# WMU Studies in Maritime Affairs

Volume 6

More information about this series at http://www.springer.com/series/11556

Aykut I. Ölçer • Momoko Kitada •
Dimitrios Dalaklis • Fabio Ballini
Editors

# Trends and Challenges in Maritime Energy Management

Springer

*Editors*
Aykut I. Ölçer
World Maritime University
Malmö, Sweden

Momoko Kitada
World Maritime University
Malmö, Sweden

Dimitrios Dalaklis
World Maritime University
Malmö, Sweden

Fabio Ballini
World Maritime University
Malmö, Sweden

ISSN 2196-8772 ISSN 2196-8780 (electronic)
WMU Studies in Maritime Affairs
ISBN 978-3-030-09026-5 ISBN 978-3-319-74576-3 (eBook)
https://doi.org/10.1007/978-3-319-74576-3



Printed on acid-free paper

This Springer imprint is published by the registered company Springer International Publishing AG part of Springer Nature.
The registered company address is: Gewerbestrasse 11, 6330 Cham, Switzerland

# Foreword

## The World Maritime University and Maritime Energy Management

Humanity is currently experiencing an era of unprecedented climate change, calling for urgent and coordinated action in order to ensure a sustainable future. Within academia and the international scientific community, it is widely accepted that greenhouse gases (GHGs) are the main drivers of climate change, contributing to the increase of global temperatures. The earth will continue to experience sea level rise, droughts, floods, increased heat, intense storms, and hurricanes (as experienced in 2017) despite our best efforts to significantly reduce GHG emissions. We all have an important responsibility in relation to future generations and our home planet to allocate sufficient resources and dedication to minimize the negative effects of climate change.

The shipping industry plays an essential role in the facilitation of world trade, being the most fuel-efficient mode of mass cargo transport. However, the expected growth of world trade represents a challenge to meet future emission targets that are required to achieve stabilization of global temperatures. According to the International Maritime Organization's (IMO) Third GHG study (2014),[1] the total annual amount of $CO_2$ emitted from international shipping is reported as 2.7% of the total $CO_2$ emissions produced worldwide. Without changes, the negative externalities of shipping will increase. This study predicts that CO2 emissions will increase by between 50% and 250% by 2050, depending on future economic and energy developments. Under a business-as-usual scenario, and if other sectors of the economy reduce emissions to keep global temperature increase below 2 °C, shipping could by 2050 represent 10% of global GHG emissions. Therefore, measures have to be taken to secure a sustainable future for mankind and ensure a competitive

[1]IMO. (2014). *Third IMO Greenhouse Gas Study*. London: International Maritime Organization.

maritime transport industry. Given the importance of international rules and regulations for ensuring sustainable shipping on clean oceans, as well as the importance for the maritime industry to remain competitive, the question arises as to how the maritime industry can best move forward to ensure a low carbon and energy efficient future.

On a global level, IMO is addressing air pollution through the International Convention for the Prevention of Pollution from Ships (MARPOL), particularly its Annex VI which limits emissions from ships, including sulfur oxides, nitrogen oxides, ozone-depleting substances, and volatile organic compounds. Measures are also in place for more energy efficient future ships as outlined in MARPOL Annex VI Chapter 4, including the Ship Energy Efficiency Management Plan (SEEMP) and Energy Efficiency Design Index (EEDI) which entered into force on 1st January 2013. The European Union (EU) has also adopted a regulation on monitoring, reporting, and verification (MRV) of $CO_2$ emissions, which will enter into force in January 2018. These regulatory measures will increase costs related to shipping operations, providing an important incentive toward energy efficient solutions. In recognition of this shift, the industry must work closely with researchers and innovators who can deliver cutting-edge solutions needed to comply with the new legal requirements. Those who do will be ahead of the competition.

Low carbon shipping provides three interrelated routes that can reduce GHG emissions:

- Increasing the energy efficiency level of ships
- Employing renewable energy on-board ships to propel fully (or, at least partially) a commercially sized merchant ship, and
- Using cleaner fuels or emission abatement technologies on-board ships such as LNG or scrubbers, respectively.

For the shipping industry, energy management is thus a key priority for energy efficient and environmentally friendly shipping that enhances profitability while operating within a tightening regulatory framework. It is very clear that significant action, including market-based measures, needs to be taken by the maritime industry in order to be able to reach a fossil fuel free industry by the end of this century.

The global importance of energy management was recognized at the 2015 COP 21 Climate Change Conference in Paris, where the IMO presented a report on its extensive efforts to address GHG emissions from shipping. The world will benefit from reduced GHG emissions with IMO's adoption of the only global, legally binding energy efficiency measures that will require ships built as of 2025 to be 30% more energy efficient than they are today. Energy management is also essential for the achievement of UN Sustainable Development Goals (UNSDGs), and in particular Goal 7: ensure access to affordable, reliable, sustainable, and modern energy for all; Goal 12: ensure sustainable consumption and production patterns; Goal 13: take urgent action to combat climate change and its impacts; and Goal 14: conservation and sustainable use of the oceans, seas, and marine resources for sustainable development.

The World Maritime University (WMU) has a strong commitment to the UNSDGs and works to support the achievement of a sustainable and energy efficient maritime and ocean industry. WMU was founded in 1983 by IMO. The fundamental objective of the university is to provide the international maritime community, and in particular developing countries, with a center for advanced maritime and ocean education, research, scholarship, and capacity building and an effective means for the sharing and transfer of technology from developed to developing maritime countries, with a view to promoting the achievement, globally, of the highest practicable standards in matters concerning maritime safety and security, efficiency of international shipping, the prevention and control of marine pollution, including air pollution from ships, and other marine and related ocean issues.

WMU aims to build knowledge, skills, and competences to enhance shipping efficiency, maritime safety and security, and the prevention of marine pollution. Since 1983, WMU has grown substantially, today offering seven specializations within the Malmö-based MSc program, two MSc programs in China, and a fast growing PhD program. In 2016, the new well-received specialization in Maritime Energy Management was launched. In line with UNSDG 7, the key pillars of the specialization are energy efficiency, renewable and clean energy, research, technology, and innovation. The programme provides a comprehensive understanding across the spectrum of maritime energy management, from on-board ships to onshore facilities such as ports and shipyards; it covers theoretical and practical aspects of maritime energy management as well as the relevant regulatory framework. Through cutting-edge research and dissemination of insights into the profitable management of alternative forms of marine and ocean energy, WMU contributes significantly to the objective of affordable, reliable, and sustainable modern energy for all.

It is clear that the demands for the shipping industry to become more energy efficient will have a broad effect across the maritime and oceans fields. The above-mentioned high-level initiatives demonstrate that a significant interest has developed in energy efficiency, renewable energy, and alternative fuels, creating an upward momentum across the shipping industry with a wide range of technical, operational, and commercial measures already implemented or under development. These demands concern actors from the private and public sectors as much as stakeholders of international organizations, NGOs, and academia. WMU is a committed partner and go-to place for informed discussion, capacity building, exchange of ideas, and applied research.

I invite you to read the chapters of this book which contains insights and analysis on how the maritime and ocean industry can achieve an energy efficient and low carbon future. Please visit our website (www.wmu.se) to get more acquainted with the World Maritime University and what it offers.

World Maritime University
Malmö, Sweden

Cleopatra Doumbia-Henry

# Acknowledgments

This book is a selection of peer-reviewed papers from the International Conference on Maritime Energy Management (MARENER 2017), held in Malmö, Sweden, 24–25 January 2017, and organized by the World Maritime University (WMU). The editors would like to take advantage of this opportunity to acknowledge the contribution and assistance of numerous individuals, such as maritime professionals, academics and researchers, MSc students of the Class of 2017 specializing in Maritime Energy Management (MEM) at WMU, and many other contributors, whose efforts led to the successful execution of MARENER 2017 and, in turn, made the publication of this book possible. MARENER 2017 provided a platform for all relevant maritime stakeholders to identify and discuss trends, opportunities, and challenges in the field of maritime energy management with the aim of achieving an energy efficient and low carbon future for the maritime industry.

The editors would first like to extend their thanks to the keynote speakers of MARENER 2017, in order of their presentations during the conference: WMU's President Dr. Cleopatra Doumbia-Henry; Secretary General of the International Maritime Organization (IMO) Mr. Kitack Lim; Mr. Cem Melikoğlu, CEO of Türk Loydu Foundation; Mr. Metin Kalkavan, Chairman, Turkish Chamber of Shipping; Mr. Carl-Johan Hagman, Head of Shipping and Ferries Stena AB; Mr. Tom Sundell, Vice President, NAPA Group; Dr. Jan Otto de Kat, Director, Energy Efficiency and Vessel Performance, ABS; Dr. Hermann J. Klein, President, German Society for Maritime Technology; Ms. Katharina Stanzel, Managing Director INTERTANKO; Ms. Anna Johansson, Minister for Infrastructure, Sweden; Mr. Johan Röstin, CEO, Copenhagen Malmö Port; Rear Admiral Paul Thomas, Assistant Commandant for Prevention Policy, U.S. Coast Guard; and Professor Dr. Mehmet Atlar, University of Strathclyde.

Our sincere appreciation is extended to the sponsors of MARENER 2017, in alphabetical order: Copenhagen Malmö Port, International Maritime Organization (IMO), MITIGATE (European Union-(EU) funded research project under the Horizon 2020 Framework), and Türk Loydu Foundation. Their generous support greatly enhanced the success of the event. Our appreciation and thanks are also extended to the MARENER 2017 media partners, in alphabetical order: Baltic Transport Journal;

Baltic University Network; DNV GL; GoLNG (EU-funded research project under the Interreg Baltic Sea Region); Harbour Review; International Association of Maritime Universities (IAMU); International Windship Association (IWSA); Motus Foundation; On-the-MoS-Way (Digital Multichannel Platform); and Ocean Renewable Power Company.

The editors would further like to thank Mr. Craig Eason, Editorial Director of Fathom, for moderating the keynote sessions, and all session chairs for their professionalism, including, in alphabetical order: Associate Professor Dr. Michele Acciaro, Kühne Logistics University; Mr. Gavin Allwright, International Windship Association; Professor Dr. Karin Andersson, Chalmers University of Technology; Professor Dr. Mehmet Atlar, University of Strathclyde; Dr. Fabio Ballini, WMU; Dr. Dimitrios Dalaklis, WMU; Dr. Jan Otto De Kat, Director, Energy Efficiency and Vessel Performance, ABS; Former Associate Professor Patrick Donner, WMU; Professor Dr. Massimo Figari, University of Genoa; Dr. Edmund Hughes, Head of Air Pollution and Energy Efficiency, IMO's Marine Environment Division (MED); Dr. Momoko Kitada, WMU; Professor Dr. Francesco Munari, University of Genoa; Professor Dr. Aykut I. Ölçer, WMU; Professor Dr. Harilaos N. Psaraftis, Technical University of Denmark; and Professor Dr. Osman Turan, University of Strathclyde.

A special thanks is extended to the members of the Scientific Committee who reviewed abstracts and full papers of the conference, in alphabetical order: Dr. Michele Acciaro, Dr. Karin Andersson, Dr. Özcan Arslan, Dr. Mehmet Atlar, Dr. Fabio Ballini, Dr. Baris Barlas, Dr. Raphael Baumler, Dr. Riccardo Bozzo, Dr. Helen Cabalu, Dr. Monica Canepa, Dr. Dimitrios Dalaklis, Dr. Jan Otto de Kat, Dr. Aref Fakhry, Dr. Massimo Figari, Mr. David Haynes, Dr. Lawrence Henesey, Dr. Mustafa Insel, Dr. Henning Jessen, Dr. Larry Hildebrand, Dr. Momoko Kitada, Dr. Dimitris Konovessis, Dr. Cristiano Morini, Dr. Francesco Munari, Dr. Takeshi Nakazawa, Dr. Nikitas Nikitakos, Dr. Aykut I. Ölçer, Dr. Harilaos N. Psaraftis, Dr. Burak Seker, Dr. Osman Turan, and Dr. Anne Marie Warris. The editors also express deep gratitude to Ms. Anne Pazaver, WMU Lecturer, who kindly proofread all the chapters in this cumulative research effort.

The editors further appreciate WMU's President Dr. Cleopatra Doumbia-Henry, for providing the leadership and vision to support the MARENER 2017 conference. Additionally, a very heavy workload was carried by the respective Organizing Committee (Aykut I. Ölçer (Chair of the Committee), Momoko Kitada, Dimitrios Dalaklis, Fabio Ballini, Josefin Madjidian, Mia Hedin, Neil Bellefontaine, Raphael Baumler, Megan Drewniak, Burak Şeker) and relevant WMU supporting staff (Erik Ponnert, Katharina Serrano, Maia Brindley Nilsson, Chris Hoebeke, Ammar Jaber, Saul Isaacson, Solveig-Karin Erdal, and René Engström). We also extend our thanks to the MSc Class of 2017 specializing in Maritime Energy Management for their assistance during the sessions and for their support of the event; they are Mark Philip Cassar, Olumide Bolarinwa Ajayi, Quang Khanh Bui, Yaser Bayoumy Abdelwahab Farag, Reza Karimpour, Maryam Mohamed Kidere, Hyuntae Kim, Edwin Nii Oduah Mensah, John Churchill Omondi, Oyegbenga Afeez Oyewunmi, Eko Maja

Priyanto, Md Abu Hasan Rony, Mohamed Said S.H. Rowihil, Wandifa Saidyleigh, Joshua Rafale Tolin, and Jing Yu.

We also express our deepest gratitude to the more than 300 participants whose valuable input and opinions contributed to active discussions and enhanced the quality of the MARENER 2017 conference. The conference looked into different topics related to Maritime Energy Management including Regulatory Framework, Energy Efficient Ship Design and Operation, Energy Management in Ports, Economics of Energy Efficiency, Social and Human Dimensions of Energy Management, Theoretical Aspects of Maritime Energy Management, Alternative Fuels and Marine Renewable Energy Including Ocean Energy.

Last but not least, the editors thank the authors of each chapter in this book for their insights and efforts in bringing trends and challenges affecting the advancement of maritime energy management to the forefront of discussions. These contributions can serve as a stimulus for further development and study of maritime energy management.

Malmö, Sweden
November 2017

Aykut I. Ölçer
Momoko Kitada
Dimitrios Dalaklis
Fabio Ballini

# Contents

**Introduction to Maritime Energy Management** . . . . . . . . . . . . . . . . . . . 1
Aykut I. Ölçer

**Part I Regulations: Challenges and Opportunities**

**MARPOL Energy Efficiency: Verging on Legal Inefficiency?** . . . . . . . . . 15
Aref Fakhry and Belma Bulut

**Analyzing Approaches to Set Greenhouse Gas Reduction Targets in Anticipation of Potential "Further Measures" for International Shipping** . . . . . . . . . . . . . . . . . . . . . . . . . . . . . . . . . . . . . . . . . 29
Seijiro Morimoto

**An Analysis of Non-conformities with the Objective of Improving Ship Energy Efficiency: Case Studies of Turkish Shipping Companies** . . . . . . 41
Özcan Arslan, Elif Bal Beşikçi, Fırat Bolat, and Koray Yaş

**Real Time Awareness for MRV Data** . . . . . . . . . . . . . . . . . . . . . . . . . . 53
Nikitas Nikitakos, Dimitrios Dalaklis, and Petros Siousiouras

**Overcoming the Challenges to Maritime Energy Efficiency in the Caribbean** . . . . . . . . . . . . . . . . . . . . . . . . . . . . . . . . . . . . . . . . 65
Vivian Rambarath-Parasram, Sukhjit Singh, and Deniece Aiken

**Energy Efficient Operations of Warships: Perspective of the Indian Navy** . . . . . . . . . . . . . . . . . . . . . . . . . . . . . . . . . . . . . . . 83
Amit Batra and Rohit Prakash

**Mexico's Reorganisation of Maritime Security Regime: A New Role for the Navy and Emphasis on Energy Related Infrastructures** . . . . . . . 95
Adriana Avila-Zuñiga Nordfjeld and Dimitrios Dalaklis

**Part II Energy Efficient Ship Design**

**Numerical Studies on Added Resistance and Ship Motions of KVLCC2 in Waves** . . . . 111
Mingyu Kim, Osman Turan, Sandy Day, and Atilla Incecik

**An Investigation of Fuel Efficiency in High Speed Vessels by Using Interceptors** . . . . 127
Ahmet G. Avci, Baris Barlas, and Aykut I. Ölçer

**A Decision Support System for Energy Efficient Ship Propulsion** . . . . 143
Mustafa Insel, Serhan Gokcay, and Ahmet Ziya Saydam

**Energy Integration of Organic Rankine Cycle, Exhaust Gas Recirculation and Scrubber** . . . . 157
Fredrik Ahlgren, Marcus Thern, Magnus Genrup, and Maria E. Mondejar

**Lighting Standards for Ships and Energy Efficiency** . . . . 169
Efe Akyurek

**Part III Energy Efficient Ship and Port Operation**

**An Integrated Vessel Performance System for Environmental Compliance** . . . . 185
Soren Hansen and Stamatis Fradelos

**Energy Efficient Ship Operation Through Speed Optimisation in Various Weather Conditions** . . . . 199
Tong Cui, Benjamin Howett, Mingyu Kim, Ruihua Lu, Yigit Kemal Demirel, Osman Turan, Sandy Day, and Atilla Incecik

**Underlying Risks Possibly Related to Power/Manoeuvrability Problems of Ships: The Case of Maritime Accidents in Adverse Weather Conditions** . . . . 213
Nikolaos P. Ventikos, Konstantinos Louzis, and Alexandros Koimtzoglou

**Simulation-Based Support to Minimize Emissions and Improve Energy Efficiency of Ship Operations** . . . . 231
Michael Baldauf, Syed Raza Ali Mehdi, Michele Schaub, Knud Benedict, Gerd Milbradt, Georg Finger, and Sandro Fischer

**Fuel Saving in Coastal Areas: A Case Study of the Oslo Fjord** . . . . 245
Karina Hjelmervik, Halvor Schøyen, Hao Wang, and Ottar Osen

**A Bayesian Belief Network Model for Integrated Energy Efficiency of Shipping** . . . . 257
Onder Canbulat, Murat Aymelek, Osman Turan, and Evangelos Boulougouris

**Smart Micro-Grid: An Effective Tool for Energy Management in Ports** . . . . 275
Monica Canepa, Giampaolo Frugone, and Riccardo Bozzo

**Energy Manager Role in Ports** . . . . 295
Fabio Ballini and Aykut I. Ölçer

**Part IV Economics and Social Dimensions of Maritime Energy Management**

**The Impact of SECA Regulations on Clean Shipping in the Baltic Sea Region** . . . . 309
Eunice O. Olaniyi, Gunnar Prause, and Jan Boyesen

**Life Cycle Assessment of Marine Coatings Applied to Ship Hulls** . . . . 325
Yigit Kemal Demirel, Dogancan Uzun, Yansheng Zhang, and Osman Turan

**The Human and Social Dimension of Energy Efficient Ship Operation** . . . . 341
Martin Viktorelius

**The Need for Education and Training in Maritime Energy Management in Myanmar** . . . . 351
Thet Hlaing Swe, Momoko Kitada, and Aykut I. Ölçer

**The Role of Maritime Transport from the Perspective of Energy and Gender: The Case of the Pacific Islands** . . . . 367
Momoko Kitada, Kuini Rabo, Ore O. Toua, and Thierry Nervale

**Part V Alternative Fuels and Wind-Assisted Ship Propulsion**

**Developing a Strategy for Liquefied Natural Gas Powered Transport Corridors in the Baltic Sea Region** . . . . 383
Josefin Madjidian, Dimitrios Dalaklis, Vytatutas Paulauskas, Lawrence Henesey, Aykut I. Ölçer, Fabio Ballini, and Momoko Kitada

**LNG Fueled Barge for Cold Ironing: Feasibility Study for the Emission Abatement in the Port of Genoa** . . . . 401
Michele Laviola, Massimo Figari, Marco Altosole, and Stefano Savio

**Decision Framework for Shipowners to Comply with Air Emission Reduction Measures: A Case Study of Methanol as a Fuel** . . . . 415
Aditya Srivastava, Aykut I. Ölçer, and Fabio Ballini

**Commercial Wind Propulsion Solutions: Putting the 'Sail' Back into Sailing** . . . . 433
Gavin Allwright

**Part VI Marine Renewable Energy**

**A Multipurpose Marine Cadastre to Manage Conflict Use with Marine Renewable Energy** . . . . . 447
Séverine Michalak

**Ocean Energy: Seeking the Balance Between States' Exclusive Rights of Exploitation and Marine Biodiversity Conservation** . . . . . 463
Wellington Tebar

**Learning from Humpback Whales for Improving the Energy Capturing Performance of Tidal Turbine Blades** . . . . . 479
Weichao Shi, Mehmet Atlar, and Rosemary Norman

**CFD Simulation of a Passively Controlled Point Absorber Wave Energy Converter** . . . . . 499
Minghao Wu, Weizhi Wang, Johannes Palm, and Claes Eskilsson

**A Framework to Improve the Coexistence of Maritime Activities & Offshore Wind Farms** . . . . . 513
Raza Ali Mehdi, Jens-Uwe Schröder-Hinrichs, Aykut I. Ölçer, and Michael Baldauf

**Editorial Conclusion** . . . . . 527
Aykut I. Ölçer, Momoko Kitada, Dimitrios Dalaklis, and Fabio Ballini

**Annex** . . . . . 533

# Introduction to Maritime Energy Management

**Aykut I. Ölçer**

## 1 Environmental Protection

Considering that the seas and oceans of our planet are associated with the most fuel-efficient method of dealing with humanity's transport needs, it is a rather self-evident fact that international shipping is crucial to world trade and the normal operation of the global economy. However, seaborne transportation is also responsible for production of greenhouse gas (GHG) emissions (like all other modes of transportation) along with the emissions of various air pollutants such as sulphur oxides ($SO_x$), nitrogen oxides ($NO_x$) and particulate matters (PMs). Hence, the maritime transport industry's environmental footprint -especially in regard to air pollution, including both GHGs and air pollutants- is not seen as very positive. Although less than 3% of the total global $CO_2$ emissions are associated with international shipping, shipping emissions could continue to increase by 50–250% by 2050, if everything follows the so-called "Business As Usual" scenario (IMO 2014).

It is also true that GHGs are produced by both natural procedures and man-made activities. However, an important dimension of the problem is that anthropogenic (man-made) GHGs are the main cause of global warming and temperature rise, responsible for climate change, which has been resulting in weather pattern changes, increased sea-level rise, and more frequent floods (IEA 2015). According to the latest report published by WMO (2017), $CO_2$ concentration in the atmosphere reached 403.3 parts per million (ppm) in 2016, which is the highest amount in the last 800,000 years. It should be noted that the figure was 400.00 ppm in 2015.

A. I. Ölçer (✉)
World Maritime University, Malmö, Sweden
e-mail: aio@wmu.se



A. I. Ölçer et al. (eds.), *Trends and Challenges in Maritime Energy Management*, WMU Studies in Maritime Affairs 6, https://doi.org/10.1007/978-3-319-74576-3_1

Within the maritime energy management context, there is no technology that can entirely remove GHGs resulting from the operations of the maritime industry. Medium to large-scale ships will indisputably continue to burn fossil fuels, including "*cleaner*" ones (like LNG) in the foreseeable future. Therefore, the vast majority of efforts in the maritime industry have been directed towards decreasing GHGs rather than eliminating them completely. Energy management, also referred to in the literature as energy efficiency or energy conservation, is accepted at IMO and within the wider maritime community to be the main mechanism to serve the purpose of reducing GHGs. Regardless of what it is called, managing energy in an optimal way or increasing the energy efficiency of a ship generally constitutes the reduction of its fuel consumption.

Concepts such as "how to reduce fuel consumption" or "how to become more energy efficient" can be dealt with during ship design, but also during regular ship operation. The biggest potential lies in the design phase, whereas ship operation also has room for improvement. It is indicative that an optimised hull form or a good coating decided during design reduces total ship resistance. The same ship can further reduce its resistance with effective hull maintenance (operational measure). Selection of the best propeller and/or propulsion system during design will lead to having more thrust available with the same input (or the same thrust with less input). Similarly, propeller cleaning (operational measure) might contribute further to increasing the propeller's efficiency. Numerous other examples can be included in the same equation; according to IMO, the combined $CO_2$ reduction potential of ship design and operation is between 50 and 75% (IMO 2009).

## 2 IMO Response and Drivers

In order to fully explore the above-mentioned potential, IMO came up with two measures (technical and operational) within Chap. 4 of the International Convention for the Prevention of Pollution from Ships (MARPOL) Annex VI, which entered into force on 1 January 2013. The one to be complied with during new ship design is an index called Energy Efficiency Design Index (EEDI); the one requiring a plan for ship operations, namely Ship Energy Efficiency Management Plan (SEEMP), is to be kept on-board existing ships. This only applies to certain ship types of certain gross tonnage. Chap. 4 of the MARPOL Annex VI is the only international binding regulation covering both design and operation of ships for GHGs resulting from international shipping.

It should be noted here that emission control areas (ECAs) introduced in Chap. 3 of MARPOL Annex VI contribute to reduction of air pollutants such as $SO_x$ and $NO_x$ emissions while contributing to GHG reduction. For Sulphur-ECAs, this is accomplished by the use of alternative fuels such as LNG/methanol, as well as using low sulphur fuel or emission abatement technologies.

In addition to the above-mentioned compulsory measures, measures of an economic nature, called market based measures (MBM), were also discussed, but have not yet come into force. Market based measures is a proposed mechanism based on the principle of "polluter pays". Prior to the future implementation of this mechanism, the European Union (EU) and IMO took an initiative on a scheme called MRV (Monitoring, Reporting and Verification) of ships' CO2 emissions; that regulation entered into force on 1 January 2018 by EU and will be in force on 1 January 2019 by IMO, with a different name, Data Collection System (DCS).

Over and above mitigating the negative impacts of GHGs, energy management can also be driven by numerous economic and business drivers. Volatility of fuel oil price, rising energy demand and high prices, scarcity of the fossil fuel trio (gas, coal, oil) are considered to be a few of those economic drivers. Potential benefits resulting from applying energy management, such as decreasing energy costs, and thereby operational costs, and increasing profits or reducing waste, are considered business drivers (Oung 2013).

The United Nations (UN) released its 2030 agenda, called UN2030 agenda, to promote sustainable development in the world while focusing on a set of actions for the 3Ps (People, Planet and Prosperity). The agenda has 17 sustainable development goals (SDGs), each of which has a number of targets, totaling 169 targets for 17 the SDGs. The SDGs have become drivers for the maritime industry to become more environmentally friendly and energy efficient in order to be able to meet the targets set under each SDG, in particular SDGs 7 (affordable and clean energy), 12 (responsible consumption and production) and 13 (climate action) (UN 2015). Needless to mention, the world population, which can be considered as the main driver of all these initiatives, continues to expand and to push the planet's boundaries (UN DESA 2015).

Those above-mentioned legislative and non-legislative motivational factors (or drivers) are the bases for energy management or energy efficiency. All these drivers have led the maritime industry to become (more) energy efficient in the way ships are designed, as well as in the way seagoing vessels are operated/maintained; producing energy for maritime transport and managing its consumption in an optimal way is simply called maritime energy management (Ölçer and Ballini 2018). This includes employing renewable energy, increasing energy efficiency through optimal design and application of "good" operation/maintenance practices, installing and using energy efficient machines and equipment as well as new innovative technologies, creating awareness for on/shore personnel, setting an energy management strategy and its objectives/targets, having top management commitment, and a combination thereof. Non-legislative drivers and compulsory technical and operational measures accompanied by the UN2030 Agenda are schematically presented in Fig. 1.

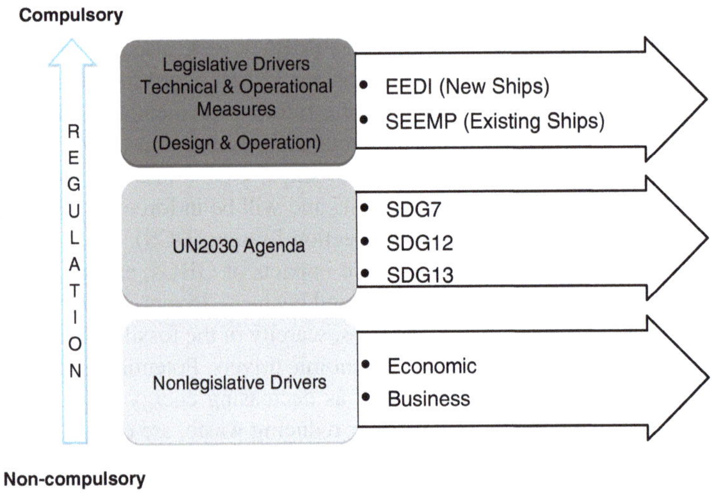


**Fig. 1** Legislative and non-legislative drivers of maritime energy management

## 3 Technical and Operational Measures

A ship can operate with the execution of its functions, including cargo loading, power generation, manoeuvring, providing hotel facilities for people on-board and many more, which require the transformation of energy sources (chemical, renewable, nuclear etc.), shown in Fig. 2, into various different forms of energy such as electricity, heat or mechanical energy to be used by energy consumers (like pumps or cargo handling equipment) (Woud and Stapersma 2002). Surely this transformation will have losses since there is no 100% efficient technology or system. The efficiency of a main propulsion engine will be around 40% while the rest (around 50%) is lost as heat and exhaust (Fig. 3), which can be recovered by using technologies like WHR (Waste Heat Recovery) (IMO 2009). The power released to shaft ($P_B$) and

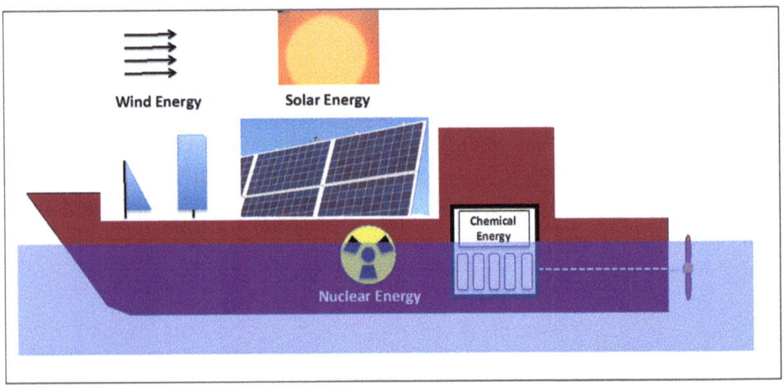


**Fig. 2** Main energy sources onboard a ship

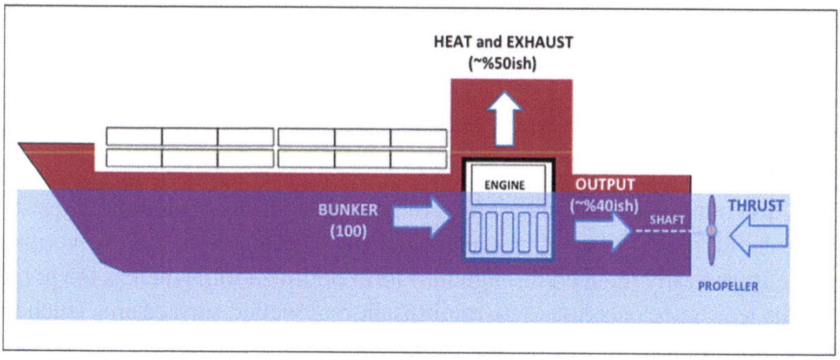


**Fig. 3** Efficiency of a main propulsion engine

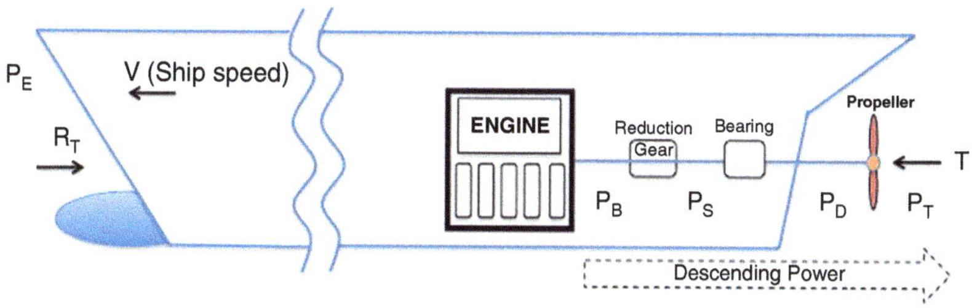


**Fig. 4** Descending power from $P_B$ to $P_T$

then propeller ($P_D$) will have further losses including mechanical losses and those resulting from propeller-hull interaction. Power losses from engine to propeller are in descending order of $P_B > P_S > P_D > P_T$ as shown in Fig. 4 where

$P_B$: Break power of main engine
$P_S$: Shaft power
$P_D$: Delivered power
$P_T$: Thrust power.

Every improvement in this chain in terms of decreased losses will contribute to increasing the ship's overall energy efficiency. Technical and operational measures are meant to be the means serving that goal.

Technical measures primarily target improvements in ship resistance and propulsion areas during the ship design phase. In very simple terms, less ship resistance or increased propulsion efficiency generally constitute less required engine power and, thereby, less fuel consumption (see Eq. 1 Molland et al. 2011):

$$P_B = P_E/(\eta_D \times \eta_{TR}) = (R_T \times V_S)/(\eta_D \times \eta_{TR}) \quad (1)$$

where

$P_E$: Effective power
$R_T$: Total ship resistance
$V_S$: Ship service speed
$\eta_D$: Quasi propulsive coefficient ($P_E/P_D$)
$\eta_{TR}$: Transmission efficiency

Reduction of total ship resistance ($R_T$) relies on the efforts of two major resistance components: (a) wave-making resistance; and (b) viscous resistance. Wave-making resistance is generally reduced through hull form optimization, whereas the potential for viscous resistance decrease is on minimisation of wetted surface area (such as air cavity or air film) or modification of the boundary layer (such as riblets or compliant surfaces) (Gokcay et al. 2004; Yang 2009). It is obvious from the above-relation Eq. (1) that lower speed is another important contributor to the reduction of fuel consumption. Speed optimization in ship design and slow steaming in ship operation are the two main methods of speed reduction.

As stated above, increased propulsion efficiency will also contribute to engine power reduction (see Eq. (1)). Quasi propulsive coefficient is the function of open-water propeller efficiency, hull efficiency and relative rotative efficiency as shown in the Eq. (2) (Molland et al. 2011). Obviously, increase in each efficiency element in the Eq. (2) will contribute to increased $\eta_D$ and so propulsion efficiency.

$$\eta_D = \eta_O \times \eta_H \times \eta_R \tag{2}$$

Where

$\eta_O$: Open-water propeller efficiency
$\eta_H$: Hull efficiency
$\eta_R$: Relative rotative efficiency

In this regard, energy efficient ship propulsion starts with the selection of an efficient propeller, which is referred to as open-water propeller efficiency. Open-water propeller efficiency is used to assess propeller performance without the presence of a hull. Torque and thrust characteristics of a propeller are measured and plotted against advance coefficient (J), hence $K_T$-$K_Q$-J diagrams are produced, which are then used to select the best propeller giving the highest efficiency ($\eta_O$). Here $K_T$ is thrust coefficient and $K_Q$ is torque coefficient. Contributing parameters to this efficiency are propeller characteristics such rake, skew, pitch, number of blades, diameter, profile sections and so on.

A propeller with the presence of a hull meaning "a propeller operating behind a hull" has different efficiency and hydrodynamic characteristics than an open-water propeller. This is an outcome of the wake gain, thrust deduction, and relative rotative efficiency, which is also referred to as propeller-hull interaction in the literature. The hull efficiency ($\eta_H$) is the function of wake fraction (w) and thrust deduction (t) factors, and expressed as in the Eq. (3) (Molland et al. 2011).

$$\eta_H = (1-t)/(1-w) \tag{3}$$

Many propulsion improvement devices have been developed to increase hull efficiency. A few indicative examples are thrust augmentation devices, wake equalizing and flow separation alleviating devices, pre-swirl devices, post-swirl devices, and propeller bos cap fins (Prins et al. 2016).

Some of these technical measures are more mature than others; similarly, some are more (or less) costly to apply, or more (or less) environmentally friendly to consider. Hull form optimization is a mature and traditional design tool, whereas air lubrication or sharkskin mimicked coating is a new technology, which requires more research in order to become a commercially viable option in the market.

It should be emphasized here that even the most energy efficient ship design has potential for improvement during the operational life of the vessel. Fuel consumption during ship operation can be reduced through three main ways (Ölçer and Ballini 2018): Optimal handling of ships; Voyage optimisation or fuel-efficient journey; Good hull, engine (main and auxiliary engines) and propeller maintenance.

There are many operational measures that can be employed during a ship's journey. Here, a few samples are provided in order to familiarise the reader with their nature and wide spectrum. For example, to reduce stresses (safety) and/or to reduce fuel consumption through trim change (energy efficiency), ships are required to deal with ballast exchange processes, which have potential for fuel saving and hence need to be handled in an optimal way. If a ship uses a controllable pitch propeller, having the optimal combination of pitch and rpm reduces fuel consumption marginally. Environmental conditions of the route that a ship takes might have significant impact on its fuel consumption. Therefore, the Captain's role in this case is to find the best route, which will cause the least fuel consumption. There are already algorithms and on-board software developed for weather routing, which takes weather and sea conditions into consideration prior to the journey. Another measure is just in time arrival (or virtual arrival), which is based on effective communication between ship and port of call; this can achieve significant fuel saving and decrease off-hire time. Roughness formed due to fouling (marine organism growth) over the years can increase fuel consumption significantly, up to 30% (or more) in severe cases. Advanced paints and periodic dry-docking as an operational measure can help reduce the roughness of a hull surface.

As highlighted before, the list is long. Nevertheless, it is necessary to bear in mind that safety should never be compromised while trying to increase energy efficiency of ships during both design and operation phases. Since energy efficiency of ships is all about reducing GHGs, every little (from pitch optimisation to slow-steaming) measure can help. Last but not least, it is important to highlight that every technical or operational measure has a different impact that varies from one ship type to another.

## 4 Barriers and Trade-Off

Despite the fact that there is a wide variety of technical and operational measures available and with different maturity levels, certain barriers exist that prevent their effective implementation. These barriers can be categorised into individual, behavioural barriers created by people, and organizational, barriers at the company level. In addition, financial and technological barriers can hinder implementation of those measures. The important thing is to eliminate all the barriers to be able to benefit from implemented measures (Kitada and Ölçer 2015; Thollander and Palm 2012).

Up to now, the GHG reduction mitigation measures applicable to ships have been explained only from an environmental point of view. However, each technical and operational measure will also create a socio-economic impact (Ballini et al. 2017). Without taking these dimensions (socio-economic and environmental) into consideration, simultaneously, it will be very difficult to make rational multiple-criteria decisions; for example externality cost versus payback time versus GHG reduction of an LNG fuelled ship retrofitted with a dual fuel engine. Therefore, the economic and social dimensions of maritime energy management must be included in the selection and implementation of technical and operational measures, in addition to the environmental dimension. However, making an evaluation in a three-dimensional domain is an analysis under a trade-off environment, where each dimension's aspects are conflicting with each other (Ölçer and Ballini 2015).

The trade-off analysis or impact assessment requires operational research methodologies to be applied, spanning from multiple criteria decision making, to monte-carlo simulation, to payback calculation, to life-cycle cost evaluation, to externality modeling and simulation, to multiple objective optimization and so on. In order to be able to increase efficiency, called maximization of efficiency in the optimization terminology, an optimization algorithm needs to be selected and employed. Sometimes, a decision has to be taken under conflicting requirements, which entails a multiple attribute decision-making methodology to be applied. Or, fuel consumption prediction will be a probabilistic estimation due to the uncertainties associated with ship and environment related parameters. Therefore, theoretical ground is inherent to maritime energy management.

The previous sections dealt with energy efficiency of ships. From here, the topics of energy efficient port operations and Marine Renewable Energy (MRE) are touched upon, which are also within the coverage of the maritime energy management field as well as this book. MRE is about producing electricity by using marine resources such as offshore wind, waves or ocean currents to be used in homes, buildings and land industry. Once electricity is produced, it can be transferred to shore grids with the use of appropriate transformers. The family of MRE is rather expanded and includes offshore wind energy and ocean energy. Ocean energy involves employing advanced technologies that convert wave energy, current energy, tidal energy, ocean thermal and salinity gradient into electricity. Among renewable energy technologies, ocean energy is the least mature one apart from tidal energy, which follows a similar development cycle to hydropower. Nevertheless, it is a relatively young and growing renewable energy sector and has great potential to decarbonise electricity production.

Energy efficient port operations use similar principles to energy efficient ship operations. Energy efficiency at ports can be achieved through using energy efficient equipment, having good operational practices, lean management and improved port processes, employing renewable energy options to produce electricity, setting port energy management strategy and its objectives/targets, having top management commitment and a combination thereof (Ölçer et al. 2017).

## 5 Motivation and Layout of the Book

In addition to some of the above-mentioned challenges, it is important to appreciate and foresee trends in the maritime energy management field. Technology is not the applications and equipment/systems that humanity relied upon one century, or even two decades ago. Those technologies have been changing at an exponential rate since the beginning of the industrial revolution. New and advanced innovative technologies solve problems or overcome challenges that humankind was not capable of easily resolving in the past. Renewable energy options started to find applications both onboard ships and at ports. The challenge as of today for renewable energy is that its capacity is not able to propel a commercially sized merchant ship. However, renewable energy, batteries, fuel cells, or their combination, depending on their technological developments, without traditional powering can be options to propel merchant ships in the future. The potential synergy between autonomous ships and the disciplines of energy efficiency and safety is not yet entirely known. Other important questions are the following: is it correct that maritime industry devotes effort toward autonomous ships rather than overcoming the huge GHG emissions related challenges? Shouldn't a zero emission ship be better than an autonomous ship emitting pollution? In the near future, should our focus be placed on carbon capture and treatment on-board a ship in addition to all the effort placed on the developed GHG emission mitigating measures? These answers are definitely not straightforward or easy to find.

The Editors hope that this book can help to identify answers to the questions previously asked, or of a similar nature, and can serve the purpose of familiarizing the reader with trends in the maritime energy management field as well as finding solutions to overcome the related challenges. That is the main motivation behind this book. As of today, there is no book in the literature of the maritime academic community with similar content, or even one with a title including "maritime energy management". This book will position itself as a pioneering work in the field of maritime energy management where the focus is given to trends and challenges. The layout of the book comprises six parts, whose descriptions are given below:

Part 1 (Regulations: Challenges and Opportunities) broadly discusses the regulatory framework of maritime energy management. The starting point is the identification of the "appropriate" way forward in relation to GHG reduction. Examination and evaluation of the impact of rather recent IMO initiatives, such as the EEDI and the SEEMP, also takes place. Additionally, the EU's MRV system of $CO_2$ emissions is discussed, as it is part of the same equation. Finally, under a forward-looking approach, challenges and appropriate responses to ensure energy efficient and optimized performance for the wider maritime industry are put forward.

Part 2 (Energy Efficient Ship Design) focuses purely on naval architecture and marine engineering related energy efficiency improvement areas from the ship design point of view. This part starts with numerical studies conducted during the ship design phase as well as studies on integration of emission abatement technologies into design. Computational Fluid Dynamics (CFD) results are demonstrated for computationally expensive processes such as (added) resistance assessment. Model tests as the most commonly used traditional validation tool are presented to substantiate the numerical results of a particular energy efficiency measure. Decision support systems, which are necessary to make rational decisions, after the settlement of numerical assessments along with their validation, are also presented for a real case study in this part.

Part 3 (Energy Efficient Ship and Port Operation) offers an analysis of energy efficiency in shipping and port operation focusing on the energy impact of those operations while complying with relevant regulations. It begins with studies related to the operational aspects of ship energy efficiency such as the minimisation of fuel consumption and risk analysis of ship performance. Several case studies are discussed and analysed with a focus on ship navigation and optimisation under particular weather conditions. Ship-port interface and micro-grids approach to energy management in ports as well as the role of the Energy Manager in ports are also covered in this part.

Part 4 (Economics and Social Dimensions of Maritime Energy Management) begins with economic and environmental implications of fuel consumption and savings from a life-cycle perspective. A case study in the Baltic Sea Region examines the impact of a particular regulation on Sulphur Emission Control Area (SECA). Human and social dimensions of energy efficient ship operation offer a good introduction to this aspect of study. Capacity building efforts to support maritime energy management in Myanmar is discussed. Gender implications of maritime transport and energy to support rural women entrepreneurs are also touched upon for Pacific Island States.

Part 5 (Alternative Fuels and Wind-Assisted Ship Propulsion) examines alternative fuels, with a special focus on their environmental benefits as well as an outlook for wind-assisted ship propulsion. Discussion of LNG and Methanol, two very promising future alternative fuels, along with brief examination of their application to ships and ports is also taking place.

Part 6 (Marine Renewable Energy) deals with renewable energy applications in the marine industry including offshore wind farms. MRE is discussed in detail, in particular Ocean Energy applications from wave energy converters to tidal energy. Besides, legal framework in EU countries with regards to MRE as well as the States' rights of exploitation and marine biodiversity conservation are discussed.

## References

Ballini, F., Ölçer, A. I., Brandt, J., & Neumann, D. (2017). Health costs and economic impact of wind assisted ship propulsion. *Ocean Engineering, 146*(1), 477–485.

Gokcay, S., Insel, M., & Odabasi, A. Y. (2004). Revisiting artificial air cavity concept for high speed craft. *Ocean Engineering, 31*(3–4), 253–267.

IMO. (2009). *Second GHG Study 2009.* London, UK: International Maritime Organization (IMO). April 2009; Buhaug, Ø., Corbett, J. J., Endresen, Ø., Eyring, V., Faber, J., Hanayama, S., Lee, D. S., Lee, D., Lindstad, H., Markowska, A. Z., Mjelde, A., Nelissen, D., Nilsen, J., Pålsson, C., Winebrake, J. J., Wu, W., Yoshida, K.

IMO. (2014). *Third GHG Study 2014.* London, UK: International Maritime Organization (IMO). April 2015; Smith, T. W. P.; Jalkanen, J. P.; Anderson, B. A.; Corbett, J. J.; Faber, J.; Hanayama, S.; O'Keeffe, E.; Parker, S.; Johansson, L.; Aldous, L.; Raucci, C.; Traut, M.; Ettinger, S.; Nelissen, D.; Lee, D. S.; Ng, S.; Agrawal, A.; Winebrake, J. J.; Hoen, M.; Chesworth, S.; Pandey, A.

International Energy Agency (IEA). (2015). Energy and climate change, World energy outlook special report.

Kitada, M., & Ölçer, A. (2015). Managing people and technology: The challenges in CSR and energy efficient shipping. *Research in Transportation Business & Management, 17*, 36–40.

Molland, A. F., Turnock, S. R., & Hudson, D. A. (2011). *Ship resistance and propulsion: Practical estimation of propulsive power.* Cambridge: Cambridge University Press. ISBN: 9781107641259.

Ölçer, A., & Ballini, F. (2015). The development of a decision making framework for evaluat-ing the trade-off solutions of cleaner seaborne transportation. *Transportation Research Part D: Transport and Environment, 37*, 150–170.

Ölçer, A., & Ballini, F. (2018). Energy management in the maritime industry. In L. L. Froholdt (Ed.), *Corporate social responsibility in the maritime industry* (Chap. 8). Heidelberg: Springer.

Ölçer, A., Baumler, R., Ballini, F., & Kitada, M. (2017). Maritime energy management. In I. D. Visvikis & P. M. Panayides (Eds.), *Shipping operations management* (Chap. 8). Heidelberg: Springer.

Oung, K. (2013). Energy management in business the manager's guide to maximis-ing and sustaining energy reduction, part 1, ISBN: 9781409452454.

Prins, H. J., Flikkema, M. B., Schuiling, B., Xing-Kaeding, Y., Voermans, A. A. M., Müller, M., et al. (2016). Green retrofitting through optimisation of hull-propulsion interaction – GRIP. *Transportation Research Procedia, 14*, 1591–1600.

Thollander, P., & Palm, J. (2012). *Improving energy efficiency in industrial energy systems an interdisciplinary perspective on barriers, energy audits, energy management, policies, and programs.* New York: Springs International Publishing. ISBN: 9781447141617.

UN General Assembly. (2015). Transforming our world: The 2030 agenda for sustainable devel-opment, 21 October, A/RES/70/1. Available at: http://www.refworld.org/docid/57b6e3e44.html. Accessed 29 Nov 2017.

United Nations, Department of Economic and Social Affairs, Population Division. (2015). World population prospects: The 2015 revision, key findings and advancetables. Working Paper No. ESA/P/WP.241.

WMO. (2017). The state of greenhouse gases in the atmosphere based on global observations through 2016, Greenhouse Gas Bulletin, No. 13 | 30 October 2017, ISSN 2078-0796.

Woud, H. K., & Stapersma, D. (2002). *Design of propulsion and electric power generation systems, Chapter 4*. London: Institute of Marine Engineers.

Yang, S. Q. (2009). Drag reduction in turbulent flow with polymer additives. *Journal of Fluids Engineering, 13*, 051301.

# Part I
# Regulations: Challenges and Opportunities

# MARPOL Energy Efficiency: Verging on Legal Inefficiency?

**Aref Fakhry and Belma Bulut**

## 1 Introduction

This chapter analyses legal aspects pertaining to ship energy efficiency measures adopted as part of the International Convention for the Prevention of Pollution from Ships 1973, as amended by the Protocols of 1978 and 1997 thereto (MARPOL) (UNTS 1983; UK 1999). The measures discussed in this work consist of the Energy Efficiency Design Index (EEDI) and the Ship Energy Efficiency Management Plan (SEEMP). These two measures were introduced pursuant to the International Maritime Organization's (IMO) Resolution MEPC.203(62), adopted on July 15, 2011, as part of a new Chapter 4 titled "Regulations on Energy Efficiency for Ships" within MARPOL's Annex VI (IMO 2011a). The latter Annex deals with air pollution from ships. The new Chapter came into force on January 1, 2013.

The chapter presents a historical recount of the development of MARPOL leading up to the adoption of the energy efficiency standards in Annex VI. EEDI and SEEMP are then described and dissected as regards their scope, legal nature, effectiveness and enforceability. A discussion of the relevance of EEDI and SEEMP in ship chartering, sale and building contracts ensues.

A. Fakhry (✉)
World Maritime University, Malmö, Sweden
e-mail: af@wmu.se

B. Bulut
Maritime Law Expert, Istanbul, Turkey



A. I. Ölçer et al. (eds.), *Trends and Challenges in Maritime Energy Management*, WMU Studies in Maritime Affairs 6, https://doi.org/10.1007/978-3-319-74576-3_2

## 2 Historical Background

MARPOL is a living instrument that has evolved over the years. The original International Convention for the Prevention of Marine Pollution from Ships, 1973, as modified by the Protocol of 1978 relating thereto, contained five Annexes numbered I to V and covering pollution by oil, noxious liquid substances in bulk, harmful substances carried in packaged form, sewage and garbage.

In the late 1980s, scientific studies verifired that ships were a source of air pollution even though the effects of ships' exhaust gases on human health and ecosystems were not immediately visible and surfaced over a long term (IMO 1998). With the aim of reducing ships' emissions and their contribution to global air pollution, IMO's Marine Environment Protection Committee (MEPC) was invited to urgently develop legally binding measures (IMO 1991).

In 1997, at an international conference of Parties to MARPOL, a new Annex VI entitled "Regulations for the Prevention of Air Pollution from Ships" was added to the Convention (IMO 1997a). The new Annex entered into force on May 19, 2005. As of November 7, 2017, 88 States representing 96.16% of world tonnage were party to the new Annex (IMO 2017).

The 1997 international conference adopted Resolution 8 entitled "$CO_2$ Emissions from Ships" inviting MEPC to identify and develop greenhouse gas emission strategies (IMO 1997b). In 2000, an IMO Study on Greenhouse Gas Emissions from Ships showed that 1.8% of the world's total $CO_2$ emissions were caused by ships (IMO 2000). Although it was highlighted that shipping was the most energy-efficient mode of cargo transport, the Study suggested that further measures could be taken to reduce $CO_2$ emissions (IMO 2000). The Study recommended resort to operational measures, such as optimal utilisation, and technical measures, including improved hull shape or propeller design, as potential solutions for the reduction of emissions.

MEPC was later urged to identify and develop measures to limit or reduce greenhouse gas emissions from ships (IMO 2003a). In 2009, MEPC developed technical and operational measures consisting of EEDI and SEEMP (IMO 2009a). In 2011, those measures were adopted and incorporated as a separate Chapter 4 in Annex VI of MARPOL (IMO 2011a). The Regulations on Energy Efficiency for Ships entered into force on January 1, 2013.

## 3 Energy Efficiency Measures Under MARPOL

This section describes EEDI and SEEMP. Both measures mandatorily apply to all ships of 400 gross tonnage and above that are engaged in voyages going beyond waters subject to the sovereignty or jurisdiction of the state the flag of which the ship is entitled to fly (MARPOL Annex VI, Reg 19).[1]

### *3.1 What Is EEDI?*

EEDI is a technical measure that aims to promote the use of energy-efficient ship designs with the ultimate aim of reducing emissions from ships. EEDI does not address operational measures; it only focuses on design.

EEDI applies to: new ships; new ships that have undergone a major conversion; and new or existing ships which have undergone an extensive major conversion and are regarded as newly-constructed ships by the Administration[2] (MARPOL Annex VI, Regs 20 and 21). Existing ships that have not undergone an extensive major conversion are left outside the scope of application of EEDI regulations.

Each ship must have its own specific EEDI showing the estimated performance of the ship for energy efficiency purposes. Two EEDI values are provided for: (a) the attained EEDI, which is defined as the EEDI value achieved by an individual ship as per Regulation 20; and (b) the required EEDI, which is defined as the maximum value of the attained EEDI allowed under Regulation 21 for the specific ship type (for example, container ship or tanker) and size. In other words, the attained EEDI must be equal to or below the required EEDI.

As will be explained later, EEDI is in the nature of goal-based standards,[3] giving freedom to shipbuilders, ship designers and other interested parties to adopt the most economical and suitable energy-efficient ship design. Nevertheless, MEPC has introduced guidelines to assist stakeholders in the smooth and uniform implementation of the mandatory regulations in Chapter 4 of MARPOL Annex VI. Those guidelines should be considered when EEDI is calculated. According to the 2014 Guidelines on Survey and Certification of the Energy Efficiency Design Index (EEDI), the verification process is carried out in two phases: a pre-verification phase and a final verification phase (IMO 2014, 2015).

---

[1]According to Article 3(3) of MARPOL, warships, naval auxiliaries and other ships owned or operated by a state and used for the time being only on government non-commercial service are excluded from the scope of application of the Convention.

[2]Article 2(5) of MARPOL defines the term "Administration" as "the Government of the State under whose authority the ship is operating." The provision further states that with respect to a ship the term "Administration" means the government of the ship's flag state.

[3]See Sect. 4.2 below.

Pre-verification takes place during ship design, whereas final verification is carried out at sea trials. The 2014 Guidelines contain elaborate technical information on the verification stages; therefore, only a brief explanation will be provided here. At the design stage, that is, before the start of ship construction, an application for an initial survey is submitted to a verifier in the form of an EEDI Technical File along with other relevant background documents and information. The verifier may request additional information if necessary. In essence, the verifier verifies the attained EEDI and issues a report of pre-verification. At the sea trials stage, final verification is effected of the ship's condition and speed, as well as the revised EEDI Technical File. This is followed by the issuance of "the certificate."[4] It should be added that in the event of a major conversion, both a preliminary and a final verification are conducted.

As seen above, according to the 2014 Guidelines, the entire EEDI verification procedure is conducted by a verifier, which is defined as an Administration[5] or an organisation duly authorised by it (IMO 2014, 2015). In short, under MARPOL, only the flag state has the authority for preliminary and final EEDI verification.

## 3.2 *What Is SEEMP?*

Unlike EEDI, SEEMP is an operational measure. As a management mechanism for companies and ships, SEEMP aspires to better energy efficiency in ship operations and, ultimately, a reduction of ship emissions. SEEMP is ship-specific, but of course it may be linked to the company's energy management policies. A well-designed SEEMP should be seen as beneficial for both the environment—because it contributes to the reduction of emissions, as well as the company—by helping reduce fuel consumption (IMO 2009a).

SEEMP applies to all ships—whether new or old—falling within the purview of Regulation 19 of Annex VI (MARPOL Annex VI, Reg 22.1).

By virtue of Resolution MEPC.278(70), new regulations on a data collection system for fuel oil consumption were introduced into MARPOL Annex VI, and are expected to enter into force on March 1, 2018 (IMO 2016a). Under the new provisions, ships of 5000 gross tonnage and above are required to include in their SEEMP a description of the methodology that will be used to collect fuel oil consumption data, which must then be reported to the ship's Administration (MARPOL Annex VI, Regs 22.2 and 22A).

---

[4]It is noteworthy that the 2014 Guidelines do not name the certificate. Pursuant to Regulation 6.4 of Annex VI, an International Energy Efficiency Certificate (IEEC) is issued following survey and verification that the ship's attained EEDI is in accordance with Chapter 4 and that the SEEMP is on board.

[5]See footnote 2 above for the definition of "Administration."

Similar to EEDI, SEEMP is in the nature of a goal-based standard; thus, stakeholders are free to adopt their own SEEMPs by taking into account the 2016 Guidelines for the Development of a Ship Energy Efficiency Management Plan (SEEMP) (IMO 2016b). The 2016 Guidelines provide for methodologies that may be adjusted in accordance with the needs of companies and ships. According to the Guidelines, the SEEMP should consist of two Parts, as follows:

- Part 1 of the SEEMP contains the ship management plan aimed at improving energy efficiency. It delineates four steps that need to be followed for ultimate ship performance optimisation, namely: planning; implementation; monitoring; self-evaluation and improvement (IMO 2016b).
- Part 2 of the SEEMP contains the ship fuel consumption data collection plan. This Part only applies to ships of 5000 gross tonnage and above. It describes the methodologies to be followed for collecting data and the processes for reporting it to the ship's Administration, that is, the flag state (when the new amendments on the data collection system enter into force).

# 4 General Legal Analysis

Having described EEDI and SEEMP, this section carries an analysis centred in law. It is noteworthy that the promotion of energy-efficient ships is relatively new to the *public law and policy of shipping*. Historically, energy efficiency was driven largely by market factors (Faber and 't Hoen 2015; Faber et al. 2016). However, since EEDI and SEEMP entered into force in 2013, energy efficiency has become a compulsory requirement for ship design and operation. The discussion will proceed from four angles, namely:

- the meaning of energy efficiency
- the legal nature of EEDI and SEEMP
- the effectiveness of EEDI and SEEMP
- the enforceability of EEDI and SEEMP

## *4.1 Meaning of Energy Efficiency*

EEDI and SEEMP come under Regulations on Energy Efficiency for Ships, which is the heading of Chapter 4 of Annex VI of MARPOL. Questions may arise as to the meaning and scope of the term "energy efficiency." For instance, does it cover sustainable consumption of resources, aside from the issue of pollution?

The term "energy efficiency" is nowhere defined in MARPOL.

Nevertheless, the objective of the new MARPOL provisions is stated as follows in the originating IMO Resolution MEPC.203(62):

> to improve energy efficiency for ships through a set of technical performance standards, which would result in reduction of emissions of any substances that originate from fuel oil and its combustion process, including those already controlled by Annex VI.... (IMO 2011a)

In the dictionary, "efficiency" is defined as follows:

> The ratio of output to input power, often expressed as a percentage. For example, an efficiency of a fossil fuel power plant of 40 per cent would mean that the ratio of the electrical power output to the heat flow into the plant was 40 per cent. (Jelley 2017).

Returning to IMO Resolution MEPC.203(62), it appears from the wording quoted above that the MARPOL provisions are not focused on a diffuse or general concept of efficiency—such as the meaning conveyed by the dictionary definition. MARPOL's energy efficiency is focused explicitly on "reduction of emissions" from "fuel oil and its combustion process." Energy efficiency is thus the means of attaining the ultimate goal of the reduction of emissions.

## 4.2 *Legal Nature of EEDI and SEEMP*

Apart from the meaning of energy efficiency in MARPOL, it is useful to examine the legal nature of the measures put in place. It should be stated at the outset that MARPOL does not prescribe the way in which the ship must meet the minimum requirements for increasing energy efficiency. EEDI and SEEMP are in the nature of goal-based standards (IMO 2008a). The concept of goal-based standards was introduced with the aim of removing any possible competition between classification societies by giving a greater role to IMO in determining the fundamental standards to which new ships are built (IMO 2002, 2003b). Goal-based standards differ from prescriptive ones in respect of their methodology and aim. Prescriptive standards stipulate the specific means of achieving compliance and mandate what to do or not to do, whereas goal-based standards set objectives and allow alternative methods for achieving compliance (Hoppe 2006).

Goal-based standards provide greater freedom and flexibility to apply the most suitable and cost-effective methods in the development of solutions to rapid technological changes. Even if the ways of achieving goals may change as a result of new technological developments, the rules created in the form of goal-based standards would not fall behind.

Currently, in every sector, including shipping, technology is quickly changing with the result that regulations may have difficulty keeping up. Whenever technologies change, prescriptive standards also need to be changed to keep pace and this causes an extra regulatory burden on, inter alia, IMO, as well as instability in practice. Of course, goal-based standards may need to be modified from time to time, but not as frequently as prescriptive standards.

Goal-based standards consist of five different tiers: the actual goals (Tier I); functional requirements (Tier II); verification of conformity (Tier III); rules and regulations (Tier IV); industry standards and practices (Tier V) (IMO 2011b). When new regulations and rules are adopted in the form of goal-based standards, those five tiers have to be considered with caution. If the goals and functional requirements are not properly identified or the link among the tiers is not adequately articulated, then implementation, effectiveness and enforcement problems are likely to arise. While it may be said that the adoption of ship energy efficiency rules in the form of goal-based standards gives freedom to stakeholders to benefit from new technological developments by applying the most cost-effective and suitable mechanisms, some of the deficiencies in EEDI and SEEMP identified in the following two sections of this commentary would point to flaws in the design of the measures.

## 4.3 *Effectiveness of EEDI and SEEMP*

The high-level goal set for MARPOL energy efficiency measures is to provide a coherent and comprehensive mandatory framework for reducing emissions from ships (IMO 2008b, 2011a). However, because of the shortfalls discussed below, it is submitted that the MARPOL energy efficiency measures may fail to achieve the set goal.

First, the limited scope of application is noteworthy. Under Regulation 19 of MARPOL Annex VI, the energy efficiency measures apply to ships of 400 gross tonnage and above engaged in voyages going beyond waters subject to the sovereignty or jurisdiction of the flag state. However, pursuant to Regulation 20, EEDI applies only to: new ships; new ships that have undergone a major conversion; and new or existing ships which have undergone an extensive major conversion and are regarded as newly-constructed ships by the Administration. Excluding existing ships that have not undergone an extensive major conversion from the scope of application of EEDI would surely affect adversely the achievement of the emissions reduction goal.

Furthermore, being goal-based standards, EEDI and SEEMP mean that shipbuilders, fitters, repairers, owners, classification societies, IMO and the maritime community at large have to come up with technical solutions to attain the set result. This is a huge effort and one that cannot be taken for granted.

Interestingly, the UK has not yet given effect to EEDI and SEEMP into its law (MCA 2012), which may be seen as evidence of the difficulty of implementation.

## 4.4 *Enforceability of EEDI and SEEMP*

Apart from effectiveness, an essential feature of good law is that it should be easily enforceable. Some key features of the current system would buttress the view that MARPOL's energy efficiency rules are significantly unenforceable both in practice and theory.

First, there is the issue of the lack of a control mechanism. In contrast with EEDI, which, as explained above,[6] has a certain level of control carried out by a verifier—at the ship's design and sea trial stages, SEEMP has a vital weakness insofar as it lacks an equivalent control mechanism. Unlike EEDI, none of the SEEMP development stages involve any verification conducted by the Administration or any other authority. This means that there is no mechanism to control whether a SEEMP is planned, implemented, monitored or evaluated properly.

An example of this deficiency may be seen in relation to SEEMP Part 2.[7] Under a new regime which is expected to enter into force on March 1, 2018, fuel consumption data will have to be sent to the ship's Administration and thereafter transferred to the IMO Ship Fuel Oil Consumption Database. Surprisingly, there is no provision describing what happens next. It seems that the Administrations will collect data for the purpose of collecting data.

Another major flaw is the ambiguity regarding port state control of energy efficiency measures. Regulation 10.5 of MARPOL Annex VI reads:

> In relation to chapter 4, any port State inspection shall be limited to verifying, when appropriate, that there is a valid International Energy Efficiency Certificate on board, in accordance with article 5 of the Convention

Under MARPOL Annex VI, an International Energy Efficiency Certificate (IEEC) is issued following survey and verification that the ship's attained EEDI is in accordance with the requirements of Chapter 4 and that the SEEMP required by Regulation 22 is on board (MARPOL Annex VI, Regs 6.4 and 5.4). MARPOL Annex VI further states that the IEEC shall be issued or endorsed by the Administration, that is, the government of the flag state, or an organisation duly authorised by it (MARPOL Annex VI, Reg 6.5; MARPOL Art 2(5)).

According to Article 5 of MARPOL, port state control is limited to verifying the existence of a valid certificate on board. Nevertheless, when "there are clear grounds for believing the condition of ship or its equipment does not correspond substantially with the particulars of that certificate," a more thorough inspection may be conducted and some sanctions, such as detainment of the ship, may be applied by the port state.

It is noteworthy that, on the one hand, Regulation 10.3 of MARPOL Annex VI provides that the "[p]rocedures relating to the port State control prescribed in Article 5 of the . . . Convention shall apply to this regulation." Furthermore, Regulation 10.4 provides that nothing therein shall be construed to limit rights and obligations regarding port state control. Reference is also made, on the other hand, to Regulation 10.5, which states that in relation to Chapter 4 containing the energy efficiency measures, port state control is limited to verifying the existence of a valid IEEC on board.

---

[6]See Sect. 3.1 above.

[7]See Sect. 3.2 above.

It is to be observed that Article 5 of the Convention and Regulation VI/10.5 may be inconsistent with the result of rendering the enforceability of the energy efficiency measures uncertain to say the least. Which provisions should be given priority: Article 5 of the Convention or Regulation VI/10.5? Although MPEC has adopted guidelines on port state control under the revised Annex VI, there is no mention of this issue (IMO 2009b).

Analytically, it is observed that, while Article 5 is located in the main part of the Convention and addresses general port state control matters for all the Annexes, Regulation VI/10.5 is restricted to port state control touching on Chapter 4. The upshot of this inconsistency may be resolved by resort to the common legal axiom to the effect that *lex specialis derogat legi generali* (the special or specific law supersedes the more general provisions).

In conclusion, it appears from the wording of MARPOL that the only party that can enforce EEDI is the flag State (or an organisation authorised by it) in the following circumstances:

- before a new ship is put in service and before the IEEC is issued;
- after a major conversion.

As for the port state, it is empowered to verify that there is an IEEC on board the ship. It appears, however, that the port state is not permitted to check the vessel's compliance beyond ascertaining the documentary requirements.

The resulting bar on the port state to verify EEDI compliance is somehow striking in international maritime law, and certainly the practices followed in IMO conventions.

As far as SEEMP is concerned, it is quite clear from the wording of MARPOL that no enforcement is expected beyond the verification that the SEEMP is on board.

## 5 EEDI and SEEMP in Private Maritime Transactions

Having considered general legal questions arising from EEDI and SEEMP, this section of the commentary examines the responsibility for compliance in the context of private legal relationships concerning the ship. The analysis will cover charter-parties,[8] and ship sale and purchase agreements.

[8]It is noteworthy that charterparties have for some time included incentives for energy efficiency. An example is the slow steaming clause incorporated in charterparties (Nikaki 2016).

## *5.1 Charterer's Rights*

The charterer has an interest in ascertaining that the ship possesses an International Energy Efficiency Certificate. This interest is notionally the same as that of ensuring that the ship meets all statutory requirements found in MARPOL, SOLAS and similar instruments. If the ship was found not to be in compliance in any respect with such instruments, it would risk being detained or delayed by government inspectors carrying out random or periodic controls, whether on behalf of the flag state or the port state.

Under a charterparty agreement, the owner's duty to fit out the ship for the contractual service means that the ship must meet technical standards, including regulatory prescriptions found in such instruments as MARPOL. A typical clause in this regard is GENCON's Owners' Responsibility Clause, which requires for any liability to be incurred by personal want of due diligence on the part of the Owners or their Manager to make the Vessel in all respects seaworthy and to secure that it is properly manned, equipped and supplied, or personal act or default of the Owners or their Manager (BIMCO 1994). Such a clause has been interpreted as covering MARPOL. In *Golden Fleece Maritime Inc v ST Shipping and Transport Inc* [2008] EWCA Civ 584, [2009] 1 All ER (Comm) 908, [2008] 2 Lloyd's Rep 119, the court upheld a claim by time charterers against owners that an exemption from MARPOL's tanker construction standards as provided for in Regulation 13H (5) had to be secured under the terms and conditions contained in the charter. There was a discussion about physical as opposed to legal fitness of the vessel. In effect, the latter was just as much required as the former. Legal fitness covered such things as the certification and documentation required by law for the ship to trade. It would not be difficult to conclude from this judgment that EEDI documentation is covered by the legal fitness duty.

Another reason that the charterer may have in ascertaining compliance of the chartered ship with the EEDI and SEEMP requirements will typically be based on its desire to secure a good image of its chartering business, by publicising its hiring of ships with good energy efficiency records.

An analogy could be made to the SIRE system operating in the tanker industry. SIRE is a database of ship particulars concerning safety and environmental soundness. Individual ship reports are shared within the industry, allowing prospective charterers to ascertain the condition of a particular vessel (OCIMF 2017). SIRE has reportedly developed as a result of the pressure felt by segments in the industry to maintain a good public image (de la Rue 2009). It is not clear whether SIRE already includes EEDI and SEEMP data.

There is, however, one fundamental difference between EEDI and SEEMP as contrasted with other statutory requirements: Compliance with EEDI and SEEMP is not likely and, actually, will never be verified—except at the launch of the vessel or following a major conversion, as explained in Sects. 3 and 4.4 above. This is unlike most other statutory requirements which undergo periodic surveys and inspections. It may thus be argued that the charterer has no real incentive or interest in ensuring that the chartered ship *actually* meets the applicable EEDI and SEEMP standards.

## 5.2 Ship Buyer's Rights

Similarly to the charterer, as seen from the previous section, the ship's buyer will have a legitimate interest and certain rights in demanding that the ship comply with the EEDI and SEEMP requirements given their mandatory nature. These rights typically derive from the ship's sale and purchase agreement.

Nevertheless, two situations need to be distinguished. Where a ship is being built or is undergoing a major conversion, MARPOL requires the carrying out of a survey. In the case of the purchase of a second-hand ship (with no major conversion involved), no survey is mandated as such under MARPOL.[9]

The MARPOL-imposed survey at the time of the ship's commissioning or major conversion should provide a natural opportunity for verifying the ship's EEDI compliance. Verification at any other stage will not be necessary nor practically feasible,[10] with the corollary that the discovery of non-conformities will be quite unlikely.

Moving on to consider the buyer's rights for demanding an EEDI-compliant ship, it is useful to examine the wording of standard contracts. EEDI can be first approached as a regulatory requirement in the context of a shipbuilding contract. The Shipbuilders' Association of Japan (SAJ) (1974) form provides in its article I(3), titled "Classification, Rules and Regulations":

> The VESSEL shall also comply with the rules, regulations and requirements of other regulatory bodies as described in the Specifications in effect as of the date of this Contract. (SAJ, article I(3).)

Usually, the specifications would include reference to MARPOL or IMO standards (Curtis 2012), so conformity with EEDI requirements is dependent technically on a more or less explicit mention in the contractual documentation.

This being said, EEDI will also often feature as a requirement of the classification society overseeing the construction of the vessel. As such, it will therefore be covered by this other provision of the SAJ form:

> The VESSEL, including its machinery, equipment and outfittings shall be constructed in accordance with the rules (the edition and amendments thereto being in force as of the date of this Contract) of and under special survey of . . . (herein called the "Classification Society"). . . . (SAJ, article I(3).)

The China Maritime Arbitration Commission (CMAC) Standard Newbuilding Contract (2011) is more explicit as to the need for the ship to be built to standards set forth by the regulatory authorities (and the classification society) (CMAC, article II, titled "Classification, Rules and Regulations").[11]

---

[9]See Sect. 3.1 above.

[10]See Sect. 4.4 above.

[11]See also article XIV(3), which provides: "The VESSEL shall comply with the above rules and regulations which are in force at the time of signing the Contract."

Article III of the CMAC contract deals with the ship's design specifications. It starts by requiring the specifications to "determine . . . compliance with the rules and regulations of CLASSIFICATION SOCIETY and relevant authorities" (sic).[12] The same article goes on to distinguish between two situations, namely, where on the one hand "the BUYER assigns ___ design company from outside of China to design the VESSEL" and, on the other hand, "the design company is assigned by the BUILDER". It is unclear what assignment signifies in this context. Crucially, in the former situation, the builder will undertake no responsibility for "any deficiency of speed, deadweight, fuel oil consumption and hold capacity due to non-compliance with the relevant regulations and rules," whereas that responsibility will attach in the latter situation.

There appears to be hardly any case law touching specifically on compliance with regulatory requirements in the ship sale and purchase contract.

Aside from featuring as a regulatory requirement, EEDI compliance may also be approached as a matter of ship design. For a newbuilding, the primary workmanship warranty rests with the builder. This rule was evidenced in *M Gillespie & Co v James Howden & Co* (1885) 12 R 800. The case makes it clear that the contractual design is very binding and the builder cannot argue that it was not feasible! The case is hornbook law. In situations where the design is provided by the buyer, the rule is that there is no warranty by the builder; however, the case will remain that EEDI is a MARPOL requirement, so the matter will fall back to the category of regulatory and class requirements discussed above.

## 6 Conclusion

In conclusion, it may be said that that EEDI and SEEMP are partly ineffective as they do not reach the ultimate goal of reducing emissions. Furthermore, the new rules are partly unenforceable as practically no one checks the situation behind the required documents, namely, IEEC and SEEMP.

By definition, a goal-based standard leaves room for creativity and flexibility—which may be welcome in a fast evolving technological environment, but it may also lead to diversity in the means of achieving the required goal. It would have been arguably quite unfathomable to expect port state control inspectors to verify a ship's actual EEDI methodology and characteristics, simply because of the sheer diversity in designs that may be adopted by individual ships. Another obvious reason is the complexity of EEDI calculations and verifications. For the same reason, while SEEMP must be present on board, it would have been unrealistic to expect port

[12]See also article XI(1)(2), which commits "the BUYER [to] provide the BUILDER with the necessary documentation including specifications, drawings, instruction books, manuals, test reports and others required to comply with all applicable rules and regulations."

state control officers to check the exact underpinnings of the methodologies and processes adopted in the SEEMP for each particular ship.

While it is admitted that the likelihood of verification of actual and real compliance with the EEDI and SEEMP requirements is almost non-existent—something of an aberration in law, at least in theory, the upshot of the argument is that the whole system for ship energy efficiency relies on trust in the EEDI-related documentation generated by the flag State or the recognised organisation acting on its behalf at the time indicated for the ship's certification, or the SEEMP generated and kept by the ship. Workable as it may be, this system strikes at legal efficiency.

The impending regime for data collection is intended to palliate the current non-verifiability of real compliance with the standards. By requiring the collection of energy efficiency data for each ship and its transmission to the flag state and ultimately IMO, it is expected that the latter will be placed in a monitoring stance and thence in a position to refine the effectiveness and enforceability of the current rules.

## References


BIMCO. (1994). Uniform general charter (as revised 1922, 1976 and 1994 (Code Name: "GENCON").

China Maritime Arbitration Commission (CMAC). (2011). Standard new building contract (Shangai Form).

Curtis, S. (2012). *Law of shipbuilding contracts*. Oxon: Informa.

de la Rue, C. (2009). *Shipping and the environment*. Oxon: Informa.

Faber, J., & 't Hoen, M. (2015). Historical trends in ship design efficiency. http://www.cleanshipping.org/download/CE_Delft_7E50_Historical_trends_in_ship_design_efficiency_DEF.pdf. Accessed 18 November 2017.

Faber, J., 't Hoen, M., Vergeer, R., & Calleya, J. (2016). The impact of full form on efficiency (Historical trends in ship design efficiency). https://www.transportenvironment.org/sites/te/files/publications/2016_CE_Delft_Historical_Trends_in_Ship_Design_Efficiency.pdf. Accessed 18 November 2017.

Golden Fleece Maritime Inc v ST Shipping and Transport Inc. (2008). EWCA Civ 584, [2009] 1 All ER (Comm) 908, [2008] 2 Lloyd's Rep 119.

Hope, H. (2006). Goal-based standards- a new approach to the international regulation of ship construction. *IMO News*. http://www.imo.org/en/MediaCentre/MaritimeNewsMagazine/Documents/2006/IMONews12006.pdf. Accessed 5 November 2017.

IMO. (1991). Resolution A.719(17) Prevention of air pollution from ships. In *IMO Assembly seventeenth session*.

IMO. (1997a). Protocol of 1997 to amend the International Convention for the Prevention of Pollution from Ships, 1973, as modified by the Protocol of 1978 relating thereto, September 26, 1997, IMO Doc MP/CONF. 3/34.

IMO. (1997b). Resolution 8-CO2 emissions from ships. In *IMO Doc MP/CONF.3/35*.

IMO. (1998). MARPOL-25 Years. Focus on IMO.

IMO. (2000). Report on the outcome of the IMO study on greenhouse gas emissions from ships. In *IMO Doc MEPC 45/8*.

IMO. (2002). Consideration of the strategy and policy of the organisation including the report of the working group, IMO strategic plan. Submitted by Bahamas and Greece (C 89/12/1).

IMO. (2003a). Resolution A.963(23) IMO policies and practices related to the reduction of greenhouse gas emissions from ships. In *IMO Doc A 23/Res.963*.

IMO. (2003b). Consideration of the strategy and policy of the organisation including the report of the working group, IMO strategic plan- New build standards. Submitted by Bahamas and Greece (MSC 77/2/5).
IMO. (2008a). Prevention of air pollution from ships (Section 4). In *IMO Doc MEPC 57/21*.
IMO. (2008b). Report on the working group on greenhouse gas emissions from ships. In *IMO Doc MEPC 57/WP.8*.
IMO. (2009a). Report of marine environment protection committee on its fifty-ninth session. In *IMO Doc MEPC 59/24/Add.1*.
IMO. (2009b). Resolution MEPC.181(59) 2009 Guidelines for port state control under the revised MARPOL Annex VI. In *IMO Doc MEPC 59/24/Add.1*.
IMO. (2011a). Resolution MEPC.203(62) Inclusion of regulations on energy efficiency for ships in MARPOL Annex VI. In *IMO Doc MEPC 62/24/Add.1*.
IMO. (2011b). Generic guidelines for developing IMO goal-based standards. In *IMO Doc MSC.1/ Circ.1394*.
IMO. (2014). Resolution MEPC.254(67) 2014 Guidelines on survey and certification of the energy efficiency design index (EEDI). In *IMO Doc MEPC 67/20*.
IMO. (2015). 2014 Guidelines on survey and certification of the energy efficiency design index (EEDI), as amended (Resolution MEPC.245(67), as amended by Resolution MEPC.261(68)). In *IMO Doc MEPC.1/Circ.855/Rev.1*.
IMO. (2016a). Resolution MEPC.278(70) Data collection system for fuel oil consumption of ships. In *IMO Doc MEPC.70/18/Add.1*.
IMO. (2016b). Resolution MEPC.282(70) 2016 Guidelines for the development of a ship energy efficiency management plan (SEEMP). In *IMO Doc MEPC.70/18/Add.1*.
IMO. (2017). Status of Treaties 2017. http://www.imo.org/en/About/Conventions/StatusOfConventions/Documents/Status%20of%20Treaties.pdf. Accessed 18 November 2017.
International Convention for the Prevention of Pollution from Ships. (1983). 1973, Nov. 2, 1973, 1340 U.N.T.S. 61.
Jelley, N. (2017). Efficiency. In *A Dictionary of energy science*. doi: https://doi.org/10.1093/acref/9780191826276.001.0001.
MCA. (2012). Pollution-Entry into force of the energy efficiency design index. https://www.gov.uk/government/uploads/system/uploads/attachment_data/file/441041/MGN_462.pdf. Accessed 18 Nov 2017.
M Gillespie & Co v James Howden & Co. (1885). 12 R 800.
Nikaki, T. (2016). Slow steaming clauses and international sales contracts: A successful marriage? In B. Soyer & A. Tettenborn (Eds.), *International trade and carriage of goods* (1st ed., pp. 15–55). Oxon: Informa.
OCIMF. (2017). Ship inspection report programme. https://www.ocimf.org/sire/about-sire/. Accessed 18 November 2017.
Shipbuilder' Association of Japan (SAJ form). (1974).
UK. (1999). Miscellaneous Series 029/1999: Cm4427.

## *Further Reading*

Protocol of 1978 relating to the International Convention for the Prevention of Pollution from Ships. (1973). Feb. 17, 1978, 1340 U.N.T.S. 61.

# Analyzing Approaches to Set Greenhouse Gas Reduction Targets in Anticipation of Potential "Further Measures" for International Shipping

**Seijiro Morimoto**

## 1 Introduction

Reduction of greenhouse gas (GHG) emissions from ships has been a major challenge at the International Maritime Organization (IMO). This specialized agency is recognised as the forum to discuss measures to address GHG emissions from international shipping as implied by article 2.2 of the Kyoto Protocol. The Marine Environment Protection Committee (MEPC) of the IMO has discussed a range of measures, including technical, operational and market-based measures (MBMs). During the 62nd session of the MEPC (MEPC62) in 2011, an amendment to Annex VI of the International Convention for the Prevention of Pollution from Ships (MARPOL) was adopted to introduce regulations on the Energy Efficiency Design Index (EEDI) for new ships and the Ship Energy Efficiency Management Plan (SEEMP) for all ships. That regulation entered into force on 1 January 2013.

Considerations on possible MBMs were suspended at MEPC65 in 2013 after years of deliberation (IMO 2013), and it was agreed to progress the matter based on the 'three-step approach', which consists of data collection, data analysis and decision-making on further measures. In 2016, discussions on the Data Collection System (DCS) for fuel oil consumption of ships culminated into the adoption of the amendment to MARPOL Annex VI to incorporate regulations on the DCS at MEPC70 (IMO 2016a). In 2015, the European Union adopted Regulation (EU) 2015/757 on the monitoring, reporting and verification (MRV) of carbon dioxide ($CO_2$) emissions from maritime transport, which "*should serve as a model for the implementation of a global MRV system*," according to the preambular paragraph (34) of the regulation.

S. Morimoto (✉)
Japan Maritime Center, Tokyo, Japan
e-mail: s-morimoto@jpmac.or.jp



A. I. Ölçer et al. (eds.), *Trends and Challenges in Maritime Energy Management*,
WMU Studies in Maritime Affairs 6, https://doi.org/10.1007/978-3-319-74576-3_3

Although progress has thus been made to deal with climate change in the international shipping sector, the IMO is now facing an ever-increasing challenge to set the GHG emissions reduction target, due to developments in other fora: the ICAO agreed on the global aspirational goals at the 37th Assembly in 2010 to improve fuel efficiency of aircrafts by 2% on an annual basis and to achieve carbon neutral growth from 2020 (CNG2020) with a view to implementing a global MBM (ICAO 2010); the Paris Agreement was adopted at the twenty-first session of the Conference of the Parties to the UNFCCC (COP21) in 2015, and entered into force in November 2016 with 190 Parties communicating their intended nationally determined contributions (INDCs) in accordance with relevant COP decisions. Many of the INDCs included quantified emission reduction targets up to 2030, which may become formal targets under the nationally determined contributions (NDCs) and will be subject to the global stock-taking process every 5 years for further review and revision in view of contributing to the objective of the UNFCCC.

The IMO, which had considered the issue of target-setting at MEPC59 and MEPC60 but deferred the consideration thereafter, resumed the consideration on target-setting at MEPC68 in 2015 (IMO 2015b). Following the consideration on how to progress the issue, the "Roadmap for developing a comprehensive IMO strategy on the reduction of GHG emissions from ships" was approved at MEPC70 (IMO 2016a). According to the roadmap, the initial IMO Strategy is scheduled for adoption in spring 2018, and deliberations for the development of the strategy will likely entail considerations on target-setting. It is against this backdrop that this paper aims to evaluate the strengths and weaknesses of various approaches in setting the GHG[1] reduction target for international shipping, taking into account the specific circumstances relating to the shipping industry and developments in other sectors, in order to find the appropriate approach for the shipping sector.

Setting a reduction target requires consideration on various elements. Parties to the UNFCCC have agreed that INDCs may include information such as reference point/base year, time frames and/or periods for implementation, scope and coverage, planning processes, assumptions and methodological approaches, and how the Party considers that its INDC is fair and ambitious, in light of its national circumstances, and how it contributes to achieving the objective of article 2 of UNFCCC (UNFCCC 2015). These information elements, which most Parties addressed in their INDCs (UNFCCC 2016), may also be relevant to the international shipping sector, but as a starting point, this analysis will consider various types of targets and approaches or methodologies for target-setting.

The remainder of the content is organized as follows: Sect. 2 considers what would be the appropriate type of target for international shipping, taking into account the characteristics of different types of targets as well as roles and features of international shipping; Sect. 3 analyzes the strengths and weaknesses of potential

[1]Hereinafter, the term 'GHG' will be replaced by '$CO_2$' in the context for international shipping, since carbon dioxide is considered as the most important GHG emitted by ships in terms of quantity and of global warming potential (IMO 2009a).

approaches to set the GHG reduction target for international shipping, in order to find the appropriate way forward; Sect. 4 summarizes the arguments and provides concluding remarks.

# 2 Consideration on Types of Targets

## *2.1 Absolute Target and Intensity Target*

The quantified emission reduction targets communicated by Parties to the UNFCCC in their INDCs can be categorized in four main types: (a) absolute emission reduction target, which is expressed as an emission reduction relative to the level in a specified base year; (b) emission reduction target relative to business-as-usual (BAU) level; (c) peaking target, which specifies the year or time frame in which emissions are expected to peak; and (d) intensity target, expressed as reduction of emission per gross domestic product (GDP) or per capita relative to the level in the base year. According to UNFCCC (2016), 32% of the Parties that submitted INDCs have set (a), 45% have set (b), 2% have set (c), and 4% have set (d). The international aviation sector's CNG 2020 and 2% annual efficiency improvement goals could be categorized as (c) and (d) respectively, although the base year for the latter is rolling instead of fixed.

The first three types have commonalities in that they all show the development of absolute emission level; the difference lies in that (a) shows the reduction relative to the base year level, whereas (b) shows the reduction relative to the projected BAU level, and (c) expressing a trajectory for multi-year period in which emissions following the target year do not exceed the level of the said year, instead of showing a target for one specific year.

In contrast with these absolute emission targets, intensity target only shows the development of intensity (emissions per output) level and does not in itself provide the absolute emission level. However, an intensity target can be converted to an absolute emission target, once activity level is taken into account, and conversely an absolute emission target can be converted to an intensity target. To which specific type of absolute emission target an intensity target could be converted would depend on the stringency of the intensity and the assumed level of activity.

## *2.2 Characteristics of Absolute Target and Intensity Target*

Literature review on assessment of target options provides some noteworthy characteristics of these two major types (of targets): absolute (volume) and intensity (efficiency) targets.[2] First and foremost, an absolute target has higher certainty of

[2]For example, the work of Ellerman and Wing (2003) explained the differences and similarities of absolute versus intensity-based emission limits.

Absolute target ↓ (CO2 Emission)     Intensity target ↓ (CO2 Emission / Activity)

$$CO2\ Emission = Activity \times \underbrace{\frac{CO2\ Emission}{Activity}}_{(Carbon\ Intensity)}$$

$$= Activity \times \underbrace{\frac{Energy}{Activity}}_{(Energy\ Intensity)} \times \underbrace{\frac{CO2\ Emission}{Energy}}_{(Fuel\ Mix)}$$

**Fig. 1** Scope of absolute target and intensity target. The formula expresses how the level of $CO_2$ emission is determined as a function of two contributing factors, namely the activity level and carbon intensity. Note that factors influencing carbon intensity are energy intensity and fuels consumed

environmental outcome in terms of $CO_2$ emission level, because it aims to control the overall emission which factors in both the activity and carbon intensity, while intensity target, which aims to control the carbon intensity only, has lower certainty of the overall emission because it would be affected by the activity level which is uncontrolled by the intensity target (see Fig. 1).

On the other hand, if assessed against the certainty of environmental integrity in terms of carbon intensity, absolute target has lower certainty, because the unexpected change in the activity level would alter the required level of intensity to achieve the target; for example, if the absolute emission target is achieved simply due to unpredicted low growth of activity, it may not necessitate the level of intensity to be improved as initially expected.[3] The tradeoff between these two (un)certainties stems from the difference in whether the target factors in the activity level or not.

Another feature pertinent to this tradeoff is that certainty of abatement costs to achieve the target level would be higher for intensity targets, since the costs for complying with the absolute targets may be influenced by changes in activity levels, while intensity targets may reduce such economic uncertainty by reducing the impact of changes in activity levels. For example, the work of Rietbergen and Blok (2010) pointed out that the major disadvantage of absolute targets is the high uncertainty of the costs related to achieving the target, while intensity targets can reduce uncertainty in compliance costs in case of unexpected growth of activity. Herzog et al. (2006) also argued that intensity targets may reduce the economic uncertainty associated with particular targets by adjusting to economic changes.

[3]Rietbergen and Blok (2010) have indicated that the quality of environmental outcome of measures with absolute targets is not necessarily satisfactory since there are cases in which emissions are reduced not as a result of abatement technologies, whereas environmental integrity of intensity targets is much more certain because the commitment level remains the same even if output (activity) fluctuates.

## 2.3 Appropriate Type of Target for International Shipping

These characteristics as well as the implications of adopting specific type of target, in light of the distinctive roles and features of international shipping, should be taken into account when considering the reduction target for international shipping.

International shipping serves as the backbone of global trade and plays a vital role for the sustainable development of world economy, in particular developing countries. World seaborne trade volumes accounted for over 80% of merchandise trade and developing countries' share of seaborne trade in terms of goods loaded and unloaded have both exceeded 60% in 2015 (UNCTAD 2016). In value terms, seaborne trade accounted for 67% of merchandise trade in 2015, excluding intra-EU trade (IHS Markit 2016). The share of merchandise trade in world GDP has increased, averaging 48% in the last decade, and the magnitude has been higher for least developed countries, whose share has averaged 53% in the same period (World Bank 2016). The importance of transport has also been recognized in the context of Sustainable Development Goals (SDGs), as freight volumes by mode of transport, including maritime transport, has been proposed to be used as a measure of progress towards achieving target 9.1 of the SDGs[4] (UNCTAD 2016).

International shipping is also the most carbon efficient mode of transport to which modal shift from less environmentally friendly mode of transport would be an effective measure to reduce global $CO_2$ emission. In contrast, if cargoes are shifted from maritime transport to other transport modes, for example due to climate policy resulting higher price for shipping, this would increase total emissions because other modes have lower transport efficiency (IMO 2009a).

Considering these characteristics, it seems that intensity target would be more appropriate to accommodate growth of trade[5] and modal shift, which would also conform to the idea of promoting sustainable development and consistent global GHG mitigation policies. It would also be useful as a measure of progress to improve sustainability of international shipping, which would contribute to the development of efficient global supply chains.

As a way of complementing the deficiencies of intensity targets, in particular the attributed uncertainty of environmental outcome in terms of absolute emission level, it should be noted that absolute emission trajectories could be derived by applying different activity scenarios to intensity targets, which would provide a better insight into how the absolute emission would develop if intensity targets are achieved.

---

[4]Target 9.1 of the SDGs is to "*develop quality, reliable, sustainable and resilient infrastructure, including regional and transborder infrastructure, to support economic development and human wellbeing, with a focus on affordable and equitable access for all.*"

[5]IMO (2009a) projected that the annual growth rate of world seaborne trade from 2000 to 2050 ranges from 2.1% to 3.3% in the base case scenarios, while the average growth rate from 2000 to 2016 is estimated as 3.5% (Clarksons Research 2017).

# 3 Analysis on Approaches to Set GHG Reduction Target

## 3.1 Three Potential Approaches for Target-Setting

Based on literature review and discussions held at MEPC, a number of approaches (or methodologies) could be envisaged for the development of $CO_2$ emission reduction target for international shipping, which could be categorized into three main types: carbon budget approach, similar reduction approach, and efficiency based approach.

Carbon budget approach envisions setting an absolute emission reduction target, based on the carbon budget[6] estimated for the shipping sector, which is derived from the remaining cumulative global $CO_2$ emission budget to limit global warming to 2 °C, assuming shipping's share of global emissions remains constant over time. For example, Smith et al. (2015) have assumed cumulative global $CO_2$ emissions consistent with the two-degree scenario over the period from 2011 to 2100 to be 1428 Gt-$CO_2$,[7] applied international shipping's share of global emissions[8] to the cumulative global emissions in order to obtain the carbon budget for international shipping (33 Gt-$CO_2$), and depicted an emission trajectory that would develop as BAU until 2020 and linearly decrease thereafter to use up the remaining cumulative budget in 2079.

Similar reduction approach could be considered, in broad terms, as an approach to determine a target for international shipping so that the level of reduction efforts in the shipping sector would be the same as other sectors. For example, the work of Cames et al. (2015) refers to, as one potential option, an approach to aggregate GHG emission reduction pledges made by Parties to the UNFCCC and transfer this global effort to international shipping, based on the idea that emissions of the shipping sector should develop at the same pace as the world at large. Another example could be the approach proposed by Norway (IMO 2010), in which marginal abatement costs (MAC) required to achieve the global reduction target are applied to international shipping, utilizing the MAC curve developed for the international shipping sector, to determine the reduction target for international shipping, based on the idea that the economic efforts to reduce emissions (abatement costs) should be equal with those of other sectors.[9]

Efficiency based approach could be conceptualized as an approach to determine a target based on the levels of efficiency improvement that are deemed technically achievable, based on the potential analysis of efficiency improvement of ships.

---

[6]Carbon budget is the area under a GHG emissions trajectory that satisfies assumptions about limits on cumulative emissions estimated to avoid a certain level of global mean surface temperature rise, and it may be defined at the global level, national, or sub-national levels (IPCC 2014).

[7]The number falls within the range of 960–1430 Gt-$CO_2$, a scenario included in IPCC (2014) in which $CO_2$ concentrations in 2100 is estimated to be 530 ppm $CO_2$-eq with no overshoot.

[8]IMO (2015a) estimated the average share of international shipping's emissions relative to global emissions to be 2.4% from 2007 to 2012.

[9]Hereinafter, the approach proposed by Norway is referred to as 'Equal MAC approach.'

Efficiency (intensity) targets developed under this approach could derive absolute emission targets in order to have a better insight into the expected trajectories when intensity targets are achieved. Examples can be drawn from proposals made by Japan (IMO 2009b, 2016b), in which efficiency improvement of different ship types as well as the expected $CO_2$ emission reduction levels relative to BAU emissions for different transport projection scenarios are depicted.

## 3.2 *Preliminary Assessment of Potential Approaches for Target-Setting*

These three approaches could be assessed in light of the criteria provided in the Intergovernmental Panel on Climate Change (IPCC) Fifth Assessment report (AR5) for evaluating international climate policies: environmental effectiveness, which can be construed as the extent to which the policy achieves its objective to reduce the causes and impacts of climate change; economic performance, which requires consideration on economic efficiency and cost-effectiveness of specific policies; distributional and social impacts, which include aspects of fairness and equity; and institutional/political feasibility, which measures the likelihood of a policy being accepted, implemented, and complied with. An initial attempt of preliminary assessments of the three approaches against these criteria (see Table 1) is provided below.

With regard to the first criterion, the carbon budget approach seems to have higher environmental effectiveness, since it aims to allocate the science-based global $CO_2$ budget consistent with the two-degree scenario to international shipping. In contrast, the other two approaches seem to have lower environmental effectiveness: the likelihood of achieving the 'two-degree target' under the similar reduction approach would conceptually depend on the efforts of other sectors, which still seems insufficient according to UNFCCC (2016) to achieve the 'two-degree target'[10]; environmental effectiveness of the efficiency based approach would depend on the stringency of the efficiency level, which would likely be less stringent than the level of efficiency improvement required to achieve the absolute emission target derived from the carbon budget approach,[11] a case which would necessitate detailed analysis on reduction potential in international shipping in terms improving carbon intensity.

---

[10]UNFCCC (2016) pointed out that the aggregate GHG emission levels resulting from the implementation of INDCs communicated by 4 April 2016 would be higher than the emission levels under least-cost two-degree scenarios based on IPCC AR5 scenario database.

[11]According to Cames et al. (2015) and Smith et al. (2016), the estimated required carbon intensity improvement in 2050 to achieve the reduction targets developed under the carbon budget approach would be over around 80% relative to the intensity levels in recent years.

**Table 1** Preliminary assessment of approaches to set reduction target for international shipping

| Assessment criteria[a] | Carbon budget approach | Similar reduction approach | Efficiency based approach |
|---|---|---|---|
| Environmental effectiveness | Higher | Lower (depends on efforts of other sectors) | Lower (depends on stringency of efficiency level) |
| Economic performance | | 'Equal MAC approach' aims for cost-effectiveness | |
| | Depends on economic efficiency and cost-effectiveness of measures | | |
| Distributional and social impacts | Less equitable (capability is not duly taken into account) | 'Equal MAC approach' aims for fair distribution of costs | More equitable (reduction potential is taken into account) |
| Institutional and political feasibility | Depends on environmental effectiveness, economic performance, and distributional and social impacts | | |

[a]Assessment criteria are drawn from IPCC (2014)

According to IPCC (2014), assessment on the second criterion requires consideration on whether a policy maximizes net benefits (economic efficiency) and whether it attains a specific level of environmental performance at least cost (cost-effectiveness). Although the 'Equal MAC approach' seems to conform with an element of cost effectiveness, which requires that marginal compliance costs should be equal among regulated parties, since the approach aims to set a reduction target that would equalize the marginal abatement costs of the shipping sector with other sectors, the overall assessment of the three approaches may require evaluations of specific measures on whether they meet the conditions mentioned above.

The third criterion, which includes aspects of fairness and equity, relates to how costs and benefits are shared among parties, and it is considered that such distributional equity and fairness are important attributes and requirements for effective implementation of climate policies. Regarding this criterion, the similar reduction approach would seem somewhat equitable because the basic idea is that the shipping sector should aim for similar reduction efforts with other sectors. The 'Equal MAC approach' clearly aims for fair distribution of costs among sectors based on the MAC analysis, however, it may not necessarily take into account the ability-to-pay of the shipping sector. Efficiency based approach would seem more likely to be equitable than the carbon budget approach, because the latter seems to derive the reduction target basically from the scientifically required reduction level in a top-down manner without duly regarding the capability of the shipping sector, while the former takes into account the reduction potential of international shipping based on the analysis of intensity improvement.

Assessment on the fourth criterion, in particular political feasibility, may require comprehensive evaluation of the first three criteria; environmental effectiveness, economic performance, and distributional and social impacts. It may also depend on the preference on which criteria is prioritized over another. In this regard, it is worth noting that there may be conflicts or tradeoffs between and among these criteria; for example, unfair distribution of costs and benefits could prove politically challenging, because efficiency may be gained at the expense of equity objectives (IPCC 2014).

Thus, strengths and weaknesses of each approach may vary depending on the criteria, however, in order to progress the consideration in a constructive manner to reach a feasible target for international shipping in view of achieving the objectives of the UNFCCC, a balanced approach that would benefit from the strengths of each approach could be more useful. In other words, a hybrid approach, in which consideration is based on the analysis of the reduction potential in terms of improvement of carbon intensity and the extent to which the stringency of the intensity could be strengthened to meet the 'two-degree target', taking into account the reduction efforts of other sectors and characteristics of international shipping, would seem to be appropriate to set a fair, ambitious and achievable target.

## 3.3 *Issues Regarding "Further Measures"*

According to the roadmap referred to in Sect. 1, the adoption of the initial strategy will be followed by the data collection, data analysis and decision making phases. In case the initial strategy includes a reduction target for international shipping, considerations on whether the target assumes reductions via offsetting may be needed in order to keep consistencies, because further measures, in particular MBMs, that may be considered in the later stage may or may not include offsetting mechanisms.

Offsetting, which means purchasing offset credits from non-shipping sectors as a means of nominally reducing emissions from international shipping, has cost saving potential, but it should also be noted that concerns have been raised regarding validity of baselines and additionality of credits obtained from Clean Development Mechanism (CDM) projects (IPCC 2014) as well as risks of double-counting with respect to offset credits that may be used for the international aviation sector (Hermwille 2016; FERN 2016), any of which may undermine the environmental effectiveness of offsetting.[12] Uncertainties also remain as to future price of offset credits, which may also affect its economic performance.

It would therefore seem to be more appropriate not to assume reductions made via offsetting as a major component in the reduction target, until its environmental effectiveness and economic performance are duly substantiated, so as to ensure a deliberate decision-making process for the development of the initial strategy. It should also be noted that such a measure would not be a sustainable solution in the long term, because simply buying emission credits from other sectors will not lead to actual improvement of carbon intensity in the international shipping sector.

[12]The environmental integrity of Joint Implementation (JI) offsets during the first commitment period of the Kyoto Protocol is also questioned, according to Kollmuss et al. (2015).

## 4 Conclusion

Based on considerations on different types of targets made above, this analysis puts forward that intensity target seems to be more appropriate than absolute target for international shipping, since the former type would better accommodate growth of seaborne trade and modal shift to international shipping, which would also conform to the idea of promoting sustainable development and consistent global GHG mitigation policies. Also, based on the preliminary qualitative analysis on potential approaches for target-setting, another important finding is that considerations to set the reduction target for international shipping should be based on the analysis of the reduction potential in terms of improvement of carbon intensity and the extent to which the stringency of carbon intensity could be strengthened to meet the 'two-degree target,' in order to set a fair, ambitious and achievable target.

Although considerations on target-setting, if launched in the process of developing the IMO strategy, may entail deliberations on many different aspects, which could not be thoroughly discussed here, findings from literature review and analysis made above may have some relevance for the consideration to set a reduction target for international shipping. In particular, the identification of the importance of carefully analyzing the reduction potential may seem to indicate the high value of increasing sector-wide discussions on identifying specific areas which have the potential to significantly improve carbon intensity, including measures for improving energy intensity and for adopting low-carbon energy including renewable energy, and assessing the feasibility of implementing those measures.

It therefore seems urgent that such deliberation takes place, aiming for the most ambitious level of efforts while taking into account on what would be the most appropriate way forward for the industry, so that the international shipping sector could take meaningful action towards contributing to the global efforts to address climate change and for the enhancement of sustainable development.

**Acknowledgements** The author would like to thank Akira Okada, Koichi Yoshida, Hideaki Saito, Hiroshi Matsui, Kazuya Nakao, Tamio Kawashima, Tomoo Kitayama, Keiji Tomoda, Yoshiharu Tsuno, and Takuma Matsuda for their valuable comments. Opinions expressed, are of the author's alone.

## References

Cames, M., Graichen, V., Faber, J., & Nelissen, D. (2015). *Greenhouse gas emission reduction targets for international shipping* (Discussion paper on behalf of the Federal Environment Agency). Accessed December 19, 2016, from http://www.oeko.de/oekodoc/2241/2015-023-en.pdf

Clarksons Research. (2017). *Shipping review and outlook*, Spring 2017.

Ellerman, A. D., & Wing, I. S. (2003). *Absolute versus intensity-based emission caps*. Accessed December 19, 2016, from http://citeseerx.ist.psu.edu/viewdoc/download?doi=10.1.1.297.7192&rep=rep1&type=pdf

EU. (2015). Regulation 2015/757 of the European Parliament and of the Council of 29 April 2015 on the monitoring, reporting and verification of carbon dioxide emissions from maritime transport, and amending Directive 2009/16/EC (Text with EEA relevance), from http://data.europa.eu/eli/reg/2015/757/oj

FERN. (2016). *Cheating the climate: The problems with aviation industry plans to offset emissions.* Accessed January 10, 2017, from http://www.fern.org/sites/fern.org/files/Fern%20Cheating%20the%20climate%20briefing%20paper%208pp.pdf

Hermwille, L. (2016). *Offsetting for international aviation. The state of play of market-based measures under ICAO.* JIKO Policy Brief No. 1/2016. Accessed November 19, 2016, from http://www.carbon-mechanisms.de/publikationen/details/?jiko%5Bpubuid%5D=451

Herzog, T., Baumert, K., & Pershing, J. (2006). *Target: Intensity. An analysis of greenhouse gas intensity targets.* World Resources Institute. Accessed December 8, 2016, from http://pdf.wri.org/target_intensity.pdf

ICAO. (2010). *Resolution A37-19: Consolidated statement of continuing ICAO policies and practices related to environmental protection – Climate change.* Accessed February 24, 2016, from https://www.icao.int/environmental-protection/37thAssembly/A37_Res19_en.pdf

IHS Markit. (2016). *World trade service.* Accessed January 16, 2017, from https://diw.ihs.com/index.aspx

IMO. (2009a). *Second IMO GHG Study 2009.* International Maritime Organization (IMO), London, UK, April 2009; Buhaug, Ø., Corbett, J. J., Endresen, Ø., Eyring, V., Faber, J., Hanayama, S., Lee, D. S., Lee, D., Lindstad, H., Markowska, A. Z., Mjelde, A., Nelissen, D., Nilsen, J., Pålsson, C., Winebrake, J. J., Wu, W., Yoshida, K.

IMO. (2009b). *Consideration of appropriate targets for reducing CO2 emissions from international shipping proposed by Japan.* IMO doc. MEPC 59/4/35.

IMO. (2010). *Alternative emission caps for shipping in 2020 and 2030 proposed by Norway.* IMO doc. MEPC 60/4/23.

IMO. (2013). Report of the Marine Environment Protection Committee on its Sixty-Fifth Session. IMO doc. MEPC 65/22.

IMO. (2015a). *Third IMO GHG Study 2014.* International Maritime Organization (IMO), London, UK, April 2015; Smith, T. W. P., Jalkanen, J. P., Anderson, B. A., Corbett, J. J., Faber, J., Hanayama, S., O'Keeffe, E., Parker, S., Johansson, L., Aldous, L., Raucci, C., Traut, M., Ettinger, S., Nelissen, D., Lee, D. S., Ng, S., Agrawal, A., Winebrake, J. J., Hoen, M., Chesworth, S., Pandey, A.

IMO. (2015b). Report of the Marine Environment Protection Committee on its Sixty-Eighth Session. IMO doc. MEPC 68/21.

IMO. (2016a). Report of the Marine Environment Protection Committee on its Seventieth Session. IMO doc. MEPC 70/18.

IMO. (2016b). *GHG emissions reduction target of international shipping based on energy efficiency proposed by Japan.* IMO doc. MEPC 70/7/3.

IPCC. (2014). *Climate change 2014: Mitigation of climate change.* Contribution of Working Group III to the Fifth Assessment Report of the Intergovernmental Panel on Climate Change. Accessed December 27, 2016, from http://www.ipcc.ch/report/ar5/wg3/

Kollmuss, A., Schneider, L., & Zhezherin, V. (2015). *Has joint implementation reduced GHG emissions? Lessons learned for the design of carbon market mechanisms.* Stockholm Environment Institute Working Paper No. 2015-07. Accessed January 4, 2017, from https://www.seiinternational.org/mediamanager/documents/Publications/Climate/SEI-WP-2015-07-JI-lessons-for-carbon-mechs.pdf

Rietbergen, M., & Blok, K. (2010). Setting SMART targets for industrial energy use and industrial energy efficiency. *Energy Policy, 38*, 4339–4354. https://doi.org/10.1016/j.enpol.2010.03.062.

Smith, T., Raucci, C., Haji Hosseinloo, S., Rojon, I., Calleya, J., Suárez de la Fuente, S., et al. (2016, October). *CO2 emissions from international shipping. Possible reduction targets and their associated pathways*. Prepared by UMAS, London. Accessed November 17, 2016, from https://www.danishshipping.dk/en/press/news/reducing-shippings-co2-emissions-the-danish-shipowners-way-forward/

Smith, T., Traut, M., Bows-Larkin, A., Anderson, K., McGlade, C., & Wrobel, P. (2015). *CO2 targets, trajectories and trends for international shipping*. Accessed September 13, 2016, from http://www.tyndall.ac.uk/sites/default/files/co2_targets_trajectories_and_trends_for_international_shipping.pdf

UNCTAD. (2016). *Review of maritime transport 2016*. Accessed November 29, 2016, from http://unctad.org/en/PublicationsLibrary/rmt2016_en.pdf

UNFCCC. (2015). *Decision 1/CP.20. Lima call for climate action*. Available online at: http://unfccc.int/resource/docs/2014/cop20/eng/10a01.pdf#page=2.

UNFCCC. (2016). *Aggregate effect of the intended nationally determined contributions: An update*. Available online at: http://unfccc.int/resource/docs/2016/cop22/eng/02.pdf.

World Bank. (2016). *World development indicators*. http://data.worldbank.org/indicator/. Accessed 16 January 2017.

# An Analysis of Non-conformities with the Objective of Improving Ship Energy Efficiency: Case Studies of Turkish Shipping Companies

**Özcan Arslan, Elif Bal Beşikçi, Fırat Bolat, and Koray Yaş**

## 1 Introduction

Maritime transport has become the backbone of global trade due to its capacity, connectivity and reliability. According to the United Nations Conference on Trade and Development (UNCTAD 2017), 80% of percent of world goods trade by volume is carried by sea and handled by ports worldwide.

The normal operations (and growth) of the shipping industry due to the volume of world trade are resulting into increased fuel consumption and greenhouse gas (GHG) emissions. To mitigate these problems, the International Maritime Organization (IMO) introduced a series of emission control regulations under MARPOL Annex VI. These regulations contain two important issues to improve ship energy efficiency, following their entry into force, on 1 January 2013: Energy Efficiency Design Index (EEDI) (for new ships) and Ship Energy Efficiency Management Plan (SEEMP) (for all ships) (IMO 2011). Whilst the primary aim of the EEDI is to reduce carbon emissions from ships over time with innovative technologies or measures, SEEMP is focusing on the improvement of "operational conditions", by implementing existing technologies on board a ship. In addition, a huge part of the ship operating cost is associated with fuel prices, which form approximately 35–70%

Ö. Arslan (✉) · E. Bal Beşikçi
Istanbul Technical University, Istanbul, Turkey
e-mail: bale@itu.edu.tr

F. Bolat
Republic of Turkey Ministry of Transport Maritime Affairs and Communications, Ankara, Turkey

K. Yaş
DITAS Marine Operations and Tanker Management Co., Istanbul, Turkey


A. I. Ölçer et al. (eds.), *Trends and Challenges in Maritime Energy Management*, WMU Studies in Maritime Affairs 6, https://doi.org/10.1007/978-3-319-74576-3_4

of the total ship operating cost (Wigforss 2012). Thus, reducing energy consumption can lower management costs; that eventually maintains competitive power of the shipping company and reduces the respective environmental pollution (Perera and Mo 2016). As a result, improving ship energy efficiency has become one of the most important issues for the shipping industry in terms of both environmental and economic concerns.

The major oil companies established the Oil Company International Marine Forum (OCIMF) in 1970 in response to increasing public awareness of marine pollution after the Torrey Canyon incident, which was the largest pollution incident in shipping history. The main aim of OCIMF is to offer operational recommendations and guidelines with the aim of ensuring high operating and safety standards (OCIMF 2014). In 1993, OCIMF launched a Ship Inspection Report Programme (SIRE) to address concerns about sub-standard tanker operation and maintenance and to provide a standardized inspection format for ship inspections. This study scrutinizes the results of SIRE inspections conducted on tankers, which occupy an important place in world trade, with a continuous rise observed in their development in terms of energy efficiency. SIRE inspection results have been used in numerous studies in literature to indicate the deficiencies of ships mainly related to safety issues and to demonstrate the effectiveness of inspections to reduce the likelihood of casualties (Howell and Bhattacharya 2016; Kececi and Arslan 2017; Knapp and Franses 2009). However, no research has investigated these SIRE results in terms of ship energy efficiency, especially from the point of ship engine and steering compartments.

It is intended to analyze the results of inspections of Turkish tanker companies, which are conducted under the Ship Inspection Report Programme (SIRE), within the scope of Chapter 11 (engine and steering compartments) and discuss the points that affect the energy efficiency of the vessels, both directly and indirectly. The aim is to investigate the deficiencies regarding energy efficiency and the root causes thereof. Flag State and Port State Inspections as well as other industrial inspections are excluded from the assessment.

## 2 SIRE (Ship Inspection Report Programme)

The SIRE programme is a tanker risk assessment tool used by ship-owners, vessel operators, port and terminal authorities and government agencies. Large oil companies worldwide have developed a system called 'OCIMF' (Oil Companies International Maritime Forum), for the purpose of investigating tanker accidents and responding appropriately for future avoidance. Large oil companies publish the results as regards the tankers they assess under "SIRE". Tanker companies and operators can transport the cargo of large oil companies based on the fact that they have undergone inspections, with the least deficiency. If the vessels fail to get approval in such inspections, they do not qualify to transport the cargo of the large oil companies, which in turn leads to major financial losses.

Inspection guidelines, which are observed in the inspection to ensure a standard inspection procedure, are included in OCIMF-SIRE inspection manual. The assessment results should be recorded in a database with all details in order to achieve a proper assessment. The content of the database is as follows:

Ship name: Recording which vessel is inspected will ensure that the performances of the vessels within the company are compared and the details of that ship are tracked.

Date of inspection: Recording the inspection dates will ensure that the performance of the fleet and the change in the performance of the fleet in time is monitored.

Inspection number: Designating a number for each inspection will prevent double records or missing records; and ensure that the inspection results are examined in a more detail way.

Name of OCIMF inspecting company: It will ensure that the assessment results of the inspecting oil company are compared with other inspecting companies and thus help to determine which conditions the inspecting companies consider more important.

Name of Inspector: It will ensure that the findings of the inspectors are classified.

The OCIMF-SIRE inspection manual specifies the criteria used to identify the deficiency found in the inspection. The manual is composed of 13 chapters and sub-classes of these chapters. These chapters are (SIRE 2016):

- Chapter 1. General Information
- Chapter 2. Certification and Documentation
- Chapter 3. Crew Management
- Chapter 4. Navigation
- Chapter 5. Safety Management
- Chapter 6. Pollution Prevention
- Chapter 7. Structural Condition
- Chapter 8. Cargo and Ballast Systems
- Chapter 9. Mooring
- Chapter 10. Communications
- Chapter 11. Engine and Steering Compartments
- Chapter 12. General Appearance and Condition
- Chapter 13. Ice Operations

Chapter 11 of OCIMF-SIRE manual was utilized since this study examines only the discrepancies and deficiencies stemming from the ship engine and steering compartments as well as the impacts thereof on energy efficiency.

# 3 Method

## 3.1 *Assessment of Research Findings*

The assessment of the non-conformities identified in vessel inspections can be used as a mean to determine the points that might affect the energy efficiency of the vessels. In the study, technical and engine room non-conformities and deficiencies, which were identified in SIRE inspections carried out on nineteen tanker operators in Turkey, were reviewed and assessed statistically. The data recorded consists of the inspection results for 19 companies between 2006 and 2016. The total number of inspections is 269.

The result of the study carried out according to the companies inspected is given in Table 1. In this context, company no. 1 is the most inspected company, with 42 inspections and the rate of finding/inspection is 1.6. Company no. 16 had 1 inspection and the rate of finding/inspection is 1.00. The average rate of finding/inspection for the companies is 1.4.

The distribution of the number of inspections/findings by years is given in Fig. 1. It is pointed out that the rate of inspection/finding based on this study is 1.3 between 2011 and 2016. The highest rates were observed prior to 2010. The average rate of inspection/finding identified by years is 1.5.

**Table 1** The rate of finding/inspection

| Firm no | Inspection number | Findings | Finding/inspection rate |
|---|---|---|---|
| 1 | 42 | 67 | 1.6 |
| 2 | 13 | 15 | 1.2 |
| 3 | 2 | 2 | 1.0 |
| 4 | 23 | 28 | 1.2 |
| 5 | 25 | 40 | 1.6 |
| 6 | 10 | 12 | 1.2 |
| 7 | 5 | 5 | 1.0 |
| 8 | 4 | 7 | 1.8 |
| 9 | 5 | 9 | 1.8 |
| 10 | 24 | 37 | 1.5 |
| 11 | 15 | 17 | 1.1 |
| 12 | 23 | 41 | 1.8 |
| 13 | 40 | 98 | 2.5 |
| 14 | 12 | 20 | 1.7 |
| 15 | 2 | 2 | 1.0 |
| 16 | 1 | 1 | 1.0 |
| 17 | 2 | 2 | 1.0 |
| 18 | 18 | 21 | 1.2 |
| 19 | 3 | 4 | 1.3 |
| Total | 269 | 428 | 1.4 |

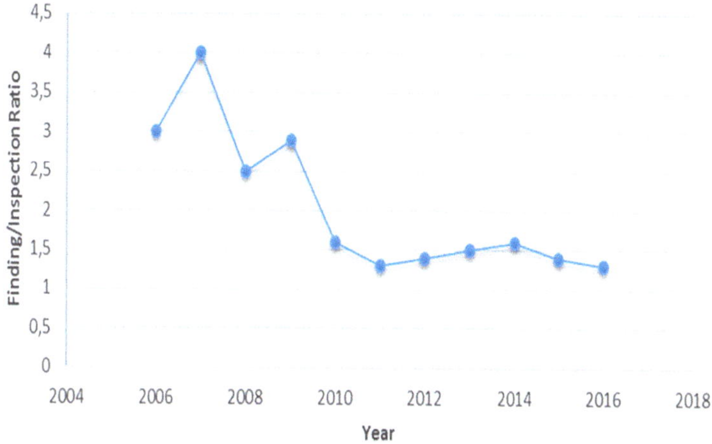


**Fig. 1** The distribution of the number of inspection/finding by years

## *3.2 Analysis of Root Causes to the Findings*

It is of importance to investigate the root causes to eliminate the deficiencies identified in the inspections. Should the root cause be identified, it will be set out clearly how to eliminate the deficiency.

The deficiencies identified in the vessels can be discussed under 6 main topics.

R1: System failure: This root cause indicates that the deficiency is caused by a lack of procedure and policy.

R2: Implementation error—Company: This root cause indicates that the deficiency is caused because the procedures are not implemented by the office.

R3: Implementation error—Vessel: This root cause indicates that the deficiency is caused due to the procedures not implemented by the vessel.

R4: Technical error: This root cause indicates that the deficiency is caused because of reasons such as malfunction or dysfunction of equipment.

R5: Design and construction error: This root cause indicates that the deficiency emerged when the vessel was built.

R6: Personal errors: This root cause indicates that the deficiency was caused by persons not related to the vessel, such as port officials, or inspection teams and classification society, among others.

Table 2 illustrates the analysis of root causes for deficiencies on a company basis.

Figure 2 illustrates the distribution in percentage, based on the root causes to a total of 428 deficiencies given in Table 2.

Table 2 The analysis of root causes for deficiencies on a company basis

| Firm no | Inspection number | R1 | R2 | R3 | R4 | R5 | R6 | Total findings |
|---|---|---|---|---|---|---|---|---|
| 1 | 42 | 17 | 3 | 24 | 18 | 5 | 0 | 67 |
| 2 | 13 | 3 | 1 | 7 | 4 | 0 | 0 | 15 |
| 3 | 2 | 1 | 0 | 1 | 0 | 0 | 0 | 2 |
| 4 | 23 | 10 | 0 | 13 | 4 | 1 | 0 | 28 |
| 5 | 25 | 8 | 2 | 13 | 10 | 7 | 0 | 40 |
| 6 | 10 | 4 | 2 | 4 | 0 | 2 | 0 | 12 |
| 7 | 5 | 0 | 0 | 2 | 2 | 1 | 0 | 5 |
| 8 | 4 | 2 | 0 | 2 | 2 | 1 | 0 | 7 |
| 9 | 5 | 2 | 0 | 6 | 1 | 0 | 0 | 9 |
| 10 | 24 | 11 | 3 | 17 | 5 | 1 | 0 | 37 |
| 11 | 15 | 4 | 1 | 6 | 1 | 5 | 0 | 17 |
| 12 | 23 | 1 | 1 | 30 | 6 | 3 | 0 | 41 |
| 13 | 40 | 40 | 1 | 45 | 4 | 8 | 0 | 98 |
| 14 | 12 | 2 | 0 | 8 | 6 | 4 | 0 | 20 |
| 15 | 2 | 1 | 0 | 0 | 0 | 1 | 0 | 2 |
| 16 | 1 | 1 | 0 | 0 | 0 | 0 | 0 | 1 |
| 17 | 2 | 1 | 0 | 0 | 1 | 0 | 0 | 2 |
| 18 | 18 | 2 | 1 | 8 | 6 | 4 | 0 | 21 |
| 19 | 3 | 0 | 0 | 1 | 1 | 2 | 0 | 4 |
| Total | 269 | 110 | 15 | 187 | 71 | 45 | 0 | 428 |
| % | | 25.7% | 3.5% | 43.7% | 16.6% | 10.5% | 0.0% | |

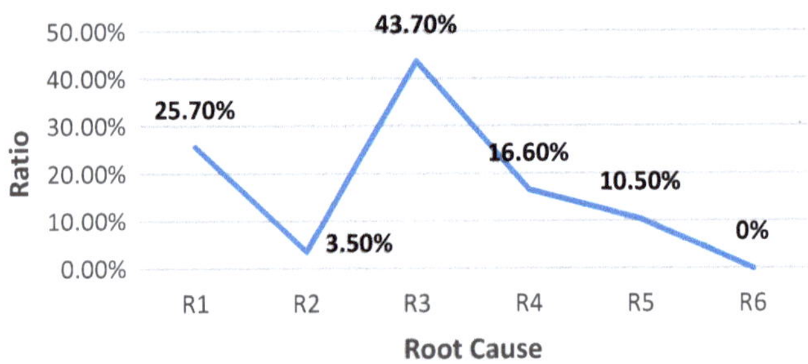


Fig. 2 The distribution of root causes in percentage

## 3.3 *Investigation of Findings Affecting Ship Energy Consumption*

The SIRE manual was analyzed within the scope of Chapter 11 (Ship engine and steering compartments) and the sub-classes of Chapter 11 mentioned below are considered to affect ship energy efficiency directly and indirectly.

Subclass 11.10: Does the operator subscribe to a fuel, lubricating and hydraulic oil testing programme, and is there a procedure in place to take into account the results?

Subclass 11.14: Is a planned maintenance system being followed and is it up to date?

Subclass 11.15: Is a comprehensive and up to date inventory of spare parts being maintained?

This item does not affect energy consumption directly, but leads to increase in consumption if necessary backups are not available in case of any failure. To give an example, failure to repair the eva ejector pump due to unavailability of necessary parts, in the event of a malfunction and supplying it by operating the fire pump which is a bigger electric pump.

Subclass 11.41: Is all of the machinery in good order and does it appear to be well maintained?

Existing findings are analyzed within the scope of the Chapter 11 of the SIRE manual. 66 of 428 findings affect the energy consumption directly and indirectly and make up 15% of total findings.

Table 3 displays the distribution of the root causes to the findings, which affect ship energy efficiency.

In addition, Figs. 3 and 4 include a graphic demonstration of Table 3.

According to Table 3, there are:

**Table 3** The distribution of the root causes to the findings which affect ship energy efficiency

| Chapter and items | Total inspection | R1 | R2 | R3 | R4 | R5 | R6 | Total findings |
|---|---|---|---|---|---|---|---|---|
| 11.10 | 17 | 2 | 3 | 14 | 0 | 0 | 0 | 19 |
| 11.14 | 7 | 3 | 1 | 5 | 0 | 0 | 0 | 9 |
| 11.15 | 14 | 10 | 0 | 4 | 0 | 0 | 0 | 14 |
| 11.41 | 24 | 1 | 0 | 7 | 11 | 5 | 0 | 24 |
| Total | 62 | 16 | 4 | 30 | 11 | 5 | 0 | 66 |
| Ratio (%) | | 24% | 6% | 45% | 17% | 8% | 0% | |

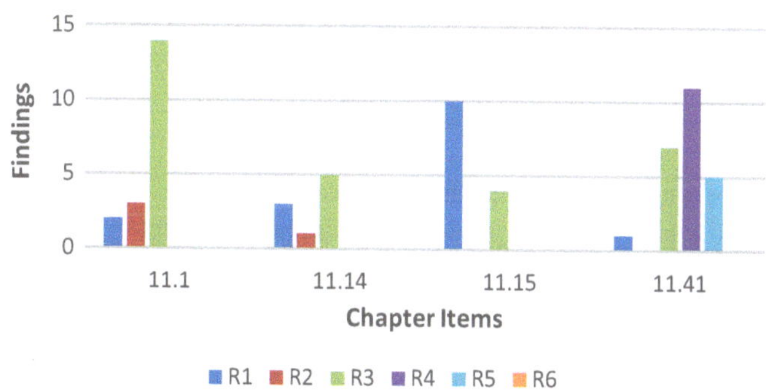


**Fig. 3** The analysis of root causes to the findings which affect ship energy efficiency

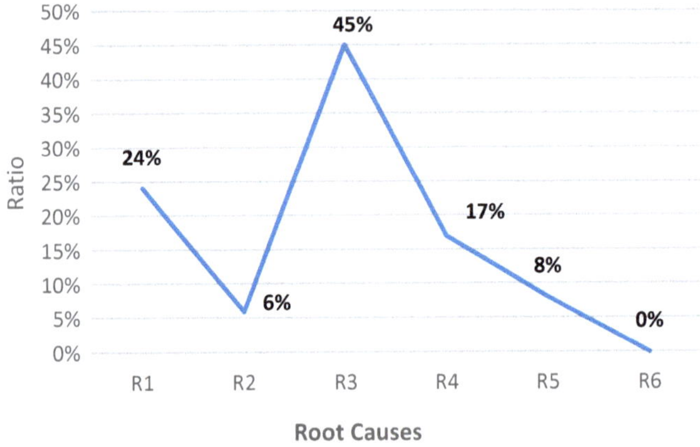


**Fig. 4** The distribution of root causes in percentage which affect ship energy efficiency

- deficiencies relating to 11.10 (related to fuel, lubricating and hydraulic oil), 19 ea.
- deficiencies relating to 11.14 (related to the planned maintenance system), 9 ea.
- deficiencies relating to 11.15 (related to inventory of spare parts), 14 ea.
- deficiencies relating to 11.41 (related to machinery status), 24 ea.

Sampling from the identified findings is provided next.
11.10

- Pending result was not sent along in the new sample although Stern Tube and Main Engine oil analyses contain the term, "caution".
- The operator failed to take a fuel oil sample and send it for analysis for each fueling.
- The oils were not sent for analysis although the SMS manual specifies 6 months as the oil analysis interval for Bow thruster, Steering Gear and Thermal oils.

There are successful oil analysis techniques such as spectrometric (Wei 2010) and occasional oil analyses to be used for acquiring engine maintenance. Such analyses may play an important role in deciding if the ship engine operates in normal conditions or not and guarantee detection of fault under certain conditions of the engine and the faulty threshold values of the selected variables (Basurko and Uriondo 2015).
11.14

- History of engine room maintenance plan cannot be accessed.
- 10 outdated jobs are present in planned maintenance system.
- Although the engine room is inspected continually according to ongoing survey method, the programme and computer used for this job is not class-certified.

The main engine performance which is directly associated to ship energy efficiency should be monitored by a planned maintenance system for performance monitoring and performance evaluation (Kökkülünk et al. 2016). A planned maintenance system aims to help ship engineers display the performance change and degradation of the main engine (Gay et al. 2004).

The results show that a planned maintenance system is not used efficiently to detect engine performance. Some duties related to engine maintenance are not followed in due time. It can be understood that the tanker company and the crew on board are not aware of the importance of the engine planned maintenance system.

11.15

- A spare parts inventory covering the critical equipment spares is not available.
- Spare part inventories must be updated once every 6 months as required by the company procedure. However, an up-to-date inventory is not available on board the vessel.

Inventory management is a significant component of engine maintenance that guarantees the readiness of spare parts and tools in the right balance of stock and stock-outs at the right time (Al-Turki et al. 2014). Lowering maintenance costs can be achieved by spare part inventories (Lazakis et al. 2016). For this purpose, Condition Based Maintenance (CBM) may also be used to increase machinery availability, reliability, efficiency and safety by lowering maintenance costs through maintained spare part inventories (Mechefske 2005).

11.41

- The existing shaft generator is not operational on board the vessel.
- Leakages are present on the oil filter to the Diesel Generator no. 3.
- Sea water pumps leak through the mechanical seal.

Figure 5 displays the repetition times (frequency) of the items. Based on this figure, 11.41 was repeated 24 times, 11.10 19 times, 11.15 14 times and 11.14 9 times.

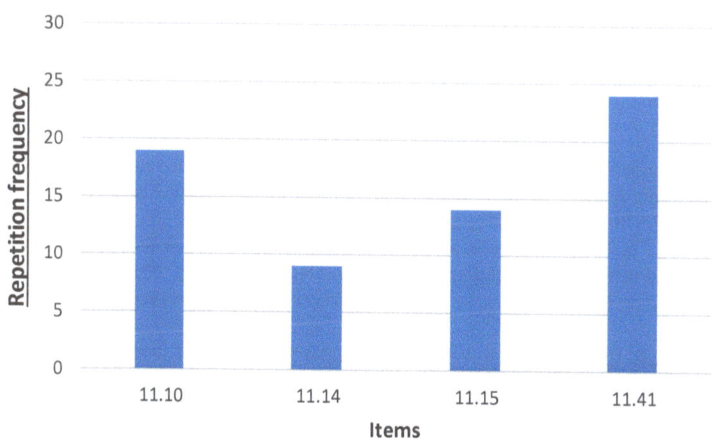


**Fig. 5** The repetition times (frequency) of the items

## 4 Conclusion and Recommendations

Maritime transportation is an essential component for global trade, representing by far the most effective means of transportation of goods. Therefore, vessel energy efficiency has become an important issue and interest in reducing fuel costs and achieving benefits in the environmental aspect has increased.

In this study, technical and engine room deficiencies and non-conformities (chapter 11), in SIRE inspection results obtained from 19 maritime companies, are examined and assessed statistically.

The data recorded between the years 2006 and 2016 is classified according to the related chapters in the SIRE inspection manual and scrutinized statistically one by one. The root causes of the deficiencies were analyzed and, in particular, those related to the chapter on the ship engine and steering compartments were evaluated in terms of ship energy efficiency.

By examining the root causes of deficiencies, it can be seen that vessel-based implementation error is considered to be the most important factor in these deficiencies. For this reason, vessels should be monitored and inspected frequently by their companies to detect deficiencies on board and identify personnel training needs associated with ship energy efficiency implementation.

This study analyzed only the deficiencies concerning ship engine and steering compartments and the impacts thereof on vessel energy efficiency. Future studies may focus on investigating other areas (i.e. navigation and ship's hull structure) where deficiencies affecting energy efficiency on board vessels occur and determining the root causes for which corrective actions need to be identified. This will help provide financial and environmental benefits by increasing energy efficiency on board the vessels if relevant corrective actions for such deficiencies are identified and applied.

## References


Al-Turki, O., Ayar, T., Yılbas, B. S., & Sahin, A. Z. (2014). Integrated maintenance planning in manufacturing systems. *Springer Briefs in Manufacturing and Surface Engineering*. https://doi.org/10.1007/978-3-319-06290-7_2.

Basurko, O. C., & Uriondo, Z. (2015). Condition-based maintenance for medium speed diesel engines used in vessels in operation. *Applied Thermal Engineering, 80*, 404–412.

Gay, R. R., Palmer, C. A., & Erbes, M. R. (2004). *Power plant performance monitoring*. Rsquared Publishing.

Howell, K. E., & Bhattacharya, S. (2016). Functional and territorial jurisdictions: Regulating a globalized shipping industry. *Journal of Transport Geography, 55*, 92–100.

IMO, Resolution MEPC.203(62), MEPC 62/24/Add.1 Annex 19, 2011. Amendments to the Annex of the Protocol of 1997 to Amend the International Convention for the Prevention of Pollution from Ships, 1973, as Modified by the Protocol of 1978 Relating Thereto.

Kececi, T., & Arslan, O. (2017). SHARE technique: A novel approach to root cause analysis of ship accidents. *Safety Science, 96*, 1–21.

Knapp, S., & Franses, P. H. (2009). Does ratification matter and do major conventions improve safety and decrease pollution in shipping? *Marine Policy, 33*, 826–846.

Kökkülünk, G., Parlak, A., & Erdem, H. H. (2016). Research Paper: Determination of performance degradation of a marine diesel engine by using curve based approach. *Applied Thermal Engineering, 108,* 1136–1146.

Lazakis, I., Dikis, K., Michala, A. L., & Theotokatos, G. (2016). Advanced ship systems condition monitoring for enhanced inspection, maintenance and decision making in ship operations. *Transportation Research Procedia, 14*, 1679–1688.

Mechefske, C. K. (2005). *Machine condition monitoring and fault diagnosis*. Boca Raton, FL, USA: CRC Press, Taylor & Francis Group.

OCIMF. (2014). *Sire inspector training and accreditation guideline*. Available online https://www.ocimf.org/media/54903/SIRE-Inspector-Accreditation-Guidelines-January-2014-Final.pdf. Accessed 08 April '17.

Perera, L. P., & Mo, B. (2016). Review: Emission control based energy efficiency measures in ship operations. *Applied Ocean Research, 60*, 29–46.

Ship Inspection Report (SIRE) Programme. (2016). Oil Companies International Marine Forum, 2014 Edition Rev 1.2.

UNCTAD. (2017). *Review of maritime transport*. United Nations Publication.

Wei, H. J. (2010). Research of marine diesel engine condition detecting base on BP neural network and spectrometric analysis. In *International Conference on Multimedia Technology (ICMT)*, Ningbo, China, 2010 (pp. 1–4).

Wigforss, J. (2012). *Benchmarks and measures for better fuel efficiency*, Report No. NM-12/29. Chalmers University of Technology, Göteborg, Sweden.

# Real Time Awareness for MRV Data

**Nikitas Nikitakos, Dimitrios Dalaklis, and Petros Siousiouras**

## 1 Introduction

European Union's (EU) Regulation 2015/757, dealing with monitoring, reporting and verification of carbon dioxide ($CO_2$) emissions in relation to maritime transport, entered into force on July 1, 2015. The objective of this regulation, also known as the MRV Regulation, is to gain a better understanding of fuel consumption and $CO_2$ emissions related to shipping activities within Europe. The MRV framework is under continuous improvement; Delegated Regulation 2016/2072 has amended the initial Regulation and it is to be read in conjunction with Delegated Regulation 2016/2071 and the Implementing Regulations 2016/1927 and 2016/1928. Briefly, irrespective of the vessel's flag, the regulation under discussion applies to ships greater than 5000 GT undertaking one or more commercial voyages into, out of or between European Economic Area (EEA) related voyages ports. The MRV Regulation requires per-voyage and annual monitoring of $CO_2$ emissions; other parameters are also included, such as quantities of cargo carried and miles travelled.

Furthermore, the annual disclosure of aggregated data on a ship-by-ship basis is required. Ships that are scheduled to perform more than 300 voyages per year or operate solely within Europe during the annual monitoring period are exempt from monitoring the parameters on a per-voyage basis. Instead, they can report on an annual basis in order to reduce the administrative burden. The regulation applies to commercial voyages only; this type of activity is defined as calling at an EEA port to offload or load cargo or passengers for commercial purposes. Voyages calling for non-commercial reasons, such as bunkering, crew relief, supply stops, dry-docking,

N. Nikitakos · P. Siousiouras
Department of Shipping, Trade and Transport, University of the Aegean, Chios, Greece

D. Dalaklis (✉)
World Maritime University, Maritime Education & Training Specialization, Malmö, Sweden
e-mail: dd@wmu.se



A. I. Ölçer et al. (eds.), *Trends and Challenges in Maritime Energy Management*,
WMU Studies in Maritime Affairs 6, https://doi.org/10.1007/978-3-319-74576-3_5

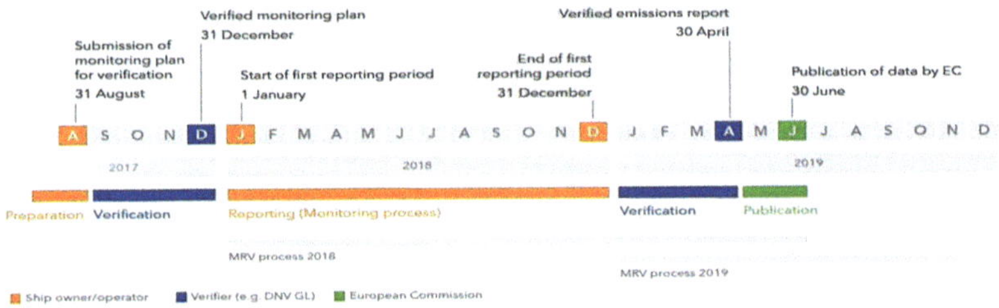


**Fig. 1** EU MRV regulation timeline [source DNV GL (2017)]

maintenance or safe harbor are exempt. According to this new regulation, vessels should monitor fuel consumption, distance traveled, time spent at sea and finally loaded cargo. All these parameters provide the so-called "Energy Efficiency Indicators"; the issue of energy efficiency is very high on the contemporary political agenda, as a rather extended number of governments around the world focus heavily on how to reduce wasteful energy consumption, strengthen energy security and reduce/cut down greenhouse gas emissions.

It is necessary to note that ships larger than 5000 GT represent about half of the total number of those that are calling at EU ports; they are also the source of about 90% of greenhouse gas emissions. As a result, the decision that vessels moving to and from EU ports must be monitored and report $CO_2$ emissions starting from January 2018 is fully justified. Before the beginning of the first reporting period, all ships are required to have a ship-specific monitoring "plan". A third party verifier must assess this and during each annual reporting period (calendar year), the ship will monitor fuel consumption and other parameters in accordance with the agreed monitoring plan. At the end of the annual reporting period, a ship-specific emissions report must be prepared and submitted to a third party verifier for verification of accuracy. Upon satisfactory verification, a Document of Compliance will be issued; the verified emissions report will be submitted to the European Commission (EC). All the important dates for the operationalization of the MRV Regulation are summarized via Fig. 1.

## 2 MRV and the Shipping Industry

It is a rather self-explanatory fact that vessels engaged in maritime transport have a certain impact on the environment, mainly because of their reliance on internal combustion engines and the related emissions. Considering that ships engaged in international trade remain the only means of transportation not currently included within the EU's commitment to reducing greenhouse gas emissions, the MRV regulation is now covering the gap. Therefore, the specific regulatory framework is the first step of a staged approach for the inclusion of maritime transport's emissions into the European framework of greenhouse gas reduction, alongside all other modes of transport. The overall strategy consists of three consecutive steps: (a) Monitoring, reporting and verification of carbon emissions from ships; (b) GHG reduction targets for the maritime transport sector; and (c) Further measures, including Market-Based Measures (MBM). The MRV framework is expected to reduce $CO_2$ emissions by up to 2%, compared with the "business as usual" current situation, according to the Commission's impact assessment. This framework could also reduce net costs to owners by up to €1.2 billion per year in 2030. In addition, it will provide useful insights into the performance of individual ships, as well as their associated operational costs and even allow determination of their potential resale value. All the above will benefit ship owners, who will be better equipped to take decisions on major investments, and will be a very useful tool available to support acquisition of the corresponding finance.

Shipping companies, in order to maintain their licenses to operate in Europe, will need to comply with that rather extensive string of Regulations. In the MRV Regulation, the following fuel-consumption methodologies are all deemed as acceptable: (a) Bunker Fuel Delivery Note (BDN) and periodic stock-takes of fuel tanks; (b) Bunker fuel tank monitoring on board; (c) Flow meters for applicable combustion processes; and (d) Direct emissions measurements. Ships can use a combination of these methodologies if this approach results in an improvement in the accuracy of monitoring (eur-lex.europa.eu, 2017). The shipping company shall define, in the monitoring plan, exactly which monitoring methodology will be used to calculate fuel consumption for each ship type under its responsibility and ensure that once it has been chosen, it is consistently applied. It is also interesting to note that when selecting a monitoring methodology, the improvements from greater accuracy shall be balanced against the additional costs.

Making the right choices from the very beginning is the most effective way to reduce the associated "administrative" burden; a seamless transition and continuity of the company's already established mode of operations is also crucial. The monitoring method to be chosen is not prescribed in fine detail; however, it must be robust enough to ensure that data gaps are minimized and errors/omissions are avoided. A good fitting of the MRV method with the company's existing processes will allow effective compliance and reduce the impact upon the standard way of doing business. Every vessel's data will be made publically available; this approach will provide clients with transparent information about operations and environmental performance. Needless to point out, only selecting the most appropriate efficiency measures for vessels will pave the way towards maximizing the return on investment and minimizing, simultaneously, the commercial risk. Compliance will be checked between January and April 2019, during the first calendar year that the monitoring data will be submitted; after the initial launch of the scheme, this will become an annual requirement.

From a ship owner's point of view, being able to monitor and manage all aspects of the ship's operations while the vessel is at sea can provide a very powerful business tool. Making well-informed decisions about the deployment of the ship, the respective maintenance/procurement needs and even the optimization of certain functions such as bunkering needs provide indicative examples of rising opportunities. The monitoring framework can also provide a tool to evaluate the performance of the power plant/engine in case upgrades or maintenance work has taken place. As such, the presence of a powerful monitoring methodology is of paramount importance, as it can help reduce operating costs, prevent potential damage to the associated machinery equipment and help the timely detection of inefficiencies. It is, however, necessary to note that MRV alone will not lead to a significant reduction of emissions or inefficiencies; on the other hand, it is a very valuable awareness tool for improving the ship's operation. Additionally, benchmarking will certainly be needed; only the creation of a database that describes performance data and emissions can provide the basis for comparative analysis. This will provide a realistic picture of $CO_2$ emissions from shipping. Along with the introduction of innovative monitoring technologies, (such as Internet of Things (IoT) initiatives, like stream reasoning) an awareness tool for corrective actions will be created. The purpose of the analysis at hand is to facilitate real time decision making. A conceptual framework to allow for MRV anomaly detection in order to help ships to early identify and correct arising risks is, therefore, presented.

## 3 Real Time Decision (RTD)

According to Power (2002 and 2011) decision support is a broad concept that prescribes using computerized systems and other tools to assist in individual, group, and organizational decision-making. In practice, any system that processes and stores data or can report the associated information as it is happening should be considered as (an on-line) real-time system. More specifically, real time translates into the concept wherein informational inputs to decision-making processes are available as soon as there are changes in the environment that alter those exact informational inputs. It should be clarified at this point that real time means "near real time" in practical terms because there is always some form of latency between (a) the actual state change, (b) the reflection of that state change in data in one or more systems of record and (c) the availability of the changed data to decision makers. Real time is not the same for every decision task. There are factors such as data availability, data capture, and transmission speeds, which are less of a problem; recent technological advances provide new tools that are very effective in receiving and accessing real-time data. For example, the recently emerged concept of "Internet of Things" (IoT) can facilitate real-time decision-making and support innovation.

In summary, this "real-time" decision-making must be supported by six (6) major capabilities: 1. Monitoring activity is conducted by using decision rules to create/send alerts. In a very simplistic description, a sensor or data-gathering instrument captures actions/data and individual data elements, or groups of observations which are processed by using rules and messages; the necessary alerts are then sent. It is also useful to highlight that it is not necessary to capture and store (all) the relevant data used. 2. Quick exchange and transmission of data towards the relevant systems and decision makers; the necessary information may be exchanged among multiple systems, or towards numerous people. 3. Information is available to a decision maker as soon as a change occurs. For that purpose, data must be updated and processed (almost) at the same time a change occurs; this information must be immediately accessible by the decision maker. For reasons of effectiveness, the decision-maker must always be able to access and use the latest data; this is achieved by storing and continuously updating all necessary information. 4. Processing-storing the data and reporting raw or summarized data as the events are happening. This way, new data is directly streamed (or, "transmitted") to the relevant decision makers. 5. Provision of analytical tools or/and creation of knowledge at the point of use. With that function available, and irrespective of the data's rate of change, the decision-maker can perform analyses or obtain advice immediately and as needed. 6. Provision of synchronous exchange of data from the point of origin to the end-user. This exchange may be supplemented by using voice, video and/or text messages. The exchange may be initiating and confirming a transaction or supporting collaboration.

In the wider context of optimizing the conduct of shipping operations, the MRV data can clearly play an important role. For example, using the available data could facilitate (early) anomaly detection. To elaborate more, deviation(s) from the vessels' expected behavior could be noticed immediately and the root cause identified by further investigation; this could also improve the "situational awareness" in relation to the overall state of the vessel. It is also important to note that relying on data handling in order to detect anomalies is not a new endeavor for the maritime industry. Vespe et al. (2009) have already put forward a methodology for automatic identification system (AIS) data anomaly detection; several other methods and algorithms have been proposed to tackle this issue, mainly correlating the "historical" data with "new" data provided by the AIS receiver. These existing methods are mainly based on neural networks using either supervised and unsupervised learning, or statistical/probabilistic models. The analysis that takes place in the next section is inspired by relevant research in air traffic control developed under the framework of EU research project SERSCIS (Kostopoulos et al. 2012), in conjunction with the authors' previous work on AIS (Nikitakos 2016; Pallikaris et al. 2016). A new concept to enhance MRV data handling based on complex event processing and particularly on data stream reasoning is discussed; identification of risks related with anomaly detection is achieved and the appropriate actions to deal with the problem are then initiated.

## 4 Stream Reasoning Technology

According to researchers such as Valle et al. (2008) and Stuckenschmidt et al. (2010), stream reasoning is an approach that can be used if information arrives as a stream of (time stamped) inputs. The specific approach has two very important features that could be helpful in data handling: (a) the knowledge base can be continuously updated and reasoning goals continuously re-evaluated as new assertions arrive; (b) the analysis considers events from a finite time window, and not only at a single instant. Therefore, introducing stream reasoning in the MRV data processing could help to overcome some of the current limitations. Potential areas of improvement include allowing the concrete system model to be continuously updated, which should be faster than generating a completely new model each time an update is needed; reducing the time lag between the evolution of the real system and that of the concrete system model, making it possible to resolve recent and rapid changes in the real system is also possible. Furthermore, representing protracted as well as instantaneously observed behaviors in the model by including information over an extended time window is feasible. Finally, allowing reasoning algorithms to take account of system changes during the time window can help to target, specifically, the instantaneous system composition and status. On the other hand, it is important to consider that research efforts on stream reasoning are still at a very early stage. More time and effort will be needed until a sound architecture is in place; the state of the art in relation to stream reasoning is presented in a fine level of detail via Fig. 2, consisting of four (4) main processing steps.

The first step (SELECT) in the stream reasoning is necessary in order to select relevant data from input streams by exploiting load-shedding techniques. This can be achieved by introducing sampling policies that probabilistically drop stream elements to deal with bursts of streams that may have unpredictable peaks. In the next step presented (ABSTRACT), the sampled streams are fed into a relevant data schema that continuously generates aggregate events by enforcing aggregation queries. Outputs of the abstract step are then consolidated as RDF (Resource Description Framework) streams, an unbounded bag of pairs $<\rho, \tau>$ where $\rho$ is a RDF triple and $\tau$ is the timestamp that denotes the logical arrival time of RDF statement. This step entails the development of aggregate query language and system for query RDF data in the form of data streams. According to Pellet (an open-source Java based OWL 2 reasoner) there are various routes to follow, such as Streaming SPARQL, Continuous SPARQL (C-SPARQL) and Time-Annotated SPARQL; all extend SPARQL methodologies can help to handle both static RDF graphs and transient streams of RDF triples.

The third step (REASON) is creating the necessary associations; RDF streams are injected into background knowledge in order to perform reasoning tasks. Given that

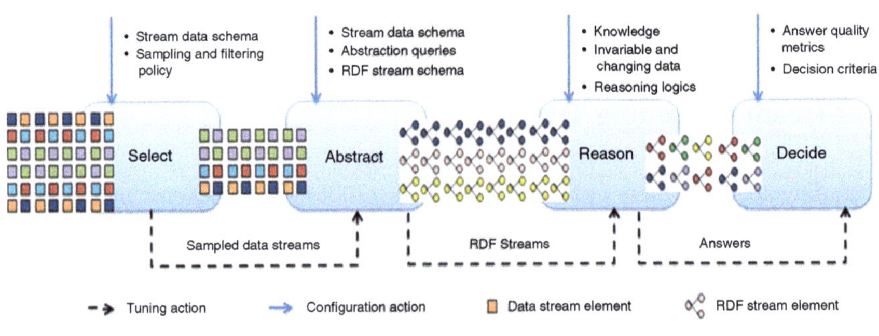


**Fig. 2** Information processing steps in stream reasoning

the reasoning process is not aware of expiration time, the reasoning results remain valid until the next update. A pre-reasoning process is used to generate the current system snapshot and it is responsible for maintenance of incremental materialization of RDF snapshots. The efficient incremental materialization of RDF snapshots is a research challenge under investigation. Finally, the fourth step (DECIDE) is very essential. Before producing answers to the reasoning tasks, the answering process must be facilitated via a "decision step", where quality metrics and decision criteria defined by an application developer are used to evaluate whether the quality of the answer is good enough (or not) and otherwise adapt the behavior of each previous step. After some consideration, the conclusion is that, for the time being, it is not possible to use stream reasoning in a simple way to address MRV real time intelligence implementation. However, the underlying concepts can be used to enrich the proposed run-time architecture and to provide more flexibility.

## 5 Proposed Real Time Framework

In this section, the necessary conceptual framework is discussed in detail. This framework must be designed in order to exploit semantic system models and enable the use of machine reasoning to support the end user in making and implementing decisions at the real-time mode (run-time). In simple terms, this translates into creating a semantic model of the running system based on the available monitoring data and then using it to "capture" the exact status of the system. Furthermore, it is necessary to present the information from this model to the user in the appropriate way to help him/her to understand and address the identified risks (enhancement of situational awareness). The tools developed to support machine-assisted system modeling allow a structure and the respective properties to be described before the actual system is created (by dynamic runtime composition). This model can be termed an "abstract system model", since it describes the structure of the system but not its actual composition. The proposed conceptual framework is presented in Fig. 3.

Avoiding further details, which would be beyond the scope of the current analysis, it is important to highlight that there are two separate reasoning processes taking place: The first one is "semantic reasoning" in order to identify potential threats of anomaly behavior classification based on whether these are addressed by the controls present in the running system; the second one is a "Bayesian inference for likelihood estimation" that each threat is currently being carried out. As Valle et al. (2008) have already pointed out, within the proposed framework of Fig. 3, the user can have simultaneous access to three different types of information (presentation): 1. What exactly are the system vulnerabilities, or what exact types of threats is the system unable to manage?; 2. What is the current likelihood/probability of each threat to be materialized?; and 3. A prediction of what will be the threat impact on the maritime emissions performance. Moreover, the relevant outcome can be classified into three different classes. The first one is termed as "Blocked threat/activity", in the

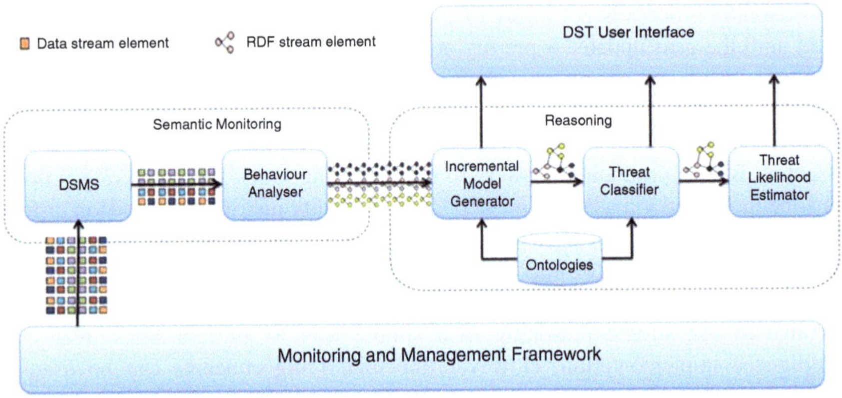


**Fig. 3** Conceptual framework incorporating stream reasoning concepts

case that the system has the appropriate control to prevent the abnormal behavior from creating a problem. For the second case, "Mitigated activity" describes a situation wherein the abnormal behavior cannot be prevented, but the system controls provide a response that will (at least) counteract its effect on maritime safety and security. In the third case, the term "Vulnerability" describes the situation wherein the system does not have any means of preventing the abnormal behavior or counteracting its effects on the targeted system asset.

The main objectives of the monitoring and decision support tool are the following: 1. To perform "Risk Classification" (indicative categories for use include "low", "medium", "high" according the potential impact and also "blocked", "mitigated", "vulnerabilities" depending on how well the potential threats are addressed by controls); 2. A "Periodic Assessment" (the Decision Support Tool (DST) should refresh, in a periodic timeframe, the model and dynamically reduce the involved risk factors); 3. Provide "Threat Explanations" (the DST should provide explanation and details in relation to those identified threats, which should help the operator in the loop to understand the system and to take appropriate actions if needed); and 4. Perform "Propositions" (the DST should allow the operator to revert to past model versions when required allowing the user to make "what – if" tests on the model by adding controls and comparing the results with the original model). So the fault monitoring DST tool provides continuous feedback and suggests new control actions that can be useful while providing the capability to test their effect on "what – if" scenarios. It is necessary to note that the user should be presented with the three vulnerability classifications: the good ones are placed on the left (blocked and mitigated threats) and the most troubling threats (vulnerabilities) are positioned on the right. As already briefly mentioned, the core semantic language is OWL, the Web Ontology.

The semantic language used in the specific effort has as a requirement that the models in the DST must be in OWL format. The version of the OWL language is OWL2. This support tool is built on JAVA 1.6 and SWT 3.738. It is no coincidence that most web semantic projects are built on JAVA because of its flexibility; JAVA is also used in the systems' DST. The reasoner holds a very important role in the DST; the one used is "Hermit 1.3.5". There are other reasoners, which could be used for the same purpose, but it is questionable if they would be able to handle very large volumes of data (and in the necessary "real-time" condition). Though "Hermit" can, for the time being, handle very well the associated volume of data, a new reasoner should be designed in order to adapt reasoning to Bayes inference used in the approach under discussion. Conclusively, semantic models have proved very useful in the application area of security and risk management of several critical infrastructures including maritime ones; the conceptual tool presented in this analysis provides all the necessary facts in a clear and easy to understand manner to the end users and the involved decision makers.

## 6 Conclusions

On a daily basis, ships of different sizes and capabilities carry huge quantities of cargo and a very large number of passengers cost effectively, cleanly and safely. Bauldauf et al. (2016) have already emphasized that through the regulatory efforts of the International Maritime Organization (IMO), several maritime Conventions stipulate both the operational environment and the training requirements for mariners/seafarers. It is a self-explanatory fact that the importance and influence of MARPOL-International Convention for the Prevention of Pollution from Ships, STCW-Standards of Training Certification and Watch-keeping and SOLAS-the International Convention for the Safety of Life at Sea, among others, are widely known throughout the maritime community and many outsiders. IMO's efforts, along with numerous other national (and often regional) legislations/regulations create a rather complex operating framework for international shipping. In any case, the issue of improving the environmental "footprint" of vessels transiting the seas and oceans of our planet is now regarded as one of the most urgent tasks to which the shipping industry needs to (effectively) respond. It is indicative that, according to the 2014 IMO's Green-House Gas (GHG) Study, ships emit approximately 1000 million tonnes of carbon dioxide ($CO_2$) on an annual basis; for the period 2007–2012, on average, the maritime transport sector accounted for approximately 3.1% of annual global $CO_2$ and approximately 2.8% of annual GHGs.

The EU introduced Regulation (EU) 2015/757 provides guidance on Data Collection, Monitoring, Verification and Reporting; the specific arrangement is also widely known as the EU MRV Regulation. The MRV Regulation applies to ships above 5000 GT and its epicenter is the level of $CO_2$ emissions released during voyages from/to EEA ports carried out after January 1, 2018. More importantly, ships are subject to the MRV Regulation regardless of their flag. According to the MRV framework, by August 31, 2017, shipping companies must submit a monitoring plan to their verifier for each of their ships. The plan will indicate the method chosen to monitor and report $CO_2$ emissions, as well as any other relevant information. For ships falling under the scope of MRV for the first time after August 31, 2017, the concerned company must submit a monitoring plan to its verifier without undue delay and no later than two months after each ship's first call in a port under the jurisdiction of an EU/EAA Member State.

In the current analysis, a stream reasoning architecture was presented in order to provide a "warning signal" to shipping companies related with measurements performed under the MRV's framework. The main advantage of the proposed method is its real time application in the detection of potential anomalies. The concept of stream reasoning is useful for MRV data intelligence to improve current limitations. However, there are still many research challenges. To begin with, the improvement of the Behavior Analyzer component (by using appropriate algorithms) is considered important. The associated threat classifier should also be extended, in order to handle different types of classifications, but still defined by SWRL rules. Additionally, the creation of a specialized classifier that would be more effective/faster than the general-purpose reasoner should be considered in order to improve performance. Furthermore, the "Bayesian threat likelihood estimator" implementation should be optimized, under the notion to reduce processing time. Finally, the abnormal activity hypothesis sampling should take into account any secondary effects. All these are not simple tasks: evaluating performance of the conceptual framework in real conditions, as well as implementing the necessary improvements can be achieved only via extensive future research activities. Therefore, further investigation of the specific research questions is expected to take place in a "real-life" situation, as part of future activities, following the implementation of the methodology under discussion.

## References

Bauldauf, M., Dalaklis, D., & Kataria, A. (2016). *Team training in safety and security via simulation: A practical dimension of maritime education and training*. Paper presented at the 10th International Technology, Education and Development Conference (International Academy of Technology, Education and Development (IATED), Valencia, Spain, 7 March 2016.

http://eur-lex.europa.eu/legal-content/EN/TXT/?uri=celex%3A32015R0757. Accessed January 2017.

https://www.dnvgl.com/maritime/eu-mrv-regulation/index.html. Accessed June 2017.

Kostopoulos, D., Leventakis, G., Tsoulkas, V., & Nikitakos, N. (2012). *An intelligent fault monitoring and risk management tool for complex critical infrastructures*. Paper presented at the SERSCIS Approach in Air-Traffic Surface Control, UKSim 14th International Conference on Modelling and Simulation, 28–30 March 2012.

Maritime Trajectory Mining for Anomaly Detection-NATO RTO-MP-SCI-247; North Atlantic Treaty Organization (NATO), 2009; Vespe, M.; Pallotta, G.; Visentini, I.; Bryan, K.; Braca, P.

Nikitakos, N. (2016). *Real time intelligence of AIS data*. Paper presented at Maritime Knowledge Discovery and Anomaly Detection Workshop, JRC-Ispra, 5–6 July 2016.

Pallikaris, A., Katsoulis, G., & Dalaklis, D. (2016). *Electronic navigation equipment and electronic chart display and information systems (in Greek)*. Faliro: Eugenides Foundation.

Power, D. J. (2011). Challenges of real-time decision support. In F. Burstein, P. Brézillon, & A. Zaslavsky (Eds.), *Supporting real time decision-making: The role of context in decision support on the move* (pp. 3–11). Cham: Springer.
Stuckenschmidt, H., Ceri, S., Valle, E. D., & Harmenlen, F. V. (2010). *Towards expressive stream reasoning*. Paper presented at the Dagstuhl Seminar on Semantic Aspects of Sensor Networks, Dagstuhl, Germany, 24–29 January 2010.
Third IMO GHG Study 2014; International Maritime Organization (IMO) London, UK, April 2015; Smith, T. W. P.; Jalkanen, J. P.; Anderson, B. A.; Corbett, J. J.; Faber, J.; Hanayama, S.; O'Keeffe, E.; Parker, S.; Johansson, L.; Aldous, L.; Raucci, C.; Traut, M.; Ettinger, S.; Nelissen, D.; Lee, D. S.; Ng, S.; Agrawal, A.; Winebrake, J. J.; Hoen, M.; Chesworth, S.; Pandey, A.
Valle, E., Ceri, S., Barbieri, D. F., Braga, D., & Campi, A. (2008). *A first step towards stream reasoning*. Paper presented at the Future Internet Symposium, Vienna, Austria, 28–30 September 2008.

## Further Reading

Barbieri, D. F., Braga, D., Ceri, S., Della Valle, E., & Grossniklaus, M. (2009). *C-sparql: Sparql for continuous querying*. In: Proceedings of the 18th international conference on World Wide Web, Madrid, Spain, 20–24 April 2009.
Community's 7th Framework Programme; Grant Agreement No. 225336.
Guidance on the application of AIS binary messages; International Maritime Organization (IMO), SN/Circ.236, 2004.
http://www.efta.int/Legal-Text/EEA-Agreement-1327. Accessed January 2017.
https://ec.europa.eu/clima/policies/transport/shipping_en#tab-0-3. Accessed January 2017.
ISO/IEC 27005:2011; Information technology -- Security techniques -- Information security risk management, International Organization for Standardization, 2011.
Pellet: OWL 2 Reasoner for Java. http://pellet.owldl.com. Accessed 15 January 2017.
Recommendation ITU-R M.1371-3, 2007; Technical characteristics for an automatic identification system using time division multiple access in the VHF maritime mobile band.
SERSCIS: Semantically Enhanced Resilient & Secure Critical Infrastructure Services; European.
W3C OWL Working Group. OWL 2 Web Ontology Language: Document overview. http://www.w3.org/TR/owl2-overview/. Accessed 21 January 2017.

# Overcoming the Challenges to Maritime Energy Efficiency in the Caribbean

**Vivian Rambarath-Parasram, Sukhjit Singh, and Deniece Aiken**

## 1 Introduction

Maritime Energy Efficiency is an imperative for the small island developing States of the Caribbean. Failure to implement the necessary measures to promote energy efficiency and the uptake of low carbon technologies will exacerbate the effects of global warming on these territories. With limited research and development into maritime technology being undertaken in the region, technology transfer from developed States to the Caribbean is a key requirement for facilitating compliance with the applicable international standards. However, technology transfer is not an end itself. States of the region must be vested with the appropriate legal and institutional framework, to create the enabling environment to promote technological uptake, and drive compliance and enforcement. The inadequacy of the regional and national legal framework is often difficult to overcome as other legislative priorities often dominate the respective legislative agendas. The number of Draft Marine Pollution Bills across the region represents an array of unrealized marine environmental protection measures. The inability to implement such a legislative agenda is an indicator that, for plans for the promotion of energy efficiency to move ahead, the implementation will have to be premised on a legal framework that is far from ideal and the enabling environment will need to be created in the context of existing legislation and institutional players that are unconventional when compared to developed States.

In the context of the above, the analysis at hand will:

V. Rambarath-Parasram (✉) · S. Singh
The University of Trinidad and Tobago, Chaguaramas, Trinidad and Tobago
e-mail: vivian.r-parasram@utt.edu.tt

D. Aiken
Caribbean Maritime University, Kingston, Jamaica



A. I. Ölçer et al. (eds.), *Trends and Challenges in Maritime Energy Management*, WMU Studies in Maritime Affairs 6, https://doi.org/10.1007/978-3-319-74576-3_6

1. Briefly highlight key literature that describes the enabling environment for technology transfer
2. Provide the methodological approach to data gathering and analysis
3. Outline the barriers to maritime energy efficiency peculiar to the region
4. Outline the opportunities for overcoming these barriers
5. Describe the enabling framework to facilitate uptake of energy efficiency
6. Recommend an approach for implementing relevant international standards based on an analysis of existing legislative frameworks.

## 2 Summary of Key Literature

The enabling environment for technology transfer involves the complex interface of human factors, institutional knowledge, institutional capacity and machinery. Hagood (2006) stated that the failure of the technology transfer process is usually due to the failure of more than one of these interfaces and invariably that failure contains elements of each (Kranzberg 1986). The overall failure may be attributed to the singular failure, weakness, or absence of a given component, but they are always linked to each other. Technological knowledge and technological transfer, key aspects of the enabling environment, are further explained below.

Technological knowledge is also a component of the enabling environment. It is the combination of empirical information, familiarity with processes and materials, and the know-how required in making an object. Explicit knowledge can be stored, copied, and transferred by impersonal means but the tacit knowledge "*which cannot be formulated explicitly is vital for a successful technology transfer*" (MacKenzie and Spinardi 1995, p. 45). Therefore, this transfer of knowledge accounts in part for the important elements of absorption of knowledge and adaptation of technology.

Technology transfer is a process involving assessment, agreement, implementation, evaluation and adaptation, and repetition. Institutional barriers and policies influence the transaction process, as well as the efficiency of the transfer process, in particular in the adaptation and repetition stages (Worrell et al. 2001). In most cases, some (or significant) adaptation, modification, or redesign of commercial technologies or products is needed to make these useable in local contexts. If a technology does not meet the needs of local consumers or is not optimized for local operating conditions, it will have limited uptake (Ockwell et al. 2015). Most existing policies and programmes for transfer of environmentally sound and greenhouse gas abatement technologies are tailored to national needs and are rarely internationally focussed. However, technologically advanced countries with policies in place that promote technology transfer usually have vested interests in the promotion of such transfers (Worrell et al. 2001).

Ray (2012) linked the economies of technology transfer to patent regimes. Developing countries face a challenge when making a choice between encouraging the transfer of the latest technology and the diffusion of such technologies. Weak patent regimes often found in developing countries, therefore, do not provide the requisite safeguards to technological innovators and this effectively creates a disincentive and a barrier for technology providers. This inadvertently leads to an adverse effect on the welfare of the developing country, as donors transferring technology to a developing country may not transfer their best available technology if that country has a weak patent regime. An adequate patent regime is, therefore, an essential part of the legal, institutional and regulatory framework.

While intellectual property regimes must be significantly strengthened in developing countries, the real bottleneck to low carbon technology diffusion is the lack of a stable policy and regulatory framework and a coherent investment environment in the energy sector (Rai et al. 2014). The most successful energy efficiency programmes around the world are statute based (IEA 2010). The legal and institutional framework provides the following four components of the enabling environment:

- Policies that reflect Governmental approaches to technology transfer
- Enabling legislation, rules and regulations
- Assigning institutional responsibility for the development of relevant regulations and/or implementing programmes. In some instances, this may involve the development of existing and new institutional capacities.
- Providing and accessing funding to facilitate related activities.

Although technology transfer is often seen as a private interaction between two companies or trade partners (Ray 2012), institutional barriers and policies influence the transaction process, as well as the efficiency of the transfer process. Under perfect market conditions, all additional needs for energy services are provided by the lowest cost measures, whether through energy supply increases or energy demand reductions. However, evidence suggests market barriers exist which does not allow investments even when they are lower in cost than marginal energy supply (Singh 2017). The uncertainties of shipping often lead to higher perceived risks, and therefore demand a more supporting environment when compared to other sectors.

# 3 Methodology

Research into the uptake of energy efficiency in the Caribbean is embryonic and, therefore, exploratory research is utilized in the current analysis. Saunders et al. (2012), Thomas (2006) and Wilson (2014) suggest an approach of inductive reasoning when published research is lacking or unavailable. Inductive reasoning is based on learning from experience. Patterns, resemblances and regularities in experience are observed to reach conclusions. Inductive reasoning for the current analysis began with detailed observations of the globally existing barriers to energy efficiency technology transfer and then moved towards more abstract generalisations and ideas. The intent was to explore the complex maritime environment in relation to factors causing barriers to energy efficiency with an objective of presenting the perspectives applicable to the region. Thus, the literature is used to frame the

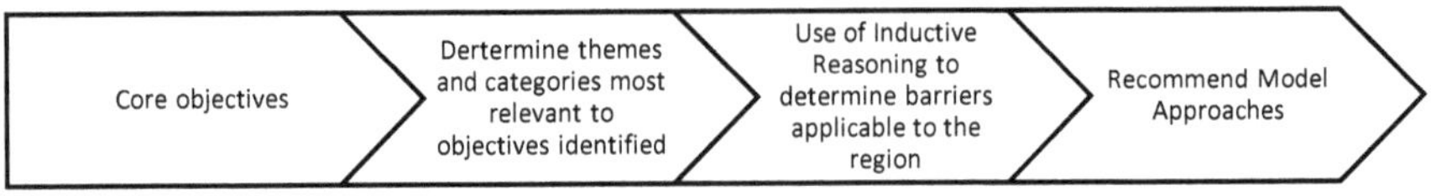


**Fig. 1** Methodological approach (adapted from Thomas 2006; Wilson 2014)

problem and to understand the relationship of various contributing factors (Creswell 2009; Saunders et al. 2012) (Fig. 1).

In implementing an inductive analysis approach the authors:

- Condensed varied raw text data into summary formats;
- Established clear links between the research objectives and the summary of findings derived from the raw text data to ensure that these links are both transparent and justifiable; and
- Suggested a model approach based on the underlying experience and processes that are evident in the text data.

The inductive component of the analysis was carried out through multiple readings and interpretations of the raw data. Although the findings are influenced by the evaluation objectives or questions outlined under the research on the topic, the findings arose directly from the analysis of the raw data. Inevitably, the findings were shaped by the assumptions and experiences of the evaluators conducting the study and carrying out the data analyses. For the findings to be practical, the evaluator suggested and made recommendations about what is more important and less important from the data. The existing data in the literature was observed and reviewed for similarities and differences. These observations were then generalized in an attempt to form conjectures. Therefore, the conclusion provides open views based on such generalizations. As the conjectures formed are not proven by any new data, they are validated based on the summary of existing textual data. Hence, this provides a transparent and justifiable link to differentiate and highlight the factors valid for the Caribbean region.

## 4 The Barriers to Energy Efficiency in the Caribbean

There are known barriers to the adoption of energy efficiency that debilitates the uptake processes in the Caribbean. These barriers exist even though there are clear incentives and economic benefits that promote energy efficient technology adoption. Therefore, these barriers, especially in the realm of institutional and technological capacities of the region, can be defined as the mechanisms inhibiting the diffusion of energy efficient measures. Figure 2 captures the energy efficiency barriers from existing research based on organizational theory, behavioural perspectives, economies and theoretical concepts (Sorrell et al. 2004, 2011).

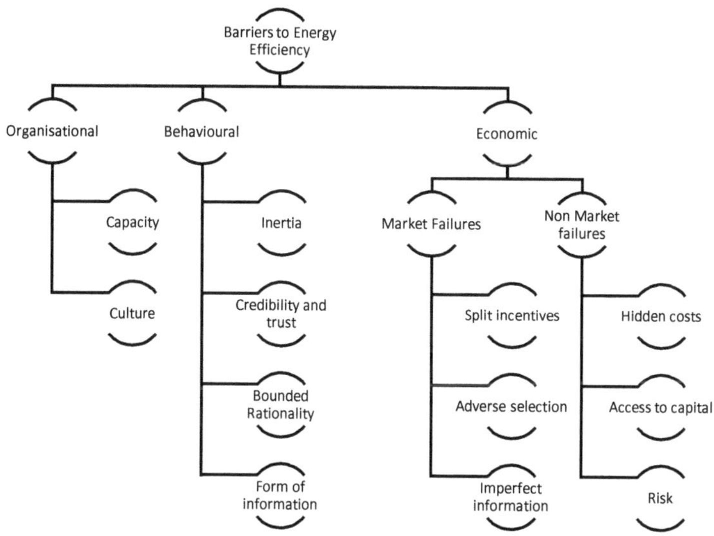


**Fig. 2** Classification of barriers to energy efficiency (adapted from Singh 2017)

The Intergovernmental Panel on Climate Change (IPCC 2013) divided the barriers to energy efficiency into four major categories—limited capital access, lack of skilled personnel, lack of relevant information and other barriers. However, given that the causes of barriers are often interlinked, it is difficult to find the exact reason for the lack of action. Hence, to provide a distinctive view of the challenges within the Caribbean region, it is imperative to understand two drivers of energy efficiency: regulatory compliance and the business case for energy efficiency. Regulatory compliance is fuelled by the requirements of Annex VI of the International Convention for the Prevention of Pollution from Ships (MARPOL 73/78). MARPOL Annex VI aims to reduce shipboard air pollution and Chapter IV of said annex details the energy efficiency requirements for vessels more than 400 gross registered tonnes. The business case for the uptake of energy efficiency can only be made when operating costs are reduced or capital costs are reduced when compared to existing operational conditions.

While financial considerations and regulatory compliance are noted drivers for energy efficiency, the inadequacy of the enforcement of international standards in the region creates an environment where ship operators do not perceive the need to comply. Ockwell et al. (2008) highlighted a central role for both national and international policy interventions in achieving low carbon technology transfer. At the national level, domestic policies that provide incentives for the use of low carbon can play a strong role in overcoming cost barriers and developing markets for new low carbon technologies. National level efforts are also required in developing national systems of innovation, actively engaging with international collaborative initiatives and ensuring appropriate infrastructure is in place to foster technological development and change. Maritime technologies are very capital intensive in nature and given the current status of shipping, ship owners and operators are capitally constrained. Many banks, globally, have reduced their ship financing commitments, restricting ship-owners capital access and credit.

Given the current volatility of the market and the absence of such support, it is difficult for a ship owner to self-finance for new technologies especially given the long payback periods. In addition to developing market frameworks and financial mechanisms favouring low carbon technologies, nations are also required to exert more direct influence over technology transfer processes. Stulgis et al. (2014) suggest a Self-Financing Fuel Saving Mechanism (SFFSM) to facilitate the adoption of fuel efficient technology through tripartite contractual agreement between technology provider, fuel payer and the financer. An Emission Compliance Service Agreement (ECSA) is suggested for alternative fuels, specifically LNG dual fuel engines. The financier's payback and return are generated by future fuel cost savings through a hedge against the price spread between low-sulphur fuel and LNG. There are no clear indications by prospective charterers of any preferences for vessels utilising alternative fuels. Therefore, any such investment is left as the last resort until there is a regulatory demand for such investments. Further, lack of any incentives for going beyond the minimum requirements does not promote voluntary use of alternative fuel technologies. Government intervention is likely to play a key role in facilitating low carbon technology transfer by developing a suitable policy (Shujing 2012).

### *4.1 Inadequacy of Existing Legal and Institutional Framework*

The IMO's Regional Maritime Advisor-Caribbean highlighted the poor state of implementation of maritime conventions in the region at a recent regional conference (Young 2016). In the presentation, he outlined the status of 12 conventions in the Caribbean. Only 3 of the 12 Conventions that were reported on reflected implementation levels above 50% across the region. With respect to MARPOL Annex VI, there are 10 reported ratifications (71% of the region), with only 2 territories having implemented the requirements into their domestic law, resulting in a 20% rate of implementation for the region in respect of this Annex of the MARPOL Convention.

In recognition of the key roles regional and local regulatory compliance and enforcement play in facilitating the uptake of energy efficiency and low carbon technologies, the gaps within the legal landscape must be examined. Monitoring of Maritime traffic in the Caribbean is undertaken by Flag States and Port States with occasional intervention under the umbrella of Coastal State jurisdiction. Maritime administrations in the region typically lack both the technical capacity and the requisite human resources to be effective monitors of maritime traffic. The relatively small economies with limited resource base, therefore, require maximum use of existing maritime surveillance institutions for enforcement of international standards.

### 4.1.1 Regulatory Control

Flag States in the region are supported in their implementation and enforcement by the Caribbean Memorandum of Understanding on Port State Control (CMOU). The CMOU comprises 17 members, including the Cayman Islands and these countries aim to inspect 15% of the vessels visiting their respective ports. It must be noted that many of the vessels plying the Caribbean area are non-convention vessels and, therefore, adherence to the standards of the Caribbean Cargo Ship Safety (CCSS) Code and the Code for Safety of Small Commercial Vessels (SCV) is being championed. In selecting energy efficiency and emission reduction methods, compliance with current and future emission regulations has been an important driver for ship owners and operators. With enhanced support from the Caribbean states, the CMOU could be one of the most significant drivers for implementation of MARPOL Annex VI.

### 4.1.2 Maritime Administration

The Maritime Administration and its enabling legislative frameworks often reside within Ministries where its functions are often overlaid by other priorities of these Ministries. It has also been observed that countries with their Maritime administrations constituted as Statutory Authorities or special purpose companies are more efficient in their implementation of conventions. Some countries have established maritime authorities while others, like that of Trinidad and Tobago, remain housed in a Ministry typically responsible for Transport.

Marine and maritime surveillance is currently facilitated by a myriad of Governmental actors and these typically include the following: Ministries responsible for National Security, Finance, Environment and Transport. Maritime administrations across the region, if not constituted as statutory authorities are divisions within their Ministries responsible for Transport. Such designations make it difficult for maritime legislative agendas to surface as priorities for legislative intervention. As a consequence, the legal frameworks for many Caribbean territories are not able to effectively facilitate low carbon initiatives.

The legislation governing maritime administration across the region and those of the Commonwealth Caribbean are very similar. Given this circumstance, the legal framework governing maritime administration in Trinidad and Tobago is being used as a case study to demonstrate the institutional gaps that adversely affect maritime administration. Table 1 identifies the Ministries, their sub-divisions and core functions as related to maritime administration.

The summary in Table 1 illustrates a fair amount of jurisdictional overlap, which often results in the presumption that responsibility for certain matters may rest with

**Table 1** Case study: summary of institutional framework impacting Maritime Administration in Trinidad and Tobago

| Ministry | Division | Legislation | Core function |
|---|---|---|---|
| National Security | Trinidad and Tobago Coast Guard (TTCG) | Defence Act 1962 Chap 14:01 | Responsible for maritime defence |
| | Immigration | Immigration Act 1969 Chap 18:01 | Implementation of immigrations law |
| | Marine Interdiction Unit | Customs Act 1938 Chap 78:01 | Joint agency between Customs and TTCG for surveillance and enforcement of counter drug trafficking |
| Ministry of Finance | Customs | Customs Act 1938 Chap 78:01 | Implementation of duties, taxes and tariffs |
| Ministry of Work and Transport | Maritime Services Division | Shipping Act 1987 No 24 | Implementation of Shipping Act 1987 and the administration of maritime matters for the State |
| | Port Authority of Trinidad and Tobago | Port Authority Act 1961 Chap 51:01 | Oversight of the operations of the Port of Port of Spain |
| | Harbour Master | The Transfer of Functions (Harbour Master) Order 1988 | Functions undertaken by the Director Maritime Services and no longer active unit for physical surveillance of Harbours and Ports |
| Ministry of Trade and Industry | InvestTT | http://www.investt.co.tt/about-us | Investment facilitators |
| | Maritime Industry Development Steering Committee | Formed by Cabinet in 2016 | Inform Governmental policy for interventions for development in the Maritime Cluster |
| Ministry of Planning and Development | Environmental Management Authority (Air Pollution Rules) | Environmental Management Act 2000 | Management of the environment; coordination and implementation of environmental laws |
| | | Air Pollution Rules 2014 | |
| Ministry of Agriculture Land and Marine Resources | Institute of Marine Affairs | Institute of Marine Affairs Act 1976 Chap 37:01 | Responsible for research on marine resources and advisors to the Government and Wider Caribbean Region on marine environmental protection |
| Ministry of Energy and Energy Affairs | Bunkering | Petroleum Taxes Act 1974 Chap 75:04 | Oversee the implementation of National Energy initiatives |

other governmental actors and, therefore, gives effect to another tragedy of common responsibility. This uncertainty of responsibility directly impacts the effectiveness of implementation of international maritime standards. The legal and institutional framework is also impacted by the outreach and initiatives of many international organizations.

### 4.1.3 Response of International Organizations

International organizations positively impact the legal and institutional framework of the Caribbean in facilitating the adoption of international standards. These international initiatives backstop many legislative agendas and provide technical and financial resources to effect international objectives (Slaughter and Burke-White 2006). Two international organizations affecting maritime governance are the International Maritime Organization (IMO) and the United Nations Environment Programme. One notable initiative by IMO is the creation of Maritime Technology Cooperation Centres across five developing regions. These MTCCs form a Global MTCC Network (GMN) that works towards promoting ship energy-efficiency technologies and operations, aimed at the reduction of harmful emissions from ships. The University of Trinidad and Tobago has been awarded the contract for the MTCC-Caribbean. This initiative is being implemented between January 2017 and December 2019.

UNEP, through the Regional Marine Pollution Emergency, Information and Training Centre—Caribe (RAC/REMPEITC) provides *inter alia* training and technical support for the implementation of internal conventions in the Wider Caribbean Region. Despite this international support, many States of the region are still lagging in terms of legal, technical and administrative requirements for effective implementation of multi-lateral environmental agreements, in general, and the MARPOL Annex VI, in particular.

## *4.2 Absence of Market Based Measures*

Lack of market based measures (MBM) presents a major barrier towards uptake of energy efficient and low carbon technology. As defined by UNEP, market based measures (MBM) "are policy approaches that encourage behaviour changes through their impact on market signals rather than through explicit directives." MBM's are based on the fact that organizations and individuals are cost driven and any modification is likely to influence behaviours. Therefore, setting a cost to carbon emissions in the Caribbean region will create an incentive to cut such emissions ether using new technology or increasing energy efficiency. Positive financial considerations constitute a major driver for investment into energy efficiency.

Ship type, adaptability, power rating and economic issues are some other major considerations. For the process of technology adaption to be success, a bespoke approach based on the investment evaluation must be constructed. As in the case of any other investment, technology transfer and adoption are also facilitated by incentives. A clear incentive for such an investment is to comply with the MARPOL 73/78 requirements required by a positive compliance enabling environment.

## 4.3 Inadequate Baseline Data

As a first requirement for an efficient and economic use of low carbon technology, it is vital to prepare a baseline of the existing technology in the region. In the absence of a baseline, as a region, it is vital to create one for a clear distinction of various existing technologies in use. Without such a baseline, identification and persuasion of technology adapters is a difficult task. Governmental intervention in the technology transfer also plays a key role in low carbon technology transfer by developing suitable policy. Government intervention can exert direct influence on technology adaption or transfer process by providing market frameworks and financial mechanisms. However, such appropriate Governmental interventions need to be informed by relevant data and analysis.

## 4.4 Insufficient Technology Transfer

As emphasised in the discussion on the enabling environment, technology transfer is a process by which expertise or knowledge related to some aspect of technology is passed from one user to another for economic gain (Schnepp et al. 1990). Such gains include mitigation of future costs associated with climate change along with any financial benefits involved for the providers and adapters, equally, in the process (Ockwell et al. 2015). The IPCC provides a wider definition of technology transfer that is "*a broad set of processes covering the flows of know-how, experience and equipment for mitigating and adapting to climate change among different stake holders such as governments, private sector entities, financial institutions, non-governmental organizations (NGO's) and research and education institutions*."

Developing economies have more scope and potential to make a significant contribution towards emission reductions. The transfer of low carbon technology to developing countries will assist the on-going efforts to reduce carbon emissions. Successful transfer and absorption of these existing low carbon technologies is a critical process in tackling climate change globally. Given the environmental and economic benefits, developed and developing countries can equally benefit from this process of technology transfer. However, to better understand the technology transfer, it is vital to know the aspects of the adoption process. Existing literature widely distinguishes the technology transfer as between:

- *Vertical technology transfer*: developing new innovations in low carbon technology and then offering them to the market (transfer of technologies from research and development to commercialization); and
- *Horizontal technology transfer*: the diffusion of the technology to the wider market. The transfer from one to another geographical location.

In the case of low carbon technologies, especially maritime technologies, there are elements of both vertical and horizontal transfer, especially given the need for commercialization within the new regional context. It is also imperative to state here that commercial acceptance within the maritime sector is influenced by logical business sense and mandatory requirements and compliance within the operating region. For technology transfer to be effective, the targeted receivers must be able to adopt, implement and sustain the technology. This requires investment into infrastructure and human capital.

## 5 Opportunities for Overcoming the Barriers to Energy Efficiency in the Caribbean

As was stated above, key drivers for energy efficiency include regulatory compliance and economic incentives. Studies (Sorrell et al. 2004, 2011; Ockwell et al. 2015) on the adoption of energy efficiency have frequently indicated that adoption of energy efficient technology leads to long term cost effectiveness. Legislation specifically providing incentives for compliance and promoting maritime energy efficiency in the Caribbean is non-existent and the absence of relevant baseline data creates an obstacle to targeted interventions The implementation of economic incentives will require legislative interventions for fiscal reform of maritime interests. Given the slow pace of maritime legislative reform in the Caribbean, the opportunities for driving compliance must be explored in the context of the existing legal and institutional frameworks that impact maritime operations. Consequently, compliance will need to be focussed through flag state and port state controls.

The possible financial benefit for the Caribbean of such legislative reform may be inspired by the actions of Norway and Singapore. Norway's NOx tax and NOx fund is an example of an effective local program that creates a financial scheme and business case for NOx reduction. The NOx tax is collected from participating industries and is fed into a NOx fund. The NOx fund then provides finances to those organizations that want to implement NOx reduction measures, including the shipping industry. On the basis of the scheme, a large number of ships have so far been equipped with NOx reduction technologies. This fund has also widely financed major Norwegian initiatives such as the move to LNG as fuel for ships operating in Norwegian waters. There is no clear incentive for ship-owners in the Caribbean region to adopt Maritime energy efficiency measures.

Some Administrations have taken or are evaluating the use of EEDI for differentiated registration fees or tonnage taxation. For example, such an initiative was taken by Singapore MPA (Maritime Port Authority) in 2011 under the Singapore Green Ship Programme. The MPA provides incentives to ship owners who adopt energy efficient ship designs. Accordingly, Singapore-flagged ships registered on or after 1 July 2011, which go beyond the requirements of the IMO's EEDI, enjoy a 50% reduction on the Initial Registration Fees (MPA 2017). They also enjoy a 20% rebate on Annual Tonnage Tax. Existing ships which utilise energy efficient ship designs are also part of this programme, and eligible to receive the 20% rebate on Annual Tonnage Tax. Given that the Caribbean has some of the largest shipping registries in the world, there is an opportunity to explore this model of incentives.

## 5.1 Enabling Environment to Facilitate Uptake of Energy Efficient Measures in Shipping

To create a win-win scenario for technology adopters and providers, the creation of a genuine collaborative environment is required. The process itself will be certainly exposed to various hurdles given that it will involve complex coordination and cooperation of various public and private sector organizations. This situation may be resolved given the various initiatives, specifically in the maritime sector, by IMO. In MARPOL Annex VI Regulation 23, emphasis is laid on co-operation along with transfer by promoting technical cooperation towards efficiency of ships.

While creating an enabling environment, each stage of the value chain must be considered. It is not only important to lift the existing constraints but also to fill the gaps in existing regulatory and administrative mechanisms. Any successful policy reform supporting an enabling environment requires changes in implementation that facilitate compliance for targeted stakeholders. Such intervention will need to be cognizant of the incentives that drive business decisions. The development of the maritime sector is on the legislative agenda of many Caribbean territories. However, current top down approaches have failed to achieve desired results in the absence of support from the bottom level.

Successful technology transfer requires an enabling environment and capacity building in addition to overcoming other major barriers like institutional constraints, intellectual property rights (IPR) and finance. Especially given IPR considerations, many feel uncomfortable with the term "Technology Transfer". The modern approach advocates new terminology—technology co-operation and sees the technology transfer as a two-way learning process instead of a one-time transaction. Given the other barriers to technology co-operation, the most needed actions in the Caribbean region are the creation of an enabling environment and capacity building.

The enabling environment will recognize the disparity between the available technologies and the technology in use and facilitate co-operation among regional administrations. Such cooperation should be directed towards the removal of barriers and adopt a framework which is efficient for receiving the required technology. This will require investment into institutions, infrastructure and the socio-economic environment. Capacity building is required to extract and absorb the complete benefits of the new technology adopted. Human and institutional capacities will need to be prepared accordingly.

### 5.1.1 Enhancing the Reach of Existing Institutional Actors

Despite the seeming limitations, with so many Governmental actors involved in maritime administration, there is an opportunity for Maritime concerns to be effectively addressed if there is a clear and focused designated "champion" for particular interests. With such a designation, the operations of these governmental bodies can be traced and avenues for cooperation in implementation can emerge.

The creation of the MTCC-Caribbean provides a framework from which national and regional maritime agendas promoting the uptake of low carbon technology can be leveraged. Capacity building for the implementation of MARPOL Annex VI is one of the key deliverables for all the MTCC's. Consequently, it is envisioned that the MTCC-Caribbean will play a key role in capacity building and fostering the partnerships between existing institutions for the purposes of promoting energy efficiency.

## *5.2 Financial Incentives for Efficient Fuel Operations*

In view of the complexity of maritime operations, transfer and adoption of technology must go beyond compliance and offer financial incentives, preferably through fuel efficient operations. Fuel cost benefit is also related to legislation. For example, by using shore-generated power instead of running auxiliary diesel engines, vessel operators enjoy a fuel cost benefit, the magnitude of which depends very much on the going fuel price. A legal requirement mandating a switch to Low Sulfur fuel in Emission Control Areas (ECAs) significantly improves the cost effectiveness of OPS, since it implies the use of marine gasoil (MGO) that is more expensive than heavy fuel oil (HFO). At present, there is no ECA established within the Caribbean region other than those areas falling under the jurisdiction of the United States of America.

The lack of any major shipbuilding or repair yards in the region presents a challenge to adopters in the form of high acquisition and maintenance costs. The limited access to ship yards creates an environment for monopolies. Such a monopoly in the business may distort the availability of appropriate and acceptable new technology. Given the nature of the business, where suppliers are reluctant to share information with a view to maintaining international competitiveness, the IMO Green House Gas Study (2009) highlighted three policy approaches to GHG emissions reduction. These are illustrated in the Fig. 3.

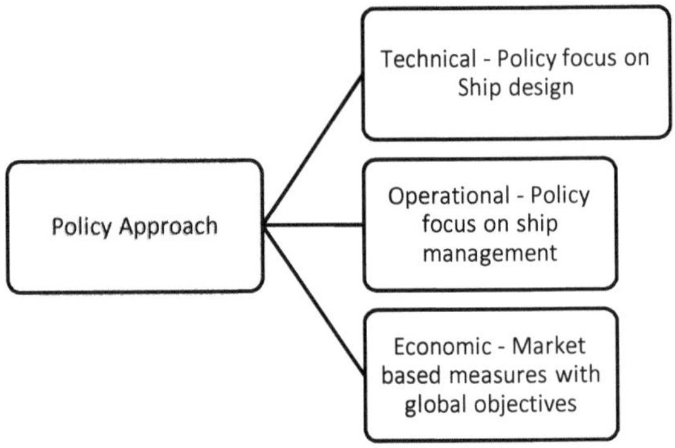


**Fig. 3** Policy approaches to GHG reduction (IMO 2009)

Through MEPC 62 and Chapter 4 of MARPOL Annex VI, some options have been implemented. These include technical[1] and operational[2] requirements that have been merged in the present state of regulations. Although these regulations emphasize compliance, which is left to the shipping companies, they fail to achieve the desired results towards energy efficiency and emission control.

# 6 An Approach for Implementing Relevant International Standards Based on an Analysis of Existing Legislative Frameworks

## 6.1 *Monitoring, Reporting and Verification*

Effective technology transfer is necessary to achieve energy efficiency and reduction of emissions. Given the existing status of administrations in the region, a *sui generis* approach will involve the innovative use of the existing institutional framework and market based measures especially involving private stakeholders. Creating the enabling environment and capacity building through a tailored approach will require incentives not only for compliance but towards a behaviour promoting efficient use of energy and a focus on energy abuse related to emissions. The aim should be not only to enable technology transfer but to sustain and absorb the associated benefits. Therefore, as highlighted through EU proposal 525/2013, a strong and accurate "monitoring, reporting and verification" (MRV) system must be created as it will provide a robust foundation for market based measures. Further research into cultural and socio-economic factors may be required in this sector to support policy decisions.

[1]Energy Efficiency Design Index (EEDI).

[2]Ship Energy Efficiency and Management Plan (SEEMP).

Regional administrations will need to enhance existing flag and port state control measures especially given that most Caribbean shipping falls outside the remit of the national registries. The absence of any existing reporting system related to use of fuel and emissions regionally, and a lack of designated focal points for collecting and analysing such data, limits development of regional MBM. MRV constitutes the first stage for the development of MBM. The MTCC's global network is a step closer to the formation of that baseline and providing a platform for such technological cooperation. Mapping the existing technology used on board vessels by major energy consumers and reporting on fuel type and consumption data, MTCC Caribbean aims to provide the much-needed baseline to the regional administrations to facilitate improved policy frameworks.

## 6.2 *Institutional Partnerships to Facilitate Implementation*

In addition to focusing on capacity building initiatives for Flag, Port and Coastal State actors, private sector partnerships also represent an untapped resource for promoting the implementation of international conventions. Key private sector actors include, Ports- private ports and enhanced ship vetting criteria for access; marinas; ship repair facilities; and shipping agents and non-vessel owning common carriers (NVOCC). Figure 4 illustrates the institutional relationships and key stakeholders required to facilitate adoption of low carbon technologies.

## 6.3 *Capacity Building Initiatives*

Leveraging the formation of an MTCC in the Caribbean, the countries of the region can harness training opportunities and identify the human and physical resources required to assist in the implementation of low carbon technology initiatives.

**Fig. 4** Institutional partnerships

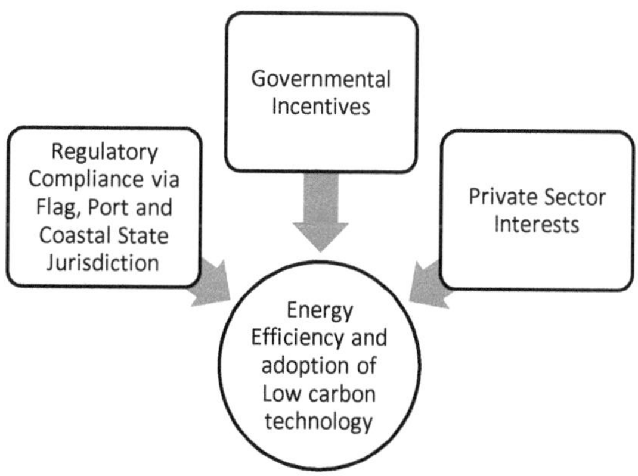

Maritime training institutions provide a ready platform as receivers of sustainable technology transfer. Information sharing, cooperation and partnerships will prevent duplication of efforts resulting in efficient use of resources.

## 7 Conclusion

Overcoming the challenges to maritime energy efficiency in the Caribbean requires the promotion of MBM and innovative use of existing legal and institutional constructs. Both individually and collectively, Caribbean nations must recognize the complexity of the sector and develop strategies to capitalize on the interlinked nature of human and institutional factors. Absorption and adaption of technology will require investment into capacity building and leveraging existing institutional actors as far as possible.

There is a need to create a baseline of the existing technology on the vessels operating within the Caribbean region. Mapping of existing capacities and technologies against contemporary technologies will identify the continuums of the existing energy efficiency gap. It is expected that the MTCC Caribbean will play a significant role in determining the existing gap through capacity building initiatives and pilot projects.

Policy and institutional drivers are major components in technology transfer. Through appropriate actions, these policies can facilitate MBM and circumvent the organisational and market barriers to technology transfer. The relation between technology transfer and IPRs must be discussed leading to a successful regional transition to sustainable development. Policy interventions that address the IPR regime are vital to the process of technology transfer. In the absence of the regulatory framework, it is difficult to influence compliance in a highly regulated sector such as the maritime industry.

In the discussions above, various state agencies and private sector stakeholders were identified as key facilitators in creating the enabling environment for promoting the uptake of energy efficient technology. Further analysis into indigenous approaches will provide opportunities for public-private partnerships and enable informed policy decisions on incentives. The implementation of Flag, port and coastal State control measures, alongside incentives for compliance and technology transfer hold the key to success. Enhanced regional collaboration between maritime administrations, the CMOU and the MTCC Caribbean will result in better MRV. Such collaborations, amongst Caribbean States and regional institutions, guided by the provisions of MARPOL Annex VI, will provide the foundation for *sui generis* regimes tailored to overcome the specific barriers and challenges of individual countries in the uptake of maritime energy efficiency.

## References

Creswell, J. W. (2009). *Research design – Qualitative, quantitative and mixed approaches* (3rd ed.). London: SAGE Publications.

Hagood, J. D. (2006). Why does technology transfer fail? Two technology transfer projects from Peronist Argentina. *Comparative Technology Transfer and Society, 4*(1), 73–98.

IEA. (2010). *Enabling energy efficiency: A comprehensive review of energy efficiency good governance*. Paris: IEA/OECD.

IMO. (2009). *Second IMO GHG study 2009*. London: IMO.

IPCC. (2013). In T. F. Stocker, D. Qin, G.-K. Plattner, M. Tignor, S. K. Allen, J. Boschung, A. Nauels, Y. Xia, V. Bex, & P. M. Midgley (Eds.), *Climate change 2013: The physical science basis, contribution of Working Group I to the Fifth assessment Report of the Intergovernmental Panel on climate change* (p. 1535). Cambridge, UK/New York, NY: Cambridge University Press.

Kranzberg, M. (1986). The technical elements in international technology transfer: Historical perspectives. In J. R. McIntyre & D. S. Papp (Eds.), *The political economy of international technology transfer* (pp. 31–45). Westport, CT: Quorum Books.

MacKenzie, D., & Spinardi, G. (1995). Tacit knowledge, weapons design, and the uninvention of nuclear weapons. *American Journal of Sociology, 101*(1), 44–99.

MPA. (2017). Green Technology Programme. http://www.mpa.gov.sg/web/portal/home/maritime-companies/research-development/grants-and-incentives/green-technology-programme.

Ockwell, D., Sagar, A., & de Coninck, H. (2015). Collaborative research and development (R&D) for climate technology transfer and uptake in developing countries: Towards a needs driven approach. *Climatic Change, 131*(3), 401–415.

Ockwell, D., Watson, J., MacKerron, G., Pal, P., & Yamin, F. (2008). Key policy considerations for facilitating low carbon technology transfer to developing countries. *Energy Policy, 36*, 4104–4115. Transition towards Sustainable Energy Systems.

Rai, V., Schultz, K., & Funkhouser, E. (2014). International low carbon technology transfer: Do intellectual property regimes matter? *Global Environmental Change, 24*, 60–74.

Ray, S. (2012). Technology transfer and technology policy in a developing country. *The Journal of Developing Areas, 46*(2), 371–396.

Saunders, M., Lewis, P., & Thornhill, A. (2012). *Research methods for business students* (6th ed.). London: Pearson Education Limited.

Schnepp, O., Glinow, V. M. A., & Bhambri, A. (1990). *United States-China technology transfer*. Englewood Cliffs: Prectice-Hall.

Shujing, Q. (2012). The analysis on barriers of low carbon technology transfer. In *Energy Procedia, 14, 2011 2nd International Conference on Advances in Energy Engineering (ICAEE)* (pp. 1398–1403).

Singh, S. (2017). *Determining the role of human factors influencing energy efficiency within a commercial and organisational framework of a shipping multinational company', MS international management (oil and gas)*. Liverpool: University of Liverpool.

Slaughter, A., & Burke-White, W. (2006). 'The future of international law in domestic' (or the European way of law). *Harvard International Law Journal, 47*(2), 2006.

Sorrell, S., Mallett, A., & Nye, S. (2011). Barriers to industrial energy efficiency: A literature review. United Nations Industrial Development Organization Working Paper No. 10/2011.

Sorrell, S., O'Malley, E., Schleich, J., & Scott, S. (2004). *The economics of energy efficiency: Barriers to cost-effective investment*. Cheltenham: Edward Elgar Publishing.

Stulgis, V., Smith, T., Rehmatulla, N., Powers, J. & Hoppe, J. (2014). CWR shipping efficiency finance report.

Thomas, D. R. (2006). A general inductive approach for analyzing qualitative evaluation data. *American Journal of Evaluation, 27*(2), 237–246.

Wilson, J. F. (2014). *Essentials of business research: A guide to doing your research project* (2nd ed.). SAGE Publications Ltd: London.

Worrell, E., van Berkel, R., Fengqi, Z., Menke, C., Schaeffer, R., & Williams, R. O. (2001). Technology transfer of energy efficient technologies in industry: A review of trends and policy issues. *Energy Policy, 29*, 29–43.

Young, C. (2016). MARPOL and the liability conventions: the implementation challenge in the region. In *Presented by Colin Young at the Annual Conference of WiMAC in The Cayman Islands. September 5–9.*

## Further Reading

EU (European Union). (2013). Proposal for a Regulation of the European Parliament of the Council on the monitoring, reporting and verification of carbon dioxide emissions from maritime transport and amending Regulation (EU) No 525/2013, COM(2013) 480 final.

UNEP. (2013). Working Group I contribution to the fifth assessment report of the intergovernmental panel on climate change.

# Energy Efficient Operations of Warships: Perspective of the Indian Navy

**Amit Batra and Rohit Prakash**

## 1 Introduction

Climate change concerns have brought energy and environment sustainability to the forefront of global discourse. The current global warming trend is of particular significance as it is human induced and increasing at an alarming rate. This particular issue has, therefore, emerged as one of the biggest challenges that the world is facing today; a challenge that can only be mitigated with an all-inclusive effort. Emissions from ships are considered to be a significant source of air pollution, both for conventional pollutants and greenhouse gases, and significant work is being undertaken globally to improve their "environmental footprint".

India, is a fast developing economy, which is at the center of global discourse, be it social or economic development. The country stands at the crossroads of a growth trajectory on one hand and a responsibility towards the issue of global warming, on the other. With 2.4% of the world's surface area and representing one sixth of the world's population, climate change is a major challenge for India, which faces large scale climate variability and is exposed to enhanced risks from climate change. The government policy has been aligned towards an ambitious climate change mitigation policy with goals to ensure that energy is available to all whilst still being environment friendly. In recognition of the growing problem of climate change, the Indian government has declared a voluntary goal of reducing emission intensity by 33–35%, over 2005 levels, by 2030 (MEFCC-GoI 2015).

IN is a maritime force with blue water capabilities and global reach, which envisages achieving a force level of 200 ships by 2027. With its expanding outreach, it continues to induct new modern energy intensive platforms which have significant carbon emissions. While the IN has been performing the military, benign, diplomatic

A. Batra (✉) · R. Prakash
Integrated Headquarters Ministry of Defence (Navy), New Delhi, India
e-mail: a.batra@cranfield.ac.uk

 
A. I. Ölçer et al. (eds.), *Trends and Challenges in Maritime Energy Management*, WMU Studies in Maritime Affairs 6, https://doi.org/10.1007/978-3-319-74576-3_7

and constabulary roles over the last six decades, it has launched its own 'Green Initiatives Program' (GIP) aligned with the national objective of a 'Greener India and the World' and thus added another dimension of social responsibility to its existing roles. The GIP focuses on the key principles of reduction, diversification and use of clean technologies. With an all-inclusive approach, the program envisages an energy and resource efficient IN, which is environmentally responsible. With the implementation of the GIP, the IN aims to add a "*Green footprint to its blue water capabilities*". The analysis at hand highlights the steps being envisaged by the IN towards increasing energy efficiency for operations of warships as a part of its GIP for operational platforms, and examines the opportunities and challenges in their implementation.

## 2 Drivers and Challenges for the "Green Initiatives Program"

Owing to their strategic role, warships, despite being significant consumers of energy are excluded from any international regulations on energy and environment. Any new beginning needs to be carefully thought of, calibrated, and uncompromising on the effectiveness of operations. With this tight rope to walk, IN, in the last two and a half years, has undertaken a number of significant steps to address the energy and environmental issues over the entire spectrum of naval operations, infrastructure and administration. While the scenario of global climate change and its resultant effects is a primary driver behind the GIP towards energy efficiency and environment, there are multiple tertiary benefits that the IN is also vying for. An important consideration in deploying a fleet is the cost of operations, of which fuel accounts for the major share of the revenue expenditure. Lower consumption of fuel would directly reduce the operating costs, which in turn reduces the Total Ownership Cost (TOC) or Life Cycle Cost (LCC). Another aspect of lower consumption of fuel is an improvement in availability and reach within the same operating cost. From a purely economic aspect, though the initial investment in energy efficiency technology is high, the payback through savings in running costs should make the investment viable. Energy efficient technologies on board ships will result in lowering onboard fuel consumption, which could translate into an enhanced operational reach for the ships (and the whole fleet). This would also result in increased operational efficiency of sea-going platforms, thus improving their combat and operational effectiveness.

The GIP is not devoid of challenges. One of those challenges has been non-availability of mature technologies for direct application on warships. Most of the technological enhancements to increase efficiency involve considerable investment, which may be constrained by the overall budget of a ship design or refit project. In addition, one of the biggest challenges that this program faces is that of 'behavioral changes' which entails acceptance of 'cultural shift' to 'greener methods of operation' by the crew of warships and sensitivity towards energy use onboard.

**Fig. 1** Summary of the overall objectives of the IN Green Initiatives Program

## 3 Objectives of the 'Green Initiatives Program'

The GIP of IN aims to reduce both energy in relation to operations (used on ships) as well as support needs (shore facilities) (IDME-IN 2014). The energy efficiency and conservation measures in shore installations are largely focused on increasing the share of renewable energy. However, the current analysis focuses on the measures being or planned to be introduced on the IN fleet as a part of the GIP.

Energy conservation and energy efficiency are both concepts of reducing energy consumption without altering the functions of the organization. While the former is achieved by adopting little changes in the behavior of the system, the latter usually involves the use of technological advances in the field of efficient equipment and system design. GIP aims for energy conservation and increasing energy efficiency of the IN fleet to be implemented through the following objectives as shown in Fig. 1:

- Efficient design of warships to reduce energy use within the envisaged operational profile. This is planned for implementation by acquisition of new platforms that are energy efficient as per predefined targets and retro-fitment of developed energy efficiency modules on legacy platforms.
- Optimization of operations to minimize energy use without compromising on the operational role or efficiency. Aligned with the naval doctrine, energy usage can be optimized by planning fleet operations aided by decision making tools that incorporate the 'green enablers' as discussed later in the paper.
- Maintenance activities including monitoring to sustain the efficient design. It is envisaged that energy consumption can be effectively used as the basis for condition monitoring of the hull as well as the propulsion and power generation plants.

## 4 'Go Green' Enablers

Towards implementation of the GIP for the IN fleet, multiple enablers have been identified in the areas of design, operations and maintenance of the fleet as shown in Fig. 2. These areas are themselves deeply related and driven by technology that needs to be either developed or identified, and then inserted for each of these enablers, integrated overall by the GIP.

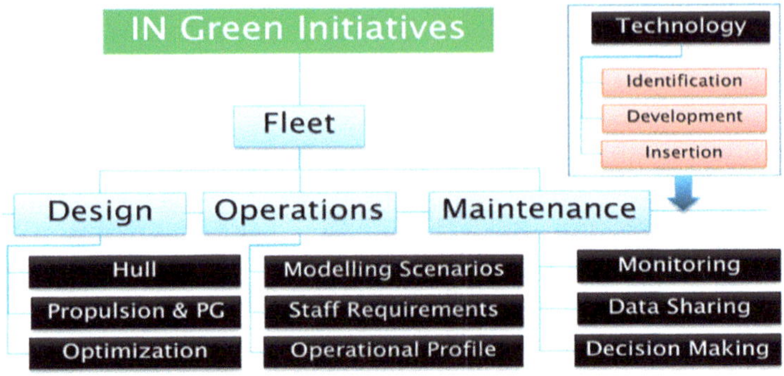


**Fig. 2** Enablers for the IN Green Initiatives Program

IN, like any other navy, maintains a fleet of ships at various stages in their life cycle to maintain the force levels as per the force level requirement and the naval doctrine. The IN is also embarked upon a large shipbuilding plan with many classes of ships currently at various levels of design and production. IN maintains its own dedicated ship design organization which is currently engaged in the design of several combatants including an aircraft carrier. The GIP of the IN seeks to infuse the design process with the requirement of reducing its emission footprint as one of the primary methods to achieve the program objectives. This capability will allow for technology identification, integration and insertion into new designs as well as retrofitting into existing platforms.

When designing and building warships, attaining high overall efficiency for propulsion and power generation plants is always a design objective, with an aim to get 'bang for the buck'. However, from the perspective of achieving low emission targets, a need is felt to better optimize designs to achieve them. Optimizing the design to meet these targets will also lead to many indirect benefits which can significantly enhance the capability of the ships by achieving higher endurance with low running costs and thus 'total cost of ownership'. Many of the measures undertaken to improve new ship designs will also have the capability for 'technology insertion' into ships already in service to improve their overall operational efficiency.

### 4.1 Design for Higher Efficiency

Warship design is largely driven by required operational capabilities, which drive the choice of architecture for the propulsion and power generation plants. The propulsion plant is required to operate at varying speed requirements, from very low speed operation for 'loitering' to high speed ('sprint') requirements. It is common for propulsion plants of frigates and destroyers to spend maximum time at less than half the rated power. This varying requirement of power is often met in warships by using a combination of equal powered engines adding power based on power requirements through a gearing arrangement, or using a combination of lower powered 'cruise' engines for lower speeds and higher powered 'boost' engines for higher speeds also through a gearing arrangement.

### 4.1.1 Plant vs Component Efficiencies

The overall efficiency of a power generation and propulsion plant needs to be viewed from the perspective of the complete range of the ship's operating profile. This overall efficiency can obviously be improved by "improving" individual component efficiencies. Of these efficiencies, targeting the efficiency of the engine to achieve overall plant efficiency is usually an obvious choice. While engine designers and manufacturers are innovating within the technological boundaries to increase the engine efficiencies, in an ever competing market scenario, naval designers face the dilemma: can the same overall plant efficiency be achieved by targeting the overall plant for a well-integrated solution as having complex arrangements in engines. An example of such a complex arrangement in an engine is the WR21 intercooled and recuperated marine aero derivative gas turbine, designed and manufactured by an international consortium, led by Northrop Grumman as prime contractor, which offers a reduction in fuel burn of 30% across the typical ship operating profile. This design, with low specific fuel consumption characteristic across the engine's operating range and having significantly higher part load efficiency, was developed to give the engine the flexibility to be used as both the 'cruise' and 'boost' engine. Despite all the advantages it offers, the WR21 currently finds application only in the Type45 destroyers of the Royal Navy (RN).

The propulsion (or power plant) designer has to make a clear choice on the overall plant architecture before the final equipment selection phase, so as to ensure that the selection remains optimal from the perspective of overall plant efficiency. High individual component efficiencies do not always translate to the best overall plant efficiency. For example, if exhaust heat recovery is planned to be used in a plant then an engine with lower efficiency would be an optimal choice for the plant so that its exhaust heat is enough to efficiently operate the recuperator. Hence, in many ways more flexibility in design is available if overall plant efficiency is targeted rather than targeting complex engine architectures; all the plant components need to be selected based on this requirement.

### 4.1.2 Integrated Electric Plants and Drives

The latest generation of warships needs more energy than ever to meet the high energy requirements of the technologically evolved weapon systems and sensors. This, along with the requirement of increased operational flexibility, with a wide range of operating scenarios from 'loitering' speeds, to 'cruise' speeds for economic transit and 'high sprint' or 'boost' speeds is moving the propulsion and power generation architecture towards electric drives, which provide more flexibility and more efficiency when operated in the scenario of wide range of operations.

The Integrated Full Electric Propulsion (IFEP) is a good option for application on larger ships, typically large destroyers and even aircraft carriers, with the advantage of having the capability to direct power from any source to propulsion, ship service or combat/launching systems. IFEP provides excellent design opportunities that include high operational flexibility with optimum definition of the size and the number of electric generators according to electric balance so as to operate them in their optimal efficiency range. A fact worth noting is that pooling of energy to meet all power requirements for a ship is not a new concept. Older ships using steam plants were 'integrated' on the steam side. However, with the transition to gas turbine and diesel engines, this integration was lost with modular energy sources. Thus IFEP brings back the "integration" of energy for ships through the electrical side for obvious benefits. IFEP is technologically intensive and does add significant complexity to the overall propulsion plant and thus its application is growing slowly with maturing of technology.

Smaller ships, typically frigates, are increasingly applying the Hybrid Electric Drive (HED) arrangement due to the inherent volume constraints that preclude full electric drive. HED cross connects the ship power generation system with the propulsion system by placing an electric motor and its associated motor drive in the drive train. At low speeds, the propulsion engines are turned off and the electric motor(s) provide the propulsion power. In new ship designs, HED can reduce the prime movers needed by adding motors. It does not eliminate space, weight, and cost, but adds efficiency. With HED, the application of propulsion diesels used as 'cruise engines' in traditional 'combination' propulsion plant arrangements can be eliminated and replaced by electric motors and their associated motor drives. This may require higher power generators. This also reduces the number of exhaust ducts required in the ship superstructure creating more volume for combat systems. The elimination of two propulsion prime movers also reduces construction and maintenance costs. The FREMM destroyers of the French and Italian Navies are an example of this type of configuration.

HED can also be a retro fitment option on existing fleet ships, especially on 'all gas turbine ships', to significantly improve fuel economy. Warships typically have a large redundancy built into the power generation system. This additional power available in the power generation system can be used to drive the motor in the propulsion drive train. Placing the propulsion load on the generator sets of the power generation system and not having to operate the larger propulsion engines at low part power reduces the effective specific fuel consumption for the propulsion load. Operating the generator sets closer to their rating point also reduces the specific fuel consumption. Overall, the fuel burn can be reduced by 8% or more depending on the ship operating profile (Tuddenham 2015). In the context of the IN, the Delhi class destroyers would make an excellent case for such retrofit. While IN does lack a land based prototype facility for proving such an arrangement before the system is installed onboard, it could undertake sea trials on the similar Rajput class destroyers before they are decommissioned.

### 4.1.3 Hull and Propeller Design Improvements

Many improvements have been made to the hull and propulsion plant design which can be applied to both new designs as well as adopted into ships already in service. Addition of bow and stern end bulbs as well as addition of stern flap has shown good potential to increase overall hull efficiency. Introduction of Propeller Boss Cap Fins has also been shown to increase efficiency. Another area that can make a significant impact in increase of efficiency is measures against bio-fouling, which include improved hull paint schemes and propeller coatings. Significant ongoing work is underway towards the same topics in civilian vessels, which can be directly applied towards warships.

### 4.1.4 Development of 'Staff' and Design Requirements

Warships, like their merchant marine counterparts, are built for a length of service which usually spans 25–40 years. A properly matched propulsion plant design is absolutely vital for the ship to effectively achieve its mission for this length of service in which the operating profile may evolve to be significantly different from the originally envisaged profile. Usually ships are designed as per specific design points for speed and endurance with an operational profile used as guidance primarily for determining the maintenance requirements. Often the ship rarely operates at these design points on which the entire design has been based. Hence, it is essential that a realistic operational profile of the ship is used to design the ship. This profile can be worked out by statistically analysing the operating pattern of ships with similar characteristics, envisaging the future operating role of the ship integrated with the fleet.

The top speed requirement of a ship, is one of the most critical factors that determines the overall propulsion design. It is very important that this is determined carefully as at higher speed ranges; there is a huge price in terms of installed power that is to be paid for each knot of demanded additional speed. If this speed is rarely used while the ship usually operates at lower speed regimes, another penalty is paid of running the propulsion plant at a non-optimized load.

### 4.1.5 Multipoint Design Point vs Single Design Point Optimization

Considering the aforementioned scenario with an ever evolving operating profile, the naval ship design could immensely benefit from utilizing the Multi Design Point (MDP) method used for design of gas turbine engines in the aero industry. Though Analytic Hierarchy Process (AHP) and Decision-Matrix analysis methods are often used in naval design for making decisions related to design and selections, MDP would make a good choice for naval propulsion and power plant design considering the complexity of the process to find an optimal solution to meet multiple, often conflicting requirements with a complex operating profile.

MDP methodology is used to overcome the problem of selecting a single design point for design. Selecting one design point generally results in an 'overdesign' with reduced performance at off design conditions. The MDP methodology for aero engine design has been published in open domain (Schutte 2012) which presents a transparent on-design cycle analysis method that simultaneously meets performance requirements and constraints at numerous design points.

Analogous to the MDP for aero engine design, for the naval propulsion and power plant design, the designer can parametrically create candidate plants to populate the cycle design space (CDS), representing all of the candidate propulsion and power plant cycles that meet the performance requirements for the specified requirements and operating profile from which a "best" propulsion and power plant can be selected. MDP allows many design points to be included in the analysis and is a more elegant approach than separate on and off design analyses. Each point within the CDS represents a unique candidate engine.

The parameters which define and shape the CDS are grouped into technology rules and performance requirements. Together with the technology rules, the performance requirements shape and constrain the CDS. The performance limits and technology limits define the boundaries of the feasible CDS, while the performance targets and component performance estimates shape the CDS within those boundaries. The goal of the designer is to select the best plant from a design space of feasible candidate plants with consistent technology level and performance capability for a particular application. MDP analysis will result in an optimal solution that is fully feasible, and meets all requirements and constraints at all design points of interest.

## 4.2 *Optimized Fleet Operations*

A significant reduction in overall emissions footprint can be made by managing the naval operations better at both ship and fleet level. At ship level, the fluids on-board can be managed better to achieve the best trim for a particular loading condition by using Trim Optimization tools which have already been developed.

Management at fleet level can primarily be controlling the overall fleet speed, which is by no means an easy task considering the vast number of scenarios in which the fleet operates along with its composition. The availability of a management tool which can help in determining the optimal speed, considering the overall composition of the fleet with individual fuel consumption rates of its ships and with any present limitations, will go a long way in helping the fleet operations team to better plan their tasks at hand. Voyage Planning and Weather Routing management tools can also be used to plan ship and fleet deployment and operations.

### 4.3 Monitoring Design, Operations and Correlations

Presently, IN measures absolute consumption of energy at the input level by monitoring the fuel used, without any relation to the energy efficiency of operations. It is essential that overall efficiency of naval operations, in the various operating scenarios, is measured to initially benchmark and then for monitoring. A detailed analysis of the processes, operations, procedures, and parameters of operation is necessary to identify opportunities to conserve energy and increase energy efficiency at all levels.

With the intention to study energy efficiency, it is essential to create a system of measuring energy consumption by fleet across systems and analyze trends to identify patterns. From these analyses, benchmarks for energy consumption for various ships and submarines at equipment, system, ship and ship class levels need to be established. For new acquisition ships, these benchmark levels need to be compared with the theoretical design energy consumption parameters to define correlation coefficients. To undertake these and achieve the objective of reduced energy consumption, it is necessary to formulate a comprehensive and navy wide *Energy Management Plan*, with individual Energy Management Plans at various unit levels.

#### 4.3.1 Current Practice

Currently, IN uses fuel consumption to monitor the overall energy consumption of the fleet. This data is usually not available integrated with the factors which influence the energy consumption levels like efficiency and degradation of propulsion and power generation plants, state of underwater hull as well as scenarios in which the fleet operates.

IN follows a practice of creating a baseline for fuel consumption of ships, undertaken during the Contractor Sea Trials (CST) or Acceptance Trials Phase after major refits of ships and recorded in the ship's data book. IN, thereafter, regularly monitors the fuel on an annual basis by undertaking 'fuel consumption trials'. This captured data is correlated with the baseline data to determine the overall deterioration with ageing of ships due to factors as that of hull, propulsion and power generation plants. This data is used to undertake passage planning and deployment of IN ships.

IN also uses the Engine Exploitation Factor (EEF) to monitor the average hours of exploitation of engines onboard for each hour of hours spent underway. EEF is defined as the ratio of the sum of the running hours of all propulsion engines onboard a ship to the number of hours the ship is underway for a given period. Based on the average statistics gathered over time, EEF for each class of ships is laid out and is monitored for each ship against this specified value.

While the 'fuel consumption trials' give a measure of the overall propulsion 'degradation' of a ship from the perspective of maintenance, the EEF is used to monitor the 'operation' of the propulsion plant over an averaged period. Both the above methods have a major advantage in their simplicity and ease of reporting and monitoring, they also have their inherent flaws. The ‘fuel consumption trials’ undertaken at ship crew level, can be inconsistent in the terms of trial conditions and there is little correlation available of the overall degradation with component hull or propulsion element degradation. EEF does not capture the important factor of the operating scenario which influences the number of engines in operation. EEF also offers no separation of information on the ‘cruise’ or ‘boost’ engine hours on ships with such arrangements. Hence, there is a need to define better factors to monitor efficiency at both design and operating levels. Some navies do use the factor of ‘fuel consumed per mile run’ for monitoring efficiency in operations but, again, this factor does not take into account the operating scenario of ships and, unless averaged over large periods of time, this factor will not give the desired output.

### 4.3.2 Adaptation of IMO’s Indices for Monitoring Performance

IMO has promulgated the Energy Efficiency Design Index (EEDI) which has been in force since 01 January 2013 for all new ships of the merchant marine fleet. IMO has also recommended the use of an Energy-Efficiency Operational Indicator (EEOI) as a monitoring and/or benchmark tool which will measure the energy efficiency of each voyage of the ship.

The EEDI specifies a minimum energy efficiency level (measured in $CO_2$ emissions per tonne-mile) for different ships as per their size, set as a reference baseline, which is to be achieved in the ship design. The choice of technologies to achieve this specified level, which incrementally increases from 2013, is left to the ship owners. The EEOI enables operators to measure the fuel efficiency of a ship in operation and to gauge the effect of any changes in operation or maintenance activity. EEOI (measured in $CO_2$ emissions per tonne-mile) from operational data over time can be used for comparison of emissions from each voyage.

EEDI and EEOI cannot be directly applied in a naval scenario as naval ships operate for a significant percentage of their deployment time loitering or at high speeds depending of the scenario. These indices could, however, be adapted to include factors such as the operation mode and theatre and developed for specific classes of ships for monitoring design as well as operations. This is not a simple task as putting numbers or weightages for operation mode and theatre will need significant study and analysis.

### 4.3.3 Development of Tools for Monitoring Performance

Along with the development of indices for monitoring performance, there is also a need to develop tools which automate or assist in the capturing and analysis of data and suggest corrective and operative actions to the crew to best manage the propulsion and the power generation plant of plant. The tools should also aid in monitoring operations at the fleet level.

The present method of capturing data in various log books of the engineering and navigation departments has its disadvantages in the cumbersome tasks of collating, stitching and analysing. With the modern day digital platform based control and monitoring systems, much of the data required for monitoring indices is readily available in these systems in various formats. However, there is a requirement for an integrated tool to process this information and present it to the user for decision making. The decision making could also be facilitated by adding "intelligence" to these systems in the form of active algorithms which suggest corrective and operative actions. The tool to achieve this could be the 'Intelligent Energy Dashboard' as a real time operations monitoring and management tool. This tool is already available in many commercial forms. To be applied in the naval context at ship and fleet level, these commercial tools will need to be significantly modified. This will also require development of additional algorithms for integration into these systems.

# 5 Conclusion

The IN has developed a 'Green Initiatives Program' as a part of a national initiative for reducing emissions in line with the rising global initiatives against environmental pollution and degradation; although naval vessels are not covered by the respective international regulations, this is in line with IMO's drive to control emissions from sea going vessels. Though the IN, like any other navy, is not mandated to follow the emission norms laid down by the IMO, it has taken these initiatives towards a national responsibility aiming at a more sustainable future. The IN also recognizes the many indirect benefits of the program, which include lowering of operating costs and increasing the reach of its fleet by increasing the endurance of its ships and achieving better health of propulsion and power plants.

There is significant work being undertaken globally in all spheres of technology by commercial, non-governmental and governmental agencies to reduce the emission footprints of commercial ships. Much of what is being done is applicable directly or indirectly for application in naval ships. However, for this flow of information and technology transfer, there is a need for synergy amongst all the involved agencies. IN is endeavoring to reach out to partners globally with such agencies, with a vision to collaborate in the areas of ship design optimization tools, design of energy efficient propulsion and power generation plants, development of ship and fleet 'Energy Dashboards', ship trim optimization tools and hull fouling measurement systems amongst others. Development of such synergy could go a long way toward making our planet 'greener' and a better planet to live in for our future generations.

## References

Directorate of Marine Engineering-Indian Navy: (DME-IN). (2014). Environment Conservation Roadmap (Internal IN Document, not available in public domain).

Ministry of Environment Forest and Climate Change-Government of India (MEFCC-GoI). (2015). India's intended nationally determined contribution - Towards climate justice.

Schutte, J., Tai, J., Sands, J. & Mavris, D. (2012). *Cycle design exploration using multi-design point approach*. Proceedings of ASME Turbo Expo 2012 GT2012.

Tuddenham, R. S., & Maynard, S. (2015). *Advanced propulsion technologies for electric ship architectures*. Indian Navy Advanced Naval Propulsion Technologies Conference Proceedings 2013.

## *Further Reading*

Anderson, T., Gerhard, K., & Sievenpiper, B. (2013). *Operational ship utilization modeling of the DDG-51 class*. ASNE Proceedings 2013.

Batra, A., (2015). Warship propulsion and powering technology: The present and the future. In *Indian Navy advanced naval propulsion technologies conference proceedings 2013*.

Batra, A., & Gupta, T. (2009). Marine engine emissions and their control present and the future. *Journal of the Institution of Engineers (India), 89*, 16–24.

Dominic, C. S., & Karafiath, G. (2012). *Hydrodynamic energy saving enhancements for DDG 51 class ships*. Report by Naval surface warfare center, Carderock Division, Resistance and Propulsion Division.

Maynard, S. (2015). The Cost of speed, Naval Technology, Naval Forces IV, 2015.

McCoy, T., Zgliczynski, J., Johanson, N. W., Puhn, F. A. & Martin T.W. (2007, October). Hybrid electric drive for DDG-51 class destroyers, *NEJ*, 119(2).

O'Rourke, R. (2006). Navy ship propulsion technologies: Options for reducing oil use. Report: Background for Congress Updated December 11, 2006.

Walters, R., Gaffney, R., Knoch, S., & Plath, E. (2011). *Improving maintenance management tools and data collection to increase surface ship fuel efficiency*. ASNE Proceedings 2011.

Webster, J. S., Fireman, H., Allen, D. A., Mackenna, A. J. & Hootman, J.C. (2007, October). Alternative propulsion methods for surface combatants and amphibious warfare ships. *NEJ*, 119(2).

# Mexico's Reorganisation of Maritime Security Regime: A New Role for the Navy and Emphasis on Energy Related Infrastructures

Adriana Avila-Zuñiga Nordfjeld and Dimitrios Dalaklis

## 1 Introduction

Previous research efforts have already pointed out that the current state of implementation and level of compliance with the International Ship and Port Facility Security Code (ISPS Code) in Mexico is rather poor (Nordfjeld and Dalaklis 2016). This chapter will also investigate in a similar direction. It will first provide a summary of the latest reform to Mexico's national legislation dealing with various important issues of the maritime domain. Then, it will identify the contradictions in the relevant laws, in relation to effective implementation of national legislation and international instruments, with special focus on the ISPS Code. Subsequently, it will study the acceptance of the transferral of the National Maritime Authority from a civil institution (MCT) towards a military one (MMAR) within the four different authorities directly related to the maritime industry's operations/activities. Finally, findings in relation to field research activities in main Mexican ports and especially important energy hubs are presented, followed by the associated conclusions and recommendations with the aim to improve their security status.

The International Maritime Organization (IMO) communicates via its web site the information that thirty-five (35) different countries located in the American Continent have signed and ratified the International Convention for the Safety of Life at Sea (SOLAS), 1974; Mexico is included in the aforementioned group. The SOLAS 1974 Convention includes Chapter XI-2 concerning special measures to enhance maritime security, and especially Regulation XI-2/3 that enshrines the International Ship and Port Facilities Security Code (ISPS Code). Apart from the ISPS Code, the SOLAS 1974 Convention is used to "operationalize" a very influential safety instrument for shipping operations: the International Management Code

A. A.-Z. Nordfjeld · D. Dalaklis (✉)
World Maritime University, Malmö, Sweden
e-mail: dd@wmu.se



A. I. Ölçer et al. (eds.), *Trends and Challenges in Maritime Energy Management*,
WMU Studies in Maritime Affairs 6, https://doi.org/10.1007/978-3-319-74576-3_8

for the Safe Operation of Ships and for Pollution Prevention (International Safety Management (ISM) Code). Furthermore, in its Chapter XI-1, there are special measures to enhance maritime safety that provide the basis for carrying out inspections of foreign ships in national ports under Port State Control (PSC); the latter can also be described as an exercise of the right of protection of the coastal State, as established in Article 25 of the United Nations Convention on the Law of the Sea (UNCLOS). In the case of Mexico, there is a confusing situation after the country's last major legal maritime reform, where it was established that the National Maritime Authority shall be executed by the Ministry of Marina (MMAR), and it transferred all Master of Harbours to MMAR; however, the Port Authority remained with the Ministry of Communications and Transport (MCT).

Implementation and compliance with the ISPS Code is a topic studied extensively in the United States and Canada. The US has, indeed, been a pioneer in port security. Already in 1997, 7 years before the ISPS Code entered into force, the US Department of Transportation launched the "Port Security National Planning Guide", which was "intended to be used by port directors and officers of port security departments in fulfilling their responsibilities toward the development of an effective port security program" (US Department of Transportation 1997). The guide included requirements of a "Security Survey and Risk Assessment" and the development and implementation of a "Security Plan". Kenneth (2009) pointed out that "*the evolution of organized security processes in the maritime sector can be understood as a product of increasing governmental and commercial concerns about the criminal exploitation of seaports, [. . .] and the rising threat of global terrorism*". In the case of Canada, the ISPS Code has been implemented since it entered into force in July 2004. Rudner (2009) included maritime ports as part of the "Critical National Infrastructure" that must be protected from terrorism and provided emphasis on ports serving the energy sector. The author highlighted the need for "a national strategy/plan for the protection of Canada's critical national infrastructure against exogenous risks and threats".

Mexico, similarly to Canada, is one of the largest producers of oil and gas and has a significant number of critical infrastructures, including numerous maritime ports linked to the energy sector. Mexico's State owned company (Pemex) reported that, as of 1st January 2014, the proven reserves of Mexico's crude oil were calculated to be 9812 million barrels (MMb), whereas the proven reserves of natural gas were calculated to be 16,549 billion cubic feet (Pemex, Petroleos Mexicanos 2014). This report also mentioned that the biggest volume of proven reserves of crude oil, equivalent to 70%, are located in marine oil fields and only 30% of them are located in terrestrial fields. At the same time, 56% of the proven reserves of natural gas are located in terrestrial fields, with the remaining 44% in marine fields. It must also be highlighted that most of the Mexican ports are directly or indirectly serving the oil energy market for developed marine oil fields.

## 2 Mexico's Maritime Reform

To better understand the latest developments in Mexico in relation to the maritime domain, it is important to consider that a few decades ago, the MMAR held the control of ports and the merchant marine; but in 1977 the ex-President José López Portillo reformed the "Law of Public Federal Administration" transferring to the MCT all the activities related to the development of merchant marine, as well as the construction of port infrastructure and the respective administration/operation. However, the MCT and particularly the Harbour Masters (also called Port Captains in Mexico) were never equipped in a satisfactory manner to deliver the functions of maritime safety and security, such as port state control and inspection of vessels. Almost 40 years later, all these functions are (again) reallocated to the authorities of MMAR, with the latest maritime legal reform.

For a rather prolonged period, drug cartels in Mexico were conducting a wide array of illegal activities that resulted in extremely high levels of violence. Considering these implications, as well as less unforeseen crimes such as the theft of hydrocarbons and demands of "extortion payments" to owners and operators of port terminals (Nordfjeld and Dalaklis 2016), the Mexican government attempted the reorganization of its maritime security regime. Not only is efficient use of energy resources important, but "losing" tremendous quantities of oil due to theft is totally unacceptable and must be addressed via the right types of policies. Several existing laws were revised; a new set of regulations was introduced in 2014 that included amendments to the "Law of Ports", the regulation in relation to the establishment of the "Unified Centre for Port and Maritime Security" (called "UCMAR" thereafter), and the "Law of Navigation and Maritime Trade", among others. However, the Parliament focused on port security and port installations; it did not include in the regulation the part related to maritime security and the requirements for ships, established in the ISPS Code, such as the Ship Security Officer (SSO), Ship Security Assessment (SSA), the Ship Security Plan (SSP) and the International Ship Security Certificate (ISSC). From April 21st 2014 to June 2017, when the last updates to this chapter were made, there was no institution appointed responsible for maritime security conforming Mexico's national legislation. Additionally, the creation of the "Military Navy Unit for Port Security/Protection" ("Unidad Naval de Protección Portuaria" and represented with the acronym UNAPROP, in Spanish, called NAUPPRO thereafter) was decided, covering only the part of the ISPS Code related to protection of ports and port installations.

Concerning maritime security, as already mentioned, an important legal instrument is Chapter XI-2 of the Safety of Life at Sea Convention 1974, (SOLAS Convention) which relates to the ISPS Code. This Code is defined by the IMO as "*the comprehensive set of measures to enhance the security of ships and port facilities, developed in response to the perceived threats to ships and port facilities in the wake of the 9/11 attacks in the United States*" (International Maritime Organization 2012). Whereas part A of the Code establishes the mandatory provisions, the non-mandatory ("recommended") part B provides guidelines about how to comply with the mandatory requirements of part A. After numerous years of ISPS Code implementation, Mexico does not yet comply with its requirements at an acceptable level, although some improvement is lately recorded, mainly because of the 2014 security reforms (Nordfjeld and Dalaklis 2016).

As highlighted previously, the mandatory Part A of the ISPS Code introduced the obligation for contracting governments to appoint a Designated Authority to carry out certain maritime security duties/responsibilities established in the Code. This Designated Authority holds the responsibility for setting maritime security levels and ensuring compliance with the maritime security measures at all ports (where the ISPS Code applies) through the Port Security Assessment (PSA) and Port Facility Security Assessment (PFSA). The revision, approval and control of compliance of the Port Security Plan (PSP) and Port Facility Security Plan (PFSP), which shall be based upon the PSA and the PFSA, is also included in the respective responsibilities. The development of PSP/PFSP is also related to the activities of the Port Security Officer (PSO) and Port Facility Security Officer (PFSO).

As established by IMO, there are three different security levels: Security Level 1 (normal) requires the minimum protective security measures at all times; Security Level 2, which requires additional protective security measures for the specific period of time that the risk of a security incident is heightened; and Security Level 3, which requires specific protective security measures and may result in the suspension of commercial operations. Security response under Level 3 is transferred to the Government or other organizations responsible for dealing with significant incidents (International Maritime Organization 2012). In Mexico, it is the "Unified Centre for Port and Maritime Security" (UCMAR) that is responsible for coordination and execution of all security operations under Level 3 and it is the Commander of the Navy Military Zone of each jurisdiction where an UCMAR is established who shall be appointed as the Chairman of that respective UCMAR, as established in provision 8 of Regulation of the UCMAR published on the Official Diary of the Federation, on 21st of April, 2014. This provision also includes the notion that the Harbour Master shall be nominated as the Vice-Chairman of the UCMAR, and, in addition, three employees of each of these institutions shall be selected as advisers. A UCMAR shall be established at each of the 16 Federal Integral Port Administration (s) (FIPAs) designed to receive vessels of over 500 gross tonnage, or which receive vessels sailing international trading routes, according to Provision 6 of the referred regulation.

According to the "Law of Ports" Article 19 SECOND, the UCMAR is a group of inter-institutional coordination between the MMAR and the MCT for the application of and compliance with the ISPS Code. This part was not amended with the Decree. However, since MMAR obtained control of both Master of Harbours and the UCMAR itself, the "inter-institutional character" of this group should be re-evaluated, keeping in mind that MCT will maintain control of the "*Fideicomiso de Formación y Capacitación para el Personal de la Marina Mercante Nacional*" (represented with the acronym "*FIDENA*" in Spanish, which may be translated to "Fond of Education and Training of the National Merchant Marine", and called FETNAMM thereafter). This is the authorized institution for giving the IMO's model courses 18.1, 18.2 and 18.3 concerning the ISPS Code, and the only authorized foundation so far, for performing Port Security Assessments (PSA) and Port Facility Security Assessments (PFSA).

Article 19 THIRD, paragraph II of this law highlights that "*the UCMAR shall apply all the dispositions and response measures within the framework of the Chapter XI-2 of the SOLAS 1974 Convention, comprising the ISPS Code and ensure the establishment of a series of functions and actions for each of the respective three security levels*". This is part of the maritime security reforms that took effect on the 21st of April 2014. However, because of lack of knowledge about the reform itself, the MMAR did not fully exercise its authority and the Harbour Master was erroneously recognized as the authority (for the ISPS Code) by the PFSOs and port agencies. The UCMAR is responsible for the revision of PSA/PFSA, which covers the security risks evaluation for ports or port facilities, which shall be the base for developing the PSP/PFSP, submitted and approved by the Designated Authority. The responsible person for developing the PSA/PFSA and PSP/PFSP is the Port Security Officer (PSO) or Port Facility Security Officer (PFSO), including compliance with all requirements established in the ISPS Code and reflected in the PSP/PFSP as training, exercises, practices, inspections audits and modifications via formalised procedures to the plan.

In another research of Nordfjeld and Dalaklis (2017), it was rightly pointed out that the PSO/PFSO must attend security incidents and keep incident security records updated, which must be considered in the risk evaluation and integrated into the security plan to achieve a constant reduction of risks and continuous improvement of port and maritime security. UCMAR shall verify, among other aspects, that security incidents are properly recorded in the security incidents register; even with MCT supervising the port authority, the UCMAR is responsible for control of compliance with the ISPS Code and co-responsible for the revision and approval of PSA/PFSA and PSP/PFSP, according to its Regulation, Article 7, paragraph II, which states that: "*[UCMAR] shall participate in the evaluation of risks of maritime and port security, previous to the elaboration of the security plans and it shall propose the necessary modifications and updating to those plans*". Paragraph III of this article, mentions that once the plan has been approved, the UCMAR shall participate in the verification and control of compliance with plans, ensuring their effective implementation. Findings of the above mentioned study included inconsistencies within the data of the necessary security incident records, or even their total absence. Neither the MMAR nor the MCT has effectively addressed the issue of security incident record keeping.

The UCMAR is currently performing the responsibilities of a Port Security Committee, a committee recommended by the ISPS Code to coordinate security procedures and measures. Under the UCMAR, there is also the Port Security Assessor Committee, which is normally integrated by the UCMAR and customs-immigration authorities operating at the port; the management of the port operator and port terminals is also included. Additionally, municipal and regional Governments with associated interests in that jurisdiction are part of the committee. However, it is important to consider the need to balance the openness of an advisory committee with the need to protect the confidentiality of sensitive security information (International Maritime Organization 2012).

In 2016, the President of Mexico, Enrique Peña Nieto, presented a law initiative to amend and improve diverse provisions of the "Law of Public Administration", "Law of Navigation and Maritime Trade" and the "Law of Ports". The aim was to transfer the control, inspection, vigilance and other activities related to the merchant marine and the maritime industry, including Harbour Masters from a civil authority (MCT) to a military one (MMAR). However, a significant exclusion was port development and administration apparatus, including the port authority, which was previously exercised through Harbour Masters. This law initiative was approved by the National Congress without any change and published as a Decree on the Official Diary of the Federation (ODF) on the 19th of December 2016.

This legal document establishes the new attributions of the Ministry of Marine, which entered into force on the 17th of June, 2017. Every authorized port shall have a Master of Harbour, which will be administrated by MMAR with a delimited maritime territorial jurisdiction, which, according to Article 9 of the decree, shall have attributions to authorize bear away and bear off as well as customs clearance of naval craft; flag and register Mexican vessels and naval artefacts; grant permissions for offering maritime passenger transport services and nautical tourism within the waters of jurisdiction; regulate and monitor that the nautical routes meet the required conditions for safety, depth, marine signs, maritime traffic control and nautical assistance; and inspect vessels, among others.

The duties and responsibilities of the NAUPPRO should also be (re-)evaluated and confirmed. This unit currently has the functions of a coast guard. It is under the control of the Commander of the Navy Zone at each port, who is the President of the UCMAR. However, its guidelines concerning duties and responsibilities are not clear enough with respect to its duties under the UCMAR. Yet, these units have the responsibility of port protection under the control of the Commander of the Navy Zone, and they must ensure control and compliance with ISPS Code requirements at the port and give all necessary information to the Commander, who holds the role of Chairman of the UCMAR, may inform the rest of this group of actions taken by the NAUPPRO. The actual Guidelines of NAUPPRO are currently too general and rather vague. Due to their military discipline, they do not take actions to ensure compliance with ISPS Code requirements if they do not receive concrete and specific orders or if it is not clearly established and requested in their procedures. The development of rules/protocols with specific and concrete duties for this unit is recommended. Security duties performed by the NAUPPRO may include, among others, considering that currently they only attend the ISPS Code section related to port installations:

- Control/verification that all maritime ports and terminals are operating with an updated Statement of Compliance (SOC).
- Revision and control that all ports and terminals have employed a certified PSO/PFSO with an updated and valid certificate.

- Revision and control that all personnel from ports and terminals with security duties have been certified with the necessary training concerning the required course 18.2 as established in the ISPS Code.
- Revision and control that all personnel from ports and terminals without specific security duties have been certified with the necessary training concerning the required course 18.3 as established in the ISPS Code.
- Develop and keep the official register of maritime security incidents for that specific port.
- Revise and control that all PSO/PFSO keep their own registers updated for reporting maritime security incidents.
- Participate in the revision and analysis of Port Security Assessments (PSA) and Port Facility Security Assessments (PFSA), for terminals.
- Participate in the inspections for approval of PSA/PFSA.
- Participate in the revision and analysis of Port Security Plans (PSP) and Port Facility Security Plans (PFSP) and make observations for necessary modifications to those plans.
- Participate in inspections and audits for approval of PSP/PFSP before the issuance of the Statement of Compliance (SOC).
- Carry out random inspections on scene and general inspections to verify the compliance of PSP/PFSP.
- Plan, coordinate and carry out security drills (level 3) at a minimum interval of once a year with no more than 18 months between them.
- Participate in the evaluation of security drills (level 3).
- Participate in and make the necessary observation for the external security audit, which shall be performed with a minimum interval of once a year.
- Develop and keep updated the register of approved PSA/PFSA; PSP/PFSP; drills; inspections and other relevant operations for availability to IMO, in the case of mandatory audits.

Agreement number 039 for the creation and activation of 19 NAUPPROs at the main maritime ports of Mexico was published on the 31st of March, 2014, on the Official Diary of the Federation. However, only 14 in total have been created and activated up to this point. The activation of NAUPPROs at the ports of La Paz, Puerto Vallarta, Puerto Chiapas, Matamoros and Dos Bocas is still lacking; at this last port, the official creation of the UCMAR is also lacking.

The Presidential Decree recently reformed the country's legal framework and allowed the complete shift not only of maritime security, but also of the safety aspect (including the port-state control function) from a civil authority (MCT), to a military one (MMAR). The previous failures of the Mexican Designated Authority are associated with a large number of security incidents and various accidents which, through poor inspections, have contributed not only to loss of property at sea, but also to human casualties and significant marine pollution. Aranda (2016) pointed out that while the Chamber of Senators analysed the initiative of law, three maritime accidents that occurred that summer can be used as evidence of deficiencies in the system under MCT. The Harbour Master of Tampico Port has also been denounced for corruption and abuse against fishermen. More specifically, MCT employees were accused of corruption by the owners of the fishing vessels "Rastreador", "Banpesca", "Fenicio", "Astur" and "Tampico Star". All these argued that they were forced to hand over a shrimp carriage quota to personnel from that institution, to be able to work (Vergara 2016).

The initiative emphasized the convenience of defining one National Maritime Authority to comply with national and international obligations and the need to delimitate and redistribute the attributions that are exercised by both institutions, the MMAR and MCT. However, several duties and responsibilities are still "shared". While the MMAR is responsible for implementation and compliance with the ISPS Code, the MCT holds the port authority role and is responsible for imposition of sanctions related to violations of the Code. On the positive side, the new legal regime emphasises that MMAR currently has the human and material resources necessary to comply with the exercise of the national sovereignty and authority in the Mexican marine zones, as well as to guarantee the compliance of the port with the national (maritime) legal framework.

In summary, the reforms to Article 8 SECOND of the referred "Law of Navigation and Maritime Trade" established the new attributions of the Ministry of Marine, and empowered MMAR to approve licenses for passenger and tourism maritime transport services with small boats, authorize vessels to bear away and bear off as well as customs clearance. Flag and registration issues of vessels, administration of the national registers of maritime crews and ships, as well as inspection and verification of national and foreign vessels are included in those tasks. Additionally, compliance with international conventions as well as national legislation and official Mexican norms related to maritime safety and security is another of its tasks.

At the same time, the Decree reformed Article 8 of the "Law of Navigation and Maritime Trade", concerning the new attributions of the MCT, confirmed that MCT will continue to lead the administration of ports, as well as the training framework of the merchant marine. A significant inconsistency is that the decree modifies Article 7 of the "Law of Navigation and Maritime Trade" and establishes that the "National Maritime Authority" is exercised by the Federal Executive Power through MMAR for the exercise of national sovereignty, dealing with maritime safety and security issues, as well as exercising the Right of the State to be applied at the Mexican marine zones, while Article 9 B establishes that MCT will exercise its functions at the ports through the Offices of Services to the Merchant Marine. Additionally, the "Law of Ports", Article 16, clearly establishes that the Port Authority is exercised by the Federal Executive Power through MCT, which will be exercised through the "Offices of Services to the Merchant Marine" established at all ports.

Vessel inspections are currently performed at the dock in the ports of Mexico, while according to common practice this should be done outside in the open ocean between 12 and 24 miles, in the contiguous zone. MMAR should evaluate the allocation to Harbour Masters of a certain number of the interceptor-patrols they already have or assess the acquisition of some Ocean Patrol Vessel(s) Defender II, which is a relatively small naval vessel designed to perform coastal defence duties, but large and seaworthy enough to patrol offshore areas in the open ocean.

Concerning the offshore area of Campeche, the control centre for maritime traffic was formerly managed and operated by Pemex, a private company. To ensure the proper exercise of the maritime authority in this area, MMAR should recover the control and operation of this centre.

## 3 Research Methodology of Field Activities

The research methodology of the current study includes semi-structured interviews with Harbours Masters, Directors of Customs Maritime Units, Directors of ports, PFSOs from ports and certain number of hydrocarbons terminals, Presidents of UCMAR (1st Commander of the Navy Zone, Navy Sector or Navy Station) and Commandants of the port and maritime police (Naval Unit for Port Security, also called NAUPS) and the Gulf's Navy Force. Additionally, participant observation at the terminals and Harbours Master's installations and ports, including not only the land areas but also the maritime passages took place; visits were made on board oceanic patrols of the Mexican Navy. The visits to the mentioned ports in Mexico had an average duration of a week at each port.

The poll of data included the leaders of the mentioned institutions from the ports of Altamira and Tampico (Tamps.), Tuxpan, Veracruz and Coatzacoalcos (Ver.), Dos Bocas (Tab.), Ciudad del Carmen (Campeche), and Progresso (Yuc.) as well as another Hub Port from the Pacific in Mexico. By doing this, the researcher covered the four pillars for the operation of the port. The pool of data is deemed sufficient, based on the fact that it includes eight out of sixteen FIPAs in Mexico where the ISPS Code applies, representing 50% of them and seven out of seven (100%) FIPAs situated in the Gulf of Mexico, where oil exploration and production activities take place. One State-Owned Integrated Port Administration (Ciudad del Carmen, Campeche) where most of the oil activity is concentrated was also included in the study. Two loose interviews with top directors from MMAR were performed on key themes identified from issues prevalent in the analysis of the interviews with Presidents of the UCMAR and Federal FIPAs. From the total number of persons invited to participate, representing Master of Harbour offices at these nine cited ports, only one of them rejected the invitation. The same figures correspond to representatives from the maritime customs units (one) and directors from FIPAs (one) as well as the PFSO from that same port, who due to extreme work volume had to decline the interview.

The purpose of the study was described to the participants via an information cover-sheet letter where the research objectives were explained, clarifying that their participation was voluntary, confidential and without any economic contribution, or gifts. The Research Ethics Committee of the World Maritime University approved the study. The total of interviewed participants was 57 persons, all of whom worked in areas of maritime safety and security. Interviews were carried out in the participants' workplaces by the researcher, tape recorded and transcribed verbatim. Data was examined line-by-line, and the main categories and themes were identified and coded using thematic analysis and constant comparison of data. The researcher searched thoroughly for all divergent views to form a rich description of different factors.

## 4 Results

From the interviewed subjects with functions of Master of Harbour, 75% expressed their approval to be transferred to the MMAR and supported the law decree, while 25% strongly disapproved of the transfer between the institutions. An important conclusion is that the people interviewed were either very much in favour or very much against the change. 87.5% of Directors of maritime customs units and 75% of Directors of FIPAs supported the law decree; at the same time, 100% of Presidents of UCMAR held a positive view. The same figures correspond to Commandants from the NAUPPRO (coast guard-navy), whereas 87.5% of Port Security Officers (PSO) supported the transfer of Master of Harbours to MMAR; and 80% of the interviewed subjects functioning as Port Facility Security Officers (PFSO) from hydrocarbon terminals supported the transfer.

Twelve categories were identified that either prevent or promote the transfer of Harbours Master from MCT to MMAR: (I) poor performance of Harbours Masters from the MCT; (II) acute lack of human and material resources available for Harbours Masters to comply with their duties and responsibilities in a satisfactory manner; (III) high exposure to corruption on the part of Harbours Masters related to Mexican crime ring organizations due to death threats against them or their families; (IV) fear on the part of Harbours Master of losing their jobs in a late stage of their productive life; (V) uncertainty on the part of Masters of Harbours with respect to their salary level, working rights and pension schemes; (VI) improvement of salary and working rights for Masters of Harbours and their personnel with the transferral to MMAR; (VII) already high participation of the Mexican Navy performing duties that correspond to Harbour Masters; (VIII) fear on the part of directors of port administrations of lack of knowledge, stubbornness and unwillingness/inflexibility to negotiate on factors related to business issues with respect to authorities from the Mexican Navy; (IX) security improvement for offshore installations and vessels serving the oil market; (X) trust in a significant improvement of maritime safety and security on the part of directors of maritime customs units with the transfer of Harbour Master offices to the Ministry of Marine; (XI) bigger and better capabilities to combat crime ring organizations that are using port installations for transport of drugs, weapons, money laundering and transport of stolen hydrocarbons, and finally; (XII) need to provide juridical support and attributions to the MMAR for activities already performed by the Navy and which are currently outside their juridical attributions.

# 5 General Discussion

A number of participants' quotes are used to better illustrate the associated data. Concerning category (I) poor performance of Masters of Harbours, one of the quotes that best illustrates the situation is taken from participant number 6 who said: "*unfortunately, the master of harbours have lost a lot of authority and it is very sad that nobody takes them seriously anymore*". This situation might be explained with category (II) acute lack of human and material resources available for Harbour Masters to comply with their duties and responsibilities in a satisfactory manner. The statement that best explains this aspect is from participant 1-G, who claimed that: "*the Maritime Designated Authority somehow is suffering a separation of attributions and material resources, because since 1940 the Mexican Navy; [...], used to give all the required support to Harbours Masters [...] but when they were transferred to MCT, [...] all the naval craft and necessary equipment was left with MMAR, including navy vessels, ocean patrols, boats for search and rescue operations and helicopters. Then, with these limitations the MCT started to function since 1976 and until today*". Interestingly enough, 100% of the participants pointed out the acute lack of resources for Harbour Masters to carry out their duties in an adequate manner.

Another reason for the loss of authority of Harbour Masters, can be explained with category (III) high exposure to corruption on the part of Harbour Masters related to Mexican crime ring organizations due to death threats against them or their families; as the case of participant 36 who affirmed "*I have one year and seven months that I am working under death threats. Constant threats by phone [...] I have moved my home and my family out of the city. There is no other way. And what can I do? I have the need to work. But tell me, where is not the same*"?

Participants highlighted the already high participation of the Mexican Navy performing duties that correspond to Harbour Masters (category VII). Yet, several directors of ports manifested their fear of lack of knowledge and inflexibility to negotiate on factors related to business issues with the Mexican Navy (category VIII). To illustrate this aspect, one of them (26) is quoted: "*I believe that the marina has given courses to their people but I do not think that they have much experience and knowledge concerning the operation and administration of merchant marine*". This was a view was shared by a few Harbour Masters. One of them (18) affirmed: "*[...] one of our functions [...] is the fluidity of the port. They are accustomed to command, and to apply correctness regardless of the problem, here everything is about flexibility and judgment. They do not have it*". On the other hand, 75 per cent of as FIPAs' directors were very positive about the transferral. To illustrate their attitude, respondent 49 quoted: "[...] *I believe that the objective of this project is to strengthen the Harbour Master to be able to do their duties in an efficient and effective way concerning vigilance and control of ports.*"

Category number (X) envisions a significant improvement of maritime safety and security on the part of directors of maritime customs units with the transfer of Harbour Masters offices to MMAR; this can be explained with a quote from participant 6: "*I have had the opportunity to work in the ports where the control of the Harbour Master was already been taken by the Ministry of Marina and it has been immensely positive. There was a huge leak of information that could only be controlled with the interference of the Navy. It has been very good. [...] I am waiting for them to do so, here too*". In the agreement between MMAR and the MCT (2009), it is stated that the financial and operative capacity of crime ring organizations has allowed them to access strictly confidential information. Therefore, it is important that MMAR implement an anti-corruption programme for the transfer of MCT employees to Marina to stop old corruption and leaking of information practices.

A statement from respondent number 36 reflects category (XI): bigger and better capabilities to "fight" crime ring organizations using port installations for criminal activities: "*the infrastructure to properly exercise the maritime authority is in MMAR through the Navy. The international conventions originally signed by the MCT are already taken by the MMAR, why? It may be because we do not have the equipment and human resources that they have*".

## 6 Conclusion and Recommendations

It is a rather self-explanatory fact that the Presidential Decree under discussion has significantly influenced the Mexican legal maritime framework; it allowed the shift of maritime safety and securities responsibility, from a civil authority (Ministry of Communications and Transport, MCT) to a military one (Ministry of Marine, MMAR). The very high level of acceptance between the four pillar institutions involved in maritime safety and security operations (both at port and at sea) such as Maritime Customs Units, Military Navy Zone-with its respective UCMAR, Federal Integrated Port Administrations, maritime police-represented by the Navy Unit for Port Protection (NAUPPRO), and finally the Harbour Masters themselves is a very strong indicator.

However, Mexico has to significantly improve the current level of performance; even if the change of the designated authority may contribute to the achievement of this goal, the MMAR must ensure the application of integrity and anti-corruption programs during the transferral process, to disconnect from the main causes of past failures. Further reforms to the Law of Navigation and Maritime Trade, Law of Ports and the regulation of the UCMAR shall be enacted with regard to appointing the institution responsible for implementation and compliance with the ISPS Code, with respect to maritime security, in relation to the security requirements for the operation of vessels. The transfer of the Port Authority from MCT to MMAR should be evaluated to avoid triangulation of the activities related to the maritime authority, such as imposition of sanctions. Most FIPAs are under the control of MCT and the simultaneous execution of both roles (acting as the performer and evaluator) should be avoided.

Strengthening the maritime security regime and especially its interface with energy resources remains an action of priority in order to "avoid unnecessary casualties" as a result of illicit activities and theft. Similarly, in order to enhance the level of maritime domain awareness (MDA) in Mexico, as well as introduce a forward way of thinking about conducting maritime security operations, the MMAR could also recover control of maritime traffic centres (which are currently operated and administrated by Pemex, including the offshore and oil production area of Campeche). Better information is a prerequisite for the optimal allocation of resources. The allocation of offshore patrol vessels or "Deterrence" patrols, as well as the required personnel to operate them at the disposal of Master of Harbours could allow the latter to perform vessel inspections at the contiguity zone if they consider so necessary. Additionally, tailor made training will also be needed to build the necessary human capacity and expertise that it is not yet available; the introduction of a "no-blame" culture could also be considered. Finally, MMAR must also avoid misuse of high security trained resources in duties that are outside their competence and especially avoid allocating to the maritime police (NAUPPRO), duties and responsibilities that must be performed by staff belonging to Harbour Masters.

## References


Aranda, J. (2016). Indagan vínculo de irregularidades en capitanías de puerto con accidentes. In: La Jornada, Newspaper. http://www.jornada.unam.mx/2016/09/27/politica/013n2pol. Accessed 20 Nov 2016.

International Maritime Organization. (2012). *Guide to maritime security and the ISPS Code*. IMO.

Kenneth, C. (2009). *Port security management*. Boca Raton, FL: CRC Press Taylor & Francis Group.

Nordfjeld, A., & Dalaklis, D. (2016). *Opening of offshore oil business in Mexico and associated framework to cope with potential security risks*. Innovation Match MX 2015–2016 Conference, Guadalajara Jalisco, Mexico, 7 April 2016.

Nordfjeld, A., & Dalaklis, D. (2017). Enhancing maritime security in Mexico: Privatization, militarization or a combination of both? In P. Chaumette (Ed.), *Economic challenge and new maritime risks management: What blue growth?* MARISK Human Sea. Université de Nantes. Gomylex.

Pemex, Petroleos Mexicanos. (2014). Hydrocarbons Reservoirs to January 1st., 2014. http://www.pemex.com/ri/Publicaciones/Reservas%20de%20Hidrocarburos%20Archivos/20140101_r_h_e.pdf. Accessed 26 Jun 2016.

Rudner, M. (2009). Protecting Canada's critical national infrastructure from terrorism. *International Journal*, 775–797.

US Department of Transportation. (1997). *Port security: A national planning guide*.

Vergara, N. (2016). Pescadores acusan abusos de Capitanía de Puerto. In: Milenio News Paper. http://www.milenio.com/region/pescadores_denuncian_abusos-capitania_puerto-milenio_noticias_0_829117188.html. Accessed 16 Nov 2016.

## *Further Reading*

Arellano, S. (2015). Marina y Ejercito las que generan mas confianza. In: Milenio News Paper. http://www.milenio.com/policia/Marina_y_Ejercito_generan_mas_confianza-Dia_de_la_Armada-Pena_Nieto_0_634136618.html. Accessed 29 Nov 2016.
Cámara de Diputados del H. Congeso de la Unión, L. L. (2013). Versión estenográfica de la sesión ordinaria del martes 5 de noviembre de 2013. http://cronica.diputados.gob.mx/Estenografia/LXII/2013/nov/2013. Accessed 15 Nov 2016.
Diario Oficial de la Federacion. (2003). Acuerdo Secretarial N. 117. http://dof.gob.mx/nota_detalle.php?codigo=690229&fecha=11/09/2003. Accessed 9 Sep 2016.
Diario Oficial de la Federación. (2016). Acuerdo por el que se establecen zonas de seguridad para la navegación y sobrevuelo en las inmediaciones de las instalaciones petroleras y para el aprovechamiento integral y sustentable de los recursos pesqueros y acuícolas en zonas marinas mexicanas. http://www.dof.gob.mx/nota_detalle. php?codigo=5456197&fecha=11/10/2016. Accessed 20 Nov 2016.
H. Cámara de Diputados H. Congreso Nacional. (without date). Antecedentes históricos y condiciones de desarrollo de la Marina Mercante Mexicana. http://www.diputados.gob.mx/comisiones/marina/mercante/mamer.htm. Accessed 16 Jun 2016.
International Maritime Organization. (2016a). International Convention on Standards of Training, Certification and Watchkeeping for Seafarers, 1978. http://www.imo.org/en/OurWork/HumanElement/TrainingCertification/Pages/STCW-Convention.aspx. Accessed 8 Dec 2016.
International Maritime Organization. (2016b). ISM Code and Guidelines on Implementation of the ISM Code. http://www.imo.org/en/OurWork/HumanElement/SafetyManagement/Pages/ISMCode.aspx. Accessed 8 Dec 2016.
International Maritime Organization. (w.d.). IMO Status of Conventions. http://www.imo.org/en/About/Conventions/StatusOfConventions/Pages/Default.aspx. Accessed 10 Dec 2016.
Ministry of Communications and Transport. (2016). Economics pic-up Research. http://economistspickresearch.hktdc.com/resources/MI_Portal/Article/rp/2012/12/454833/1357614260894_454833Mexico01_454833.jpg. Accessed 27 Nov 2016.
Pope, C. (2013). The Herald of Free Entreprise tragedy. *Journal of Undergraduate Engineering Research and Scholarship*.
Vela, D. S. (2015). Ejército y Marina son bien aceptados, hasta en Tlatlaya. In: El Financiero, News Paper. http://www.elfinanciero.com.mx/nacional/ejercito-y-marina-son-bien-aceptados-hasta-en-tlatlaya.html. Accessed 26 Jun 2016.

# Part II
# Energy Efficient Ship Design

# Numerical Studies on Added Resistance and Ship Motions of KVLCC2 in Waves

Mingyu Kim, Osman Turan, Sandy Day, and Atilla Incecik

## 1 Introduction

Traditionally, only ship resistance and propulsion performance in calm water was considered at the ship design stage and during the design process even though recently the hull form has been optimised for a specific range of draught and speed ranges, considering the operational profile (Kim and Park 2015). However, when a ship advances in a seaway, it requires additional power in comparison with the power required in calm water due to weather effects and ship operating conditions. This degradation of the ship performance in a seaway, which is reported to be about 15–30% of the power required in calm water, is accounted for a "Sea Margin" to calculate the engine-propeller matching. The added resistance due to waves is one of the major components affecting ship performance in a seaway. Therefore, accurate prediction of added resistance in waves is essential to evaluate the additional power requirement, to assess the full environmental impact and to design ships with high fuel efficiency in realistic operating conditions. This can also be combined with other operational measures to ensure greater efficiency, such as voyage planning and weather routing (Zaccone et al. 2016). Additionally, correct estimation and understanding of ship motions is crucial to ensure safe navigation. Regarding international regulations, the International Maritime Organization (IMO) issued new regulations to improve the energy efficiency level of ships and to reduce carbon emissions. These regulations include the Energy Efficiency Design Index (EEDI) as a mandatory technical measure for new ships and the Ship Energy Efficiency Management

M. Kim (✉) · O. Turan · S. Day · A. Incecik
Department of Naval Architecture, Ocean and Marine Engineering, University of Strathclyde, Glasgow, UK
e-mail: mingyu.kim@strath.ac.uk



A. I. Ölçer et al. (eds.), *Trends and Challenges in Maritime Energy Management*, WMU Studies in Maritime Affairs 6, https://doi.org/10.1007/978-3-319-74576-3_9

Plan (SEEMP), which is related to ship voyage and operational efficiency as an operational measure for ships in service. In addition, the ship speed reduction coefficient (fw) has been proposed and is under discussion for the calculation of EEDI in representative sea states (IMO 2012). Moreover, guidelines for determining minimum propulsion power to maintain the manoeuvrability of a ship in adverse weather conditions (IMO 2013) have been developed for safe manoeuvring.

The added resistance and ship motion problem in waves has been widely studied by conducting experiments and numerical simulations using potential flow theory and CFD approaches. There are two major analytical approaches in potential flow methods which are used to calculate the added resistance: the far-field method and the near-field method. The far-field method is based on the added resistance computed from the wave energy and the momentum flux generated by a ship and is evaluated across a vertical control surface of infinite radius surrounding the ship. This method was first introduced by Mauro (1960) using the Kochin function, which consists of radiating and diffracting wave components. Later on, the far-field method based on the radiated energy approach was proposed by Gerritsma and Beukelman (1972) to predict the added resistance in head seas. This approach became popular in strip theory programs due to its easy implementation. Recently, Liu et al. (2011) solved the added resistance problem with a quasi-second-order approach using the hybrid Rankine Source-Green function method. Another numerical approach is the near-field method, which estimates the added resistance by integrating the hydrodynamic pressure on the body surface. This method was first introduced by Havelock (1937) who used the Froude-Krylov approach to calculate hull pressures. The near-field method was enhanced by Faltinsen et al. (1980) based on the direct pressure integration approach. Salvesen et al. (1970) introduced a simplified asymptotic method based on 2-D strip theory to overcome the deficiency of this approach in short waves. Kim et al. (2007) and Joncquez (2009) formulated the added resistance based on the Rankine panel method using a time-domain approach with B-spline functions and investigated the effects of the Neumann-Kelvin (NK) and Double Body (DB) linearization schemes on the added resistance predictions. Recently, Kim et al. (2012) formulated the added resistance using a time-domain B-spline Rankine panel method. In the present study, the 3D linear potential flow method is applied to predict ship motions and added resistance using the NK linearization scheme and near-field method in regular waves due to more accurate prediction of ship motions and added resistance of blunt ships compared to the DB method (Kim and Shin 2007).

Recently, as computational facilities have become more powerful and more accessible, CFD tools are now commonly used to predict added resistance and ship motions. It has advantages over potential codes as it can deal directly with large amplitude ship motions and with nonlinear flow phenomena such as breaking waves and green water, without explicit approximations and empirical values. Sadat-Hosseini et al. (2010) predicted the added resistance and motions for KVLCC2 based on a URANS approach. There have also been subsequent investigations on how to reduce the added resistance by modifying the hull form. Park et al. (2014) and Kim et al. (2015) modified the fore body of KVLCC2 to reduce added resistance in waves.

In the study presented in this paper, numerical simulations for the prediction of added resistance and ship motions for KVLCC2 in regular head waves are carried out using URANS and 3-D potential flow methods. The results obtained are validated with the available experimental data and, during the study, grid convergence tests are also carried out for the CFD approach. The added resistance and the vertical ship motions are examined for various wave conditions at design and operating speeds as well as in stationary condition. Unsteady wave patterns and the time history results of the resistance and vertical ship motions in waves are simulated using a CFD tool. The relationship between the added resistance and the ship motions for various ship speeds are investigated including the viscous effects and non-linear phenomena such as green water on deck.

## 2 Ship Particulars and Coordinate System

All calculations of added resistance and ship motions have been performed for KVLCC2, which represents the second variant of the VLCC-type vessel developed by the Korea Research Institute of Ships and Ocean Engineering (KRISO), which is a benchmark hull form used to study seakeeping problems by researchers. The main particulars of the KVLCC2 are given in full scale in Table 1. For CFD simulations, a model scale vessel without appendages using a scale ratio of 1/80 is employed in the calculations.

In the numerical simulations, a right-handed coordinate system x, y, z is adopted. The translational displacements in the x, y and z directions are $\xi_1$ (surge), $\xi_2$ (sway) and $\xi_3$ (heave), and the angular displacements of rotational motion about the x, y and z axes are $\xi_4$ (roll), $\xi_5$ (pitch) and $\xi_6$ (yaw), respectively, and $\theta$ angle represents the ship's heading angle with respect to the incident waves. For head seas the angle $\theta$ equals 180° and, for beam seas from the port side, the angle equals 90°.

**Table 1** Main particulars of KVLCC2

| Particulars | Full scale | Model scale |
|---|---|---|
| Length, L (m) | 320 | 4 |
| Breadth, B (m) | 58 | 0.725 |
| Depth, D (m) | 30 | 0.375 |
| Draught, T (m) | 20.8 | 0.260 |
| Displacement Volume, V ($m^3$) | 312,622 | 0.6106 |
| LCG(%), fwd + | 3.48 | 3.48 |
| VCG (m) | 18.56 | 0.232 |
| Block coefficient, $C_B$ (−) | 0.8098 | 0.8098 |

# 3 Numerical Methods and Modelling

In the present study, the 3-D linear potential flow and CFD methods are applied to predict added resistance and ship motions in regular waves.

## *3.1 3-D Linear Potential Flow Method*

3-D potential flow calculations are carried out using the PRECAL (PREssure CALculation) software developed by the Maritime Research Institute Netherlands (MARIN). The PRECAL software is based on the planar panel approach, which can calculate the seakeeping behaviour of monohull, catamaran and trimaran ships. In addition to the rigid body motions, it can also calculate the deformation modes of a ship's hull girder, internal loads, pressure on the hull and added resistance in waves. The prediction of forward speed effects is the main shortcoming in the solution of Green's functions due to the complex numerical integration process on the waterline sections. Numerical methods need to be implemented to solve the Boundary Value Problem (BVP) in the presence of forward speed and the Green's functions need to be satisfied both for the Free-Surface Boundary Condition (FSBC) and the Body Boundary Condition (BBC). PRECAL is a 3-D source-sink frequency domain code capable of solving the forward speed linear BVP using the Approximate Forward Speed (AFS) and the Exact Forward Speed (EFS) methods. In the AFS method, the BVP is solved using zero-speed Green's functions and then forward speed corrections are applied to the BVP equations. In the EFS method, exact forward speed Green's functions are used to solve the forward speed BVP, but in the PRECAL software, the Lid panel method can only be applied to the AFS formulation. In this study, forward speed ship motions are solved using the AFS formulation due to its fast and accurate results (Hizir 2015). The added resistance is calculated using the near-field method based on direct pressure integration over the mean wetted hull surface, using the second-order forces to calculate wave drift forces while the first-order forces and moments are calculated to solve ship motions. The total pressure is divided into four components. which originate from relative water height, incident wave velocities, pressure gradient and rotation times inertial terms. The added resistance force due to waves ($\Delta R_{wave}$) is calculated in the time domain as shown in Eq. (1)

$$\Delta R_{wave} = \begin{bmatrix} -\rho \int\limits_{H_0} \vec{\nabla}\phi^{(1)} \cdot \vec{\nabla}\phi^{(1)} \vec{n}^{(0)} ds \\ -\rho \int\limits_{H_0} \left(\vec{\alpha}^{(1)} \cdot \vec{\nabla}\right) \left(\frac{\partial \phi^{(1)}}{\partial t} + \vec{\nabla}\phi^{(1)} \cdot \vec{\nabla}\overline{\phi}\right) \vec{n}^{(0)} ds \\ +\frac{1}{2}\rho g \int\limits_{wl} \left(\zeta^{(1)} - \alpha_3^{(1)}\right)^2 \vec{n}^{(0)} dl + \vec{\Omega}^{(1)} \times M\ddot{\vec{X}}^{(1)} \end{bmatrix} \tag{1}$$

where the first integral is the water velocity contribution, the second integral is the pressure gradient contribution, the third integral is the relative wave height contribution and the last term is the rotation times inertia contribution. The indices stand for the order of the forces in the force contribution formulations. $H_0$ represents the mean position of the ship, $\vec{\alpha}^{(1)}$ represents the first order translation and rotation vector, $\vec{n}^{(0)}$ is the zeroth order normal vector calculated on the mean position vessel wetted surface and $\overline{\Omega}^{(1)}$is the first order rotation vector. In order to derive the added resistance equation in the frequency domain, an oscillatory description of motion and flow is introduced and the steady flow contribution is neglected. The added resistance in the frequency domain is formulated by Eq. (2):

$$\Delta \mathrm{R}_{\mathrm{wave}} = \begin{bmatrix} -\rho \int\limits_{H_0} \left|\vec{\nabla}\phi^{(1)}\right|^2 \vec{n}^{(0)} ds \\ -\rho \int\limits_{H_0} \left(\vec{\alpha}^{(1)} \cdot \vec{\nabla}\right) \left(i\omega_e \vec{\nabla}\phi^{(1)}\right) \vec{n}^{(0)} ds \\ +\frac{1}{2}\rho g \int\limits_{wl} \left|\zeta^{(1)}\right|^2 \vec{n}^{(0)} dl - \omega_e{}^2 \vec{\Omega}^{(1)} \times M\vec{X}^{(1)} \end{bmatrix} \tag{2}$$

In order to evaluate the added resistance forces, all components in the integrals are defined in perturbation series. A small parameter (ε) is introduced to define the quantities in the perturbation series. The perturbation series expansion of the relative wave height and the velocity potential can be formulated as shown in Eqs. (3) and (4):

$$\zeta = \zeta^{(0)} + \varepsilon\zeta^{(1)} + \varepsilon^2\zeta^{(2)} + O\left(\varepsilon^3\right) \tag{3}$$
$$\phi = \overline{\phi} + \varepsilon\phi^{(1)} + \varepsilon^2\phi^{(2)} + O\left(\varepsilon^3\right) \tag{4}$$

where zeroth order quantities are time independent and are assumed to be small to satisfy the linearized free-surface condition. For the same reason, time dependent parts of the series are also assumed to be small.

In added resistance calculations, only the mean values of the forces and moments are of interest. First-order quantities such as motions, velocities, accelerations etc. have a mean value of zero when the wave is given by an oscillatory function with a mean value of zero. However, second-order quantities such as added resistance have a non-zero mean value; therefore, in order to calculate the added resistance, second-order forces and moments need to be calculated. In the present study, in the calculation of added resistance, which is composed of a constant part (mean value) and a slowly oscillating part, only the constant part of the added resistance is taken into account, while the slowly oscillating part of the added resistance is trivial compared to the order of the mean value of the added resistance.

## 3.2 Computational Fluid Dynamics (CFD)

An URANS approach was applied to calculate the added resistance and ship motions in regular waves using the commercial CFD software STAR-CCM+. For incompressible flows, if there are no external forces, the averaged continuity and momentum equations are given in tensor form in the Cartesian coordinate system by Eqs. (5) and (6)

$$\frac{\partial(\rho\bar{u}_i)}{\partial x_i} = 0 \tag{5}$$

$$\frac{\partial(\rho\bar{u}_i)}{\partial t} + \frac{\partial}{\partial x_j}\left(\rho\bar{u}_i\bar{u}_j + \rho\overline{u_i'u_j'}\right) = -\frac{\partial\bar{p}}{\partial x_i} + \frac{\partial\bar{\tau}_{ij}}{\partial x_j} \tag{6}$$

where $\bar{u}_i$ is the averaged velocity vector of fluid, $\overline{u_i'u_j'}$ is the Reynolds stresses and $\bar{p}$ is the mean pressure. For Newtonian fluid under incompressible flow, the mean shear stress tensor, $\bar{\tau}_{ij}$, is expressed as Eq. (7)

$$\bar{\tau}_{ij} = \mu\left(\frac{\partial\bar{u}_i}{\partial x_j} + \frac{\partial\bar{u}_j}{\partial x_i}\right) \tag{7}$$

where $\mu$ is dynamic viscosity.

The finite volume method (FVM) and the volume of fluid (VOF) method were applied for spatial discretization and free surface capturing, respectively. The flow equations were solved in a segregated manner using a predictor-corrector approach. Convection and diffusion terms in the RANS equations were discretized by a second-order upwind scheme and a central difference scheme. The semi-implicit method for pressure-linked equations (SIMPLE) algorithm was used to resolve the pressure-velocity coupling and a standard $k-\varepsilon$ model was applied as the turbulence model. In order to consider ship motions, a Dynamic Fluid Body Interaction (DFBI) scheme was applied with the vessel free to move in heave and pitch directions as vertical motions.

Only half of the ship's hull (the port side) with a scale ratio of 1/80 and control volume were taken into account in the calculations; thus, a symmetry plane formed the centerline domain face in order to reduce computational time and complexity. The calculation domain is $-3L < x < 1.25L$, $0 < y < 2L$, $-2L < z < 1L$ where the

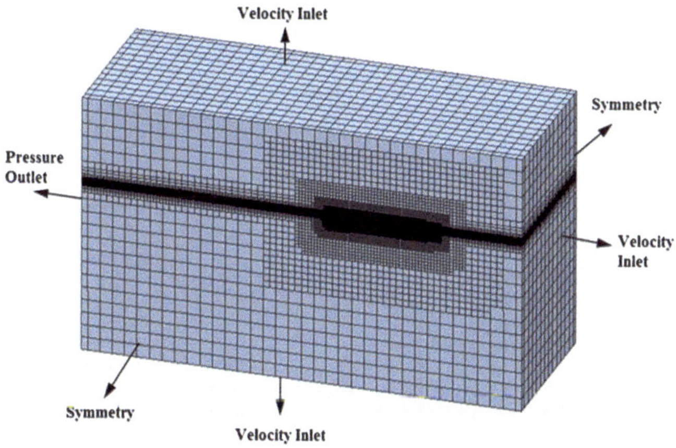


**Fig. 1** Mesh and boundary conditions

mid-plane of the ship is located at $y = 0$ and ship draught (T) is at z=0. The boundary conditions together with the generated meshes are depicted in Fig. 1. Artificial wave damping was applied to avoid the undesirable effect of the reflected waves from the side and outlet boundaries.

# 4 Discussion of Results

In this section, the simulation results using CFD and 3-D potential methods are presented and compared with available experimental added resistance (Lee et al. 2013) and ship motions data in regular head waves. Unsteady wave patterns and time history results of the resistance and vertical ship motions in waves are simulated using a CFD method. Only two degrees of freedom motions, which are heave and pitch responses, are calculated during all simulations.

## *4.1 Grid Convergence Test*

Prior to the investigation of the added resistance and the heave and pitch motions using the CFD tool, grid convergence tests were performed. The CFD simulations at 15.5 knots, which corresponds to a Froude Number (Fn) of 0.142, were carried out and the simulation results were compared with the existing experimental data. The wave length (λ/L) and the wave steepness (H/λ) are 1.2 and 1/60 respectively. This wave condition corresponds to a resonant case (Sadat-Hosseini et al. 2013). The coarse and fine mesh systems were derived by using a factor of $\sqrt{2}$ (Bøckmann et al. 2014) based on the base mesh case (Case no. C10). The simulation time step is set to

**Table 2** Test cases for grid convergence ($\lambda/L = 1.2$, $H/\lambda = 1/60$, Vs = 15.5 knots)

| Case no. | Mesh | $\lambda/\Delta x$ | $H/\Delta z$ | $T_e/\Delta t$ |
|---|---|---|---|---|
| C1C | Coarse | 70 | 14 | 181 |
| C10 | Base | 100 | 20 | 256 ($2^8$) |
| C1F | Fine | 140 | 28 | 362 |

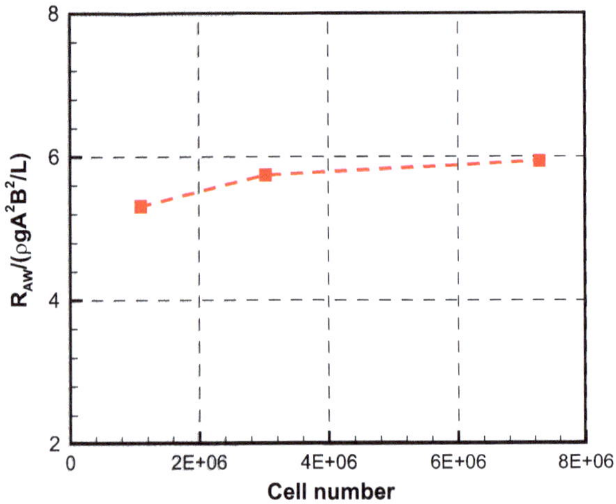


**Fig. 2** Grid convergence test for the added resistance

be proportional to the grid size as shown in Table 2 where $T_e$ represents the corresponding encountering period.

The results of the convergence tests with three different mesh systems are shown in Fig. 2 where ρ, $g$ and $A$ denote the density, gravitational acceleration and the wave amplitude parameters, respectively. As the number of cells increased, the added resistance coefficient increased, especially from the coarse mesh (case no. C1C) to base mesh system (case no. C10). However, the results of the added resistance for the base and fine mesh (case no. C1F) were almost identical. Therefore, the base mesh system was chosen for the CFD simulations in this study and the cell number and time step vary according to the wave conditions in the simulations.

The added resistance due to waves ($R_{AW}$) is obtained by Eq. (8)

$$R_{AW} = R_w - R_c \tag{8}$$

where $R_w$ and $R_c$ are total resistance in waves and resistance in calm water respectively.

## 4.2 *Added Resistance and Ship Motions at Design Speed*

Numerical calculations using the 3-D potential and CFD methods at the design speed of 15.5 knots were carried out in both calm water and wave conditions for various

**Table 3** Test cases at design speed (15.5 knots)

| Case no. | Vs [knots] | Wave length (λ/L) | Wave height (H) [m] | Wave steepness (H/λ) | fe [Hz] | Te [s] |
|---|---|---|---|---|---|---|
| C00 | 15.5 | Calm water | – | – | – | – |
| C10 | | 1.20 | 6.40 | 1/60 | 0.7560 | 1.3227 |
| C11 | | 0.50 | 2.67 | | 1.3293 | 0.7523 |
| C12 | | 0.75 | 4.00 | | 1.0186 | 0.9818 |
| C13 | | 1.00 | 5.33 | | 0.8476 | 1.1798 |
| C14 | | 1.40 | 6.40 | | 0.6872 | 1.4552 |
| C15 | | 1.60 | 7.47 | | 0.6332 | 1.5793 |

wave lengths for constant wave steepness (H/λ) ratio of 1/60. The test cases are summarised in Table 3.

The wave length is assumed to be $\lambda = gT^2/2\pi$ for deep water and the wave encountering frequency $f_e$ in Hz for model scale is calculated by $f_e = \sqrt{g/(2\pi\lambda)} + U/\lambda$ for head sea where $U$ denotes the ship forward speed in m/s.

Prior to the investigation on the added resistance, Response Amplitude Operators (RAOs) of heave and pitch motions are compared with the experimental data in regular head waves from Osaka University (Larsson et al. 2010) as shown in Fig. 3. It is a well-known fact that added resistance is proportional to relative motions, hence heave and pitch motions, and inaccuracies in the predicted motion responses may amplify the errors in the added resistance calculations. In this study, $\xi_3$ and $\xi_5$ are the amplitudes of heave and pitch motion responses, respectively, whereas $k = 2\pi/\lambda$ is the wave number in deep water. The motion responses are evaluated at the ship's centre of gravity. The zeroth and first order terms of the resistance and motion responses calculated by CFD are used for the added resistance coefficient and motion transfer functions (Shen and Wan 2013). The overestimation of the heave motion using the 3-D potential method is amplified around the resonance period ($1.0 < \lambda/L < 1.4$), while the CFD method slightly underestimates the heave motion around the resonance period for the range of λ/L from 1 to 1.4. For the pitch motions, the results obtained from both methods show good agreement with the experimental data. The overestimation of the results obtained from the 3-D potential method for the heave motions can be attributed to the AFS formulation, in which the BVP is solved using zero speed Green's functions and then forward speed corrections are applied to the boundary conditions, and also to the Neumann-Kelvin (NK) approximation where the steady wave and unsteady wave interactions are linearized. The underestimation of the results obtained from the CFD method is likely to stem from the adoption of a non-inertial reference frame in which large amplitude motion causes inaccurate capturing of the free surface. The authors will address this problem in future studies using adaptive mesh method in CFD simulations.

The numerical results of the added resistance are compared with the available experiment data (Lee et al. 2013) as illustrated in Fig. 4, which indicates that the CFD and 3-D panel methods both have reasonable agreement with the experimental data.

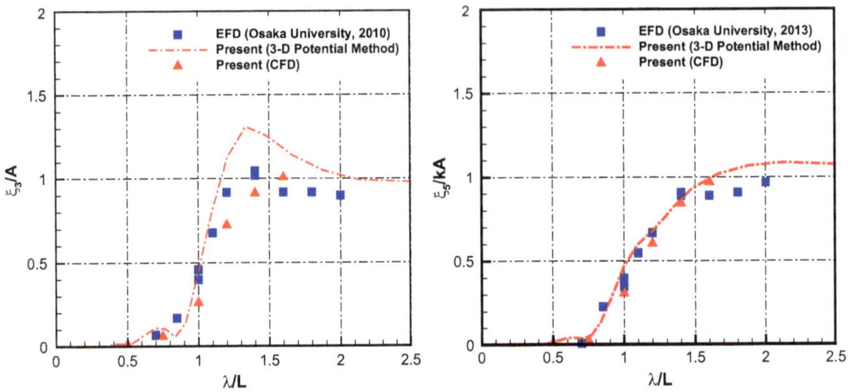


**Fig. 3** Heave and pitch responses (Vs = 15.5 knots, θ = 180°)

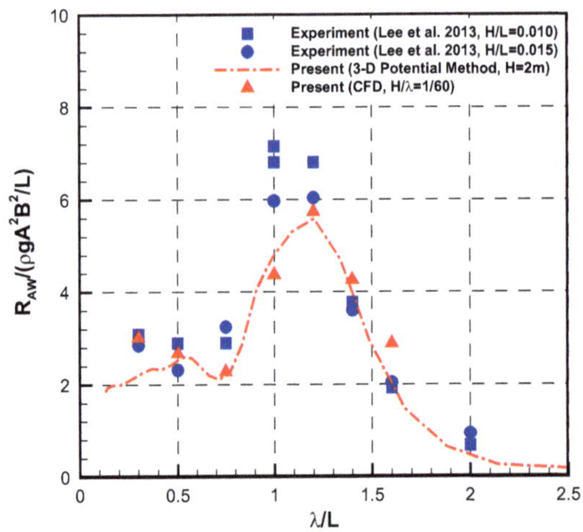


**Fig. 4** Added resistance (Vs = 15.5 knots, θ = 180°)

To visualise the ship motions and periodic wave patterns, a snapshot of the wave and the vessel motions for the C10 test case was captured with respect to the period of encounter at $\lambda/L = 1.2$ and at a vessel speed of 15.5 knots. The result displayed in Fig. 5a is the snapshot at $t/Te = 0.5$ when the ship has the largest added resistance value as is shown in Fig. 5. Additionally, the time histories of total resistance force, heave and pitch motions are displayed over an encounter period as shown in Fig. 5. The largest resistance force in waves is observed around $t/Te = 0.5$ when the bow is completely immersed and when the relative wave height is high around the bow with green water on deck, as illustrated in Fig. 5a. The position of the vessel where the maximum added resistance is recorded is when the ship has the highest immersion due to heave motion, while the pitch amplitude is almost zero.

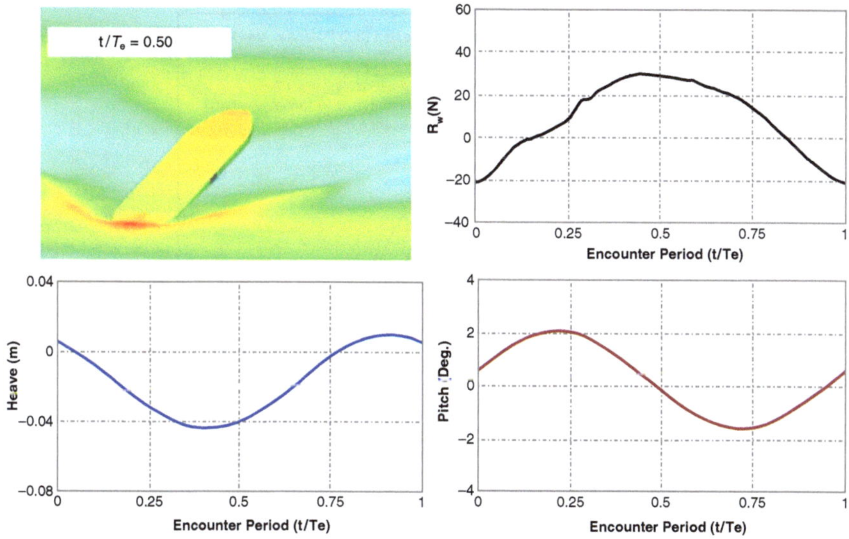


**Fig. 5** Snapshot of free surface elevation, and total resistance, heave and pitch response time histories over one period of encounter (Case no. C10)

**Table 4** Test cases at operating speed (12 knots) and at stationary conditions (0 knot)

| Case no. | Vs [knots] | Wave length (λ/L) | Wave height (H) [m] | Wave steepness (H/λ) | fe [Hz] | Te [s] |
|---|---|---|---|---|---|---|
| C71 | 12 | 0.75 | 4.00 | 1/60 | 0.9515 | 1.0510 |
| C72 | | 1.00 | 5.33 | | 0.7973 | 1.2542 |
| C73 | | 1.20 | 6.40 | | 0.7141 | 1.4004 |
| C73 | | 1.40 | 7.47 | | 0.6513 | 1.5355 |
| C81 | 0 | 0.75 | 4.00 | 1/60 | 0.7214 | 1.3862 |
| C82 | | 1.00 | 5.33 | | 0.6248 | 1.6006 |
| C83 | | 1.20 | 6.40 | | 0.5703 | 1.7534 |
| C84 | | 1.40 | 7.47 | | 0.5280 | 1.8939 |

## *4.3 Added Resistance at Stationary and Operating Speeds*

To consider slow steaming or the realistic operating speeds of the vessel, the effect of ship speed on added resistance and ship motions was investigated. In addition to the assumed operating speed (12 knots), the cases for stationary condition (0 knots) were also simulated as summarised in Table 4.

The results for the added resistance in regular waves are also compared as shown in Fig. 6. The mean added resistance estimated by the 3-D potential and CFD methods show good agreement and it is demonstrated that added resistance can be predicted reliably by using the current numerical approaches. Blok (1993) revealed

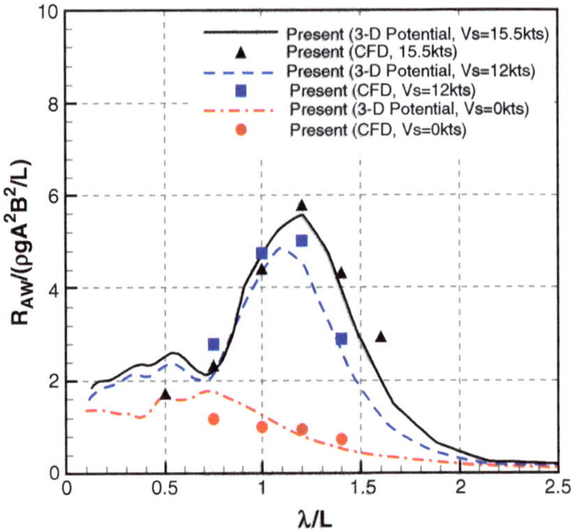


**Fig. 6** Comparison of added resistance at different ship speeds (V = 0, 12, 15.5 knots)

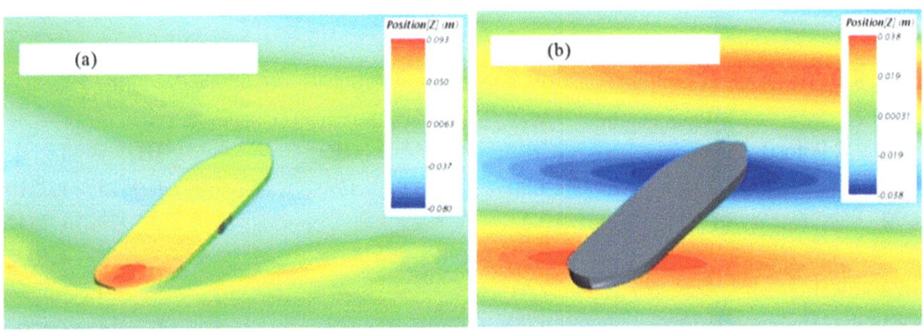


**Fig. 7** Free surface elevation at the highest added resistance instant (t/Te = 0.5, λ/L = 1.2). (**a**) Vs = 12 knots. (**b**) Vs = 0 knots

that in head seas the higher the vessel speed, the larger the mean added resistance, and the peaks of the curves shift towards longer wave periods. In the current study, Blok's observations are confirmed for the KVLCC2 because in head seas the speed dependency of the added resistance only exists for wave lengths longer than λ/L > 1, whilst in short waves the ship speed has minimal effect on the added resistance due to the governing diffraction forces.

For the added resistance resonant case at λ/L = 1.2 when the ship is at operational speed and stationary, snapshots of the wave elevation and vessel motions are illustrated in Fig. 7 at the time instant of $t/T_e = 0.5$, when the highest value in added resistance is observed. Green water on deck is observed clearly at the operational speed as shown in Fig. 7a where position (Z) refers to the free surface

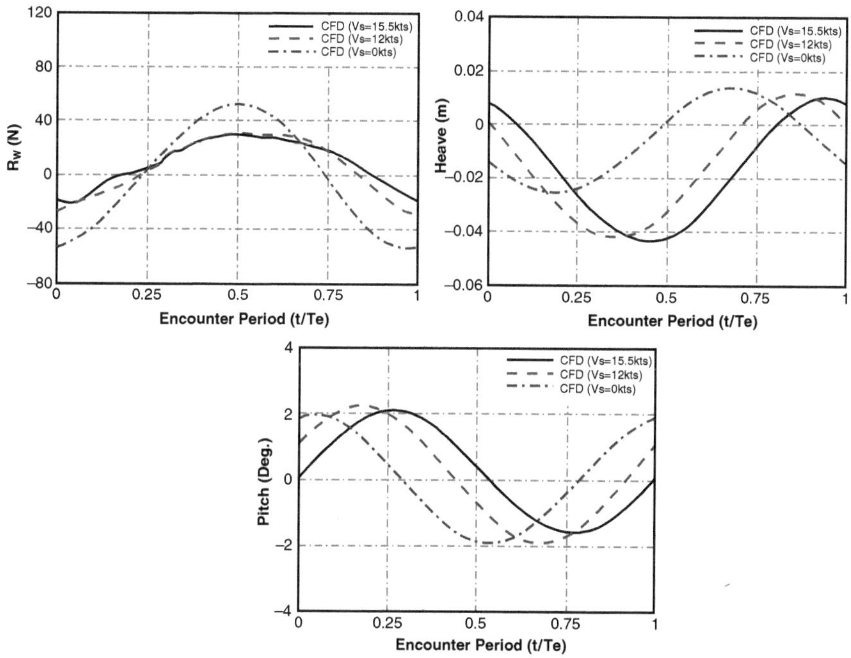


**Fig. 8** Total resistance, heave and pitch responses over one period of encounter (λ/L = 1.2)

elevation, while the relative wave elevation around the bow and stern at Vs = 0 knots are clearly observed as presented in Fig. 7b.

Time histories of the total resistance and the ship vertical motions at the encounter period are compared for the three ship speeds as shown in Fig. 8. It is observed that in the time domain the oscillation amplitudes of the total resistance force in waves at the stationary condition are higher than at other speeds even though, for the stationary condition, the mean total resistance in waves is much lower than the other ship speeds. In the stationary condition, there is no green water incidence observed as presented in Fig. 7. The heave motions at $t/T_e = 0.5$ are almost zero and the pitch motions are negative as the ship trims by stern as shown in Fig. 8. It is revealed that the oscillation of total resistance in the stationary condition is governed by the relative wave elevation. Therefore, it should be noted that the ship should be carefully operated especially in heavy weather conditions because the added resistance or drift forces at zero speed may be larger than the added resistance of a vessel advancing in waves. It should also be noted that there are concerns regarding ship maneuverability at low speed in restricted areas in adverse weather conditions (Shigunov and Papanikolaou 2015).

## 5 Conclusions

Added resistance and vertical ship motions (heave and pitch) in regular head waves were simulated using the unsteady RANS and the 3-D potential flow methods for a wide range of wave conditions at three ship speeds, which are the design speed (Vs = 15.5 knots), operating speed (Vs = 12 knots) and zero speed (Vs = 0 knots). The time histories of total resistance and ship motions in waves calculated using the CFD method were examined at zero, design and operational speed taking into account the unsteady wave patterns and viscous effects.

Firstly, the optimal mesh system was established from the grid convergence tests for the CFD simulations.

Secondly the results of the added resistance and the ship motions in regular waves using the CFD and 3-D potential methods were compared with the results of experiments at design speed and were found to have a reasonable agreement except for the heave motions which were overestimated by 3-D potential method around the resonance period ($1.0 < \lambda/L < 1.4$) and in long waves ($\lambda/L > 2$) due to the AFS formulation. The heave motions were slightly underestimated by the CFD method for the range of wave lengths ($\lambda/L$) from 1 to 1.4 because of the adoption of the non-inertial reference frame. The wave snapshots and vessel motions are investigated with the time history data of the resistance force and the vertical motions at the encounter period. For the case C10 (Vs = 15.5 knots, $\lambda/L = 1.2$), the largest added resistance in waves appeared when bow slamming was observed, at the instant when the wave elevation was high around the bow and the ship had the largest immersion even though the pitch amplitude was almost zero.

Thirdly, the zeroth order harmonic terms of the total resistance calculated by the CFD method were compared for three ship speeds. The mean total resistance forces were larger for faster ship speeds as expected. However, it was observed that the oscillation magnitude of the total resistance force in the time domain at the stationary speed was larger than at the design and operating speeds.

Finally, for the resonant test case ($\lambda/L = 1.2$) the wave elevation and the ship motions were examined at the operating and stationary speeds. When the ship had the largest added resistance, green water on deck was observed at the operating speed, while high wave elevations around the bow and stern were seen clearly for the stationary case without water on deck. From the comparison of the time histories of the total resistance and the ship motions at the resonance period, the oscillation magnitudes of the resistance force at Vs = 0 knots were higher than other ship speeds. It is likely that the resistance in waves for the ship at zero speed is affected by relative wave elevation around bow and stern since the relative wave height governs the added resistance calculations. Therefore, it should be noted that vessels in stationary condition should be carefully operated in heavy weather conditions.

**Acknowledgements** The authors are grateful to the Engineering and Physical Research Council (EPSRC) for funding the research reported in this paper through the project: "Shipping in Changing Climate. (EPSRC grant no. EP/K039253/1). The results given in the paper were obtained using the EPSRC funded ARCHIE-WeSt High Performance Computer (www.archie-west.ac.uk). EPSRC grant no. EP/K000586/1.

# References


Blok, J. J. (1993). *The resistance increase of a ship in waves*. TU Delft: Delft University of Technology.

Bøckmann, A., Påkozdi, C., Kristiansen, T., Jang, H., & Kim, J. (2014). An experimental and computational development of a benchmark solution for the validation of numerical wave tanks. In *ASME 2014 33rd International Conference on Ocean, Offshore and Arctic Engineering. American Society of Mechanical Engineers* (pp. V002T008A092–V002T008A092).

Faltinsen, O. M., Minsaas, K. J., Liapis, N., & Skjørdal, S. O. (1980). Prediction of resistance and propulsion and propulsion of a ship in a seaway. In *Proceeding of 13th Symposium on Naval Hydrodynamics, Tokyo* (pp. 505–529).

Gerritsma, J., & Beukelman, W. (1972). Analysis of the resistance increase in waves of a fast cargo ship. *International Shipbuilding Progress, 19*(217).

Havelock, T. H. (1937). The resistance of a ship among waves. *Proceedings of the Royal Society of London. Series A, Mathematical and Physical Sciences*, 299–308.

Hizir, O. G. (2015). *Three dimensional time domain simulation of ship motions and loads in large amplitude waves, naval architecture, ocean and marine engineering*. Glasgow: University of Strathclyde.

IMO. (2012). *Interim guidelines for the calculation of the coefficient $f_w$ for decrease in ship speed in a representative sea condition for trial use*. London: International Maritime Organisation (IMO).

IMO. (2013). *Interim guidelines for determining minimum propulsion power to maintain the manoevrability in adverse conditons*. London: International Maritime Organisation.

Joncquez, S. A. (2009). *Second-order forces and moments acting on ships in waves*. Ph.D. thesis, Technical University of Denmark, Copenhagen, Denmark.

Kim, B., & Shin, Y. S. (2007). Steady flow approximations in three-dimensional ship motion calculation. *Journal of Ship Research, 51*(3), 229–249.

Kim, K. H., Kim, Y., & Kim, Y. (2007). *WISH JIP project report and manual*. Marine Hydrodynamic Laboratory, Seoul National University.

Kim, K. H., Seo, M. G., & Kim, Y. H. (2012). Numerical analysis on added resistance of ships. *International Journal of Offshore and Polar Engineering, 22*(01), 21–29.

Kim, M., & Park, D. W. (2015). A study on the green ship design for ultra large container ship. *Journal of the Korean Society of Marine Environment & Safety, 21*(5), 558–570.

Kim, Y. C., Kim, K. S., Kim, J., Kim, Y. S., Van, S. H., & Jang, Y. H. (2015). Calculation of added resistance in waves for KVLCC2 and its modified hull form using RANS-based method. In *The 25th International Offshore and Polar Engineering Conference*. International Society of Offshore and Polar Engineers, Hawaii, USA (pp. 924–930).

Larsson, L., Stern, F., & Visonneau, M. (2010). Proceedings Gothenburg 2010 a workshop on numerical ship hydrodynamics. In Technology, C.U.o. (Ed.), Gothenburg, Sweden.

Lee, J. H., Seo, M. G., Park, D. M., Yang, K. K., Kim, K. H., & Kim, Y. (2013). Study on the effects of hull form on added resistance. In *The 12th International Symposium on Practical Design of Ships and Other Floating Structures, Changwon, Korea* (pp. 329–337).

Liu, S., Papanikolaou, A., & Zaraphonitis, G. (2011). Prediction of added resistance of ships in waves. *Ocean Engineering, 38*(4), 641–650.

Mauro, H. (1960). The drift of a body floating on waves. *Journal of Ship Research, 4*, 1–5.

Park, D. M., Seo, M. G., Lee, J., Yang, K. Y., & Kim, Y. (2014). Systematic experimental and numerical analyses on added resistance in waves. *Journal of the Society of Naval Architects of Korea, 51*(6), 459–479.

Sadat-Hosseini, H., Carrica, P., Kim, H., Toda, Y., & Stern, F. (2010). URANS simulation and valiation of added resistance and motions of the KVLCC2 crude carrier with fixed and free surge conditions, Gothenburg 2010: A Workshop on CFD in ship hydrodynamics.

Sadat-Hosseini, H., Wu, P., Carrica, P., Kim, H., Toda, Y., & Stern, F. (2013). CFD verification and validation of added resistance and motions of KVLCC2 with fixed and free surge in short and long head waves. *Ocean Engineering, 59*, 240–273.

Salvesen, N., Tuck, E. O., & Faltinsen, O. M. (1970). Ship motions and sea loads. *SNAME, 104*, 119–137.
Shen, Z., & Wan, D. (2013). RANS computations of added resistance and motions of a ship in head waves. *International Journal of Offshore and Polar Engineering, 23*(04), 264–271.
Shigunov, V., & Papanikolaou, A. (2015). Criteria for minimum powering and maneuverability in adverse weather conditions. *Ship Technology Research, 62*(3), 140–147.
Zaccone, R., Figari, M., Altosole, M., & Ottaviani, E. (2016). Fuel saving-oriented 3D dynamic programming for weather routing applications. In *Proceedings of the 3rd International Conference on Maritime Technology and Engineering (MARTECH), Lisbon, Portugal.*

# An Investigation of Fuel Efficiency in High Speed Vessels by Using Interceptors

**Ahmet G. Avci, Baris Barlas, and Aykut I. Ölçer**

## 1 Introduction

Nowadays, the importance of high-speed marine vessels is increasing dramatically and it is still a challenge to design high-speed, efficient marine vessels. Recently, high-speed motor yachts, military crafts, coast guard boats, rescue boats and some passenger vessels have been required. When ships advance steadily at high Froude numbers, they experience sinkage and trim motions because of the hydrodynamic forces acting on the wetted hull. The sinkage and trim motions surge the drag and resistance characteristics. Because of these sinkage and trim motions, some appendages called "interceptors" have been widely used in recent years in fast passenger ferries, high-speed motor yachts, sailing yachts, high speed fishing vessels, and military crafts in order to reduce the drag forces. Furthermore, they are used for ride, trim control and steering, as well as to deal with unexpected trim and sinkage motions. Typically, the trim interceptors are made of a thin plate extending out of the craft by a percentage generally between 0.15% and 0.4% of the LWL of the vessel (De Luca 2011). It is generally located on the bottom of the stern. Its main purpose is to utilize a positive pressure gradient at the stern part of the craft and thereby change the trim.

Subsequently, by using interceptors, lift coefficient at high angles of trim can be increased. Not only the lift coefficient increases but also the drag coefficient grows. However, hydrodynamic drag coefficient increases at a slower rate than the lift coefficient, so interceptors make the hull more efficient. Typically, after a while a

A. G. Avci · B. Barlas (✉)
Faculty of Naval Architecture and Ocean Engineering, Istanbul Technical University, Istanbul, Turkey
e-mail: barlas@itu.edu.tr

A. I. Ölçer
World Maritime University, Malmö, Sweden



A. I. Ölçer et al. (eds.), *Trends and Challenges in Maritime Energy Management*,
WMU Studies in Maritime Affairs 6, https://doi.org/10.1007/978-3-319-74576-3_10

high-speed boat starts moving, she goes into the planning regime. The rate of increase in the resistance of the planning regime (high speed) is less than the displacement regime (low speed). After achieving high speed, total resistance starts to increase again as before the planning regime. When the model is free to trim and sinkage, it can also encounter an increase in wave resistance. Yet, the main purpose of using interceptors is to utilize a positive pressure gradient at the stern part of the craft and change the trim for more speed and less resistance.

In the literature, Savitsky (2003), gave a definition of the operational characteristics and geometry of a unique hull form with the maximum speed of 50 knots. Tsai and Hwang (2004) examined the effect of the trim mechanisms of stern flap and interceptors on resistance reduction and state that the drag reduction with integrated interceptor with stern flap is better than with interceptor alone. Kohansal et al. (2010) tried to validate an appropriate algorithm solution for the hydrodynamic characteristics of planning hulls, including trim effects with interceptors and modelled the wave-making phenomenon of planning hulls, and found out the numerical results are in good agreement with experimental ones. Day and Cooper (2011) also studied interceptor effectiveness and compared the interceptors and Gurney flap, which is known as an aerodynamic device on Open 60 yacht hull, and declared a 10–18% fuel saving with reduced sinkage and trim. Ghassemi et al. (2011), studied the hydrodynamic effects of interceptors on fast speed hulls to investigate the optimal geometrical characteristics based on numerical methods and stated that the height of the interceptor blade has an important effect on efficiency. Wang et al. (2012) conducted a study of a semi-planning ship with interceptors through model tests and found that the resistance reduction effect of the interceptors is related to the loading condition and they gained a maximum 6% of reduction effect on the fully loaded condition of a semi-displacement hull. Additionally, Avci and Barlas (2015) investigated a practical way of measuring trim and sinkage for high Froude numbers including roll, pitch and heave motions using Arduinos and low cost IMU's (inertial measurement unit). They also give an example Arduino code that allows printing raw accelerometer, gyro, and magnetometer data for Eulerian Angels compatible with MPU 9150 IMU sensor.

In this work, an experimental approach is used to deal with the problem of evaluation of the effects of interceptors for fuel efficiency in high-speed vessels. This work differs from others works as follows; a fixed width interceptor is placed at the aft of a model hull in three different locations (close to keel, in the middle and close to chine). The displacement of the model in all the experiments was constant and the deadrise angle was fixed, with six different interceptor deployment heights, i.e. 1, 2, 4, 6, 8, 10 mm (the deployment heights of the interceptor blades between 0.05% and 0.5% LWL). Additionally, the trim and sinkage measurements were done by two different measuring systems, an IMU and a HSCS, to check and control the results. An extreme number of systematic experiments for a model of an 18.7 m waterjet propelled pilot boat were conducted at Ata Nutku Ship Model Testing Laboratory at Istanbul Technical University. A wooden model with a scale of 1/8.5 was manufactured and tested in the towing tank for desired loading conditions. The experiments were based on the variation of speed, trim and different location of the interceptors with different deployment depths and locations as mid of transom, near the keel and near the hull chine. These motions have also been investigated in towing tank experiments when the model is free for 2 degrees of freedom.

# 2 Experiments

## 2.1 Geometry

The tests for the current study were carried out in the towing tank of the Ata Nutku Ship Model Testing Laboratory at Istanbul Technical University. The tank measures 160 m in length, 6 m in width and 4 m in depth; for the current tests the water depth was set at 3.4 m. The carriage can travel at more than 5 m/s with a high level of accuracy in speed control and regulation. A wooden model of an 18.7 m high-speed pilot boat was used in this work with a scale of 1/8.5 without the interceptors, as shown in Fig. 1. The geometric and hydrostatic details of the model and the boat are

**Fig. 1** Bow, aft and profile views of the test model

given in Table 1. Furthermore, the sectional curves and the profile view of the high-speed pilot boat are given in Figs. 2 and 3, respectively.

## 2.2 *Experimentation Procedure*

In the investigation of the effects of trim interceptors, a vast number of parameters are taken into account to arrange the transverse position and depth of the interceptors. The length of the interceptor is 85 mm (0.043 LWL). It is positioned transversely in three places on the aft; close to chine (Fig. 4a), in the middle (Fig. 4b) and close to the keel (Fig. 4c). Additionally, six different depths were used, i.e. 1, 2, 4, 6, 8, and 10 mm (0.05%, 0.1%, 0.2%, 0.3%, 0.4% and 0.5% of LWL).

## 2.3 *Trim and Sinkage Measurements*

There are numerous systems for measuring trim and sinkage motions of a model in a towing tank, such as high-speed camera systems (HSCS), remote distance meters, LVDT, and inertial measurement units (IMU). According to the ITTC recommended procedures, sinkage fore and aft may be measured with mechanical guides,

**Table 1** The geometric and hydrostatic details of the model and the boat

| Test condition | Model ($\lambda$= 8.5) | Ship |
|---|---|---|
| Length between perpendiculars $L_{BP}$ (m) | 2.031 | 17.262 |
| Length on waterline $L_{WL}$ (m) | 1.934 | 16.439 |
| Wetted length $L_{WS}$ (m) | 1.934 | 16.439 |
| Breadth B (m) | 0.588 | 5.000 |
| Draught (amidships) T (m) | 0.109 | 0.918 |
| Draught (AP) $T_A$ (m) | 0.108 | 0.918 |
| Draught (FP) $T_F$ (m) | 0.108 | 0.918 |
| Displacement volume $\nabla$ ($m^3$) | 0.053 | 32.750 |
| Displacement $\Delta$ (ton) | 0.053 | 33.569 |
| Nominal wetted surface area $S_O$ ($m^2$) | 0.989 | 71.520 |
| Transom area $A_T$ ($m^2$) | 0.0256 | 1.855 |
| Centre of transom area $H_T$ (m) | 0.0741 | 0.630 |
| Block coefficient $C_B$ | 0.436 | 0.436 |
| Prismatic coefficient $C_P$ | 0.715 | 0.715 |
| Midship area coefficient $C_M$ | 0.611 | 0.611 |
| Waterplane area coefficient $C_{WP}$ | 0.755 | 0.755 |
| Longitudinal center of buoyancy LCB (m) (+ fwd) | −0.128 | −1.086 |
| Longitudinal center of floatation LCF (m) (+ fwd) | −0.130 | −1.108 |
| Service speed $V_S$ | 4.411 m/s | 25.0 knots |

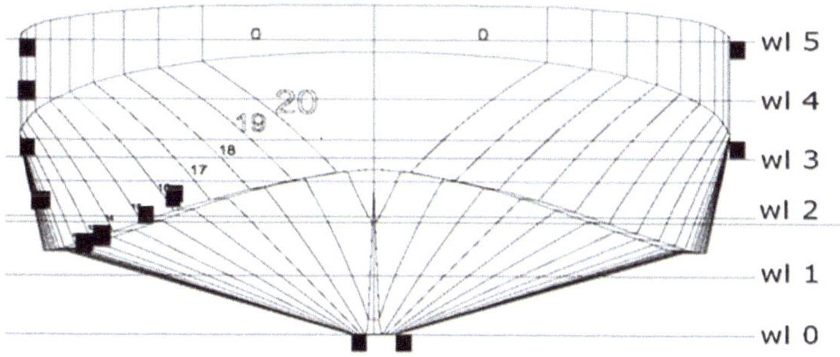


Fig. 2 Sectional curves of the test model

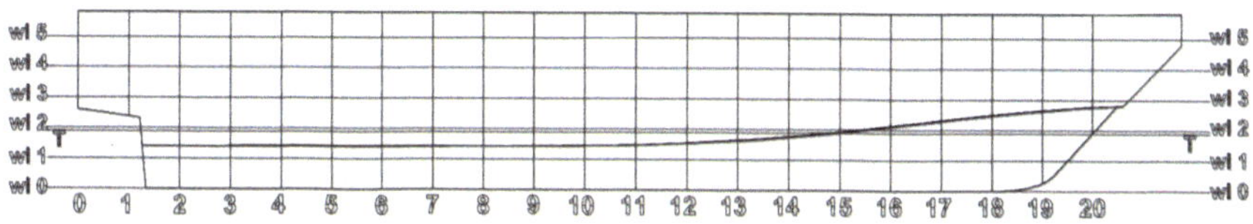


Fig. 3 Profile view of the test model

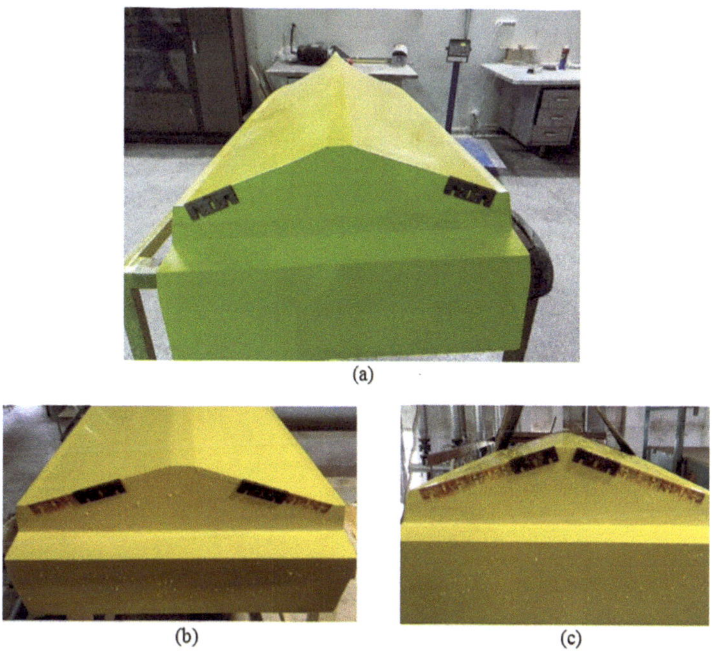


**Fig. 4** Interceptor applications of the test model, close to chine (**a**), in the middle (**b**) and close to keel (**c**)

potentiometers, encoders, LDVTs or with remote (laser or ultrasonic) distance meters. The running trim is calculated from the measured running sinkage fore and aft (ITTC 2002).

For the present study, during the resistance tests, trim and sinkage measurements were carried out by using a high-speed camera system (HSCS) as a conventional method for the first approach, photographed in Fig. 5. Although HSCS seems to be an appropriate method for trim measurement, the configuration and calibration of the system on the test model has some drawbacks. Adjusting the camera resolution and connecting cables to the computer takes time and more importantly, the LED lights, seen in Fig. 5, cannot always be captured by the high-speed camera system during the measurements, especially when the lights are on at the towing car aiming to record the wave formations as well as flow visualization of resistance tests.

The second approach used was an inertial measurement unit (IMU), which is an electronic device that measures velocity, orientation, and gravitational forces, using a combination of accelerometers and gyroscopes. An IMU can either be gimballed or strap down, outputting the integrating quantities of angular velocity and acceleration in the sensor/body frame. The general mounting of the IMU is depicted in Fig. 6. The

**Fig. 5** Trim measurements using high speed camera system (HSCS)

**Fig. 6** Inertial measurement unit device

main advantage of the IMU device is user-friendly device software. It includes on board signal processing with a Kalman Filter. In addition, the user can adjust the frequency of the system to decrease and smooth the data via interface.

For the third approach, to measure the sinkage and trim motions, simple laser distance meters were used for the vertical movement of the forepart target point and astern target point of the wooden body of the model, as seen in Fig. 7. The laser meters were also used for the present study to capture vertical movements. Motions in Z-direction were read from the laser distance meters then transformed into trim and sinkage values of the hull by the help of 3D CAD-CAM software.

As a comparison of the three different measuring devices, HSCS, IMU and laser distance meters, the trim and sinkage values of the model results are given together in Figs. 8 and 9.

For comparison, the minimum difference in trim was 0.2°; the maximum difference in trim was 0.72°, and the average difference in trim was 0.46°. In addition, as seen from the Fig. 9, the sinkage values of the model, using three different methods, were close to each other apart from some regions such as Froude numbers between 0.75 and 0.95. The trim and sinkage values measured by three different methods are in good agreement. In addition, due to the disadvantages of using the HSCS method, which was mentioned before, using an inertial measurement unit is a more accurate method and using the distance laser meters method was the easiest way to measure trim and sinkage.

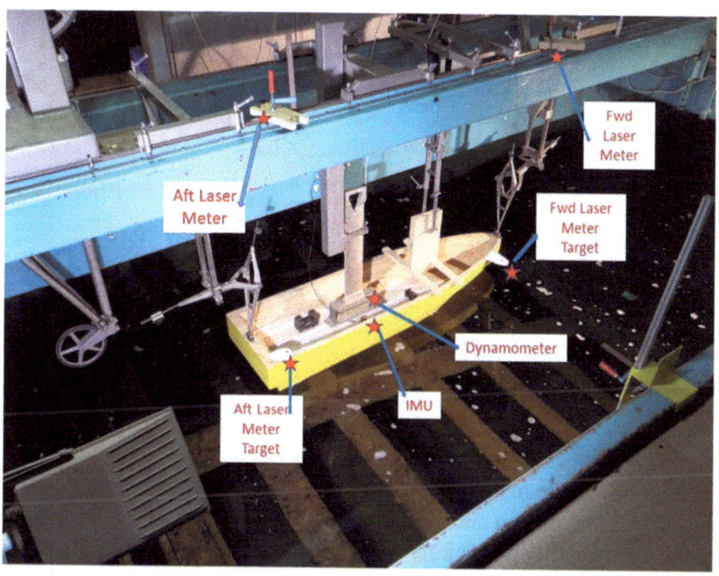


**Fig. 7** IMU device and laser meters mounted on the model

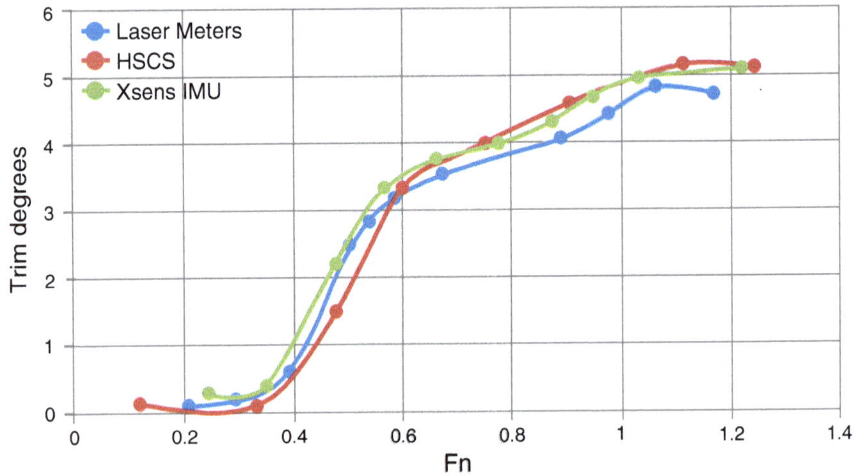


**Fig. 8** Comparison of the trim values of the model by using different methods

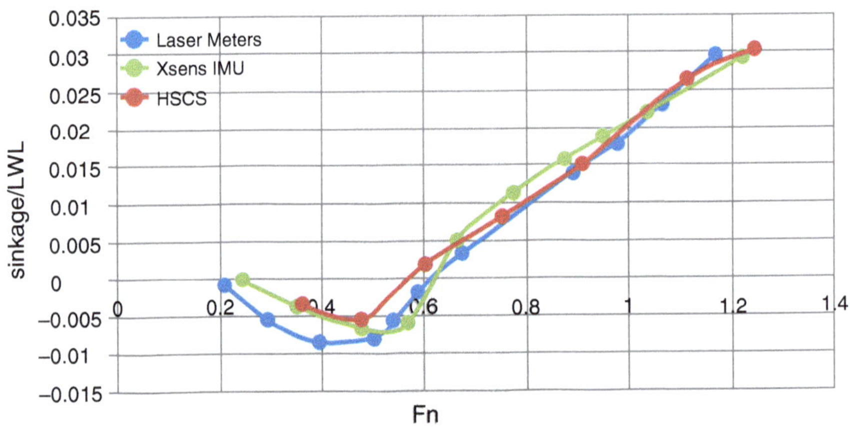


**Fig. 9** Comparison of the sinkage values of the model by using different methods

## 3 Results and Discussion

Since the ship will be thrusted by a water jet system, no appendage was introduced on the model. Therefore, resistance tests were carried out for the bare hull only. The model was tested free to trim and sink motions in calm water, though the model was fixed to heel, sway and yaw motions. Because the boat was a fast craft, form factor analysis was not carried out. The effect of air drag was not included in the analysis. Extrapolation to the full scale was carried out according to the ITTC Recommended Procedure, Testing and Extrapolation Methods High Speed Marine Vehicles

**Fig. 10** Wave formation and the trim reduction of 1 mm interceptor blade deployment in model scale at Fn about 1

Resistance Test—7.5-02-05-01 (ITTC 2002). Flow observation is given in Fig. 10 during the resistance tests.

203 experiments were conducted in the tank basin; of which 52 were experiments for close to chine placed interceptor, 59 were experiments for middle placed interceptor, and 63 were experiments for near keel placed interceptors and 29 were experiments for bare hull.

The total resistance (kg) vs model speed (m/s) measured for bare hull, 2 mm middle, near keel and close to chine interceptor positions are compared in Fig. 11. It can be seen from the figure that for Fn below 0.50 the interceptors slightly increase the drag force. Hence, when the model starts the planning motion, all the interceptors decrease the total resistance force compared to bare hull. The decrease in the total resistance force by using interceptor position close to chine is much less than the other interceptor positions. In Fig. 12, the comparison of the total resistance (kg) vs Froude number measured for bare hull, 1, 2 and 10 mm keel interceptor positions is depicted. It is interesting to note that above $Fn = 0.9$ the 10 mm keel interceptor position increases the total resistance. This is due to the fact that for high blade height deployments, and at high Froude numbers the interceptors increase the drag whilst reducing the trim. So interceptor blades must be adjusted relative to the hull, by the way; they have to be clever. Also, resistance characteristics ($C_{TM} \times 1000$) of the interceptor blade heights in h/LWL %, according to the various Froude numbers are depicted in Figs. 13, 14 and 15.

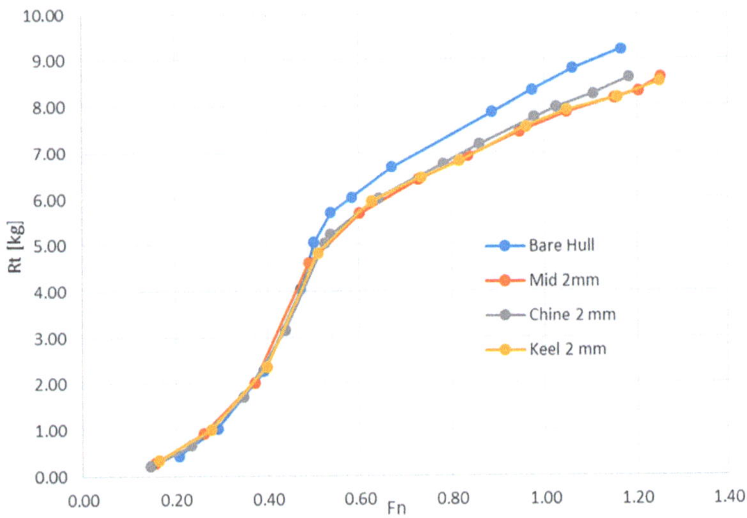


**Fig. 11** The total resistance (kg) vs Froude number measured for bare hull, 2 mm chine, middle and keel interceptor positions

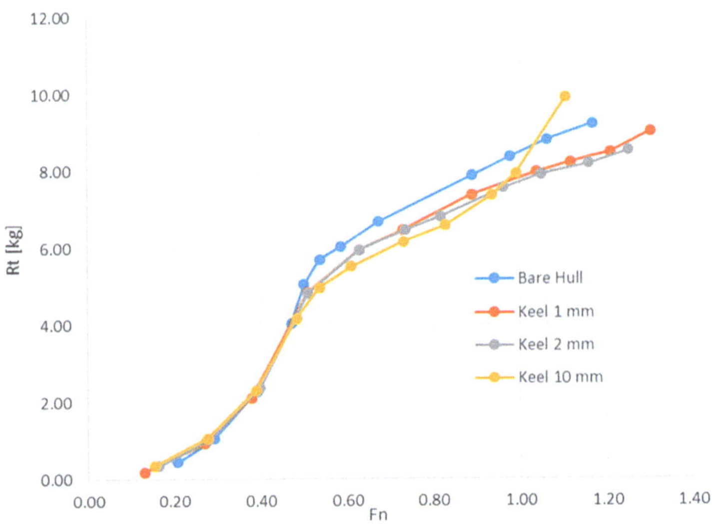


**Fig. 12** The total resistance (kg) vs Froude number measured for bare hull, 1, 2 and 10 mm keel interceptor positions

## 3.1 The Regression Model

A multiple regression analysis is utilized to make suitable predictions of the $C_T$. It is assumed that $C_T$ has a strong bond with the Froude number (Fn), interceptor area

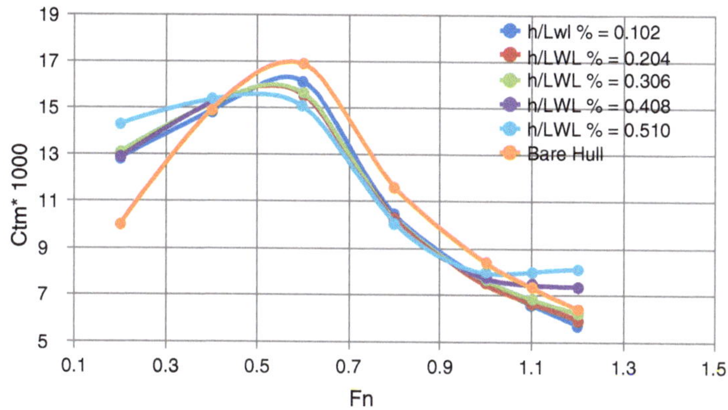


**Fig. 13** Froude number vs $C_{TM} \times 1000$ with Interceptor Heights (h/LWL %), interceptor blade is at "near keel" location

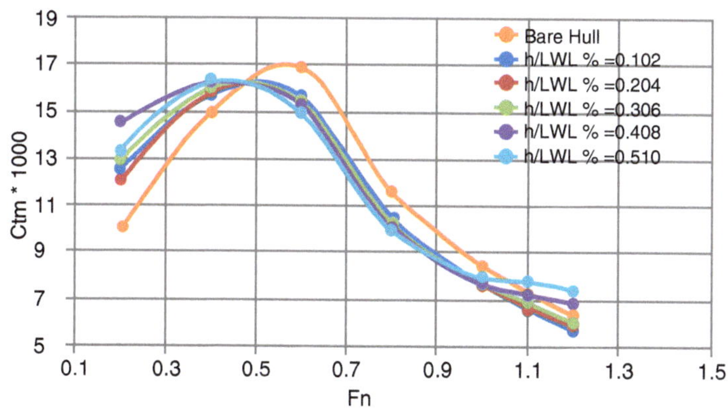


**Fig. 14** Froude number vs $C_{TM} \times 1000$ with Interceptor Heights (h/LWL %), interceptor blade is at "middle" location

(A), and trim (T). By using the well-known $C_T$ formula (Eq. 1), three different multiple regression relations are found by three different interceptor locations, i.e. close to chine, in the middle and close to keel.

$$C_{TM} = \frac{R_{TM}}{\frac{1}{2}\rho_M V_M^2 S_M} \tag{1}$$

where $R_{TM}$ is the model total resistance, $\rho_M$ is the water density, $V_M$ is the model speed and $S_M$ is the model wetted surface.

$$C_T = a_0 + a_1 Fn + a_2 A + a_3 T \tag{2}$$

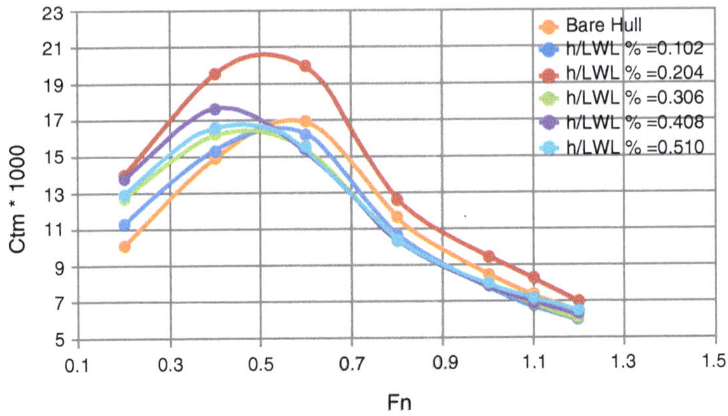


**Fig. 15** Froude number vs $C_{TM} \times 1000$ with Interceptor Heights (h/LWL %), interceptor blade is at "near chine" location

$$E = \sum_{0}^{n} \left[C_T - a_0 - a_1 Fn - a_2 A - a_3 T\right]^2 \tag{3}$$

where E is the error sum. First by taking the differentiation of the error sum with respect to parameters (Eq. 4) and then solving, the equations yield multiple regression model of $C_T$ for three different interceptor locations.

$$\frac{\partial E}{\partial a_i} = 0 \tag{4}$$

$$C_{T\,Keel} = 21.512 - 11.787Fn - 1.053A - 0.314T \tag{5}$$

$$C_{T\,Middle} = 20.103 - 11.910Fn - 0.113A - 0.028T \tag{6}$$

$$C_{T\,Chine} = 24.748 - 13.554Fn - 2.046A - 0.353T \tag{7}$$

In Figs. 16, 17 and 18, the total resistance coefficient $C_T$ vs Froude number measured for close to keel, middle, and close to chine interceptor positions compared with estimations are depicted. The suggested multiple regression model of $C_T$ for three different interceptor locations is applicable after the hull starts planning motion, i.e. Fn is greater than 0.60. It is seen from Figs. 16, 17 and 18 that the suggested model predicts the total resistance coefficient $C_T$ quite well with an average error, compared to the model experiments, of 6.9% in chine location of the interceptor, 5.2% in keel location of the interceptor and 4.2% in the middle location of the interceptor.

## 4 Conclusions and Future Work

Our conclusions and recommendations are as follows:

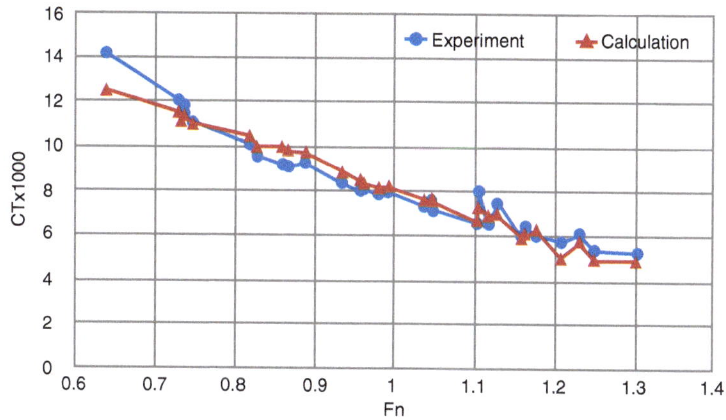


**Fig. 16** The total resistance coefficient $C_T$ vs Froude number measured for close to keel interceptor position compared with estimation. Triangle is model estimation, circle is experiment

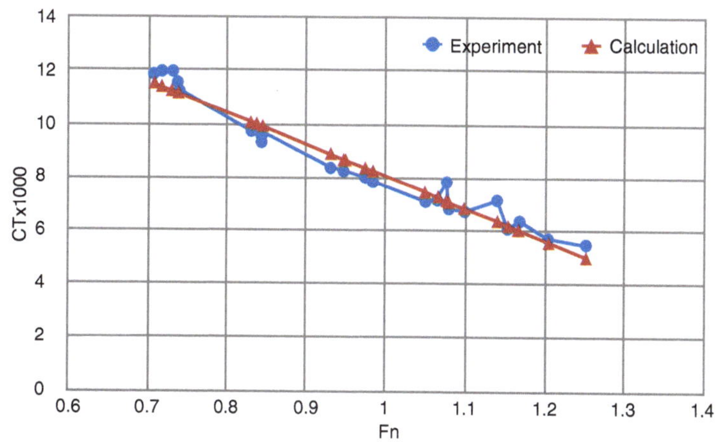


**Fig. 17** The total resistance coefficient $C_T$ vs Froude number measured for middle interceptor position compared with estimation. Triangle is model estimation, circle is experiment

- Resistance and running trim results showed characteristics common to planing craft.
- Correspondingly, the experimental results clearly state that, the interceptor systems decrease the unwanted trim angles and reduce the total resistance in high speeds, and gain approximately 14% fuel savings.
- An additional lift under the transom area generated by interceptors, led to a decrease in sinkage. When sinkage motion decreases, the wetted area and skin friction drag are also reduced.
- The interceptor systems also decrease the unwanted trim angles and supply a clear angle of sight for the boat operators.

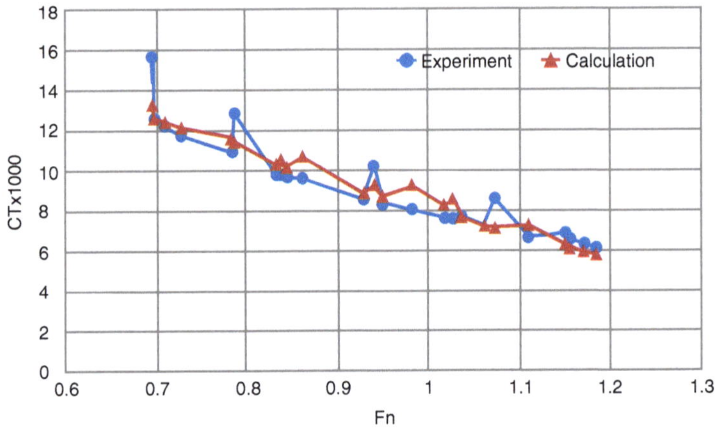


**Fig. 18** The total resistance coefficient $C_T$ vs Froude number measured for close to chine interceptor position compared with estimation. Triangle is model estimation, circle is experiment

- For all three types of interceptor locations, the interceptors increase the total resistance at slow speeds up to Fn = 0.40–0.50. After the boat starts its planning motion, the interceptor works in advance of reducing the total resistance coefficient $C_T$. After about Fn = 1, the higher blade deployments increase the total resistance coefficient as seen in Figs. 13, 14 and 15. Therefore, the authors advice using the dynamic interceptor system in high-speed boats; the interceptor blade depth is to be adjusted autonomously related to the operation speed.
- According to the study, interceptor deployment is very important regarding the advance speed of the hull. At high speeds, high depth of interceptor deployment lets the hull make a bow down motion and extreme spray, increasing the resistance. So, interceptor deployment should be arranged regarding speed and it has to be a dynamic blade deployment system with clever motions up to the speed.
- For all the interceptor locations, the optimum blade height decreases after about Fn = 1. For Fn between 0.55 and 0.85, the optimum blade heights reach the maximum values.

For future work, full transom deployment of interceptor blade experiments is planned. Additionally, it is planned to investigate the interceptor on the stepped hull. Also, CFD simulations of interceptor applications will be added.

## References

Avci, A. G., & Barlas, B. (2015). *A practical application for trim and sinkage measurements for high-speed marine vessels by using an inertial measurement unit and an arduino board.*

Paper presented at the 4th international conference on advanced model measurement technology for the maritime industry, Istanbul, October 2015.
Day, A. H., & Cooper, C. (2011). An experimental study of interceptors for drag reduction on high-performance sailing yachts. *Ocean Engineering, 38*(8–9), 983–994. https://doi.org/10.1016/j.oceaneng.2011.03.006.
De Luca, F. (2011). *Experimental study on interceptor's effectiveness*. Dissertation, Universita degli studi di Napoli "Federico II".
Ghassemi, H., Mansouri, M., & Zaferanlouei, S. (2011). Interceptor hydrodynamic analysis for handling trim control problems in the high-speed crafts. *Proceedings of the Institution of Mechanical Engineers, Part C: Journal of Mechanical Engineering Science, 225*, 2597–2618. https://doi.org/10.1177/0954406211406650.
ITTC. (2002). Testing and extrapolation methods high speed marine vehicles resistance test. ITTC-recommended procedures. In *Proceeding of 23rd International Towing Tank Conference, Venice, 2002.*
Kohansal, A. R., Ghassemi, H., & Ghaisi, M. (2010). Hydrodynamic characteristics of high speed planning hulls, including trim effects. *Turkish Journal of Engineering and Environmental Sciences, 34*, 155–170.
Savitsky, D. (2003). *On the subject of high-speed monohulls*. Paper presented at the Greek Section of the Society of Naval Architects and Marine Engineers (SNAME), Athens, October 2003.
Tsai, J. F., & Hwang, J. L. (2004). *Study on the compound effects of interceptor with stern flap for two fast monohulls*. Paper presented at the Oceans'04. MTTS/IEEE Techno-Ocean'04, (Vol. 2, pp. 1023–1028).
Wang, W., Zong, Z., Ni, S., Zhang, L., & Chen, L. (2012). Model tests of effects of interceptor on resistance of a semi-planing ship. *Chinese Journal of Ship Research, 7*(1), 18–22.

# A Decision Support System for Energy Efficient Ship Propulsion

**Mustafa Insel, Serhan Gokcay, and Ahmet Ziya Saydam**

## 1 Introduction

Environmental concerns have been taken very seriously by the maritime community and IMO (International Maritime Organization) has established Annex VI of MARPOL (International Convention for the Prevention of Pollution from Ships) which contains shipboard air emissions such as $NO_x$, $SO_x$ and $CO_2$ emissions. The designers and shipyards responded to this request by further optimizing their design through green technologies. Ship-owners have also reacted by re-evaluating their operations including trim optimization, weather routing and retrofitting their ships with green technologies. However, there is a considerably difficult task of assessing the benefits and costs of operational measures/design alternatives/retrofitting alternatives as most of the associated benefits cannot be directly measured with a reasonable level of certainty. Additionally, when there are a number of alternatives to be evaluated, complicated interaction between these technologies cannot be easily predicted or determined. Utilization of reliable numerical methods and sea trials is required to reduce such uncertainties.

Decision support systems were introduced to support ship bridge teams on decisions of route and trim optimization (Lu et al. 2015; Lin et al. 2013, 2015). These systems are either based on simplified models of ship performance or direct collection of operational data on route and system identification methodology such as neural network to generate the decision suggestions such as Besikci et al. (2016) and Petersen et al. (2012).

M. Insel (✉)
Hidroteknik Nautical Design Technologies Ltd, Teknopark İstanbul, Istanbul, Turkey
e-mail: minsel@hidro-teknik.net

S. Gokcay · A. Z. Saydam
Hidroteknik Nautical Design Technologies Ltd, Teknopark İstanbul, Istanbul, Turkey

Piri Reis University, Istanbul, Turkey



A. I. Ölçer et al. (eds.), *Trends and Challenges in Maritime Energy Management*,
WMU Studies in Maritime Affairs 6, https://doi.org/10.1007/978-3-319-74576-3_11

These systems are usually not sensitive enough for the design decisions or retrofitting option analysis, as they can simulate only the tested cases. Hence, in the current work, a decision support tool is being developed based on detailed CFD (Computational Fluid Dynamics) analysis, validated and fine-tuned by full scale trials and sea voyage data to model ship performance. Once the ship performance model is fully developed and validated, various operational measures and retrofitting options can be tested using a detailed ship performance model.

## 2 Performance Prediction Models

Performance prediction of a ship in a seaway can be separated into calm water performance prediction and speed loss in seaway due to wind, waves, swell and current, and speed loss in steering.

The calm water performance is primarily based on resistance which can be subdivided into viscous and wave resistance. Wave resistance has been the target of optimization studies since the 1970s, and highly efficient hull forms have been developed for design speed (Chun 2010). However, little gain has been achieved on viscous resistance even though it is the largest resistance component for most ship types. Newly developed techniques, include but are not limited to;

- biofouling and new paints (Schultz 2007),
- boundary layer modifying polymer injections (Yang 2009),
- air lubrication such as given in Foeth (2008) and Thill (2010)
- surface context,
  - surface patterns and shark skin (Dean and Bhushan 2010),
  - dolphin skin (Gad-el-hak 2002)

may result in resistance reduction within the range of 5–10%.

Propulsion efficiency of the ship represents probably the largest potential to gain energy efficiency. Propulsion improvement devices, PIDs such as presented by Hollenbach and Reinholz (2011), Hochkirch and Bertram (2010) and Haferman and Marzi (2011) are in the form of wake equalizing and flow separation alleviation devices, pre-swirl devices, post-swirl devices and high-efficiency propellers.

Air drag increase due to wind, and appendage drag increase due to rudder angles are well understood and should be included in resistance models. The effect of the sea waves through both resistance increase and propulsion efficiency loss results in speed loss.

A full performance prediction model was formed by combining the resistance, propulsion and weather effects as given in Fig. 1. Hence a full performance model can be given as:

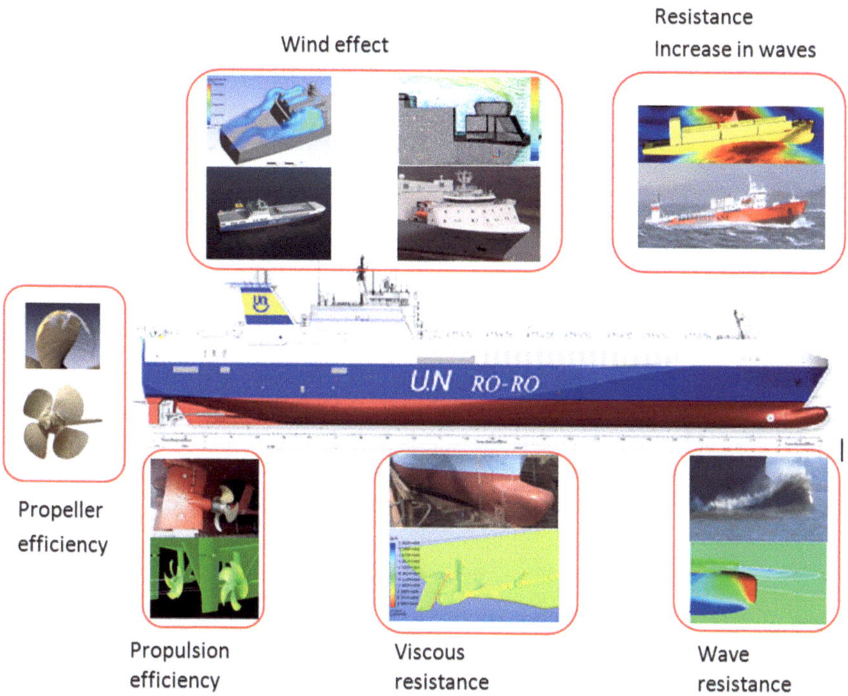


**Fig. 1** Decision support system components

$$P_B = \frac{(R_V + R_W + \Delta R_{WIND} + \Delta R_{STEER} + \Delta R_{WAVE})V}{(\eta_0 - \Delta\eta_{WAVE})\eta_H\eta_{RR}\eta_{TR}}. \quad (1)$$

where

$P_B$ Ship power requirement
$R_V$ Viscous resistance
$R_W$ Wave resistance
$\Delta R_{WIND}$ Resistance increase due to wind
$\Delta R_{STEER}$ Resistance increase due to steering
$\Delta R_{WAVE}$ Resistance increase due to waves
$\eta_0$ Propeller efficiency
$\Delta\eta_{WAVE}$ Propeller efficiency change due to waves
$\eta_H$ Hull efficiency
$\eta_{RR}$ Relative rotative efficiency
$\eta_{TR}$ Transmission efficiency
$V$ Ship speed

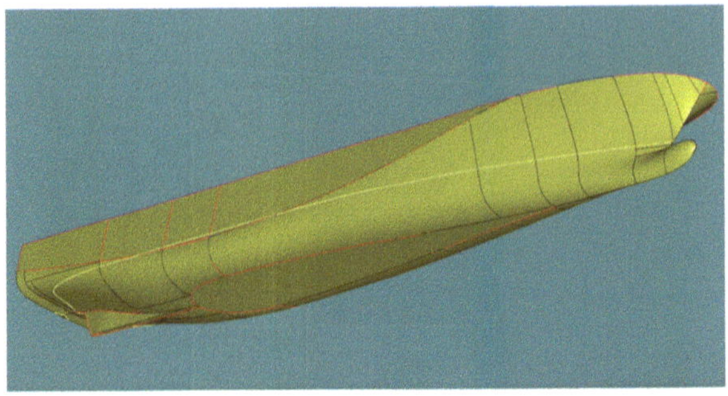

**Fig. 2** Example ship utilized in the analysis

The resistance components and propulsion efficiency can be estimated by various numerical methods. But the energy efficiency goal requires a high fidelity methodology to be able to distinguish the differences between various energy saving measures and interactions between them. A model based on CFD calculations and sea trial/voyage data for validation is proposed in the current work.

## *2.1 Numerical Model*

Numerical evaluation of prototype ship was conducted by RANS (Reynolds-Averaged Navier–Stokes) CFD analysis. Such approach has great flexibility in that any ship type and hull form can be represented including forms with high waves and spray formation. The hull form was modeled in 3D IGES form as a solid model as given in Fig. 2. The solid model has been placed in a control volume by creating two phase fluid, namely air and water. The fluid has been meshed with 2.5 million mesh elements. The bow and stern mesh configuration is demonstrated in Figs. 3 and 4.

## *2.2 Viscous Resistance*

Viscous resistance can be estimated from simple formulations such as ITTC (International Towing Tank Conference) 1957 ship-model correlation line. However, such simplistic approaches cannot determine the separation at the aft end. As propulsion improvement devices at the aft end may modify the flow, a RANS solver method was used at full ship scale without free surface. The appendages are modeled along with interactions between the appendages and energy saving measures such as PIDs.

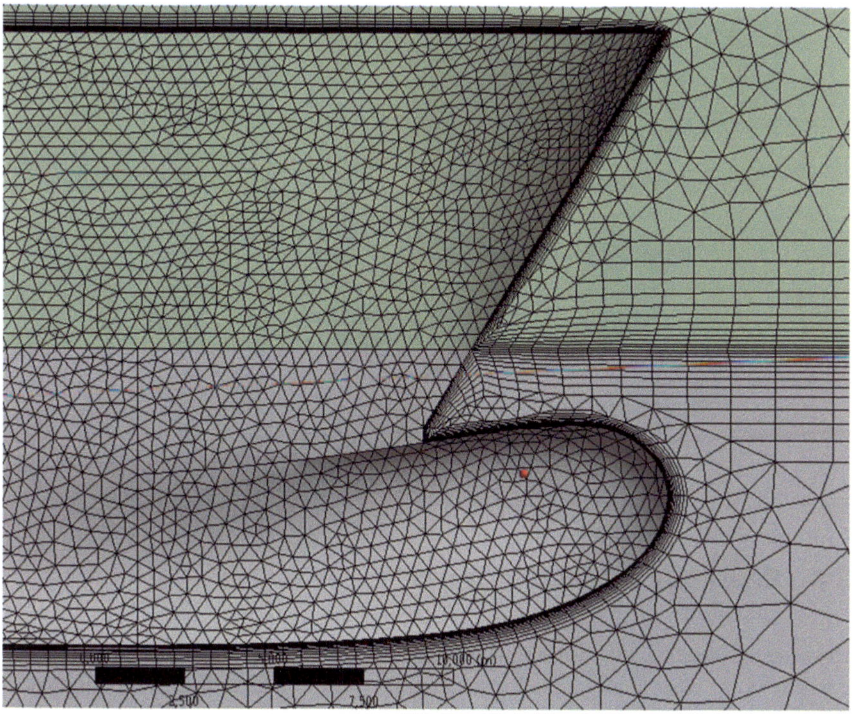

**Fig. 3** Mesh structure at the bow

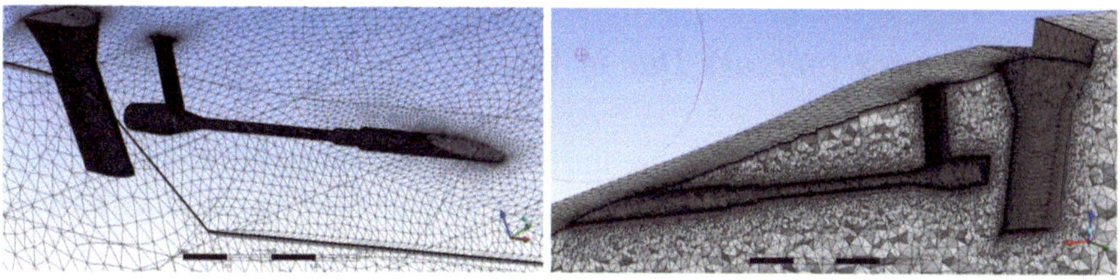

**Fig. 4** Mesh structure at the aft

## 2.3 *Wave Resistance*

Most of the current ship forms have been optimized with numerical tools, and usually potential flow tools are utilized for the wave resistance optimization. Although potential flow assumption is suitable for ship wave resistance estimation, a two phase RANS solver was used in the current method in order to capture any viscous-wave interactions due to any energy efficiency improvement measure. Figure 5 demonstrates the wave-making characteristic of the hull form.

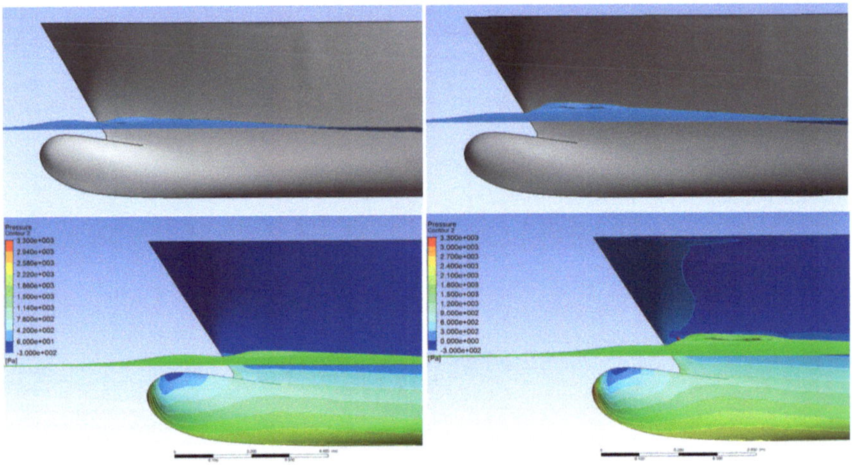

**Fig. 5** Bow waves at 19 and 21 knots ship speed

## 2.4 *Resistance Increase in Waves*

Resistance increase due to sea waves is required to estimate the real life sea conditions. A strip theory based resistance in waves methodology was adopted in the current work.

## 2.5 *Resistance Increase Due to Wind*

A RANS CFD modeling of the superstructure was built and analyzed with regard to different apparent wind angles and wind speeds to determine the resistance increase due to the existence of the superstructure. Superstructure streamlining effect on energy efficiency can be calculated with the current approach.

## 2.6 *Resistance Increase Due to Steering*

Route corrective rudder angles are required for route keeping, usually within $\pm$ 5 degrees of rudder angle. The effect of rudder angle on resistance was studied by the CFD model. This approach also allows consideration of special rudders such as twisted rudders and rudder bulbs to be tested as an energy efficiency measure.

### 2.7 Engine Dynamics

When the power requirement is estimated by using resistance and propulsion calculations, engine factory test results are utilized to convert the power requirement at a specific engine rpm to fuel consumption through specific fuel oil consumption obtained at the factory tests. Fixed pitch propeller (FPP) or Controllable pitch propeller (CPP) configuration of the shaft-propeller system is taken into account.

## 3 Validation-Sea Trials

Validation of performance prediction methodology is of prime importance as all the energy efficiency improvement measures shall be judged on this basis. Sea Trials are required to validate the performance prediction at the base configuration. These trials were conducted with equipment in the bridge for the speed determination and control parameters; meanwhile, strain gauges were installed in the shafts to measure the shaft torque and propeller rate of revolution as given in Fig. 6.

Typically pairs of runs are conducted for each speed according to ITTC standard given in ITTC (2014a, b) or according to ISO standard given in ISO 15016 (2015) and ISO 19019 (2005). A number of speeds are tested to derive the speed-power curve given in Fig. 7. Uncertainty analysis on these speed trails as shown by Insel (2008) can be up to 10% depending on weather conditions and uncertainties associated with the sea trial. Considering that some of the energy efficiency measures have effective energy savings of only about 1% of the power, such sea trials cannot be directly used to assess the effectiveness of the energy efficiency measure.

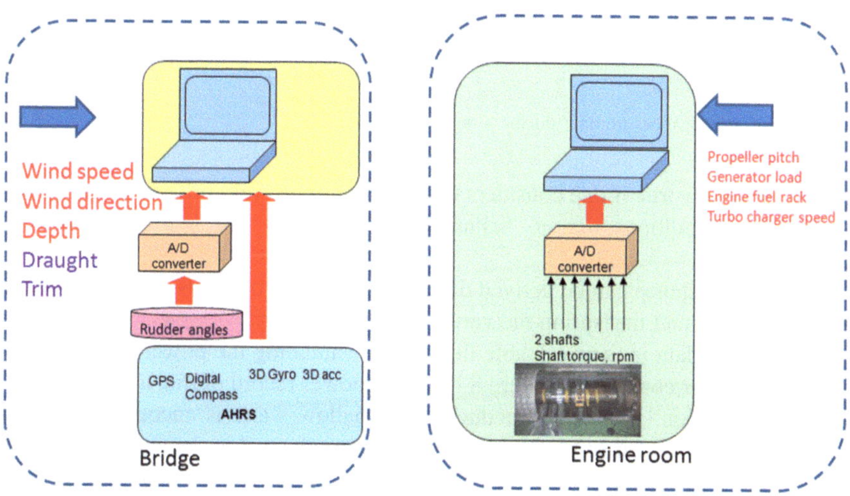


**Fig. 6** Full scale measurement system located in the bridge and engine room

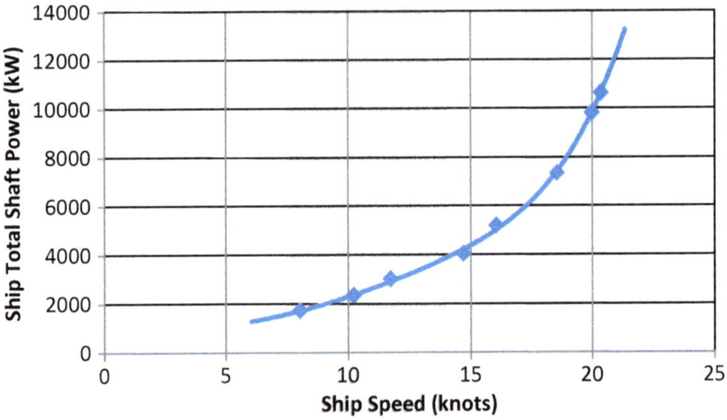


Fig. 7 Full scale ship speed versus shaft power measurement

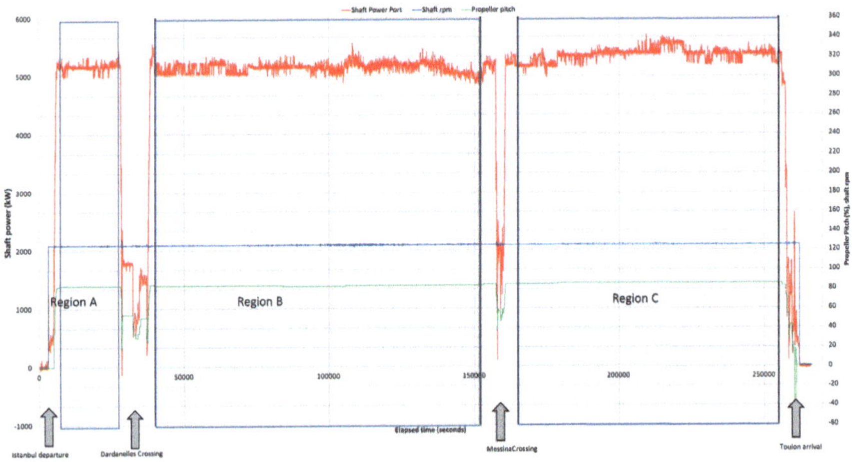


Fig. 8 Shaft power measurement during a regular sea voyage

This is especially true if one considers trial conditions are very different from actual sea voyage conditions due to ballast load instead of full load, and weather conditions.

Steering resistance can be derived during the sea trials by adjusting the rudder at different angles and measuring the variation on the power.

Sea voyage data is very valuable in terms of validating the powering change in different weather conditions. Figure 8 is a shaft power record during a voyage from Istanbul to Toulon. As speed is reduced and shallow water is encountered in the channel crossings, such as Dardanelles and Messina channel in the current case, the data on the channel crossing were eliminated to obtain real sea condition performance data. Variation of power during voyage is due to weather, speed command changes, i.e. propeller pitch setting in the current CPP case and heading changes. There is clearly a higher power requirement in comparison with sea trial, which is conducted or corrected for calm water in addition to added resistance in waves.

## 4 Decision Support System

The choice of ship configuration and operating variables form a multi-dimensional space and the ship owner has the dilemma of how to assess the ship design/retrofit options/operational measures in regular voyages. The data obtained from the numerical analysis, validated and modified based on sea trials/voyages, can now be used to simulate any ship conditions even if they have not been tested during sea trials. Combining numerical calculation and sea trial/sea voyage data improves the fidelity of the performance prediction into a suitable precision for an energy efficiency measure decision support system.

The information flow in the DSS is given in Fig. 9. There are no restrictions on ship type or hull form. The only limitation shall be energy efficiency improvement methods, which should be represented by the CFD analysis.

## 5 Case Tests

Three cases were demonstrated in the current DSS for retrofit analysis: application of a new type of paint for a containership, slow steaming of a containership, and retrofit of a high efficiency propeller for a Ro-Ro Ship.

The first example case is based on application of new paint onto two sister containerships. Sea trials were conducted to capture the effect of the paint on the viscous resistance. As both sea trials could not be conducted in the same conditions, such as wind, wave, and ship loading, the differences were corrected using the current methodology, not to equalize the conventional/new paint scenarios but also to predict the actual sea condition performance at the full load, and different sea conditions. Figure 10 demonstrates the difference between two sister ships one with conventional paint and one with new low friction paint.

The second case was based on speed reduction of a containership with CPP propeller to improve energy efficiency. In order to achieve better fuel efficiency, the engine rpm was reduced by de-rating the engine, and propeller rpm was reduced from 120 to 105 without changing the propeller. In addition to sea trails, numerical analysis was utilized to derive propeller efficiency at partial load and full load displacements. Energy efficiency gains originate from reduced resistance due to speed reduction, increased propeller efficiency at lower loading due to reduced rpm, and lower specific fuel oil consumption due to engine de-rating. Energy consumption was evaluated in sea trials and projected into operating conditions using the current methodology (Fig. 11).

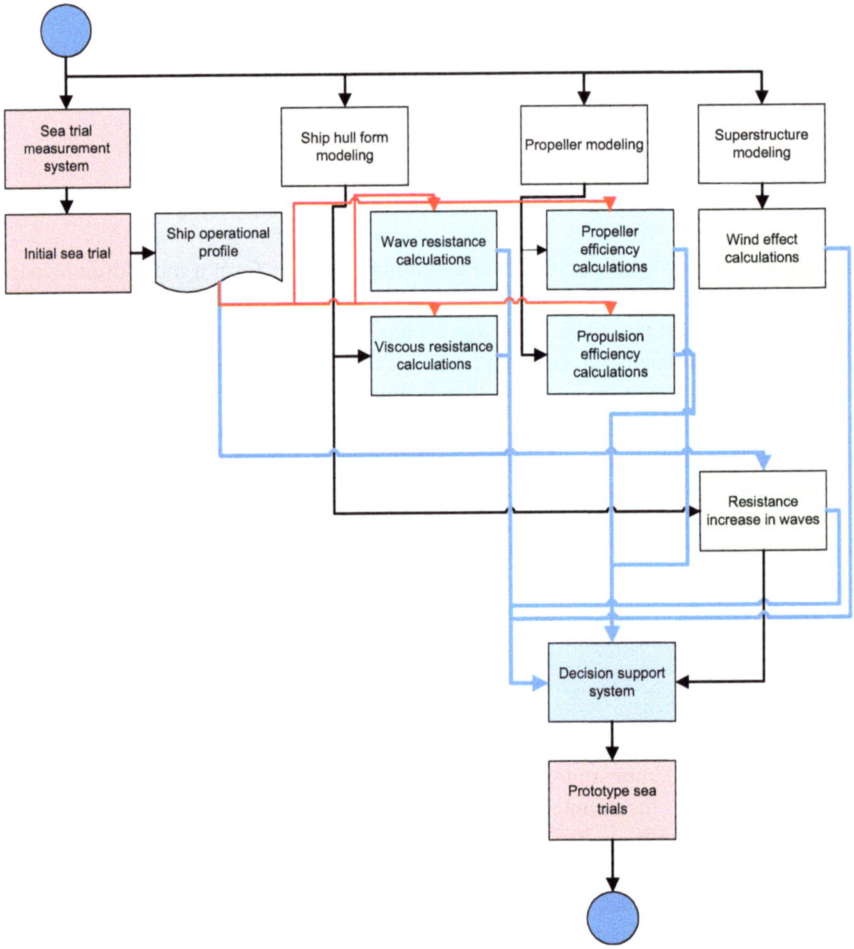


**Fig. 9** The information flow of the proposed decision support system

The last example is based on application of a high efficiency propeller onto a Ro-Ro ship. Two sister ships were tested in voyage conditions in addition to numerical analysis and compared using the current DSS methodology. The differences between ship displacements were corrected for frictional resistance, wave resistance, propeller and propulsion efficiency changes due to propeller loading. No clear gains could be obtained in this case (Fig. 12).

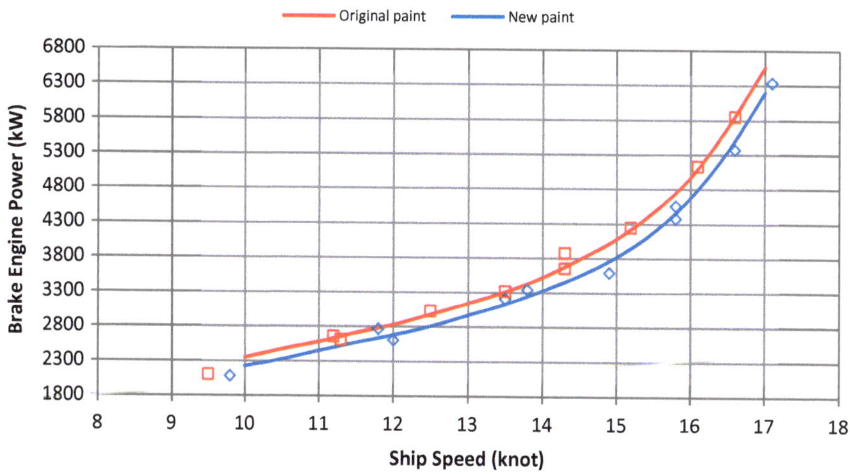


**Fig. 10** Application of a new paint

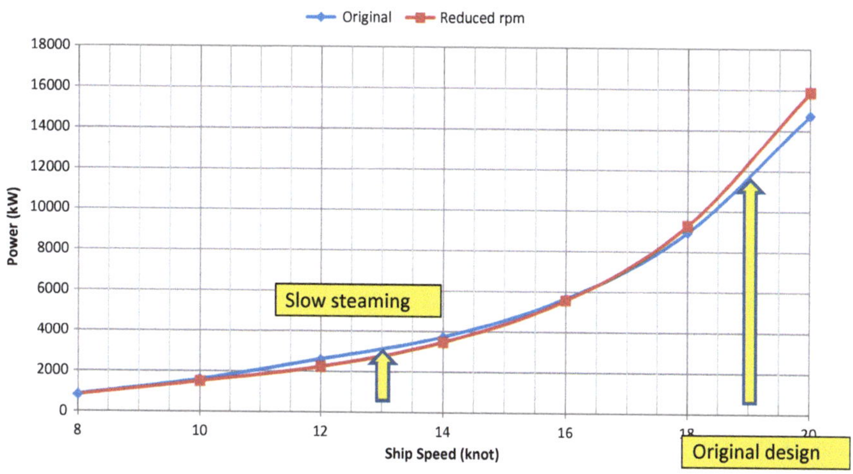


**Fig. 11** Application of slow steaming

# 6 Conclusions

Energy efficiency improvement on current hull forms may be a difficult task as they are already optimized systems. Numerous retrofit options become available as new technologies are developed. However, decision making on the efficiency improvement achievable with these technologies is a complex problem due to uncertainty associated with powering predictions. Empirical approaches are not sufficient at all to make such decisions due to even higher uncertainty involved. Sea trials alone are also not sufficient to judge the benefit of the energy efficiency improvement

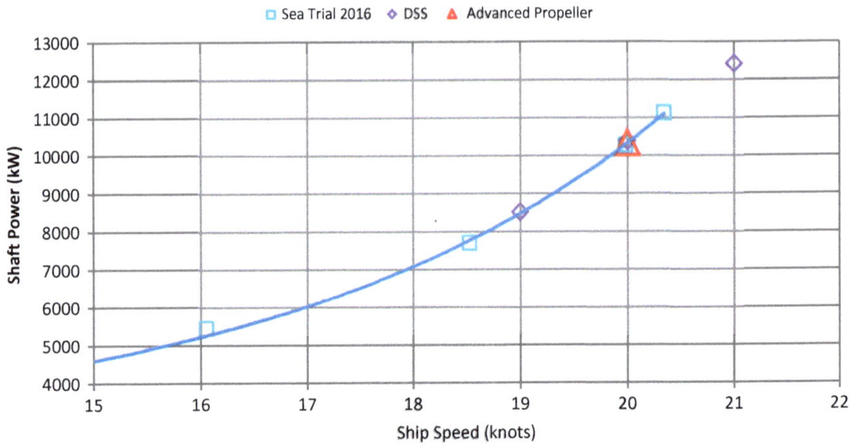


**Fig. 12** Application of high efficiency propeller

measures. High fidelity methodology is required to make rational decisions to evaluate design alternatives or retrofit options.

A methodology based on CFD calculation on the resistance, propulsion and resistance increase due to seaway and steering is now available with fast computers. When it is coupled with full scale trials and sea voyage data for validation and verification, a decision-making system becomes a promising high fidelity solution.

## References


Besikci, E. B., Arslan, O., Turan, O., & Ölçer, A. I. (2016). An artificial neural network based decision support system for energy efficient ship operations. *Computers & Operations Research, 66*, 393–401.

Chun, H. H. (2010). *Hull form parametrization technique with local and global optimization algorithms*. Paper presented at the International Conference on Marine Technology. 1–12 December 2010, Dhaka, Bangladesh.

Dean, B., & Bhushan, B. (2010). Shark-skin surfaces for fluid-drag reduction in turbulent flow: A review. *Philosophical Transactions of Royal Society A, 368*, 4775–4806.

Foeth, E. J. (2008). *Decreasing frictional resistance by air lubrication*. Paper presented at the 20th International HISWA symposium on Yacht Design and Yacht construction. Amsterdam, The Netherlands, 17–18 November 2008.

Gad-el-hak, M. (2002). Complaint coatings for drag reduction. *Progress in Aerospace Sciences, 38*, 77–99.

Haferman, D., & Marzi, J. (2011). *Propulsion enhancements for a greener environment*. Paper presented at the second symposium on marine propelsors, SMP'11, Hamburg, Germany, June 2011.

Hochkirch, K., & Bertram, V. (2010). *Options for fuel saving for ships*. Paper presented at mareforum 2010:Maritime transportation of energy, Houston, 19 February 2010.

Hollenbach, U., & Reinholz, O. (2011). *Hydrodynamic trends in optimizing propulsion*. Paper presented at the second international symposium on marine propulsors SMP'11, Hamburg, Germany, June 2011.
Insel, M. (2008). Uncertainty in the analysis of speed and powering trials. *Ocean Engineering, 35*, 1183–1193.
ISO 15016. (2015). Guidelines for the assessment of speed and power performance by analysis of speed trial data.
ISO 19019. (2005). Instructions for planing, carrying out and reporting sea trials.
ITTC. (2014a). Recommended procedures and guidelines - 7.5-04-01-01.1- speed and power trials, part 1 preparation and conduct.
ITTC. (2014b). Recommended procedures and guidelines - 7.5-04-01-01.2- speed and power trials, part 2 analysis of speed/power trial data.
Lin, Y.-H., Fang, M.-C., & Yeung, R. W. (2013). The optimization of ship weather-riuting algorithm based on the composite influence of multi-dynamic elements. *Applied Ocean Research, 43*, 184–194.
Lin, Y.-H., Fang, M.-C., & Yeung, R. W. (2015). The optimization of ship weather-riuting algorithm based on the composite influence of multi-dynamic elements II. *Applied Ocean Research, 50*, 130–140.
Lu, R., Turan, O., Boulougouris, E., Banks, C., & Incecik, A. (2015). A semi-empirical ship operational performance prediction model for voyage optimization towards energy efficiency shipping. *Ocean Engineering, 110*, 18–28.
Petersen, J. P., Winther, O., & Jacobsen, D. J. (2012). A machine learning approach to predict main energy consumption under realistic operational conditions. *Ship Technology Research, 59*(1), 64–72.
Schultz, M. P. (2007). Effects of coating roughness and biofouling on ship resistance and powering. *Biofouling, 23*, 331–341.
Thill, C. (2010). *A long road mapping drag reduction*. Paper presented at the international conference on ship drag reduction (SMOOTH-Ships), Istanbul, Turkey, 20–21 May 2010.
Yang, S.-Q. (2009). Drag reduction in turbulent flow with polymer additives. *Journal of Fluids Engineering, 131*, 051301.

# Energy Integration of Organic Rankine Cycle, Exhaust Gas Recirculation and Scrubber

Fredrik Ahlgren, Marcus Thern, Magnus Genrup, and Maria E. Mondejar

## 1 Introduction

There are regulations for shipping concerning both nitrogen oxides and sulphur oxides in certain regulated areas. The areas are called emission control areas and there are today several sulphur emission control areas (SECA) and only two nitrogen control areas (NECA), the North American Sea and US Caribbean Sea. Still, this has to be addressed if ships are to be permitted to sail in all waters, as regulations are getting stricter. The International Maritime Organisation (IMO) (2014) resolution MEPC.251(66) states that nitrogen oxide emissions are limited for engines operating in NECAs. The requirements apply to all marine diesel engines over 130 kW power. Tier standards are IMO engine emission standards that set increasingly restrictive nitrogen oxides ($NO_x$) emission limits as a function of the engine operating speed (IMO 2015). The standards are based on when the ship was built and newer ships need to conform to higher Tier standards. Tier III applies to ships built after 1 January 2016, and states that a low speed diesel engine (<130 rpm) cannot emit more than 3.4 g/kWh $NO_x$. This requirement can be met either by selective catalytic reduction (SCR), which is an after-treatment technology, or exhaust gas recirculation (EGR). SCR requires an additional input of urea and sufficiently high temperature for the process to work satisfactorily (MAN Diesel and Turbo 2012). An EGR works on the principle that part of the exhaust gas stream is recirculated to the combustion

F. Ahlgren (✉)
Linnaeus University, Kalmar, Sweden
e-mail: fredrik.ahlgren@lnu.se; fredrik@ahlgren.nu

M. Thern · M. Genrup
Lund University, Lund, Sweden

M. E. Mondejar
Danish Technical University, Lyngby, Denmark



A. I. Ölçer et al. (eds.), *Trends and Challenges in Maritime Energy Management*, WMU Studies in Maritime Affairs 6, https://doi.org/10.1007/978-3-319-74576-3_12

process which lowers the oxygen ratio and, therefore, reduces the amount of nitrogen that reacts with oxygen.

Sulphur emissions are a concern for the environment in terms of sulphur oxide ($SO_x$) particles which are formed when fuel containing sulphur is combusted. IMO regulations state that the maximum sulphur content of the fuel must not exceed 0.1% weight fraction in SECAs (IMO 2010). The fuel used today is still most commonly heavy fuel oil (HFO), which has an average of about 2.7% sulphur (European Environment Agency 2013). However, the scrubber is a complex system, which means that both pumps and auxiliary equipment require extra space and energy, which implies an increase in both capital and operational costs.

There was some concern that the sulphur directive could be an economic burden on shipping companies due to the higher costs of fuel or cleaning technologies (Sweco 2012), and most of the ships operating in the SECA today are using a distillate fuel (Kalli et al. 2009). An explanation for this is most probably that oil prices have been markedly lower than previously predicted (Ölçer and Ballini 2015), which meant that the price for switching to a cleaner fuel did not present an extra cost. The main criteria for installing a scrubber instead of using a low sulphur fuel would be the price difference between the distillate fuel and the residual fuel. Rising fuel prices are an incentive for using cheaper fuel, and this could lead to an increased use of HFO with exhaust gas cleaning techniques in the near future. Considering its price advantage in comparison with distillate fuels, there is an incitement to use high sulphuric HFO and clean the exhaust gases from $SO_x$ (Kalli et al. 2009). The investment cost of scrubber installation is high, and the most important factor is when the system is economical. Studies demonstrate that when the cost of HFO is significantly lower than low sulphur fuel there is an economic incentive for retrofitting (Jiang et al. 2014; Panasiuk and Turkina 2015).

A scrubber is an end-of-pipe solution, which means that the cleaning is done at the end of the chain rather than in a fuel distillery. A number of research studies on power plants with seawater scrubbers (Georgopoulou et al. 2015) demonstrate that the technology works well but with the drawback of higher energy consumption. End-of-pipe solutions, such as scrubbers are considered to have a negative impact on the climate in comparison with using a distillate fuel, as they reduce the overall energy efficiency of the system (Lindstad and Eskeland 2016). These additional technologies for cleaning exhaust gases come with an added cost, often called a fuel penalty, in that they increase $CO_2$ emissions but with a lower amount of $SO_x$ and $NO_x$ (Panasiuk and Turkina 2015). A way to compensate for this increase in $CO_2$ emissions is the use of waste heat recovery (WHR) technologies, such as the organic Rankine cycle (ORC). Studies of these systems have been presented by Wang et al. (2017) and Yang (2016) where an ORC was used to recover waste heat from a diesel engine. The authors of this study have demonstrated a significant efficiency gain with WHR using ORC on ships using exhaust gases (Ahlgren et al. 2015; Mondejar et al. 2016). A large two-stroke ship engine with an open loop scrubber was modelled with a waste heat recovery device by Nielsen et al. (2014), indicating an increased efficiency of 2.6% while still cleaning the exhaust gases.

The objective of this work is to make a first evaluation of the potential benefit of using an ORC to compensate for the power consumption of a scrubber. To this end, we simulated the integration of an ORC unit into an EGR and scrubber system. The potential power generation from the system was analysed for different configurations of the ORC and the available heat sources, and compared with the consumption of the scrubber.

# 2 Method

We developed a model of an ORC to simulate three different cases of heat source integration with technology for reducing both $NO_x$ and $SO_x$ emissions based on data from a two-stroke low-speed MAN engine. The characteristics of the model are specified as follows.

## *2.1 Engine Model*

The data used for the simulations were collected from the specifications of the low speed two-stroke MAN 6G50ME-C95 engine. This engine has a nominal power output of 10 MW at 100 rpm, although the dataset used in this study considers a derated engine. Derating consists of tuning and optimising the engine control, turbocharger, fuel system and other auxiliary systems for running at higher efficiency at lower load conditions. As fuel-economy is dependent on ship speed, it is common to use slow steaming as an effective way of reducing fuel consumption. The main heat sources from the diesel engine are exhaust gas, jacket cooling water, lubrication oil and the charge air cooler. If an EGR and a scrubber are installed for $NO_x$ reduction and $SO_x$ reduction, there will be additional heat sources, as the EGR consists of an exhaust gas recirculation string, where the gases are scrubbed and cooled before re-entering the cylinder. As the system becomes more complex there are several heat sources that can be utilised for WHR, and the way to integrate these is essential for an energy efficient design.

In this study we have considered an open loop wet scrubber used for $SO_x$ removal, and an EGR for $NO_x$ reduction. An engine system, which meets both the Tier III standards and the sulphur directive, has been considered.

Figure 1 presents a system overview of heat sources available from the system and the EGR, excluding the scrubber for $SO_x$, which is not shown in this figure. After the engine the exhaust gases go to the exhaust receiver. The exhaust gas receiver has one outlet-stream that goes into the turbocharger and one outlet-stream that goes to EGR. The turbocharger drives the air compressor, which feeds the combustion with fresh air, and, as the air compression raises the incoming air temperature, the air is cooled in the Charge Air Cooler (CAC). After the compressor, a by-pass valve can be opened at low engine loads. The EGR stream recirculates

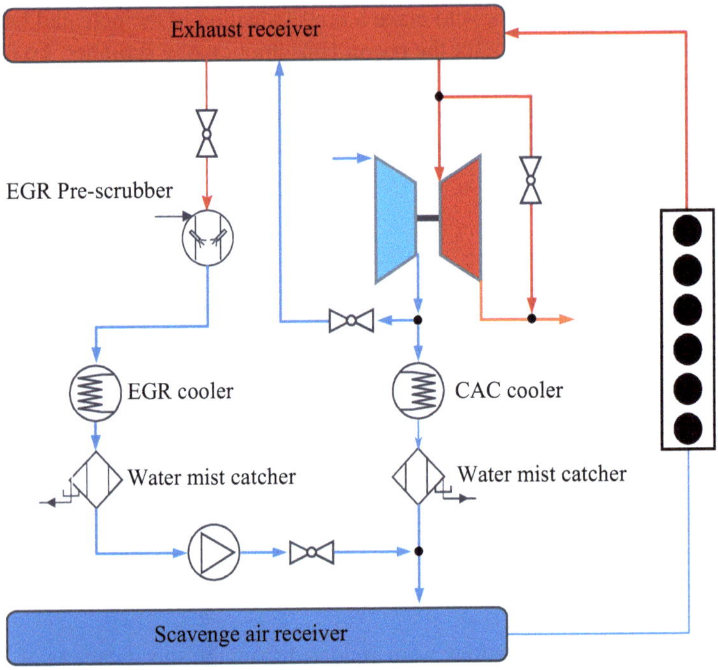


**Fig. 1** Engine EGR schematics

exhaust gases via a cleaning and cooling process. When the combustion environment is of a lower oxygen level it forms less nitrogen oxide ($NO_x$). In the EGR-stream the exhaust gases first enter the EGR pre-scrubber, which is a water sprayed into the gases. An excess of water is sprayed and part of the water is evaporated, where some of the water is still in condensed form. Later in the process, the gases are cooled in the EGR cooler; condensed water is removed in the Water Mist Catcher (WMC) before being pressurized with an EGR-fan and recirculated to the combustion process.

The heat sources considered in this study are the Exhaust EGR, Exhaust gases after the turbo, charge air cooler (CAC) and the FW Jacket cooling water.

The temperatures presented in Fig. 2 are from the manufacturers engine test data, which is used as the base in this study. As seen in Fig. 2 the temperature of all heat sources besides FW Jacket cooling vary with the engine load. The temperature and mass flow are important factors when designing an ORC for waste heat recovery as the temperature, mass flows and pressures in the cycle are linked to the efficiency and power production of the ORC. As the ORC needs to be designed for a certain point with a specific fluid, running at a varying operation (off-design) can be a challenge for efficient operation of the ORC (Mondejar et al. 2016).

In this study we optimized the integration of the ORC for one engine load, which was considered as the design point. This choice considered that the engine was already optimised for low load conditions, such as slow steaming. The design-point

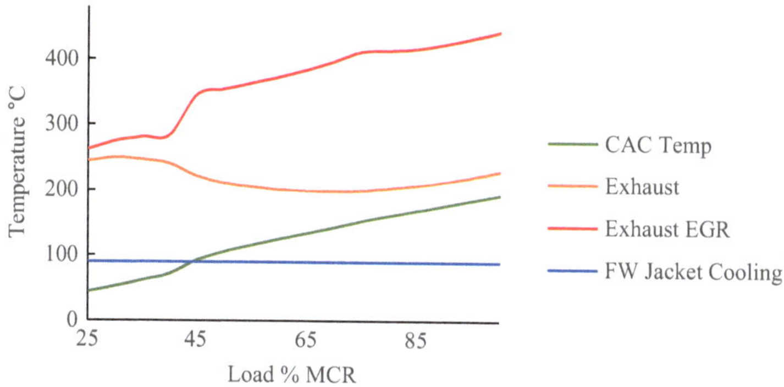


**Fig. 2** Temperature of heat sources MAN 6G50ME-C95 with varying load

load condition was set to 85%, which corresponds to a power of 5202 kW at ISO conditions.

## 2.2 *Simulation Software*

The software used for the simulations is IPSEpro (SimTech GmbH n.d.), and is a software used for calculating heat balances and simulating energy processes. It has an extensive library of standard components for thermal power, and also the ability to add components in a model development kit. The software is an equation solver, and can solve steady state simulations. IPSEpro is limited to single compound fluids and water and, in order to be able to use binary fluids in the simulation, an additional library was required. This library used Refprop (Lemmon et al. 2010) for fetching property data into the simulation using a DLL developed by Mondejar and Thern (2014). In order to optimise the integration, a Matlab (version 2013b, The Mathworks, Natick, MA) code was used to invoke the simulation via a COM-interface invoking the IPSEpro simulation.

## 2.3 *ORC Model*

An ORC is a Rankine cycle, which uses an organic fluid as the working fluid instead of water. As many organic fluids have lower boiling temperatures, the cycle can be used for lower temperature applications and, therefore, achieve more efficient cycles for lower temperature ranges (Lecompte et al. 2015). The basic system is comprised of a pump, an expander, condenser and an evaporator. The fluid is evaporated from condensate (liquid) phase to gas phase in the evaporator, and in the turbine (expander), mechanical power is delivered to the generator, which produces

**Table 1** Heat sources used in each case

| | Case 1 | Case 2 | Case 3 |
|---|---|---|---|
| Exhaust EGR | X | X | X |
| Exhaust after Turbo | X | X | X |
| Air CAC | X | X | – |
| FW Jacket water | X | – | – |

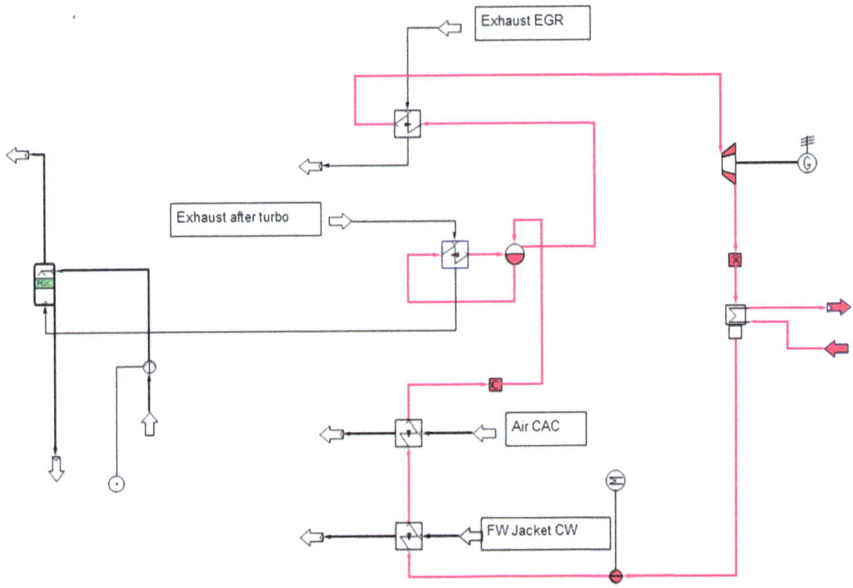


**Fig. 3** ORC Model. Case 1 with 4 heat exchangers

electrical power. The ORC evaporator pressure was optimized for six different fluids at an engine design point of 85% MCR.

Three different cases were considered: Case 1, utilising all available heat flows; Case 2, the FW Jacket water was excluded and, in Case 3, only the exhaust gas after the turbo and EGR were utilised.

When working with different heat source temperatures in the cycle, they can be integrated in different ways for each part of the cycle. As shown in Table 1, three cases were considered for this study, where all available heat sources were used in Case 1, ranging to using only exhaust gas EGR and exhaust gas in Case 3.

In Fig. 3, Case 1 is shown with the following heat sources; Exhaust EGR, Exhaust after the turbo, Air CAC and FW Jacket water. Therefore, four heat exchangers are transporting thermal energy from the sources to the ORC working medium. The heat exchangers were ordered in temperature range, with the lowest temperature first, up to the highest. The evaporation was set to happen in the heat exchanger from the exhaust gases after the turbocharger, and the two heat exchangers before heating the condensate fluid.

The last heat exchanger is for superheating, and is connected to the EGR flow, which is the source with highest temperature.

Both the condenser pump and scrubber pump isentropic efficiencies were set to 0.4, since this is a common and realistic value for industrial pumps (Yang 2016; Larsen et al. 2016). The heat exchangers for charge air and exhaust gases were set to a pinch point of 10 K, whilst the heat exchanger for jacket water was set to a 30 K temperature difference between input jacket water and output working fluid. The pinch point of 10 K was considered a reasonable value for the trade-off to keep the low heat exchanger area and an optimal mean temperature difference. The 30 K difference constraint was set due to limiting the amount of heat flow from the HT FW Jacket cooler, so not returning to low return-temperature to the cooling system, which could influence cold cylinder lining corrosion (MAN Diesel and Turbo 2016b). The pressure drop of all components was set to 0.01 MPa, which is a value in the standard order of magnitude for these systems.

The condenser pressure was set to constant 0.17 MPa in all simulation cases, as the sink temperature is normally constant. The condenser sub-cools the working fluid 4 K and is cooled by seawater with a temperature of 288.15 K (15 °C).

The net power output ($W_{net}$) is calculated from the difference between the turbine generator electrical output and the required pump work ($W_{pump_C}$) for the condenser pump and the sea water scrubber pump ($W_{pump_S}$) as shown in Eq. (1). The generator work ($W_{gen}$) is the electrical work from the generator coupled to the turbine, which has an isentropic efficiency of 0.75. The power consumption of the pump ($W_{pump}$) was calculated for an isentropic efficiency of 0.4. The heat flow from each individual heat exchanger ($Q_{HEX_n}$) is calculated with the mass flow ($m$) and difference in enthalpy between the inlet and the outlet ($h_{in} - h_{out}$) for the working liquid as shown in Eq. (2). All fluid properties for the simulations are calculated with Refprop (Lemmon et al. 2010).

$$W_{net} = W_{gen} - W_{pump_C} - W_{pump_S} \tag{1}$$

$$Q_{HEX_n} = m\left(h_{in} - h_{out}\right) \tag{2}$$

The ORC net efficiency ($\eta_{cycle}$) is obtained, as expressed in Eq. (3), by dividing the net power output of the cycle and the total heat input for all heat exchangers.

$$\eta_{cycle} = \frac{W_{net}}{Q_{HEX1} + Q_{HEX2} + Q_{HEX3} + Q_{HEX4}} \tag{3}$$

The temperature at the outlet of the turbo was limited to 287.15 K (140 °C) so sulphur condensation would not occur (Mondejar et al. 2016). The scrubber was modelled as a flue gas condenser, using seawater as the medium, spraying the exhaust gases. The seawater flow to the scrubber process was estimated at 75.89 kg/s, using the guidelines from the ABS Exhaust Gas Scrubber Advisory (ABS 2013).

The evaporator pressure was optimized using Matlab for maximum net power production.

### 2.4 Simulation Setup

Selecting the working fluid for a specific ORC application is essential for optimizing its performance. The fluids studied in this paper focus mostly on the thermodynamic perspective although some environmental aspects are also considered. In this sense the fluid must be chosen so that the characteristic temperatures required for the ORC operation are below the critical temperature of the fluid. This means that some of the fluids that were tested did not meet these requirements.

The following organic fluids were used in the simulation; isobutene, isopentane, toluene, R1233zd, R1234yf and R245fa. The fluids were chosen as initial candidates for this study based on previous work from Yang and Yeh (2015), Zhou et al. (2013) and Suárez de la Fuente and Greig (2015). The fluids R1233zd and R1234yf are considered more environmentally friendly due to their low GWP compared to R245fa.

## 3 Results and Discussion

Figure 4 represents the comparison of the net power output in cases 1–3 for the fluids with highest net power output. R245fa showed the maximum amount of power produced in Case 1. In Fig. 5, the cycle efficiency is presented, and R1233zd demonstrates the highest cycle efficiency. The maximum overall efficiency does not automatically correspond to the maximum net power production when dealing with several individual heat sources due to the combination of temperatures between the flows. As the heat source is considered waste and is normally dissipated into the environment, maximizing the cycle efficiency is of lower importance than maximizing the net power production. R1233zd has a GWP of 7 compared to R234fa of 1030

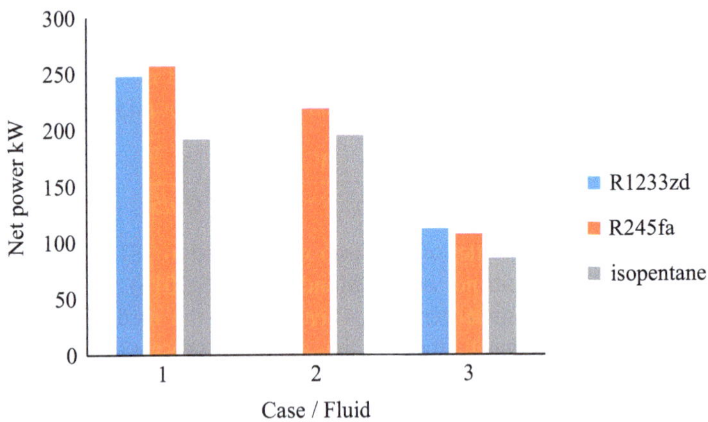


**Fig. 4** Comparison of the net power output for cases 1–3, for the fluids with highest net power output

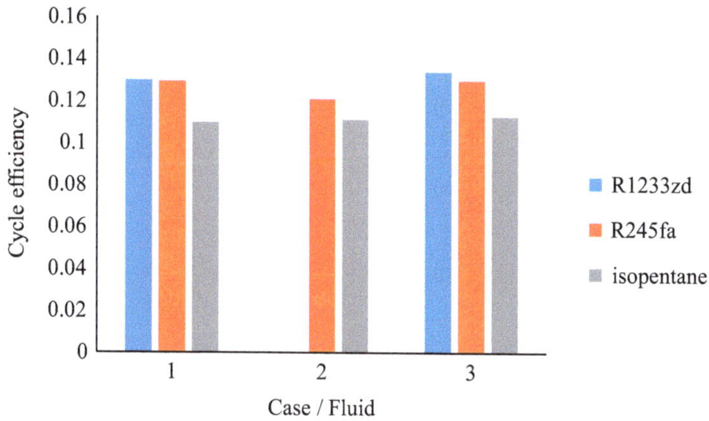


**Fig. 5** Cycle efficiency in cases 1–3

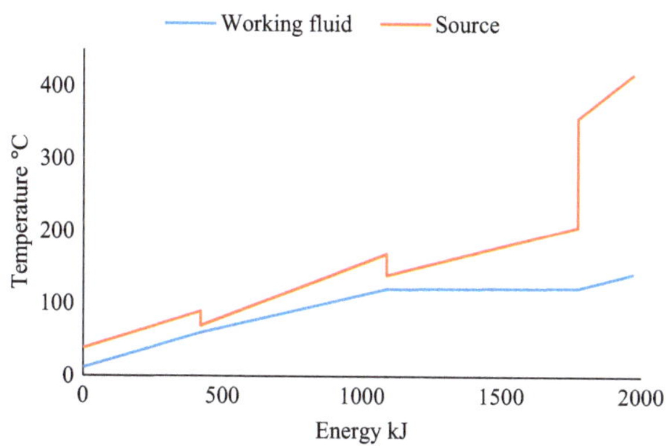


**Fig. 6** Temperature entropy diagram of heat flows, simulation case 1

(Yang and Yeh 2016) and a higher efficiency which makes it attractive for its application as working fluid of ORC.

The first heat exchanger (FW Jacket water) was a limiting factor in all simulations, as the amount of heat was limited due to risk of cold cylinder corrosion. According to the MAN data project data manual, the heat dissipation for cooling of jacket water is 1160 kW at 60 $m^3$/h (MAN Diesel and Turbo 2017), and the extracted power from the HEX 1 is below 420 kW in all cases. In all simulated cases the amount of heat was lower than the guidelines from the engine project guide (MAN Diesel and Turbo 2016a), which indicates that it is possible to take even more energy from this heat exchanger.

In Fig. 6, the temperature and enthalpy diagram is shown for Case 1. There are four different heat sources presented in the diagram, which are ordered from lowest to highest temperature; FW Jacket water, Charge Air Cooler, Exhaust gas after turbo and exhaust EGR as the highest temperature. As can be observed, there is a

Fig. 7 Scrubber, ORC and net power production

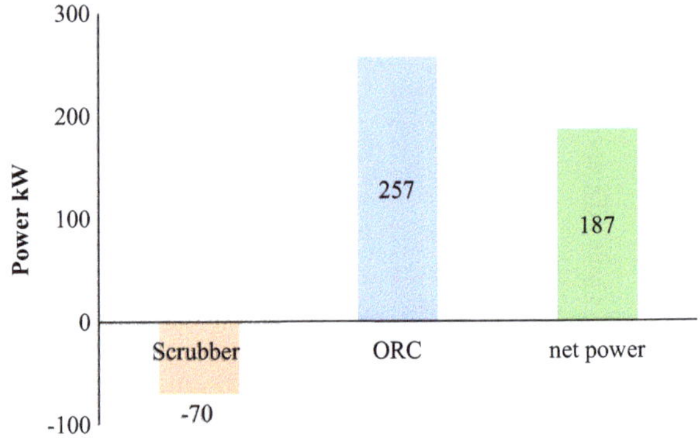


significant difference between the temperature of the exhaust gases and that of the working fluid at the turbine inlet, which means that more potential for integration is available for this heat source. In this sense, it would still be possible to use part of the exhaust heat for other heating needs on board, while converting the rest into power by means of the ORC.

The temperature after the exhaust gas heat exchanger, evaporator, was limited to 287.15 K (140 °C) so sulphur condensation would not occur, but as the scrubber removes sulphur anyway it might be possible to have a tighter integration and choose a material that can withstand the acidic environment. In this case, more power could be produced, but this needs to be addressed in a future study.

The complete design of this integration would be more complex if considering the different individual heat sources, and if the design complications and extra cost of installing additional heat exchangers into the existing engine design were considered. The study is, in this sense, only thermodynamic and theoretical, so the economical and practical issues need to be addressed in a future study.

The net power production in Case 1 with R245fa is 257 kW, and the estimated power for the scrubber operation is 70 kW based on the ABS Scrubber Advisory (ABS 2013). This implies that there is an excess of electrical power production of 187 kW, and even considering the lower ORC power production in Case 3, it still exceeds the needed power for a scrubber operation (see Fig. 7).

## 4 Conclusions

In this work, we presented a simulation of an ORC integrated with heat sources from a low-speed two-stroke marine engine equipped with exhaust gas cleaning for both nitrogen and sulphur oxides. The results showed that an ORC integrated can produce a significant amount of electrical power, and for this specific installation, an excess power of 184 kW was obtained. We can conclude that an ORC integrated into an engine with both EGR and a scrubber can cover and provide excess electrical power to the Scrubber installation, hence making the system more efficient and in fulfilment of both the Tier III standard for $NO_x$ and the sulphur directive.

**Acknowledgments** We would like to acknowledge MAN Diesel & Turbo for providing us with engine data. Acknowledgments also go to the Swedish Maritime Administration and Lund University for funding.

## References


ABS. (2013). *Exhaust gas scrubber systems advisory*. Available at: http://ww2.eagle.org/content/dam/eagle/publications/2013/Scrubber_Advisory.pdf.

Ahlgren, F., Mondejar, M. E., Genrup, M., & Thern, M. (2015). Waste heat recovery in a cruise vessel in the Baltic Sea by using an organic Rankine cycle: A case study. *Journal of Engineering for Gas Turbines and Power, 138*(1), 11702. Available at: http://gasturbinespower.asmedigitalcollection.asme.org/article.aspx?doi=10.1115/1.4031145.

European Environment Agency. (2013). *The impact of international shipping on European air quality and climate forcing*. Copenhagen. Available at: http://trid.trb.org/view.aspx?id=1248012. Accessed October 12, 2014.

Georgopoulou, C. A., Dimopoulos, G. G., & Kakalis, N. M. (2015). Modelling and simulation of a marine propulsion power plant with seawater desulphurisation scrubber. *Proceedings of the Institution of Mechanical Engineers, Part M: Journal of Engineering for the Maritime Environment, 230*(2), 341–353. Available at: http://pim.sagepub.com/lookup/doi/10.1177/1475090215571377.

IMO. (2010). RESOLUTION MEPC.176(58) (Revised MARPOL Annex VI), 176(July), pp. 1–45.

International Maritime Organization (IMO). (2014). Resolution MEPC.251(66) amendments to the annex of the protocol of 1997 to amend the international convention for the prevention of pollution from SHIPS, 1973, as modified by the protocol of 1978 relating thereto.

International Maritime Organization (IMO). (2015). Nitrogen Oxides (NOx) – Regulation 13.

Jiang, L., Kronbak, J., & Christensen, L. P. (2014). The costs and benefits of sulphur reduction measures: Sulphur scrubbers versus marine gas oil. *Transportation Research Part D: Transport and Environment, 28*, 19–27. Available at: http://linkinghub.elsevier.com/retrieve/pii/S1361920913001491. Accessed November 10, 2014.

Kalli, J., Karvonen, T., & Makkonen, T. (2009). *Sulphur content in ships bunker fuel in 2015. A study on the impacts of the new IMO*. Helsinki.

Larsen, U., Wronski, J., Andreasen, J. G., Baldi, F., & Pierobon, L. (2016). Expansion of organic Rankine cycle working fluid in a cylinder of a lowspeed two-stroke ship engine. *Energy, 119*, 1212–1220.

Lecompte, S., Huisseune, H., van den Broek, M., Vanslambrouck, B., & De Paepe, M. (2015). Review of organic Rankine cycle (ORC) architectures for waste heat recovery. *Renewable and Sustainable Energy Reviews, 47*, 448–461. Available at: http://linkinghub.elsevier.com/retrieve/pii/S1364032115002427.

Lemmon, E. W., Huber, M. L., & McLinden, M. O. (2010). NIST standard reference database 23. NIST reference fluid thermodynamic and transport properties—REFPROP, 9.1.

Lindstad, H. E., & Eskeland, G. S. (2016). Environmental regulations in shipping: Policies leaning towards globalization of scrubbers deserve scrutiny. *Transportation Research Part D: Transport and Environment, 47*, 67–76. https://doi.org/10.1016/j.trd.2016.05.004.

MAN Diesel & Turbo. (2012). Tier III two-stroke technology (p. 36).

MAN Diesel & Turbo. (2016a). G50ME-C9.5 project guide.

MAN Diesel & Turbo. (2016b). Two-stroke service experience.

MAN Diesel & Turbo. (2017). CEAS engine data report 6G50ME-C9.5 with EGR.

Mondejar, M. E., & Thern, M. (2014). A new IPSEpro® library for the simulation of binary mixtures of real fluids in power cycle analysis. *Journal of Postdoctoral Research, 2*(3), 1–6.
Mondejar, M. E., Ahlgren, F., Thern, M., & Genrup, M. (2016). Quasi-steady state simulation of an organic Rankine cycle for waste heat recovery in a passenger vessel. *Applied Energy, 119*, 1212–1220. Available at: http://linkinghub.elsevier.com/retrieve/pii/S030626191630335X.
Nielsen, R. F., Haglind, F., & Larsen, U. (2014). Design and modeling of an advanced marine machinery system including waste heat recovery and removal of sulphur oxides. *Energy Conversion and Management, 85*, 687–693. Available at: http://linkinghub.elsevier.com/retrieve/pii/S0196890414002301. Accessed August 26, 2014.
Ölçer, A., & Ballini, F. (2015). The development of a decision making framework for evaluating the trade-off solutions of cleaner seaborne transportation. *Transportation Research Part D: Transport and Environment, 37*, 150–170. Available at: http://linkinghub.elsevier.com/retrieve/pii/S1361920915000553.
Panasiuk, I., & Turkina, L. (2015). The evaluation of investments efficiency of SOx scrubber installation. *Transportation Research Part D: Transport and Environment, 40*, 87–96. Available at: http://linkinghub.elsevier.com/retrieve/pii/S1361920915001091.
SimTech GmbH. IPSEpro 6.0 U, Release 12, Build 1376. Available at: www.simtechnology.com.
Suárez de la Fuente, S., & Greig, A. (2015). Making shipping greener: Comparative study between organic fluids and water for Rankine cycle waste heat recovery. *Journal of Marine Engineering & Technology, 14*(2), 70–84. https://doi.org/10.1080/20464177.2015.1077601.
Sweco. (2012). *Consequences of the sulphur directive*. Stockholm. Available at: http://scholar.google.com/scholar?hl=en&btnG=Search&q=intitle:Consequences+of+the+Sulphur+Directive#1. Accessed November 27, 2014.
Wang, E., Yu, Z., Zhang, H., & Yang, F. (2017). A regenerative supercritical-subcritical dual-loop organic Rankine cycle system for energy recovery from the waste heat of internal combustion engines. *Applied Energy, 190*, 574–590. Available at: http://linkinghub.elsevier.com/retrieve/pii/S0306261916319055.
Yang, M.-H. (2016). Optimizations of the waste heat recovery system for a large marine diesel engine based on transcritical Rankine cycle. *Energy, 113*(August), 1109–1124. Available at: http://linkinghub.elsevier.com/retrieve/pii/S0360544216310854.
Yang, M.-H., & Yeh, R.-H. (2015). Thermodynamic and economic performances optimization of an organic Rankine cycle system utilizing exhaust gas of a large marine diesel engine. *Applied Energy, 149*, 1–12. Available at: http://linkinghub.elsevier.com/retrieve/pii/S0306261915003852.
Yang, M.-H., & Yeh, R.-H. (2016). Economic performances optimization of an organic Rankine cycle system with lower global warming potential working fluids in geothermal application. *Renewable Energy, 85*, 1201–1213. Available at: http://linkinghub.elsevier.com/retrieve/pii/S0960148115301622.
Zhou, N., Wang, X., Chen, Z., & Wang, Z. (2013). Experimental study on organic Rankine cycle for waste heat recovery from low-temperature flue gas. *Energy, 55*, 216–225. Available at: http://linkinghub.elsevier.com/retrieve/pii/S0360544213002429. Accessed August 20, 2014.

# Lighting Standards for Ships and Energy Efficiency

Efe Akyurek

## 1 Introduction

From the beginning of the nineteenth century, humankind has been under the effect of electric lighting. Electric sourced lighting is easy compared to old fashioned ways such as kerosene and candles. Besides, lighting has become more and more vital in human lives and is being researched in almost every aspect. While it is debatable whether artificial lighting has a beneficial effect on human health or not, standardized lighting probably contributes into a better morale and motivation for crew; human comfort must be included in that equation.

In general, vessels are being illuminated by generator sourced electric powered light bulbs and/or fluorescents. Nowadays, common lighting techniques are being replaced by LED (Light Emitting Diode) technology, which provides long-lasting and more efficient lighting. The lighting technology on ships also sees its share of LED technology since the maritime industry targets more efficiency in general. The shift from incandescent bulb to fluorescent light has occurred slightly and has not been researched in terms of its efficiency aspect.

A literature review reveals that shipboard lighting has been a trending topic, especially efficient lighting, since the term "green shipping" appeared. Green ships supported by LED lighting technology reduce energy consumption and require less energy for the same lighting level. LED lights deliver at least twice as much efficiency as fluorescent lights and are almost ten times more efficient than incandescent bulbs (Lighting the Way for a Greener Ship 2010).

Light and eye adaptation are associated with human comfort. A successful luminance measure of the key workstations such as the ship bridge was carried out by Wynn et al. (2012). They examined the chartroom illumination of the bridge to

E. Akyurek (✉)
Selami Ali Mah. Tank Sok, Uskudar/Istanbul, Turkey
e-mail: akyureke@itu.edu.tr



A. I. Ölçer et al. (eds.), *Trends and Challenges in Maritime Energy Management*, WMU Studies in Maritime Affairs 6, https://doi.org/10.1007/978-3-319-74576-3_13

reveal dark adaptation. The study indicated that the chartroom had two table lamps, both fitted with 40 W incandescent bulbs and the authors measured lux values in various places in the ship's bridge to determine whether lookouts were effectively dealing with the issue of "adaptation to darkness".

In addition, Herrmann (1984) developed a detailed guideline table for ship illumination for various places in 'The Ship's Lighting'. Even though the guideline is well explained with general remarks and points worthy of special attention, the source of lux value is not determined and tested on ships. As a result, a framework is drawn on lighting in general, considering the human eye, lux values and light adaptation. Despite shipboard illuminance maintaining its importance in terms of energy efficiency, human comfort and occupational safety, luminous values are still not sufficiently defined by a majority of Flag Administrations and classification societies to allow the replacement of lighting equipment onboard.

Hereby, the current analysis aims to be unique in drawing attention to how energy efficiency emerges by providing natural lighting onboard. Unfortunately, because of the lack of lighting standards for ships, this analysis initially defines optimal lux values for ships as considering human comfort by adopting ISO 8995-1:2002(en) Lighting of work places—Part 1: Indoor. The first section explains maritime based regulations on lighting, and the second section applies International Organization Standards for work places on a river–type ship. The third section gives an example of how natural lighting could be effective in ship lighting. Finally, the last section reveals the possibility and potential of energy efficiency with increased use of natural lighting onboard.

## 2 Maritime Regulations on Lighting Standards

Optimized lighting on board is an important aspect for crews' daily life, morale and eagerness to work. Early research on lighting of ships was initiated by the United States Naval Industry. The Department of Defense Naval Sea Systems Command (Department of Defense 1986) published a handbook on Lighting on Naval Ships as a pioneering study that initially designates human eye capabilities and continues by defining and determining aspects related to lighting. The handbook provides clear standards for lighting of naval vessels, dividing lighting into general lighting requirements and detailed lighting requirements. Furthermore, Elliott and Kobus (1992) carried out a study on Standardizing Shipboard Lighting: Light Fixtures and Light Bulbs on U.S. Navy Ships. The study recommended to fully standardizing shipboard lighting to optimize human performance on naval ships. Furthermore, ABS (American Bureau of Shipping) issued a guide for Crew Habitability on Ships in December 2001. The guide aims to set assessment criteria and describes the measurement methodology for obtaining a crew habitability notation. The sixth section of the guide sets several criteria for crew accommodation spaces, navigation and control spaces, service spaces, operating and maintenance spaces and red or low-level white illuminance. However, the guide covers only main source lighting lux values, excluding emergency lighting. Also, the guidance excludes the impact of daylight.

On a regulatory basis, the pioneering standards for lighting on merchant ships are mentioned in the Convention concerning Crew Accommodation on Board (ILO 92), which entered into force on 29th January 1953. The lighting requirement basically states that all crew spaces shall be adequately lighted. However, the term "adequate lighting" could be considered as subjective. In addition, the ILO Convention concerning Crew Accommodation on Board Ship (Supplementary Provisions) entered into force on 27th August 1991, without clarifying the confusion concerning adequate lighting. Finally, the Maritime Labor Convention (MLC) 2006 entered into force on 20th August 2012. Yet, neither the Standard section of the convention nor the guidelines define lighting standards for ships. Table 1 provides a comparison of all international maritime conventions regarding the lighting standards of ships.

Moreover, ILO published a Guideline (ILO 1996) as a summary of the 254th Session with 15 experts and the following remarks have been noted:

9.4.1. Areas of the ship used for loading or unloading, other work processes or transit should be adequately and appropriately illuminated.
9.4.2. Lighting should be reasonably constant and arranged to minimize glare, dazzle and the formation of deep shadows and sharp contrasts between one area and another.
9.4.3. Broken or defective lights should be reported immediately and repaired as soon as practicable.
9.4.4. It should be prohibited to enter unlighted or inadequately lighted places on the ship without safe portable lights.
9.4.5. Lights, both fixed and portable, should be checked to ensure proper operation and safe rigging prior to use. No operation should be permitted to commence or continue if lighting is insufficient (ILO 1996).

Besides, SOLAS (Safety of Life at Sea), Chapter II/1 Part D Regulation 41.2.1 states that, "*the main lighting system which shall provide illumination throughout those parts of the ship normally accessible to and by passengers or crew shall be supplied from the main source of electrical power.*" In addition to this, ships shall have an emergency lighting system that provides lighting when the main lighting system fails. However, the regulations only explain the lighting period depending on the ship type. There is no evidence or guidance for either main or emergency lighting luminance values.

Overall, on a regulatory basis, a few Administrations have minimum lighting standards for ships, such as the Isle of Man. According to the Maritime Labor Notice (MLN) 3.1 Accommodation and recreational facilities April 2013—Revision 2—Annex I, standards of artificial lighting/minimum requirements of luminance are defined for Isle of Man registered ships. However, assigned minimum illuminance requirements for some places such as passageways, sleeping rooms, and mess rooms are below the International Standards for minimum illuminance levels. Furthermore, there is no inspection procedure regarding luminance levels. Despite many studies

**Table 1** Comparison of lighting standards of ILO 92, 133, MLC 2006 and SOLAS

| ILO 92 | ILO 133 | MLC 2006 | SOLAS |
|---|---|---|---|
| Article 9 | Article 11 | Standard A.3.1 & Guidelines B3.1.4 | Chapter II/1 Part D Regulation 41.2.1 |
| 1. Subject to such special arrangements as may be permitted in passenger ships, sleeping rooms and mess rooms shall be properly lighted by natural light and shall be provided with adequate artificial light | 1. Crew accommodation shall be properly lighted | *Standard A.3.1.8.* With respect to requirements for lighting, subject to such special arrangements as may be permitted in passenger ships, sleeping rooms and mess rooms shall be lit by natural light and provided with adequate artificial light | Main lighting system shall provide illumination throughout those parts of the ship normally accessible to and by passengers or crew shall be supplied the main source of electrical power |
| 2. All crew spaces shall be adequately lighted. The minimum standard for natural lighting in living rooms shall be such as to permit a person with a normal vision to read on a clear day an ordinary newspaper in any part of the space available for free movement. When it is not possible to provide adequate natural lighting, artificial lighting of the above minimum standard shall be provided | 2. Subject to such special arrangements as may be permitted in passenger ships, sleeping rooms and mess rooms shall be lighted by natural light and shall be provided with adequate artificial light | *Guidelines B3.1.4* | Ships shall have an emergency lighting system that provides lighting when the main lighting system fails |
| 3. In all ships electric lights shall be provided in the crew accommodation. If there are not two independent sources of electricity for lighting, additional lighting shall be provided by properly constructed lamps or lighting apparatus for emergency use | 3. In all ships electric light shall be provided in the crew accommodation. If there are not two independent sources of electricity for lighting, additional lighting shall be provided by properly constructed lamps or lighting apparatus for emergency use | 1. In all ships, electric light should be provided in the seafarer accommodation. If there are not two independent sources of electricity for lighting, additional lighting should be provided by properly constructed lamps or lighting apparatus for emergency use | |
| 4. Artificial lighting shall be so disposed as to give the maximum benefit to the occupants of the room | 4. In sleeping rooms an electric reading lamp shall be installed at the head of each berth | 2. In sleeping rooms an electric reading lamp should be installed at the head of each berth | |

(continued)

**Table 1** (continued)

| ILO 92 | ILO 133 | MLC 2006 | SOLAS |
|---|---|---|---|
| Article 9 | Article 11 | Standard A.3.1 & Guidelines B3.1.4 | Chapter II/1 Part D Regulation 41.2.1 |
| 5. In sleeping rooms an electric reading lamp shall be installed at the head of each berth | 5. Suitable standards of natural and artificial lighting shall be fixed by the competent authority | 3. Suitable standards of natural and artificial lighting should be fixed by the competent authority | |

on minimum standards of illuminance for ships, on a regulatory basis, there is a gap between providing optimal lighting and energy efficiency with lighting.

On the other hand, water integrity for ships and construction is another obstacle for illumination. The aim of providing water integrity prevents construction of more openings, portholes and windows because, in ships, all portholes bring risk of water leakage.

## 3 Standardization of Lighting, Optimal Lighting and Possible Lighting Standards for a Ship

The history of lighting standards started with the oil embargo in the United States in 1973 by revealing the importance and the value of energy. For the first time in the history, President Nixon declared an energy efficiency policy via the Office of Energy Conservation within the Department of the Interior, and one of the aims was to reduce federal energy consumption by 7% (Savitz 1986). When the oil embargo was raised, President Nixon ordered government buildings (and those of the private sector, preferably) to keep in line with lighting levels of approximately 500 lux for office work; approximately 300 lux for general lighting/hallways; and approximately 100 lux for parking lots. Following the embargo, energy efficiency programs spread to other departments and most of the projects focused on the construction industry, related to more energy-efficient heating, cooling, and lighting systems (Kassakian 2013). Nowadays, standardization organizations such as ISO (International Standardization Organization), and BSI (British Standards Institution) have illumination standards depending on the space such as ISO 8995-1:2002 (CIE S 008/E:2001) Lighting of work places—Part 1: Indoor, ISO 8995-3:2006 (CIE S 016/E:2005) Lighting of work places—Part 3: Lighting requirements for safety and security of outdoor, BS EN 12193 Light and lighting—Sports lighting. This paper initiates lighting standards for merchant ships based on ISO 8995-1:2002 (CIE S 008/E:2001) Lighting of work places—Part 1: Indoor.

Based on ISO standards and previous studies, possible lighting standards could be as follows:

**Table 2** Possible lux values for ships based on ISO standards compared to present studies

| Ships place | Lux value for a river—type ship (Adopted ISO Standards) | Lux value for merchant ships(Isle of Man Circular) | Lux value for naval ships (Military Handbook) |
|---|---|---|---|
| Bridge general | 200 (in port) | – | 75.3 |
| Passages | 100 | 50 in general | 75.3 |
| Stairs | 150 | 50 in general | 75.3 |
| Sanitary areas | 200 | 100 in general | 150.6 |
| Cabins | 100 | 50 in general | 75.3 |
| Hospital | 500 | 50 in general | 301.3 |
| Ships office | 300 | 100 in general—200 at desk | 301.3 in general—451.9 at desk |
| Laundry spaces | 300 | 100 | |
| Galley | 300 | 300 | 301.3 |
| Ships stores | 100 | 100 | 32.3 |
| Mess rooms | 150 | 50 in general—150 on tables | 150.6 |
| Provision store | 100 | 100 | |
| Steering gear room | 200 | – | – |
| Engine control room | 200 | – | 226 |
| Engine room general | 200 | – | 150.6 |
| Engine workshop | 300 | – | 301.3 |
| Paint room | 100 | – | – |
| Bow & Stern Thruster Room | 200 | – | – |
| $CO_2$ room | 200 | – | – |

Table 2 indicates that not all spaces in a merchant ship are defined, despite several attempts. There are various ship types with specific spaces, such as cargo control rooms, cages for livestock carriers, etc., but the majority of ships have spaces in common, such as accommodation and engine spaces.

Optimal lighting is essential for human physiology. According to Dr. Ott (Ott Biolight Systems, Inc. 1997), the human body uses light as a nutrient for metabolic processes similar to water or food. Natural light stimulates essential biological functions in the brain and is divided into colors which are vital to people's health. On a cloudy day or under inadequate lighting conditions, the inability to perceive the colors from light can negatively affect people's moods and energy levels (Edwards and Torcelini 2002).

Working onboard has inherent risk factors; however, inadequate lighting and inaccurate color rendering of work spaces increases the risk factor and the number of accidents. Successful research has been carried out on different lighting levels and 347 accidents by Völker et al. (1995). The study investigates accidents that occurred in working spaces by type and associated them with lux levels. The result clearly indicates that 650–749 lux band results in burn and scald types of injuries, but the number of wounded persons by variety of injury types is irregular under different lux values. The general tendency in the injuries decreases slightly and reaches the lowest value at 950–1049 lux band. Consequently, the study clearly indicates that illuminance is important for occupational safety of work.

In addition, color rendering is important and stroboscopic effects of lighting can be dangerous in locations where it is important to accurately see moving parts of machinery. The danger is completely eliminated by high-frequency operated lighting (Bommel et al. 2002).

Related to this, lighting pollution caused by artificial light also has unwanted effects on human health. Lighting pollution could be defined as excessive, misdirected or obtrusive artificial light. Lighting pollution mostly occurs at night time, when pineal melatonin production decreases; additionally, at this time-period, the circadian clock alters, which may impact performance, alertness, sleep and metabolic disorders. Finally, lighting pollution, affecting human health, is caused by the following (Falchi et al. 2011):

- Luminaires that light directly at and above the horizontal;
- Wasted downward light flux outside the area to be lit;
- Over lighting;
- Lights are constantly lit;
- Growth of the total installed flux;
- Short wavelength 'blue' light (Falchi et al. 2011).

Another study carried out on lighting indicates not only that light pollution affects human physiology but that lack of daylight is related to depressive symptoms and poor quality of sleep (Harb et al. 2014). To extend this, increased access to natural light and window views is likely to have positive effects on psychology and well-being (Heerwagen 2000). Surely, light is the inevitable element of life and artificial light can serve the needs of the crew only up to a certain level.

## 4 Energy Efficiency Practices with Combination of Natural and Artificial Lighting

The previous sections aimed to define standards for lighting under international maritime conventions such as SOLAS and ILO. Since these conventions are not sufficient to define a detailed lighting plan that indicates minimum lux values, the study continues to explain and adopt ISO Standards and emphasize optimal lighting effects on crew. This vital section is intended to clarify how energy efficiency could be provided with electricity by means of lighting.

Lighting is a wide-spread topic discussed within case studies on ships, yet most studies only highlight the comparison between conventional lighting and advanced lighting systems such as LED, ignoring human comfort and the beneficial effects of natural lighting. Due to onboard living conditions and safety measures, effective illumination has always been a must.

Before incandescent light was invented, lighting onboard was provided by kerosene lamps at night time. Kerosene lamps were not sufficient to illuminate the designated spaces effectively and when considering output lighting energy divided by input kerosene energy, kerosene lamps were not energy efficient. Moreover, kerosene lamps were a potential fire risk that was not easy to extinguish in the marine environment. Besides, below decks of the ships also needed to be illuminated during the daytime because openings on the ship's side bring flooding risks. To eliminate this risk, ship builders fitted several skylight lobbies usually located on the main deck to provide illumination in daytime. This technology was totally energy efficient since no energy consumption was required to illuminate below decks during daytime.

Unfortunately, skylights and kerosene lamps were not stable as sources of lighting energy. For this reason, the invention of the incandescent bulb was revolutionary because it provides stable lighting when needed. This change in lighting technology radically affected life today. Since the traditional incandescent bulb is not energy efficient, it transforms most of the electricity into heat energy rather than lighting. By means of developing technology, the incandescent light bulb yielded to florescent and LED technology. Now, energy efficient lighting comes one step further because the incandescent light bulb and other sources of lighting require a continuous energy source to be lit. Electricity energy on a ship is mainly provided by generators that commonly consume fossil fuel sources. When each electricity lighting source runs, it increases induction on the generator, causing more fuel consumption. Even though lighting becomes stable, generators create more electricity energy than actually needed for lighting purposes.

At that point hybrid lighting could increase energy efficiency onboard. If the old skylight technology revives, lighting the ship during the daytime will require less electricity/energy. Even lighting obscure spaces up to standardized lux values requires less electricity energy. Moreover, lighting standards maintain their stabilization during day and night time. By the aspect of onboard lighting, if the consumed energy is fixed for lighting onboard and increases the lux value by additional skylights, energy efficiency increases.

Effort in energy efficient lighting by alternative methods is not only a matter for ships. It is also a matter for lighting technology of buildings. For instance, the following several methods could be used to minimize lighting energy demand for buildings:

- Eliminate excessive lighting by reducing the total lamp wattage in each activity area
- Substitute higher-efficiency lamps in existing fixtures
- Modify existing fixtures to reduce light trapping and/or improve light distribution
- Install skylights or light pipes (Wulfinghoff 2004).

On the other hand, it is an improper approach to turn off lighting sources to conserve energy. In this way, crews also conserve energy by turning off unnecessary lights, but this case is out of scope of this paper.

Briefly, defining lighting standards for ships is the first step to reflect current lighting regulations on ships. Monitoring hybrid lighting is the second step to provide both onboard safety and energy efficiency.

At this juncture, natural light supported skylights bring innovative lighting solutions for the ships' enclosed spaces. A river–type vessel has a glass skylight in the mess room as indicated in Fig. 1. Even though the glass skylight cannot deliver fresh air, it is a feasible solution to illuminate the room with natural light without compromising the water integrity of the ship.

River type ships shall close portholes with a deadlight to support water integrity during the voyage; thus, glass skylights continue shining with the sun. Besides, the sunlight supports the morale and health of the crew, apart from energy efficiency.

Energy efficiency could be defined as providing the same service while using less energy. With regard to this ship, with the help of a glass skylight, the mess room could be illuminated by the same lux value while using less fluorescent light. A standard 18 W T8 fluorescent lamp consumes at least 18 W per hour (excluding ballast). In this case, the ship is about 30 years old and a never turned off common 18 W T8 fluorescent light (such as mess room light, or a passageway light) consuming 4.733 kW (excluding ballast) of energy. The mess room of the ship has six pieces of fluorescent lighting where 28.400 kW are consumed over 30 years. The ship has three main generators of 105 kW each, and each generator consumes 200 L of MGO (Marine Gas Oil) per day. As an assumption, when one generator consumes 200 L of MGO constantly, a single generator should work for 270 h with the consumption of 2250 L of MGO for these six lamps during the ship's life.

Figures 2 and 3 indicate the active skylight from different perspectives. There is a designated space for the skylight that is painted white to reflect daylight. The general lux value of the mess room table is 140 lux with six pieces 18 W T8 type fluorescent lamp and one active skylight. While lights are switched off, the skylight provides 30 lux illuminance on the table in general for the mess room and provides sufficient lighting for at least general visibility. Without the skylight, another minimum 18 W fluorescent lamp would be necessary for the center of the mess room.

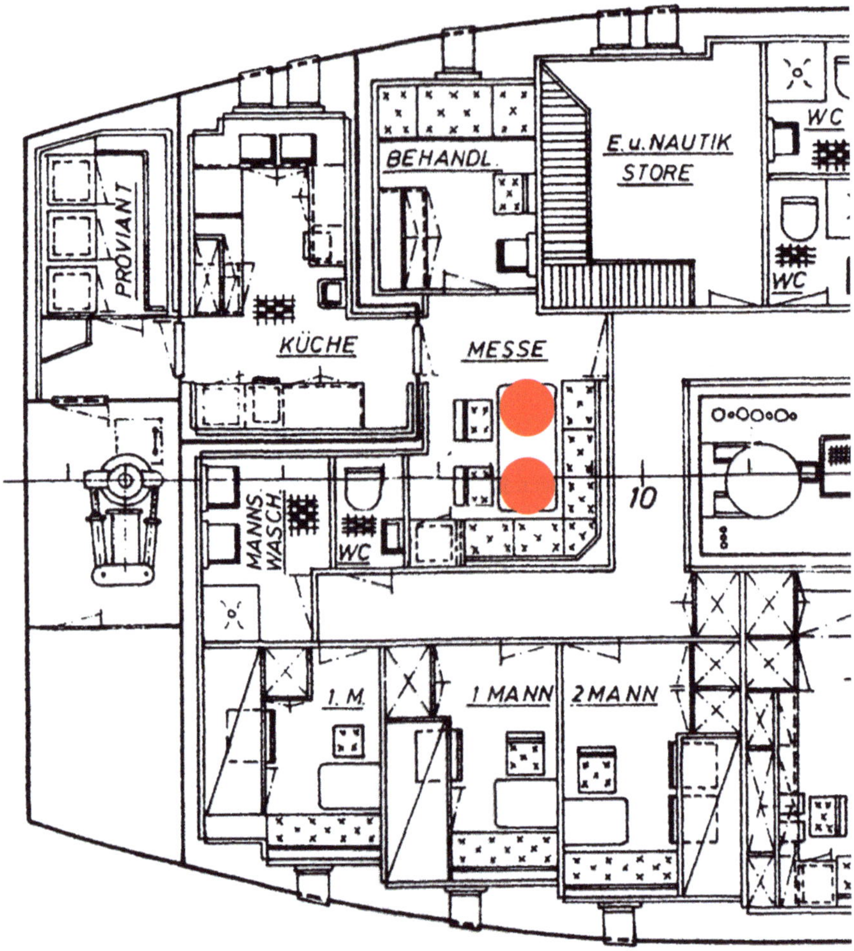


**Fig. 1** Location of portholes and glass skylights on ship (marked in red color) (Source: Original drawing of M/V Salix IMO No: 8520446)

A single skylight produces approximately 30 lux in daytime and provides continuous illuminance of about 8 h. With the help of the skylight, the reduction of one fluorescent light in the mess room during daytime provides energy efficiency while delivering approximately the same lux value with less energy. Moreover, as the skylight is placed under an artificial light, it continues working during the night time.

Regarding this, if all of the lighting fixtures are supported with an additional skylight, 18 fixtures (1 for steering gear room, 1 for galley, 1 for $CO_2$ room, 1 for

**Fig. 2** One working glass skylight in daytime (Source: Authors archive, Messroom of M/V Salix IMO No: 8520446)

**Fig. 3** Illumination of one glass skylight and one porthole in daytime (Source: Authors archive, Messroom of M/V Salix IMO No: 8520446)

**Fig. 4** The outside view of the skylight (Source: Authors archive, aft deck of M/V Salix IMO No: 8520446)

officer mess room, 1 for office, 1 for provision room, 1 for engine room store, 1 for deck store, 1 for crew mess room and 1 for common shower area plus 8 cabins) could be closed during the daytime (calculated around 8 h), which could save 28,382.4 kW over the vessel's economic life of 30 years, and support crew motivation and health.

However, skylights have some limitations as providing shipboard water integrity is an important factor. Glass may be fragile and is as expensive as 1 ton of MGO (Marine Gas Oil). Skylights also require frequent maintenance: for instance, as indicated in the example ship, the outside of the skylight is in poor condition. An average installation price for a skylight, as indicated in Fig. 4, is between €150 and €350, depending on the size and quality of the glass. During scrapping, painting and carrying goods on deck, this equipment needs to be covered and removed if necessary to protect the skylight.

Consequently, the use of skylights differs from energy conservation, which requires switching off artificial lights when not in use. However, skylight supported lamp fixtures bring energy efficiency without compromising the illuminance values. Even though skylights provide limited luminance, they save remarkable energy while increasing the morale and motivation of the crew.

# 5 Conclusion

In summary, while poor lighting brings several risks for crew, excessive lighting causes physical and psychological issues. On this basis, adequate lighting comes one step further together with the necessity of lighting standardization. This study initially searches literature and existing international maritime regulations for lighting standards as the basis of a detailed lighting proposition that indicates lux values of several common areas of ships. Since the aim is to achieve energy efficiency, it is fundamental to define the minimum energy necessary for lighting.

The International Maritime Organization (IMO) and International Labor Organization (ILO) set standards for the maritime industry and shoulder a great work load while these standards determine the minimum requirements for the industry. The aspect of shipboard illumination is also related to the SOLAS and ILO Conventions, since lighting is related to safety, security and the human element. Currently, ILO has the Maritime Labor Convention in force, and SOLAS has a lighting section that regulates lighting condition for ships. Unfortunately, the regulations related to lighting come short of the 'adequate lighting' definition from the Maritime Labor Convention. The term "Adequate lighting" does not explain minimum standards, and it is an ambiguous definition without a guideline for minimum illumination values. In addition to the critical role of lighting based energy efficiency, this paper also proposes that lighting for ships shall be regulated and inspected with the determined lux values for both work and living places.

Since sailing ships were indeed the most energy efficient ships, natural lighting was important in yesterday's technology. As soon as technology brought more control of lighting, the need for natural lighting was neglected. However, natural lighting is mostly poor for an average river-type ship because its freeboard is very low and it is hard to illuminate with portholes during open sea voyage. As a result, most of the lights onboard are constantly turned on. On this basis, skylights provide two solutions as they easily deliver natural light to designated spaces and provide water integrity.

A lux measurement has been carried out in the mess room of the ship, which is illuminated with 140 lux together with 6 florescent lights and a skylight. Additionally, a skylight alone provides 30 lux during daytime with zero electricity consumption. This can save 28,382.4 kW during the vessel's economic life of 30 years. A fixture supplied with a skylight requires less energy and illuminates the place without compromising standardized lux values. It also increases crew motivation, since it delivers natural light inside crew accommodation. Therefore, hybrid ways (combination of natural and artificial lighting) of lighting offer one of the best solutions for onboard lighting while increasing human comfort.

Energy efficiency is a trending topic and this study aims to indicate that, even with daylight, it becomes possible to provide energy efficiency on ships. For further research, lighting standards could be extended by creating a lighting plan for a ship and modelling lighting fixtures that are optimized for narrow spaces. On the energy efficiency aspect, other human comfort elements such as ventilation and thermal comforts could be researched in more detail. Besides, further research may seek a solution for placing natural lighting points on multi-decker ships.

**Acknowledgements** The author owes special thanks to Prof. Dr. Aykut I. Ölçer from World Maritime University; additionally, he would like to thank the review committee for their valuable comments and critics.

## References

Bommel, W. J. M., Beld, G. J., & Ooyen, M. H. F. (2002). *Industrial lighting and productivity*. Philips. http://www.iar.unicamp.br/lab/luz/ld/Arquitetural/interiores/ilumina%E7%E3o%20industrial/industrial_lighting_and_productivity%5B1%5D.pdf. Accessed on 22 October 2016.

Department of Defense. (1986). Lighting on naval ships. Washington, USA.

Edwards, L., & Torcelini, P. (2002). *A literature review of the effects of natural light on building occupants*. Colorado, US: National Renewable Energy Laboratory.

Elliott, F. S., & Kobus, D. A. (1992). *Standardizing shipboard lighting: Light fixtures and light bulbs on U.S. Navy Ships*. San Diego, CA: Naval Health Research Center. http://oai.dtic.mil/oai/oai?verb=getRecord&metadataPrefix=html&identifier=ADA273682, Accessed on 01 September 2016.

Falchi, F., Cinzano, P., Elvidge, C. D., Keith, D. M., & Haim, A. (2011). Limiting the impact of light pollution on human health, environment and stellar visibility. *Journal of Environmental Management, 92*, 2714–2722.

Harb, F., Hidalgo, M. P., & Martau, B. (2014). Lack of exposure to natural light in the workspace is associated with physiological, sleep and depressive symptoms. *Chronobiology International, 32* (3), 368–375. https://doi.org/10.3109/07420528.2014.982757

Heerwagen, J. H. (2000). *Green buildings, organizational success, and occupant productivity* (Vol. 28, No. 5). London, UK: Building Research and Information.

Herrmann, R. (1984). The Ship's lighting. In W. H. G. Goethe, E. N. Watson, & D. T. Jones (Eds.), *Handbook of nautical medicine*. Berlin, Heidelberg: Springer.

ILO. (1996). *Accident prevention on board ship at sea and in port. An ILO code of practice* (2nd ed.). Geneva: International Labour Office. http://www.ilo.org/wcmsp5/groups/public/@ed_protect/@protrav/@safework/documents/normativeinstrument/wcms_107798.pdf. Accessed 01 November 2016.

Kassakian, J. G. (2013). *Assessment of advanced solid state lighting*. Washington, DC: The National Academies Press. https://doi.org/10.17226/18279.

Lighting The Way For A Greener Ship. (2010). Marine log, 115(8), 30–31. Retrieved from https://search-proquest-com.dianus.libr.tue.nl/docview/758281481?accountid=27128.

Ott Biolight Systems, Inc. (1997, October). *Ergo biolight report*. Santa Barbara, CA: Ott Biolight Systems, Inc.

Savitz, M. (1986). The federal role in conservation research and development. In J. Byrne & D. Rich (Eds.), *The politics of energy research and development, Energy policy studies series* (Vol. 3, pp. 89–118). New Brunswick, NJ: Transaction, Inc.

Völker, S., Rüschenschmidt, H., & Gall, D. (1995). Beleuchtung und Unfallgeschehen am Arbeitsplatz, Zeitschrift für die Berufsgenossenschaften.

Wulfinghoff, D. R. (2004). *Energy efficiency manual*. Wheaton, MD: Energy Institute Press.

Wynn, T., Howarth, P. A., & Kunze, B. R. (2012). Night-time lookout duty: The role of ambient light levels and dark adaptation. *Journal of Navigation, 65*(4), 589–602. https://doi.org/10.1017/S0373463312000288

# Part III
# Energy Efficient Ship and Port Operation

# An Integrated Vessel Performance System for Environmental Compliance

Soren Hansen and Stamatis Fradelos

## 1 Introduction

Recent regulatory developments include the global data collection scheme for fuel oil consumption for ships agreed by IMO and the monitoring, reporting and verification (MRV Scheme) legislation approved by the EU Parliament. These regulations concern $CO_2$ emissions from ships over 5000 GT that are engaged in international and EU trade, respectively; they will soon require ship operators to demonstrate to the competent authorities that they have a monitoring plan in place and are prepared to comply with regulations.

Despite the present rather low bunker fuel prices, the 0.1% Sulphur limit within ECAs and the future 0.5% global Sulphur cap require the use of more expensive compliant fuels. Bunker fuel costs still account for a high percentage of the total voyage expenditure for a vessel and ship operators are required to put more effort into monitoring ship's fuel oil consumption, finding measures to reduce it and, thereby, also reducing GHG emissions.

This paper describes the preparations needed to be ready to comply with the EU MRV regulation and describes how a Vessel Performance Monitoring System (VPMS) can be set up to evaluate the operational efficiency of a vessel and can, at the same time, be used for environmental regulatory compliance. This can be done without introducing an additional administrative burden on the vessel's crew and without additional operational costs for owners and operators.

S. Hansen (✉)
American Bureau of Shipping (ABS), Copenhagen, Denmark
e-mail: shansen@eagle.org

S. Fradelos
American Bureau of Shipping (ABS), Athens, Greece
e-mail: sfradelos@eagle.org



A. I. Ölçer et al. (eds.), *Trends and Challenges in Maritime Energy Management*, WMU Studies in Maritime Affairs 6, https://doi.org/10.1007/978-3-319-74576-3_14

# 2 Emissions Regulations

## *2.1 EU Regulations*

The MRV Shipping Regulation adopted in April 2015 creates an EU-wide legal framework for the monitoring, reporting and verification of $CO_2$ emissions and other relevant information from maritime transport. It requires larger ships (5000 gross tons and above), irrespective of where the ship or the company is registered, calling at ports in the European Economic Area EEA (EU, Iceland and Norway) from January 1, 2018 to monitor their $CO_2$ emissions and other relevant information on journeys to, from and between EEA ports of calls, and also when in EEA ports of call.

Companies having assumed the MRV responsibility for ships visiting EEA ports will have to:

- By 31 August 2017—submit to the verifiers a monitoring plan for each ship above 5000GT visiting EEA ports, indicating the method chosen to monitor and report $CO_2$ emissions and other relevant information. Companies will need to develop a monitoring plan in accordance with Article 6 of the Regulation, using standardized electronic templates
- From January 1, 2018, monitor and report to an accredited MRV shipping verifier, data on each ships' $CO_{2,}$, fuel consumption and other parameters, such as distance, time at sea and cargo carried, so as to determine the ships' average energy efficiency.
- From 2019, by April 30 of each year submit, electronically, to the Commission satisfactorily verified Emissions reports for each of the ships concerned.
- From 2019, by June 30 of each year ensure that all ships having performed activities in the preceding reporting period and visiting EEA ports, carry on board a document of compliance issued by an accredited MRV shipping verifier. This might be subject to inspections by Member States' authorities (EU Regulation 2015).

Companies need to assess the following steps in order to demonstrate compliance with the Regulation. The following paragraphs (Sects. 2.1.1–2.1.5) will describe the above-mentioned steps.

### 2.1.1 The Monitoring Plan

The draft monitoring plan template consists of six parts (A to F) as follows (Table 1):

**Table 1** Monitoring plan elements

| | | |
|---|---|---|
| A | Record of revisions | |
| B | Basic data | Identification of the ship: Company information; Emission sources/fuel types; Emission factors; and Procedures, systems and responsibilities |
| C | Activity data | Exemptions; Fuel consumed; List of voyages; Distance travelled; Cargo carried and Number of passengers; Time spent at sea |
| D | Data gaps | Estimated gaps for fuel consumed; distance travelled; cargo carried and time spent at sea. |
| E | Management processes | Periodic reviews of Monitoring plan; IT management; Data validity; Corrections and corrective activities; Outsourced activities; Documentation |
| F | Further information | Definitions and abbreviations |

### 2.1.2 Primary Data Accuracy

This section addresses the quality of the data produced on board the vessels to ensure that it meets an acceptable level of accuracy.

Companies will be asked to provide details about their procedures for measuring data on board and quality assurance of the measurement equipment (calibration schedules, maintenance schedules, availability of necessary parts). Existing fuel measurement methods will have to be assessed against recognized standards to identify any possible causes of inconsistency in fuel measurement. Examples include if the flow meter is inaccurate due to poor calibration, if electronic data transmission fails, if lack of training leads to inaccurate meter reading, if data is logged incorrectly due to human error or if software for data collection contains bugs or other IT issues.

Whenever possible, companies will need to provide corrections for known errors, such as lack of instrument precision, calibration and zeroing, vibration or aging of the instrument. Insufficient training of onboard personnel or incomplete company procedures can also result in errors and will have to be addressed.

### 2.1.3 Data Flow and Control Activities

This section addresses the existing data flow to ensure that the data is being transferred, processed stored and retrieved in a controlled and well-organized manner.

Companies should identify the data sources and measurement/calculation activities and then create a logical sequence of data collection and processing steps (data flow). Examples for data flow activities include reading from instruments, aggregating data, converting volume to mass, transferring, processing and storage. A data flow diagram is considered to be a useful tool for analyzing and/or setting up data flow procedures.

Errors in these activities should be expected, as both onboard and shore-side staff are involved and interact with information technology systems. Therefore, companies need to assess the risks associated with this data flow and set up appropriate control measures for mitigating the identified risks.

Finally, after the application of the control measures, the company will need to assess the risks once more to ensure that the measures have been effectively applied and the remaining risks are sufficiently low for the data flow system to produce a fuel measurement report which is free from material misstatement.

The outcome will be a well-defined data flow, a set of control activities and a practical risk assessment plan. It is very important that all staff involved receive regular training and are made aware that they are responsible for implementing the procedures properly.

Companies may also choose to establish and maintain written procedures related to control activities for:

- Quality assurance of the measurement equipment, such as calibration schedules, maintenance schedules, availability of necessary parts.
- Quality assurance of the information technology system used for data flow activities. Control is to be exercised in particular regarding access to the systems backups, recovery, maintenance and security.
- Segregation of duties in the data flow activities and control activities using the four-eyes principle, meaning that the person dealing with the cross check of data is different from the person that is processing the data and produces the emission report.
- Internal reviews and validation of data, including comparison of data over time and cross-checks between different methods used for measurement of fuel consumption.
- Corrections and corrective actions such as replacement of faulty measurement instruments or improvement of control activities.
- Maintaining records and documentation, including the management of document versions.

### 2.1.4 Reporting Systems

The existing reporting system to ensure that the reports produced include all the required information defined according to the Regulation.

For each voyage and for each calendar year, companies must monitor specific parameters according to Articles 9 and 10 of the Regulation for each ship visiting EEA ports.

Companies will have to carry out a gap analysis of their existing systems against the reporting requirements of the Regulation and identify missing elements such as data required per voyage or at berth, average energy efficiency monitoring and ship specific additional information and highlight areas for improvement such as data entry quality issues. Companies will need to make sure that the data flow and

**Table 2** Emissions report elements

| Relevant part | Title |
|---|---|
| Part A | Data identifying the ship and the company |
| Part B | Identification of verifier |
| Part C | Monitoring methods and level of uncertainty |
| Part D | Annual results: Fuel consumed and $CO_2$ emitted; Distance travelled, time spent at sea and transport work; and Energy efficiency |

reporting system provides adequate data retention and control throughout the year and includes all of the required information.

For the purposes of submitting the emissions report, companies shall use the electronic version of the template available in the THETIS MRV automated Union information system operated by the European Maritime Safety Agency. The electronic version of the template of the emissions report consists of four parts (A to D) as follows (European Commission 2016) (Table 2):

### 2.1.5 Training

It is important that all staff involved receive regular training to be able to implement the procedures properly.

## 2.2 *IMO Regulations*

IMO formally adopted a mandatory data collection system for fuel consumption of ships at the Marine Environment Protection Committee (MEPC) 70 meeting in October 2016. In addition, it was agreed that an initial but comprehensive IMO strategy on reduction of GHG emissions from ships should be adopted and enter in force in 2018.

According to the global data collection scheme, larger ships (5000 gross tonnage and above) will be required to collect consumption data for each type of fuel they use as well as data relevant to energy efficiency (such as distance travelled and hours underway). The aggregated annual data will be reported to the flag state after the end of each calendar year. Having determined that the data has been reported in accordance with the requirements, the flag state will issue a statement of compliance to the ship. Flag states will be required to subsequently transfer this data to an IMO ship fuel consumption database. IMO Secretariat will be required to produce an annual report to the MEPC summarizing the data collected. Data will be anonymized so individual ship data will not be recognized.

IMO will develop guidelines regarding the implementation of the mandatory data collection scheme (for instance the verification of the collected data and the management of the IMO ship fuel consumption database). The global system will start applying in March 2018, which in practice will mean that ships will have to start collecting data for the 2019 calendar year.

## 2.3 EU and IMO Regulations

The global IMO data collection system is still under preparation and it is to be seen if this system can be combined with the EU system. The key differences between the two systems are indicated in Table 3 EU Regulation (2015) and IMO (2016):

So far, the official statement from the Commission is that it welcomes the adoption of such a mandatory data collection system. A global scheme is in line with the approach set out in the European Strategy for low-emissions mobility adopted in July 2016. Global action to reduce emissions from maritime transport is the preferred approach due to the global nature of the sector. At the same time, the global data collection system needs to be sufficiently robust to deliver reliable data. To ensure a reasonable level of robustness, the Commission and the EU Member States will continue to actively contribute to the development of IMO Guidelines required for the implementation of the system.

**Table 3** IMO & MRV comparison

| IMO DCS | EU MRV |
|---|---|
| Monitoring starting on 1 January 2019 | Monitoring starting on 1 January 2018 |
| SEEMP amendments to include the data collection and reporting methodology. | Monitoring plans and emission reports using European Commission (EC) electronic templates |
| Reporting of fuel consumption, hours underway and distance travelled | Reporting of additional information, such as cargo carried, emissions while at berth, total transport work and average energy efficiency. |
| Data verification by the Administration or R.O.s | Data verification by Verifiers accredited by the European national accreditation bodies. |
| Verification activities as defined by the Administrations taking under consideration the IMO Guidelines currently under development. | Detailed verification activities based on ISO14064 part 3 principles |
| Data will be transferred and stored into an IMO database for further analysis in support of ongoing IMO decisions. | Data will be transferred and stored into THETIS MRV automated Union information system. |
| Identification of a specific ship and Administration will not be possible. Parties shall have access to the anonymized data strictly for their analysis and consideration. | By 30 June each year, the EC shall make publicly available the information on $CO_2$ emissions and other relevant information. |

# 3 Vessels Performance Management System (VPMS)

## 3.1 *Purpose of the VPMS*

The monitoring of vessel condition, performance, and maintenance activities is vital to the effective management of vessels. Each voyage offers the opportunity to optimize speed, find the safest route through calm seas, and make sure the ship is sailing at the best draft and trim and tuned to keep course efficiently. Selected maintenance cycles impact the resistance created by the hull and propeller. Accurate and regular energy use monitoring across the fleet can highlight inefficiencies and provide a mechanism for continual improvement. Data logged in the system can be extracted and used in reports for regulatory compliance. See Fig. 1.

A Vessel Performance Management System (VPMS) will be the placeholder for the fuel efficient operations overview and performance. The VPMS system is implemented in a three tiered process (ABS 2016a):

1. Performance Monitoring
2. Performance Management
3. Performance Optimization

### 3.1.1 Vessel Performance Monitoring

Identifying key data and analyzing that data is critical to cost effective technical and operational management. Logging of data can be manual (by vessel's crew), semi-automated or automated. Integrating data collection with the vessel's control and monitoring systems can reduce crew burden and simplify the process of qualifying and analyzing condition and performance data into a standardized process for maintenance planning and decision making related to vessel operations.

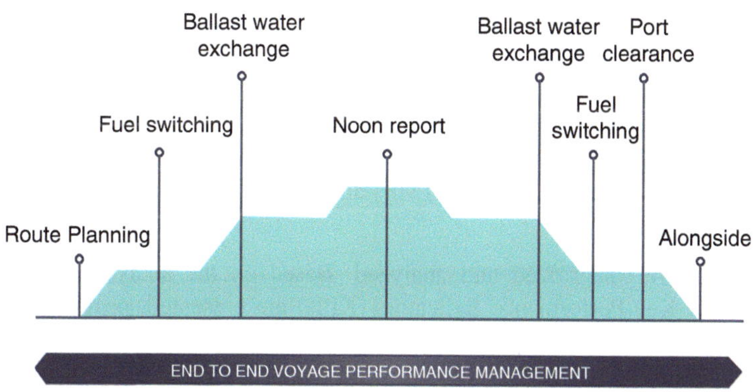


**Fig. 1** VPMS logging of events (ABS 2016c)

A full scale vessel monitoring system monitors the ship's structural integrity, functionality and performance in a real-time base and will enhance the vessel and crew's safety and fuel efficiency. In addition, the accumulated ship operational data, such as wave conditions, hull structural loads and vessel's operational performance parameters enable the owner, designer and classification to conduct data-driven analysis and support maintenance and operational decisions.

The philosophy these monitoring systems would follow is measuring, recording, monitoring, analyzing, reporting, and enhancing. These efforts speak directly to the goals of IMO Guideline on Ship Energy Efficiency Management Plans (SEEMP), a top down framework that captures the corporate commitment to energy conservation.

Many ship owners/operators have processes in place to measure, analyze and manage vessel performance. The processes should include a monitoring system to capture all events on a voyage in order to comply with environmental regulations e.g. the EU MRV system or the 2020 0.5% Global Sulphur cap and to capture all performance related activities.

### 3.1.2 Data Logging

Onboard data logging is one of the main functions of VPMS. The logged data will assist crew to track and evaluate the vessel's performance, investigate possible root causes for performance degradation and help crew to make informed decisions and correction plans. The onboard VPMS uses dedicated display units or consolidated displays with other systems as the user interface to show the real-time measurements from sensors and devices, together with historical data, data statistics, trending and prediction.

The commonly monitored parameters include

- Navigation, such as position, speed, course, rudder angle, ship motion and acceleration
- Environmental conditions, such as wind direction and speed, dominant wave direction and wave height, current direction, sea temperature and density
- Ship operation, such as drafts (fwd, aft, starboard and portside) and trim, cargo loading
- Machinery status, such as temperature and pressure of main engine components
- Fuel flow readings for main consumers, such as ME, AEs and Boilers
- Power output from Main Engine (ME) and Auxiliary Engines (AEs), RPM and shaft torque readings

The data are categorized and analyzed. Based on the analysis, a variety of performance Key Performance Indicators (KPIs) are presented, providing insights to the vessel's performance.

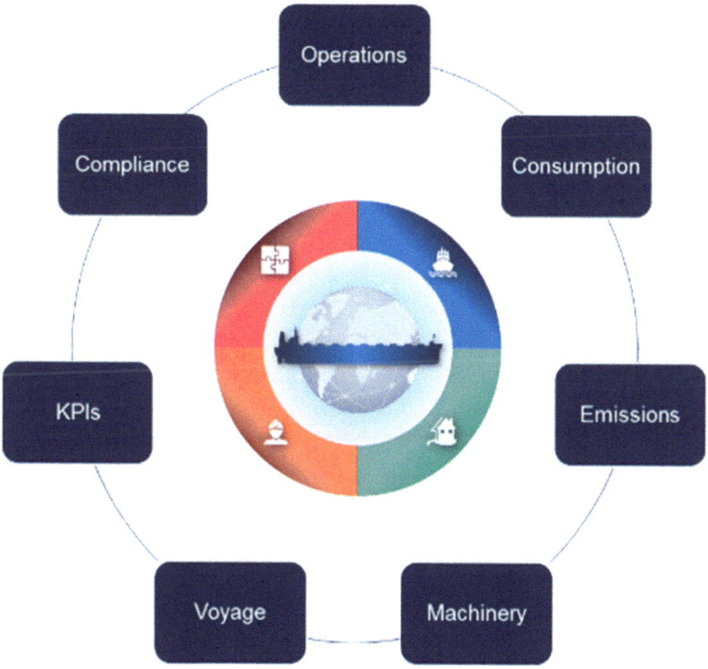


**Fig. 2** Data Interpretation (ABS 2016b)

### 3.1.3 Data Interpretation

The amount of data to be logged in a modern VMPS can be quite extensive. If the data are autologged the amount of data can be huge and once the data are logged into the VPMS, they should be sorted and arranged in relevant categories for the user of the system. See Fig. 2.

To perform a vessel performance analysis, it is necessary to have an overview of the vessel's operational profile. Area of operation, Operation modes, Weather, Speed, loading condition and ME load are important performance influencing factors. Consumption should be categorized by consumers, fuel grades and properties. Emissions are derived from fuel grades consumed and should be categorized in relevant categories for different emissions reports. The categorization of data will give the user a quick overview of the data and ease the decisions to be made for the efficient operation of the vessel.

### 3.1.4 Vessel Performance Management

The Vessel Performance Management tier consists of a detailed analysis of the most important performance parameters and results of the analysis are provided through dashboards with KPIs for decision support.

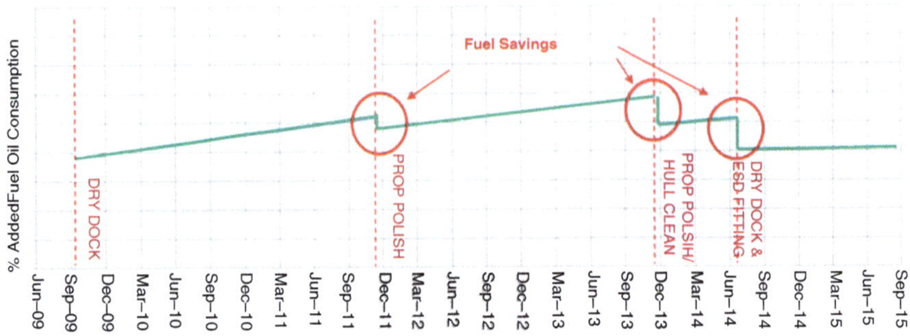


**Fig. 3** Hull/Propeller performance (ABS 2016c)

Hull/Propeller Performance

Fouling of the hull and propeller will create added resistance for the vessel and therefore added fuel oil consumption in order to maintain vessel speed. The propulsive performance losses due to fouling can be monitored by a KPI and the change in this evaluated towards a reference baseline established at sea trial/delivery event or within a pre-determined period. Key events such as dry docking, hull cleaning and propeller polishing partially restore propulsive performance and changes are noticeable and effects of events on daily fuel oil consumption are given in speed/draught—fuel tables. See Fig. 3.

Main Engine Performance

The Main Engine of the vessel is the prime consumer and monitoring/analyzing the state of this will save money and avoid breakdowns of equipment. Main Engine Performance tests provide valuable information on the state of the engine and, by establishing indexes and trends, it will be possible to predict and avoid breakdown and perform maintenance in due time. Combining Main Engine Performance with the state of the Hull and the Propeller gives an overview of the main causes of added fuel oil consumption over time. A set of dashboards provides an overview of the state of the main parameters and decision support for fuel consumption improvement actions to be taken. See Fig. 4.

Base Load Performance

In addition to the Main Engine, auxiliary engines and machinery consume energy. The Base Load performance analysis can e.g. comprise of:

- Electrical Load Analysis
- Cargo Loading/Discharging performance

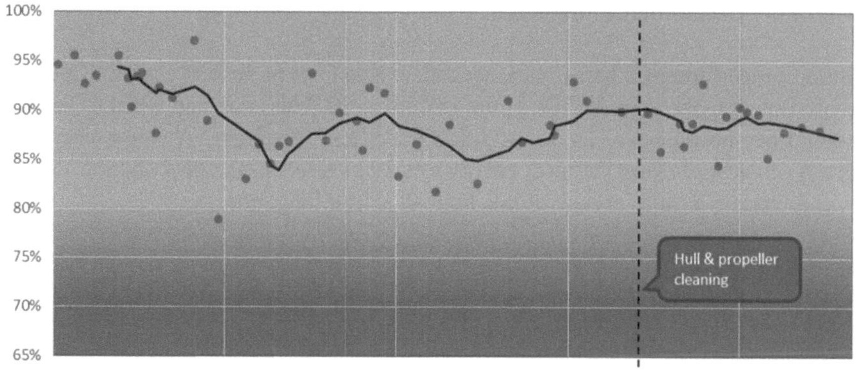


**Fig. 4** Main Engine Performance Index (ABS 2016c)

- Cargo Heating Analysis
- HVAC performance

Performance of the base load is continuously monitored and analyzed to optimize performance.

Fleet Benchmarking

Performance KPIs are also issued on a fleet level and benchmarking will identify which vessels are good performers and in which areas. Best practices of good performance can be rolled out to the whole fleet and by sharing performance KPIs among vessels, performance awareness by the crew will be enhanced for the benefit of the fleet performance.

Environmental Performance

Reporting and performance towards environmental regulations and reporting schemes are a key part of the VPMS. $CO_2$ emissions reporting to the MRV system, the IMO reporting system and fuel switching monitoring in ECA areas are part of the system. The SEEMP is the document to verify improvements in performance and a VPMS will deliver information for an active SEEMP. Ballast water management is also included in the VPMS. Since all relevant data are logged in the System it is an easy task to generate the necessary reports for the various reporting schemes.

### 3.1.5 Vessel Performance Optimization

The VPMS establishes performance KPIs for the various areas in the vessel and optimization of the performance can then be based on these KPIs.

Fleet/Pool Utilization/Composition

Based on performance benchmarking, a fleet or a pool of vessels can be evaluated with respect to the composition of the vessels in the fleet/pool, and whether vessels should be substituted or moved to other trade areas or routes. A ship-owner will always look for having the most energy efficient vessels in the fleet and the VPMS will provide a clear overview of this both short and long term.

Operational Benchmarking

Operational benchmarking can be applied not only to the performance KPIs in the fleet but also across fleets and across industry in same trading areas or vessels of the same size and type. Including external information in the VPMS provides valuable data to the ship owner on best practices in vessel operations and can provide valuable knowledge internally and help improve the operational performance of the fleet. See Fig. 5.

Voyage Optimization

Traditional voyage planning and weather routing aim to determine a route that guarantees the arrival schedule while avoiding severe weather and assuring the ship's safety. The voyage optimization in VPMS adds one more target to the optimization goals: using least fuel. Besides the optimum voyage calculated before voyages, onboard VPMS updates the optimum speed and heading continuously and refreshes the optimum route throughout the whole voyage based on the updated weather forecasts. The speed profile for a voyage plays a critical role in ship's fuel

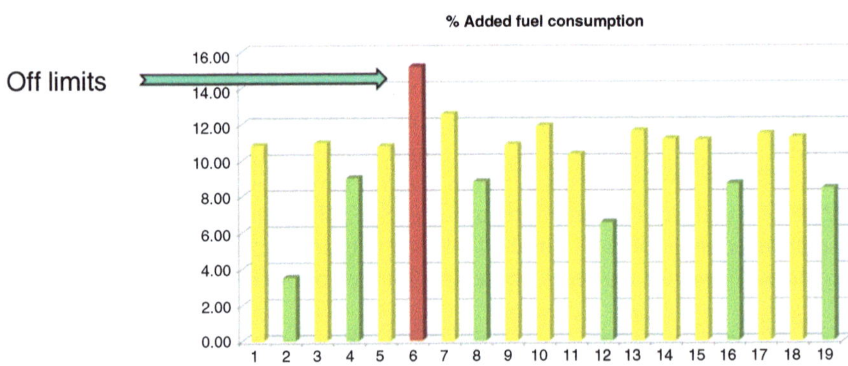


**Fig. 5** Operational benchmarking, %Added fuel oil consumption from base model (ABS 2016c)

consumption and slow steaming has been accepted as the common practice to reduce the overall fuel consumption in the shipping industry. In addition to slow steaming, voyage optimization aims to sail the ship at the most economical speed for each leg of the voyage and arrive at the destination right on time.

Trim Optimization

An optimum trim can reduce the total ship resistance and lead to less fuel consumption under the same loading condition. In the design phase, the ship's bow and stern are usually optimized for even keel conditions, which may not always be the case during real operations. A trim table is the basis for finding the optimum trim, which gives the required propulsion power for various draft, trim and speed. The optimum trim should be considered when conducting cargo planning before voyages. The optimum cargo distribution should be achieved to reduce ballast water usage and ballast and de-ballast operations during voyages. An important consideration for trim optimization is that optimum trim should not jeopardize the ship's integrity and stability. When calculating the optimum trim, the structural safety and stability limitations should be fully considered and strictly satisfied.

## 4 Conclusion

Increased costs and new regulations and requirements are the primary drivers for owners and operators to improve vessel performance. As the regulatory framework gets more complex, ship charters' requirements expand and the volume of data required to satisfy regulatory reporting increases; owners and operators are looking to leverage data to better understand their fleet's performance.

As an example of the complexity in reporting requirements, this paper presented the challenges that ship operators and owners are facing concerning establishing a Monitoring Plan for the EU MRV regulation and setting up a system for the data flow and the verification of data. It also presented a similar IMO based system and a description of the differences and similarities in the EU and IMO defined systems.

By setting up a transparent VPMS that always reflects the latest regulatory requirements and includes all the relevant logged data, it is also shown that this system can be expanded to include both Vessel and Environmental Performance matters including the required reports.

By implementing and using the VPMS as a consolidated and common logging platform, functionalities can be added without increasing the administrative burdens to the vessel's crew.

# References

ABS. (2016a). ABS Vessel Performance. Available at http://ww2.eagle.org/en/what-we-offer/asset-performance/vessel-performance.html.

ABS. (2016b). Nautical Systems Fleet Management Software/MRV solution. Available at: https://ww2.eagle.org/en/what-we-offer/Nautical_Systems_Fleet_Management_Software/solutinos_nautical_systems_fleet_management/MRV.html.

ABS. (2016c). ABS VPMS software (screenshots from software).

EU Regulation. (2015). REGULATION (EU) 2015/757 OF THE EUROPEAN PARLIAMENT AND OF THE COUNCIL of 29 April 2015 on the monitoring, reporting and verification of carbon dioxide emissions from maritime transport, and amending Directive 2009/16/EC. Available via: http://eur-lex.europa.eu/legal-content/EN/TXT/PDF/?uri=CELEX:32015R0757&from=EN.

European Commission. (2016). Shipping Emissions Templates. Available via: https://ec.europa.eu/info/law/better-regulation/initiatives/ares-2016-3985800_en.

IMO. (2016). Marine Environment Protection Committee (MEPC), 70th session, 24–28 October 2016. Available via: http://www.imo.org/en/MediaCentre/MeetingSummaries/MEPC/Pages/MEPC-70th-session.aspx.

# *Further Reading*

Hansen, S., ABS. 2017. An integrated vessel performance system for environmental compliance. In *MARENER2017 Conference*, World Maritime University, Malmo, Sweden. http://commons.wmu.se/cgi/viewcontent.cgi?article=1029&context=marener_conference.

# Energy Efficient Ship Operation Through Speed Optimisation in Various Weather Conditions

**Tong Cui, Benjamin Howett, Mingyu Kim, Ruihua Lu, Yigit Kemal Demirel, Osman Turan, Sandy Day, and Atilla Incecik**

## 1 Introduction

Shipping plays an important role in the global economy as almost 90% of goods traded worldwide are transported by sea. Greenhouse Gas (GHG) emissions and energy consumption from the shipping industry are significant, and have become very important factors that cannot be ignored for both environmental and economic benefit (IMO 2014; IHS Global Insight 2009; Oxford Economics 2014). The modern shipping industry is paying more attention to energy saving and emission reduction. In recent years, as the main regulatory body for shipping, the International Maritime Organisation (IMO) has raised several technical and operational measures towards regulating shipping energy efficiency and thereby controlling maritime GHG emissions. One of the most important concepts is the Ship Energy Efficiency Management Plan (SEEMP).

The Ship Energy Efficiency Management Plan (SEEMP) is designed to provide the framework, against which, a ship owner can develop best practice and energy efficient operations. The IMO introduced the SEEMP as a mandatory tool under MARPOL Annex VI,[1] which entered into force on January 1, 2013.

There are many categories of energy improvement methods for potential adoption within each ship's SEEMP, including Fuel-efficient Ship Operations, Optimised Ship Handling, Hull Maintenance, Propulsion System, and Waste Heat Recovery The category of Fuel-efficient ship operations recommended by SEEMP (IMO 2012), includes:

[1]MARPOL: International Convention for the Prevention of Pollution from Ships.
Annex VI: Prevention of air pollution by ships.

T. Cui (✉) · B. Howett · M. Kim · R. Lu · Y. K. Demirel · O. Turan · S. Day · A. Incecik
Department of Naval Architecture, Ocean and Marine Engineering, University of Strathclyde, Glasgow, UK
e-mail: tong.cui@strath.ac.uk



A. I. Ölçer et al. (eds.), *Trends and Challenges in Maritime Energy Management*,
WMU Studies in Maritime Affairs 6, https://doi.org/10.1007/978-3-319-74576-3_15

- Improved voyage planning
- Weather routing
- Just in time
- Speed optimisation
- Optimized shaft power

Among them, speed optimisation and weather routing are two very common and effective approaches for energy efficient ship operations (IMO 2012, 2016; Simonsen et al. 2015).

Speed optimisation means to select an optimum speed changing strategy for a given route, which can make a great contribution to energy efficient shipping. There are many types of research in this field. Several studies have assumed that ship speed is fixed (Wang and Meng 2011, 2012), while several studies have made a speed increase or reduction at certain speed intervals to see the benefits (Norstad et al. 2011; Lu et al. 2015). Several studies have carried out speed optimisation based on Historical data and Sea current forecasts (Jussi 2012). Heuristic methods, such as genetic algorithms, have also been used for speed optimisation by Golias et al. (2010), but the optimality cannot be guaranteed. Besides, several studies have created relationship models between fuel consumption and sailing speed (Lang and Veenstra 2010; Du et al. 2011) or even between oil price and sailing speed (Ronen 2011) for energy efficient shipping. All of these studies provide very positive methods but still cannot avoid constraints, especially inability to provide accurate ship operations with various weather conditions.

Actually, weather condition is a very important factor, which can not be ignored during speed optimisation. When combined with weather routing technique, speed optimisation will become more reliable. Ship weather routing is used to determine the optimum route (course and speed) between given departure and destination ports based on weather conditions and corresponding ship performance. It always aims to minimize fuel consumption, achieve a suitable ETA (estimated time arrival) or maximum ship safety. It has been proven as a very effective tool in the shipping field (Simonsen et al. 2015; Chen et al. 1998; Buhaug et al. 2009).

In this paper, a speed optimisation model is developed for energy efficient shipping. This model is composed of several sub-modules, including ship performance calculation module, grids system design module, and weather routing module. Every module has its own task, and every one of them should link with each other so that the whole model can run well towards energy efficient shipping. With the weather routing module integrated, this model can well predict ship operations in various sea states. Additionally, it can combine very well with other technologies, such as wind assistant, and fouling influence prediction. Therefore, with so many factors in consideration, this model will certainly provide stakeholders with more reliable ship operations for a shipping task.

## 2 Speed Optimisation Model

The Speed Optimisation Model presented in this paper is not only for single fixed route optimisation but also for global optimisation, which takes direction optimisation into consideration. According to the requirements of stakeholders, the only input items are the ship model, departure time and ETA, and positions of departure and destination, and then, after complicated calculation, the model will determine an optimum route towards minimum fuel consumption. The workflow of the model is shown in Fig. 1.

It can be seen that the model generally contains five modules, each of which is introduced briefly in the following paragraphs (Sects. 2.1–2.5).

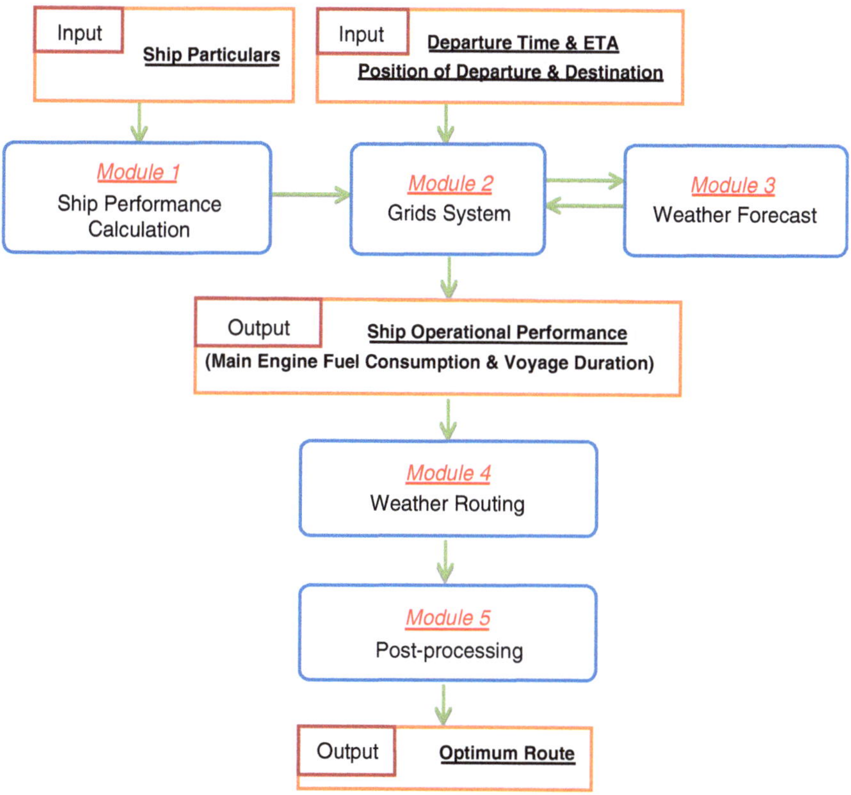


**Fig. 1** Workflow of speed optimisation model

## 2.1 Module 1: Ship Performance Calculation

Since the speed optimisation model runs towards minimum fuel consumption under various weather conditions, the ship performance prediction becomes a very important module of the whole model, which should accurately predict ship brake power under different situations. In this module, the ship particulars including ship geometry, main engine and propeller parameters should first be collected. Next, ship resistance in calm water can be predicted based on the Holtrop 84 method (Holtrop 1984), while the added resistance due to waves and wind can be calculated based on the Kwon's method (Kwon 2008) after the weather forecast module is introduced. Based on the total resistance, together with ship speed, propeller open water performance and efficiency of transmission of power, the ship engine brake power can be determined, which is necessary for fuel consumption calculation. Besides, the sea trial data, ship model test data or even noon report data can be introduced to verify the predicted ship performance. Based on the above, when weather data is given, the ship performance in this sea state can be obtained in this module, which will be used in the weather routing module afterward.

In order to obtain ship performance very quickly and conveniently, once ship particulars are decided, a ship performance database can be generated before running the whole speed optimisation model. With the given variable (sailing speed, weather data etc), ship performance can be easily determined through the database. One simple database contains these dimensions: speed, significant wave height, relative wave angle, true wind speed, relative wind angle and an output attribute: brake power (Howett 2015). The ship performance database may also include additional dimensions (e.g. draft/loading condition) or additional attributes (e.g. fuel consumption) where required. Therefore, when alternative energy saving techniques or other influencing factors are taken into account, the model only needs to generate a new ship performance database without changing anything else. This makes the speed optimisation model combine well with other external techniques.

## 2.2 Module 2: Grids System

Grids system means the waypoints distribution strategy in the shipping area. The ship will travel along the potential route formed by these waypoints. In this grids system, as presented in Fig. 2, a great circle route (The shortest distance between two points on a sphere) is taken as the reference route, which is divided into several equal stages with a certain number of nodes. Through every point, a straight line can be drawn perpendicular to its tangent line around the circle. Next, a number of nodes can be distributed along this vertical line, including upper and lower parts of the great circle. The ship in one stage can travel to any waypoint in the next stage. Here, considering that larger course deviations are not feasible and would be unrealistic for an optimum route, a limit is set for the ship in one waypoint, which can only go to the

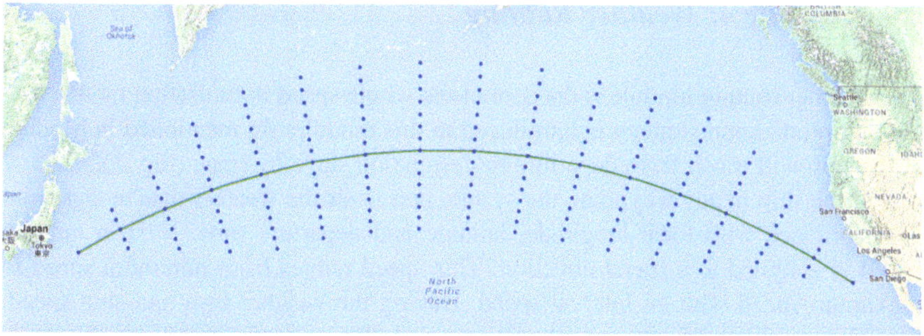


**Fig. 2** A typical grids system

nearest five waypoints in the next stage. For some special shipping tasks, such as the ship going to several fixed ports during the whole navigation, the grids system can be automatically divided into several stages based on the same principle. Besides, in most real situations, the ship travels around some islands or avoids prohibited military zones, so an area avoidance function is also developed.

It is worth noting that the speed optimisation for a single fixed route (without direction optimisation) can be regarded as a special but much easier situation within the normal global optimisation. For a fixed route, it must be formed by several certain waypoints as well. The latitude and longitude of every waypoint will be input to grids system manually. So that the grids system also has several stages but has only one waypoint, thus, one direction on each stage. The whole speed optimisation will be carried out with this simple grids system.

## 2.3 *Module 3: Weather Forecast*

This system will take only wind, and waves into account at the moment. Wave and wind data are both downloaded from the ECMWF (European Centre for Medium-Range Weather Forecasts) website. The data is constructed in gridded binary (GRIB) data form, including 10 m U wind component, 10 m V wind component, mean wave direction, mean wave period and significant height of combined wind waves and swell, and they all update every 6 h. These weather data will be converted to a Beaufort Number and relative weather direction according to ship heading for ship performance calculation. The weather forecast files are linked with three common parameters: time, latitude and longitude. With given time, latitude and longitude, the weather forecast module will provide corresponding sea conditions to the grids system.

## 2.4 Module 4: Weather Routing

The weather routing module is the core of the whole speed optimisation model. The global optimization strategy is introduced in this module. As mentioned in Module 2, the ship will travel from departure to destination stage by stage (Fig. 3).

For the ship in any waypoint, the system first reads the weather data at that point in accordance with local longitude, latitude and departure time. Next, a random speed is assigned to a travel direction. This speed ranges from minimum speed to maximum speed with an interval speed. Having the weather data and ship speed, according to ship performance calculation result, the fuel consumption on this arc is calculated under this speed and direction. Then the navigation information of this arc (fuel consumption, ship speed, navigation duration, local time and coordinates of local departure point) is stored in its arrival point. After that, the ship will travel to next waypoint based on all the information saved in this arrival point.

The whole calculation starts from the departure port. Under this simulation principal, as described above, fuel consumption modeling will be continuously utilised for all stages between departure port and destination. This process covers all the speed and direction options. In the end, the total fuel consumption and voyage duration of the different potential route with different speed set will be stored in the destination node. The recorded information will be used to select the minimum fuel consumption route at given ETA, which is regarded as the global optimisation. During the optimisation process, several smart algorithms are added to the weather routing module, which can reduce calculation time and make the whole calculation run towards the required ETA faster.

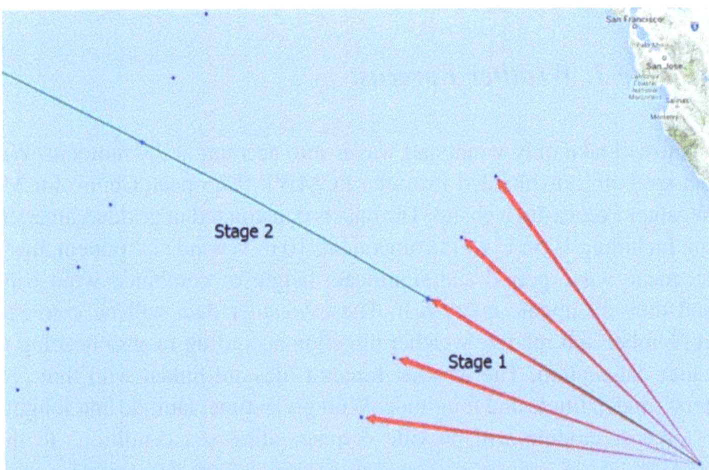


**Fig. 3** Ship routing stages

### 2.5 Module 5: Post-processing

After the calculation in the weather routing module is finished, there will be many results stored in the final destination waypoint. This module will select the minimum fuel consumption route under the required ETA for stakeholders. According to the information stored in every waypoint on this route, the speed, duration, fuel consumption and weather data of every stage will be extracted in detail. Users can see and handle them very easily.

## 3 Case Studies

A Bulk Carrier with 35,500 DWT and 175.72 m in length is taken as the ship model in the case studies. The shipping area is from Dubai to the middle of the Indian Ocean, but the ship should travel along the west coast of India. The departure and destination points are given as 25°15'N\56°51′E and 8°13'S\92°22′E, respectively. The grids system has 19 stages, and every stage has a maximum of 15 vertical waypoints with equal distance of 30 nautical miles. Ship speed ranges from minimum 9 knots to maximum 19 knots with the interval speed 0.1 knot. The required ETA is set at 242 h.

### 3.1 Case Study 1

In order to see the benefits of the model, the speed sets under six different conditions are selected for comparison as shown below:

Speed Set A—Actual route and actual voyage speed as recorded in ship noon report
Speed Set B—Actual route and optimum speed with identical ETA
Speed Set C—Global Optimisation with identical ETA
Speed Set D—Global Optimisation with 3% less ETA
Speed Set E—Global Optimisation with 3% more ETA
Speed Set F—Global Optimisation with 5% more ETA

Global optimisation means optimising both speed and direction at the same time. So the resulting route may be different from the actual route recorded in the ship noon report. All of the above have the same departure time at 2015-05-13, 23:00. The results are shown in Table 1.

As can be seen from Table 1, compared to Speed Set A, the speed optimisation (Speed Set B) for a fixed route will provide 1.47% fuel savings, while global optimisation (Speed Set C, both speed and direction optimisation) can provide 1.60% fuel savings. For Speed Set D, because it has 3% shorter ETA, the average sailing speed will increase, so it will consume 4.43% more fuel. As for Speed Set E

**Table 1** Comparison of ship operational performance of different speed sets for case study 1

| Stage | Speed Set A (knots) | Speed Set B (knots) | Speed Set C (knots) | Speed Set D (knots) | Speed Set E (knots) | Speed Set F (knots) |
|---|---|---|---|---|---|---|
| 1 | 12.5 | 12.8 | 13.0 | 13.0 | 13.2 | 12.8 |
| 2 | 12.5 | 13.2 | 13.2 | 13.2 | 12.6 | 13.2 |
| 3 | 12.7 | 12.6 | 13.0 | 13.0 | 12.6 | 12.6 |
| 4 | 12.5 | 12.6 | 13.0 | 12.6 | 9.7 | 12.6 |
| 5 | 12.4 | 10.7 | 10.1 | 11.9 | 12.6 | 9.7 |
| 6 | 13.0 | 12.8 | 12.6 | 12.8 | 11.9 | 12.6 |
| 7 | 12.8 | 12.1 | 12.1 | 12.1 | 12.1 | 11.9 |
| 8 | 12.6 | 12.6 | 12.6 | 13.0 | 12.6 | 11.9 |
| 9 | 12.4 | 13.6 | 12.6 | 13.0 | 12.1 | 12.6 |
| 10 | 12.1 | 13.0 | 12.1 | 13.0 | 11.9 | 11.9 |
| 11 | 13.2 | 12.6 | 13.4 | 13.4 | 12.6 | 11.9 |
| 12 | 12.5 | 13.0 | 12.6 | 13.2 | 12.1 | 12.6 |
| 13 | 12.4 | 9.7 | 12.1 | 12.6 | 9.7 | 9.7 |
| 14 | 12.3 | 12.1 | 11.3 | 12.1 | 11.9 | 9.7 |
| 15 | 13.1 | 12.1 | 12.1 | 12.2 | 12.6 | 11.9 |
| 16 | 12.9 | 12.6 | 12.6 | 12.6 | 13.2 | 12.6 |
| 17 | 12.4 | 12.8 | 13.2 | 13.2 | 13.4 | 13.2 |
| 18 | 11.9 | 13.2 | 13.0 | 13.2 | 13.2 | 13.2 |
| 19 | 11.8 | 12.8 | 13.2 | 13.4 | 13.2 | 13.2 |
| Voyage duration (hours) | 242 | 242 | 242 | 235 | 249 | 254 |
| Fuel consumption (tonnes) | 135.23 | 133.24 | 133.06 | 141.23 | 125.69 | 121.37 |
| Fuel savings compared to speed set A (%) | 0 | 1.47 | 1.60 | −4.43 | 7.05 | 10.2 |

with 3% longer ETA and Speed Set F with 5% longer ETA, based on the same reason, they will respectively provide 7.05 and 10.2% fuel savings.

The speed, BN (Beaufort number, which reflects wind force and significant wave height) and relative weather direction in each stage of Speed Set C are extracted in Fig. 4. To compare more clearly, the speed unit has been changed from knots to m/s. It can be seen that the ship speed has very good response to different weather conditions. When BN is increasing, the ship will reduce speed and will speed up again when the BN is low. The impact factor of BN is also higher than relative weather directions. At stages 16–19, although the ship travels in head sea or bow sea (relative weather direction from 0° to 60°) because the BN is relatively low, the ship can still travel at relatively higher speed. All of these routes are drawn on the map in Fig. 5.

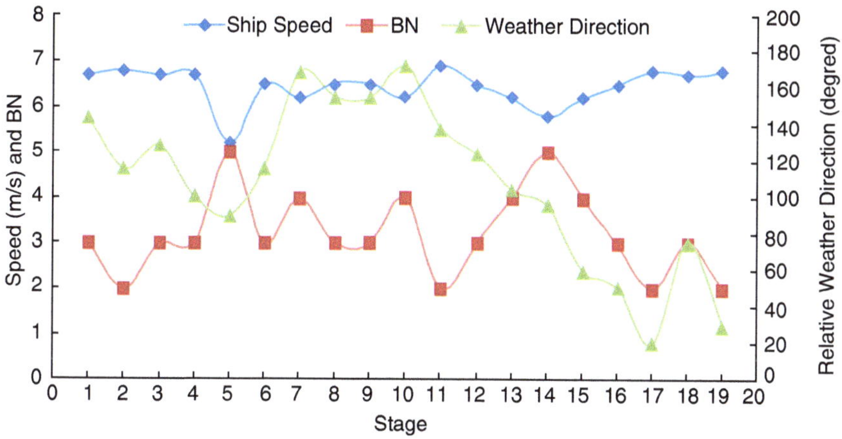


**Fig. 4** The speed and weather condition distribution of Speed Set C

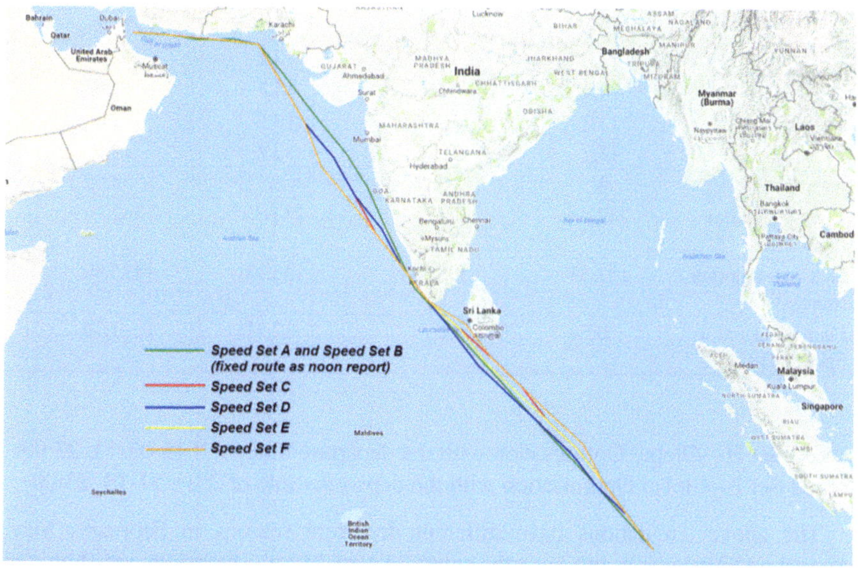


**Fig. 5** Optimal routes based on different conditions for case study 1

## 3.2 *Case Study 2*

In order to see the speed optimisation benefits under various weather conditions more clearly, another three case studies with departures at different seasons are carried out. Together with the Speed Set C, they are listed as below:

Speed Set G—Global Optimisation with the departure time of 2015-02-13, 23:00
Speed Set C—Global Optimisation with the departure time of 2015-05-13, 23:00

**Table 2** Comparison of ship operational performance of different speed sets for case study 2

| Stage | Speed Set G Feb (knots) | Speed Set C May (knots) | Speed Set H Aug (knots) | Speed Set I Nov (knots) |
|---|---|---|---|---|
| 1 | 13.2 | 13.0 | 13.2 | 12.6 |
| 2 | 13.0 | 13.2 | 13.2 | 11.9 |
| 3 | 11.9 | 13.0 | 12.6 | 11.9 |
| 4 | 11.9 | 13.0 | 13.2 | 11.9 |
| 5 | 11.9 | 10.1 | 12.1 | 13.2 |
| 6 | 11.9 | 12.6 | 12.1 | 11.9 |
| 7 | 11.9 | 12.1 | 12.6 | 11.9 |
| 8 | 12.6 | 12.6 | 12.4 | 11.9 |
| 9 | 11.9 | 12.6 | 12.6 | 11.9 |
| 10 | 13.2 | 12.1 | 12.6 | 11.9 |
| 11 | 13.2 | 13.4 | 12.6 | 13.2 |
| 12 | 13.2 | 12.6 | 12.6 | 13.2 |
| 13 | 12.6 | 12.1 | 12.1 | 13.4 |
| 14 | 9.7 | 11.3 | 12.1 | 12.6 |
| 15 | 12.6 | 12.1 | 13.2 | 13.2 |
| 16 | 13.2 | 12.6 | 12.6 | 12.6 |
| 17 | 13.4 | 13.2 | 12.2 | 12.6 |
| 18 | 13.2 | 13.0 | 10.9 | 12.6 |
| 19 | 13.2 | 13.2 | 12.2 | 12.6 |
| Average BN | 3 | 3.21 | 3.32 | 3.05 |
| Voyage duration (hours) | 242 | 242 | 242 | 242 |
| Fuel consumption (tonnes) | 132.33 | 133.06 | 136.13 | 135.40 |
| Fuel compared to Speed Set C (%) | 0.54 | 0 | −2.31 | −1.76 |

Speed Set H—Global Optimisation with the departure time of 2015-08-13, 23:00
Speed Set I—Global Optimisation with the departure time of 2015-11-13, 23:00

The above simulations have different departure seasons in February, May, August and November, but have the same ETA of 242 h. The results are shown in Table 2.

As can be seen from Table 2, with different departure seasons, the corresponding fuel consumption will change slightly. Compared with Speed Set C (May), the fuel savings are 0.54% less for Speed Set G (February), 2.31% more for Speed Set H (August) and 1.76% more for speed Set I (November). Among them, Speed set I has a different trend, which has relatively lower average BN but has more fuel consumption compared to Speed Set C. But after taking relative weather direction into consideration, it can be found that Speed Set I has relatively more head sea and bow sea situations, which will definitely add fuel consumption. Therefore, the results are reasonable. The speed, BN and relative weather direction in each stage of all the Speed Set are extracted in Figs. 6, 7 and 8.

As can be seen from these three figures, the ship speed still has very good response to different weather conditions, which proves the presented speed optimisation model works very well. Besides, all of these routes are drawn on the map in Fig. 9.

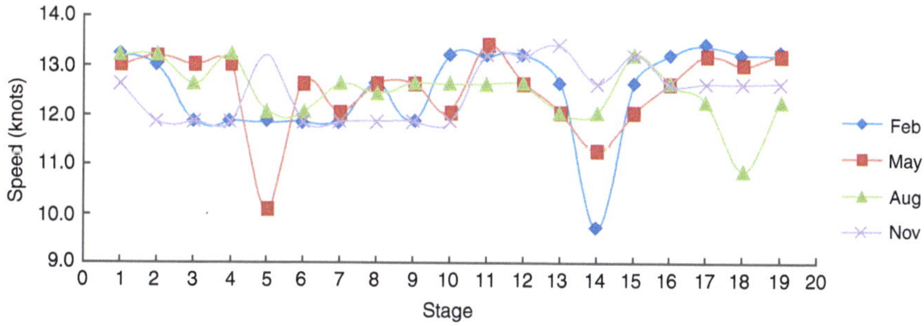


**Fig. 6** Speed distribution based on different conditions for case study 2

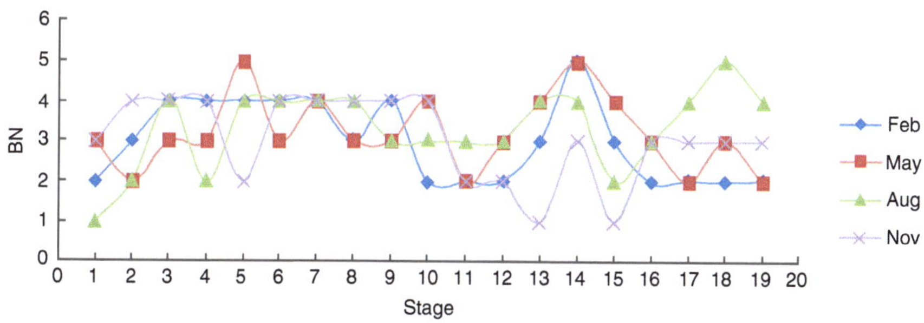


**Fig. 7** BN distribution based on different conditions for case study 2

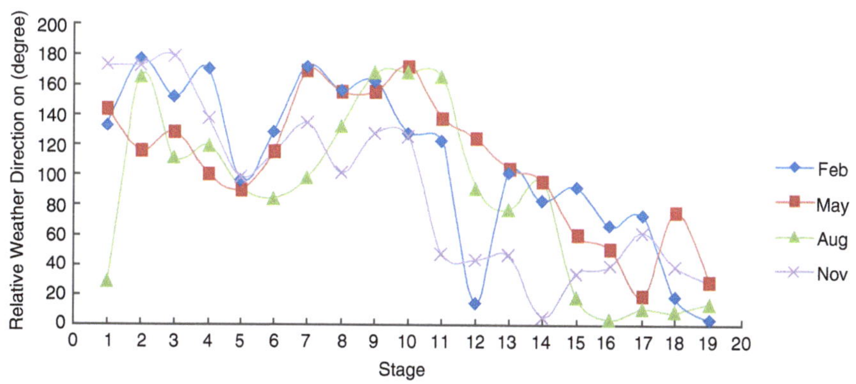


**Fig. 8** Relative weather direction distribution based on different conditions for case study 2

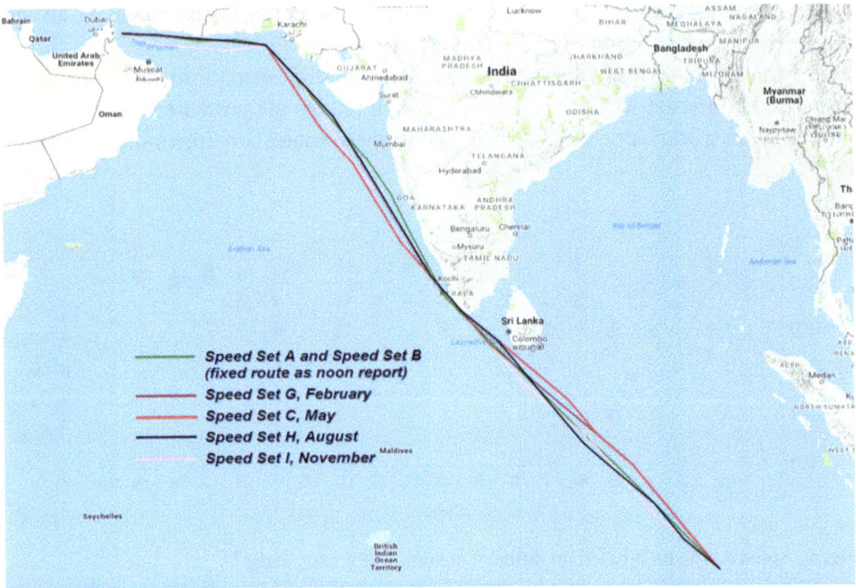


**Fig. 9** Optimal routes based on different conditions for case study 2

## 4 Conclusion

This paper presents a speed optimisation model towards energy efficient shipping. With the integration of a weather routing module, this model can provide stakeholders with optimum ship operations in various weather conditions towards minimum fuel consumption, according to the shipping schedule. Several case studies with Bulk Carriers were carried out. As can be seen from results, this model can help the case study ship save almost 1.5% in fuel consumption, which proves the model works well.

Of course, during the speed optimisation process, there are always many uncertainties such as the quality of ship performance prediction, weather forecasting limitations, service speed estimations or even operating profile of the engine, which make the final result insufficiently reliable. In the future, detailed research will be continued to further improve this speed optimisation model.

Finally, this model and the tool can be used towards: (1) Developing energy efficient ship operation strategies for shipping companies and can be used for fleet level optimisation. (2) Training crew to understand the effect of the environment on ship energy performance and develop weather routing or voyage optimisation. (3) Assessment of various technologies and operational practices on ship performance for individual voyages and annualised operations.

**Acknowledgements** This research is part of the project: Shipping in Changing Climates (EPSRC Grant no. EP/K039253/1). The author would like to express sincere thanks for the support from UK Research Council, University of Strathclyde and China Scholarship Council.

## References

Buhaug, Ø., Corbett, J. J., Endresen, Ø., Eyring, V., Faber, J., Hanayama, S., et al. (2009). *Prevention of air pollution from ships – Second IMO GHG study*. London, UK: International Maritime Organization.

Chen, H., Cardone, V., & Lacey, P. (1998). Use of operation support information technology to increase ship safety and efficiency. *SNAME Transactions, 106*, 105–127.

Du, Y., Chen, Q., Quan, X., Long, L., & Fung, R. Y. K. (2011). Berth allocation considering fuel consumption and vessel emissions. *Transportation Research, 47E*, 1021–1037.

Golias, M. M., Boile, M., Theofanis, S., & Efstathiou, C. (2010). The berth scheduling problem: Maximizing berth productivity and minimizing fuel consumption and emissions production. *Transportation Research Record, 2166*, 20–27.

Holtrop, J. (1984). A statistical re-analysis of resistance and propulsion data. *International Shipbuilding Progress, 31*(363), 272–276.

Howett, B. (2015). Report: Ship performance profile file format definition.

IHS Global Insight. (2009). Valuation of the liner shipping industry-Economic contribution and liner industry operations.

IMO. (2012). Guidelines for the development of a SEEMP. MEPC 63/23, Annex 9.

IMO. (2014). Third IMO GHG study 2014 – Exceutive summary and final report.

IMO. (2016). Guidelines for the development of a SEEMP. MEPC 70/18/Add.1, Annex 10.

Jussi, P. (2012). Overcoming the challenges in vessel speed optimization. *HANSA International Maritime Journal, 149*, 130–135. Jahrgang – 2012 – Nr.9.

Kwon, Y. J. (2008). Speed loss due to added resistance in wind and waves. *The Naval Architect, 3*, 14–16.

Lang, N., & Veenstra, A. (2010). A quantitative analysis of container vessel arrival planning strategies. *OR Spectrum, 32*, 477–499.

Lu, R., Cui, T., Turan, O., & Boulougouris, E. (2015). Speed management for energy efficient shipping. In *2015 SCC Conference, UK*.

Norstad, I., Fagerholt, K., & Laporte, G. (2011). Tramp ship routing and scheduling with speed optimization. *Transportation Research Part C: Emerging Technologies, 19*, 853–865.

Oxford Economics. (2014). The economic value of the EU shipping industry.

Ronen, D. (2011). The effect of oil price on containership speed and fleet size. *Journal of the Operational Research Society, 62*, 211–216.

Simonsen, M. H., Larsson, E., Mao, W., & Jonas, W. R. (2015). State-of-the-art within ship weather routing. In *Proceedings of the ASME 34th International Conference on Ocean, Offshore and Arctic Engineering*, St. John's, Newfoundland, Canada, OMAE.

Wang, S., & Meng, Q. (2011). Schedule design and container routing in liner shipping. *Transportation Research Record, 2222*, 25–33.

Wang, S., & Meng, Q. (2012). Liner ship fleet deployment with container transshipment operations. *Transportation Research Part E, 48*, 470–484.

# Underlying Risks Possibly Related to Power/Manoeuvrability Problems of Ships: The Case of Maritime Accidents in Adverse Weather Conditions

**Nikolaos P. Ventikos, Konstantinos Louzis, and Alexandros Koimtzoglou**

## 1 Introduction

The International Maritime Organization (IMO), in an effort to reduce Greenhouse Gas (GHG) emissions resulting from maritime operations, has introduced the Energy Efficiency Design Index (EEDI) requirement for new ships in the IMO Resolution MEPC.203(62) (IMO 2011). This new requirement was integrated in ANNEX VI of the International Convention for the Prevention of Pollution from Ships (MARPOL) as Regulation 21 in 2011 and has entered into force since 1.1.2013. The EEDI is a measure of energy efficiency of ships; it is expressed in grams of emitted $CO_2$ by the distance the cargo has been transported (g $CO_2$/t nm) and is intended to be a tool for designing new ships with lower emissions.

Because the formula for obtaining the EEDI for new ships does not provide limitations on how the required value should be met, there are concerns that this requirement will actually foster a culture of compliance rather than promote energy efficient shipping (Devanney 2011). These concerns relate to the speculation that vessel owners would be content to marginally comply with the EEDI requirement by simply installing smaller main engines on new ships. However, this could possibly lead to reduced capability of new vessels to maintain manoeuvrability in adverse weather conditions, as highlighted by the International Association of Classification Societies (IACS 2010). The potential risks for vessels related to inadequate propulsion power in adverse weather conditions include the inability to maintain or change course, i.e. drifting or loss of controllability, especially in restricted navigational areas (Shigunov and Papanikolaou 2014), and the inability to maintain ship speed.

N. P. Ventikos (✉) · K. Louzis · A. Koimtzoglou
National Technical University of Athens, School of Naval Architecture and Marine Engineering, Laboratory for Maritime Transport, Athens, Greece
e-mail: niven@deslab.ntua.gr; niven@mail.ntua.gr



A. I. Ölçer et al. (eds.), *Trends and Challenges in Maritime Energy Management*, WMU Studies in Maritime Affairs 6, https://doi.org/10.1007/978-3-319-74576-3_16

The work carried out by IACS led to the introduction of the 2013 Interim Guidelines for determining minimum propulsion power by IMO Resolution MEPC 232(65) (IMO 2013a), followed by amendments by the IMO Resolutions MEPC 255(67) (IMO 2014) and MEPC 262(68) (IMO 2015). The 2013 Interim Guidelines intend to provide a lower bound for potential reduction of the installed main engine power output, by setting a minimum propulsion requirement as a function of ship size. This minimum is determined as the power required to maintain manoeuvrability and navigate safely under adverse weather conditions, which are given specific values as a function of ship length (Papanikolaou et al. 2015).

Due to the fact that ships often operate in weather conditions that may exceed design expectations (Bitner-Gregerse et al. 2016), concerns have also been raised over the margins for wave height and wind speed values that determine what is considered as adverse weather according to the IMO 2013 Interim Guidelines. An additional concern over the IMO 2013 Interim Guidelines is related to what is considered adequate propulsion power to maintain manoeuvrability in adverse weather conditions. These disputes indicate that the defined weather conditions and adequate propulsion power may need revision, and were the basis for the SHOPERA (Energy Efficient Safe Ship OPERAtion) Project, which was conducted from 2013 to 2016 and was funded by the European Commission (FP7). As stated in Papanikolaou et al. (2016) the objectives of the Project included the specification of environmental conditions for assessing sufficiency of propulsion power and the development of simplified, yet effective, assessments for the safety margins of ships in adverse weather conditions.

This paper presents work carried out in the context of the SHOPERA Project and involves the analysis of accidents that may relate to insufficiencies in propulsion power and the application of a structured risk-based approach that aims to identify the ship types that present the highest risk of being involved in this type of accident. Section 2 provides a brief overview of the methodology that was employed to collect the relevant accident data and the risk analysis that followed. Section 3 presents key descriptive statistics, such as the distribution of ship types in the accident sample. Section 4 presents the risk analysis results, which combines the accident frequency that was calculated from the accident sample, with the severity of the consequences. Section 5 discusses the connection of the findings to the minimum propulsion power requirements, as defined in the IMO 2013 Interim Guidelines. The paper concludes the most important results from the analysis and interesting insights.

## 2 Methodology

This paper studies navigational accidents in adverse weather conditions where the sufficiency of installed propulsive power may have been a contributing factor to the development of the accident. The examined navigational accidents include manoeuvring in coastal waters and low-speed manoeuvring in restricted areas. The basic research question that this paper aims to answer is whether the risk for this type of accident is dependent on the characteristics of different ship types. Towards this end, a risk-based approach is employed to prioritize various ship types and identify the size classes that are most at risk considering this type of accident. The analysis is the next step of the work carried out in the context of the SHOPERA project (Ventikos et al. 2015) and aims to highlight the potential risk of insufficient propulsive power when navigating in adverse weather conditions.

The limitations that were faced relate to under-reporting of cases in maritime accident databases, the quantity and type of data available, the time frame for which accident frequencies could be calculated, and uncertainty regarding the underlying causes of the examined accidents.

A major concern when dealing with maritime accident data is the fact that a significant number of accidents is essentially not reported (Hassel et al. 2011), which is especially true for accidents that are flagged as "not serious". Previous studies, such as the work by Psarros et al. (2010), have highlighted the dangers in the uncritical use of maritime accident data in formal risk assessments. To maintain the integrity of the results of the analysis, only accidents that were flagged as "serious" were investigated.

Since information regarding accident consequences is, in many cases, scarce, there are limitations on the quantity and completeness of the available data. As a result, there is an inherent limitation on how many types of consequences may be considered for the consequence vector. The selected approach was to use the consequence category with the maximum amount of available information, which for this data set was hull damage suffered by the ship. Although the typical consequences of maritime accidents also include environmental impact and loss of life (Kristiansen 2005), the present analysis provides a useful estimation of the relative risk levels by considering only one of the main consequence categories (i.e. damage to property). An additional consideration for the analysis was that the available consequence data were qualitative descriptions that included indications of the severity of hull damage and its approximate location but no quantitative information, such as the extent of the damage and its exact location. This led to a semi-quantitative approach for handling accident consequences that involved structuring a dummy index with increasing levels of hull damage severity and combining the index values with the calculated accident frequencies to determine the risk level. The details of this approach are presented in the following sections.

The relative accident frequencies were calculated by dividing the number of accidents in the sample, either by ship type or accident category, by the corresponding number of ships that operated in the same time-period as the investigated accidents. The number of ships in operation is commonly used in maritime risk assessment as a basis for calculating relative accident frequencies as it represents the population of ships (fleet) that were at risk of being involved in an accident, such as in the work of Eliopoulou et al. (2016). Since the available fleet at risk number data were available for specific time periods, the accident frequencies were calculated only for the corresponding time periods. This implies that many accidents in our sample were excluded from the calculation of accident frequencies. However, the exclusion has only a small effect on the calculated frequencies since it represented only about 2% of the initial accident sample.

Uncertainty regarding the underlying causes of the examined accidents originates from the fact that, in most cases, the adequacy of installed power was not explicitly highlighted as the cause of the accident. This uncertainty translates into uncertainty regarding the relevance of the selected accidents to the context of the analysis. To address this limitation, a structured approach was implemented for inferring the relevance of the accidents by eliminating all irrelevant accident causes.

## 2.1 Data Collection

Information regarding the accidents in the sample was retrieved from the following data sources: IHS SeaWeb® Marine Casualty Database (IHS 2013); IMO Global Integrated Shipping Information System (GISIS) (IMO 2013b); and published accident investigation reports from National Authorities, which were used to retrieve missing information and for verification purposes.

The selected time frame for the analysis was accidents that occurred from January 1990 to December 2013. The analysis included accidents of ships that were built after January 1980 and up to 2011. Table 1 lists the criteria that were used for including accidents in the analysis.

As highlighted by Psarros et al. (2010), accident databases usually have limitations in capturing the causality of the recorded accidents. Therefore, the relevance of the selected accidents to the scope of the analysis was indirectly inferred by eliminating irrelevant cases following specific exclusion criteria in a structured three-stage verification process. In Stage I of this process, the collected accidents were screened according to the accident category; the excluded categories and the exclusion criteria from the following stages are listed in Table 2. In Stage II, exclusions were based on Froude number and the classification of accident severity (i.e. serious, not serious). Ships with Froude number greater than 0.5 that are considered High-Speed Crafts (HSCs), were excluded from the analysis. Since HSCs are usually equipped with main engines with high propulsion power, for achieving high speeds at sea, it is unlikely for these ships to have issues with the

**Table 1** Criteria for including accidents in the analysis

| Criterion | Value |
|---|---|
| Ship type | Container ships, Cruise ships, Ro-Ro Ferries, Ro-Ro Cargo, Pure Car Carriers, Gas Carriers, Tankers, Bulk Carriers, General Cargo |
| Year built | 1/1980–12/2011 |
| Gross tonnage | ≥400 gt |
| Accident year | 1/1990–12/2013 |
| Weather conditions | Adverse weather, as defined in the IHS SeaWeb® Marine Casualty Database |

**Table 2** Criteria used in the stages of the verification process for excluding accidents from the analysis

| Verification process | Criterion | Excluded cases |
|---|---|---|
| Stage I | Accident category | • Hull and machinery damage (incl. main engine and generator failures, blackouts, hull cracks, and failures relating to the navigational machinery),<br>• Fire/Explosion,<br>• Foundering due to ingress of water or cargo shift,<br>• Accidents in shipyard or dry dock |
| Stage II | Froude number | Fn > 0.5 (HSC) |
| | Accident severity | Not serious |
| Stage III | Relevance | Cases that were considered irrelevant based on available information from accident reports |

adequacy of installed power. Not serious accidents were excluded from the analysis because their relevance could not be safely determined, as recorded information is usually sparse and may, for example, only involve the date, location, and a vague description of consequences (e.g. minor damages). On the other hand, serious accidents are more likely to be reported in maritime accident databases (Hassel et al. 2011) and with more detailed information. In Stage III, the remaining information was cross checked against published accident reports from national maritime authorities and other sources such as the Lloyd's Casualty Week publication. The aim of this process was to minimize the false inclusion of accidents that are irrelevant to the scope of the analysis.

The following cases are specific examples of accidents that were included in the analysis. The General Cargo ship CARRIER was overwhelmed by strong winds and large waves while maneuvering on 3/4/2012 (MAIB 2013). The Bulk Carrier PASHA BULKER was rapidly approaching the shore on 8/6/2007 due to strong winds and while the Master attempted to maneuver to avoid grounding, the attempts failed due to the prevailing weather conditions (ATSB 2008).

The following cases are specific examples of accidents that were excluded from the analysis. The Ro-Ro Cargo ship CAMILLA sustained main engine breakdown on 23/1/2003 and developed a list. The LPG Carrier BENEGAS caught fire while discharging cargo on 13/7/2006.

## 2.2 Consequence Categories

Table 3 shows the consequence categories that were identified from the available information in the accident descriptions and in published accident reports.

The available information on the consequences of the accidents was scarce, with only about a third (34%) of the cases having any kind of relevant information. Hull

**Table 3** Description of available consequence data

| Consequence category | Description |
|---|---|
| Days off-hire | The number of days for the ship to re-enter service. |
| Injuries | The number of persons injured during the accident. |
| Fatalities | The number of persons that were reported dead or missing during the accident. |
| Pollution | The amount of oil that was reported to have been spilled to the sea. |
| Hull damage | A qualitative description for the hull damage the ship suffered from the accident that includes severity and approximate location. |

**Table 4** Description of categories for hull damage data

| Categories | HDI | Description |
|---|---|---|
| No | 0 | No hull damage reported. |
| Yes | 2 | Hull damage reported, with no effect on the integrity of the ship. |
| Damage above water line | 4 | Minor hull damage reported above water line, with minor effect on the integrity of the ship. |
| Damage below water line | 6 | Hull damage reported below water line that led to LOWI. |
| Total loss | 8 | Severe and extended hull damage that led to LOWI and resulted in the sinking of the ship. |

damage was selected as the consequence variable due to the greater availability of information (44%) compared to the other consequence data. Because the available hull damage data was in the form of qualitative descriptions regarding its severity and approximate location, it has been classified into discrete categories of increasing severity (see Table 3). Subsequently, the categorical severity levels were assigned to specific dummy values of a Hull Damage Index (HDI), that was constructed for the purposes of the present analysis, as detailed in Table 4. The HDI distinguishes between unspecified hull damage with no effect on the integrity of the ship, damage above and below the water line, and damage that led to total loss by Loss of Watertight Integrity (LOWI). For each accident scenario, the average HDI is used as an approximate indicator of the magnitude of the consequence severity, as these average values do not have any physical meaning. When the average HDI values are combined with the calculated frequencies for each accident scenario, the results are relative risk levels that are effective in comparing risk between the different ship types, size classes, and accident categories.

## 2.3 *Risk Triplets*

The main target of the presented analysis is to identify the ship types and the size classes that are most at risk when involved in accidents that occurred in adverse weather conditions and may have been related to insufficient propulsion power. An additional target of the analysis is to identify the types of accidents (i.e. collisions, groundings, contacts) that present the highest risk per ship type.

The risk-based methodology is based on the risk triplet idea introduced by Kaplan et al. (1981). Risk triplets are one of the many definitions of risk and are constructed in order to answer the following questions: What can happen?; How likely is it?; What are the consequences? (Kaplan 1997). A risk triplet is defined as a vector that consists of the following elements: Scenario, which is the description of the adverse/accident event; Frequency, which is the occurrence likelihood for each scenario; Consequence, which is the result of each accident scenario. The risk triplet idea has been used in the field of maritime risk assessment, such as in the work of van Dorp and Merrick (2011), who studied the risk of oil spills by generating scenarios via simulation based on a Maritime Transportation System (MTS) model, and Ulusçu et al. (2009), who studied vessel traffic risk in the Strait of Istanbul via a set of structured scenarios.

The examined accidents were classified into different scenarios, which are characterized by the following parameters: ship type, size class, accident category and installed power. It is noted that the reported Maximum Continuous Rating (MCR) is used as an indicator of installed power. For example, the case of a Handysize Bulk Carrier that was involved in a collision and had 8000 kW installed power would be codified as (Bulk Carrier, Handysize, Collision, 8000 kW).

The different scenarios are subsequently linked to average values of accident frequencies and consequence severity and, therefore, several risk triplets are constructed. In the context of this analysis, the scenario frequency is the average accident frequency that was calculated for the whole sample (i.e. that included N = 239 accident cases) and the scenario consequence is the average consequence severity. For each scenario, the following risk triplet groups were created based on the calculated frequency and consequence severity: by ship type, by ship type and size class, by ship type and accident category. Two examples of risk triplets are given in Table 5, where F1 denotes the frequency per ship type, F2 the frequency per ship type and size class, and F3 the frequency per ship type and accident category.

The risk triplets are consequently depicted as graphical risk-related curves on XY axes, where the X axis is the consequence and the Y axis is the frequency of the accident scenarios. The graphical curves that are produced are used to evaluate the different accident scenarios by ship type, size class, and accident category. The

**Table 5** Example of risk triplets in tabular format

| Scenario | | | | Frequency | | | Cons. |
|---|---|---|---|---|---|---|---|
| Ship type | Size class | Installed power (kW) | Accident category | F1 | F2 | F3 | HDI |
| Tanker | Handy-size | 14,932 | Ground. | 1.58E-04 | 7.65E-05 | 7.65E-05 | 6 |
| Bulk Carrier | Capesize | 15,225 | Ground. | 3.42E-04 | 2.61E-04 | 1.30E-04 | 8 |

different accident scenarios are prioritized by accident frequency and consequence severity. The graphical curves are produced for each ship type, which enables the comparison of the risk level between ship types, and are also broken down per size class and accident category.

## 3 Descriptive Analysis

A detailed descriptive analysis of the particular accidents in the compiled database has been presented in the work of Ventikos et al. (2015). Due to the limitations in the availability of fleet number data and information regarding the accident consequences, the risk analysis was based on a subset of the initial sample (N = 239). The subset consisted of accidents that occurred in the same time-period, as the one used for calculating the accident frequency from the initial sample, and for which at least one of the identified consequence categories was available. The first criterion was applied to ensure that the calculated frequencies, which are used in the risk analysis, correspond to the accident cases in the subset, while the second ensured that the subset is complete. Therefore, the resulting accident subset consisted of 100 accidents that may have been related to insufficient propulsion power and for which all the required consequence information was available. This section presents a brief descriptive analysis of the accident subset that was used in the estimation of the risk levels.

Figure 1 shows the percentage distribution per ship type in the original sample and the accident subset. As in the original accident sample, the majority of accidents involved General Cargo ships (38%) and Ro-Ro Ferries (28%). A chi-square test of independence showed that there are no statistically significant differences ($X^2 = 11.41$, $p = 0.179$), at a 5% level of significance, in the ship type distribution between the initial sample (N = 239) and the accident subset (N = 100). In addition, if the initial sample is assumed to be the population from which the subset is derived,

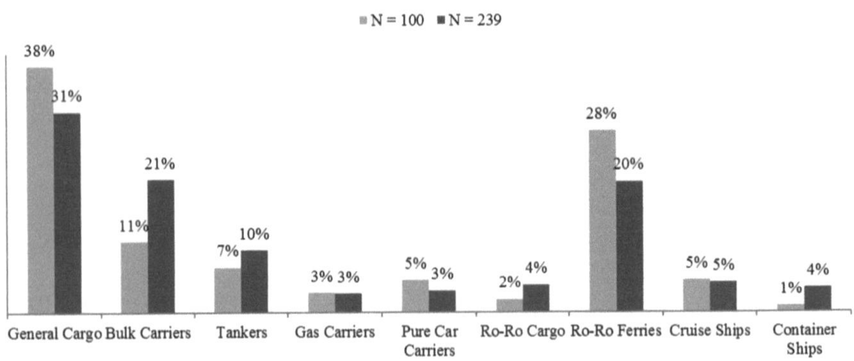


**Fig. 1** Comparison of percentage distribution per ship type between the original sample and the accident subset

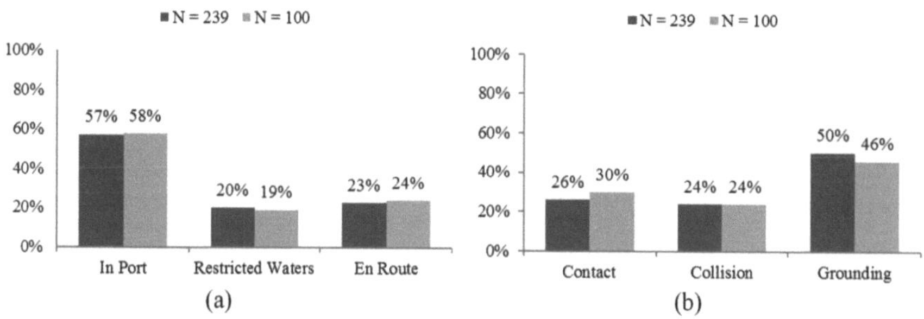


**Fig. 2** Comparison of percentage distribution per accident location (**a**) and accident category (**b**) between the original sample and the accident subset

then a statistical error of about 7.5% is calculated with a 95% confidence level. Therefore, the subset may be considered representative of the initial accidents of interest.

Figure 2 shows the percentage distribution per accident location (a) and accident category (b) for the original sample and the accident subset. It is observed that the distributions in the accident subset present only slight differences to the corresponding distributions in the original sample. Most accidents in the subset occurred in port (58%), followed by accidents while the ship was en route (24%) and those in restricted waters (19%). In addition, most accidents in the subset were groundings (46%), followed by contacts (30%), and collisions (24%).

# 4 Risk Analysis Results

This section presents the results of the risk analysis that was based on the accident subset (N = 100), which include the average accident frequencies that were used in the risk analysis and the prioritization of the identified accident scenarios.

## *4.1 Relative Accident Frequencies*

The average relative accident frequencies have been calculated from the original accident sample (N = 239) per ship type, size class, and accident category. The average accident frequencies were calculated by dividing the number of accidents in the compiled database to the number of ships in the Fleet at Risk, for the same time-period. The data for the Fleet at Risk was provided by Clarkson's Shipping Intelligence Network and was available for the time periods 1990–2013 and 1996–2013. The resulting accident frequencies per ship type are shown in Table 6. It is observed that General Cargo ships are the ship type most frequently involved in this type of

**Table 6** Fleet at risk and accident frequencies per ship type

| Ship type | Time-period | Fleet at risk | Accidents | Relative frequency |
|---|---|---|---|---|
| Container ships | 1996–2013 | 63,594 | 8 | 1.26E-04 |
| Cruise ships | 1996–2013 | 5252 | 11 | 2.09E-03 |
| Ro-Ro | 1996–2013 | 19,173 | 56 | 2.92E-03 |
| Pure Car Carriers | 1996–2013 | 9814 | 8 | 8.15E-04 |
| Gas Carriers | 1996–2013 | 22,195 | 6 | 2.70E-04 |
| Tankers | 1990–2013 | 145,159 | 23 | 1.58E-04 |
| Bulk Carriers | 1990–2013 | 143,158 | 49 | 3.42E-04 |
| General Cargo | 1996–2013 | 23,462 | 74 | 3.15E-03 |

**Table 7** Accident frequencies per ship and accident category (database)

| | Containerships | Cruise ships | Ro-Ro | Crude Oil Tankers | LNG | General Cargo | Bulk Carriers |
|---|---|---|---|---|---|---|---|
| Collision | 1.61E-02 | 4.60E-03 | 1.29E-03 | 9.35E-03 | 6.70E-03 | 5.50E-03 | 1.79E-04 |
| Contact | 3.65E-03 | 1.20E-03 | 1.36E-03 | 3.72E-03 | 2.80E-03 | 2.30E-03 | – |
| Grounding | 6.84E-03 | 9.80E-03 | 3.04E-03 | 7.49E-03 | 2.80E-03 | 7.40E-03 | – |

**Table 8** Accident frequencies as calculated per ship and accident category (FSAs)

| | Containerships | Cruise ships | Ro-Ro Ferries | Crude Oil Tankers | LNG | General Cargo | Bulk Carriers |
|---|---|---|---|---|---|---|---|
| Collision | 1.61E-02 | 4.60E-03 | 1.29E-03 | 9.35E-03 | 6.70E-03 | 5.50E-03 | 1.79E-04 |
| Contact | 3.65E-03 | 1.20E-03 | 1.36E-03 | 3.72E-03 | 2.80E-03 | 2.30E-03 | – |
| Grounding | 6.84E-03 | 9.80E-03 | 3.04E-03 | 7.49E-03 | 2.80E-03 | 7.40E-03 | – |
| Data source | MSC 83/21/2 | MSC 85/17/1 | MSC 85/17/2 | MEPC 58/INF.2 | MSC 83/21/1 | MSC 88/INF.8 | MSC 76/5/5 |

accident (3.15E-03), while accidents involving Container ships are the ones occurring least frequently (1.26E-04).

Table 7 shows the calculated accident frequencies per ship type and accident category; missing values for certain accident categories and ship types means that there were no corresponding cases in the accident database. Additionally, the Ro-Ro category contains both Ro-Ro Ferries as well as Ro-Ro Cargo ships.

Table 8 shows the corresponding accident frequencies that have been calculated in known (IMO-MSC) Formal Safety Assessments (FSAs); missing values mean that no available information was available for the specific accident category and ship type.

The frequencies calculated for the accidents in the database range from the order of $10^{-5}$ to $10^{-3}$, which is at least one order of magnitude lower than those calculated in the FSAs. This is reasonable, considering that accident frequencies in Table 8 are calculated for accidents regardless of the prevailing weather conditions, while accident frequencies in Table 7 are calculated only for accidents that occurred in adverse weather conditions.

Relative frequencies were also calculated per ship type and size class, as well as per ship type and accident category, by following the same rationale as for the accident frequencies per ship type.

## 4.2 Accident Scenarios Risk

For the prioritization of the different accident scenarios, the risk triplets are put into a tabular format and consequently sorted in order of increasing frequency and consequence severity by ship type. The accident scenarios are classified into the following two broad categories: lower frequency-higher consequence, and higher frequency-medium consequence. Two types of graphical risk-related curves were produced, each one based on the frequencies per ship type and size class and per ship and accident category. Each type of graphical risk-related curve corresponds to a different perspective and, therefore, focuses on a different aspect of the adverse situation. In addition, the graphical curves also depict a higher risk zone that is marked by a red rectangle. This region indicates the size classes, and accident categories that present the highest relative risk for each ship type. Because the higher risk region does not correspond to a quantitative risk acceptance criterion, it is used qualitatively to exhibit the difference between ship types.

The results in Table 9 focus on the ship type by comparing the accident frequency as well as the consequence severity range, where minimum and maximum consequence severity values are presented for each ship type. General Cargo ships have the highest relative risk values, i.e. highest accident frequency (3.15E-03) and consequence severity (Maximum HDI = 8), but also have the highest range of consequence severity, from HDI = 0 to HDI = 8. A wide range of consequence severity may be translated into increased overall risk due to the greater uncertainty it implies. The following accident scenarios are considered lower frequency-higher consequence: Tankers, Bulk Carriers, Gas Carriers, Ro-Ro Cargo Ships, and Pure Car Carriers. The following accident scenarios are considered higher frequency-medium consequence: General Cargo Ships and Ro-Ro Ferries.

**Table 9** Average accident frequencies and range consequence severity per ship type

| Ship type | Accident frequency | HDI | |
|---|---|---|---|
| | | Min | Max |
| Container ships | 1.26E-04 | 2 | |
| Cruise ships | 2.09E-03 | 0 | 2 |
| Ro-Ro | 2.92E-03 | 2 | 6 |
| Pure Car Carriers | 8.15E-04 | 2 | 6 |
| LNG Carriers | 5.53E-04 | 6 | |
| LPG Carriers | 1.15E-04 | 2 | 4 |
| Crude Oil Tankers | 1.58E-04 | 2 | 6 |
| Bulk Carriers | 3.42E-04 | 2 | 8 |
| General Cargo ships | 3.15E-03 | 0 | 8 |

Figure 3 shows the graphical risk-related curves that were produced per ship type and size class. It is noted that the labels indicate which ship type is represented by each line and that each point represents a specific size class of a ship type, with a corresponding average MCR, average accident frequency and average HDI. The

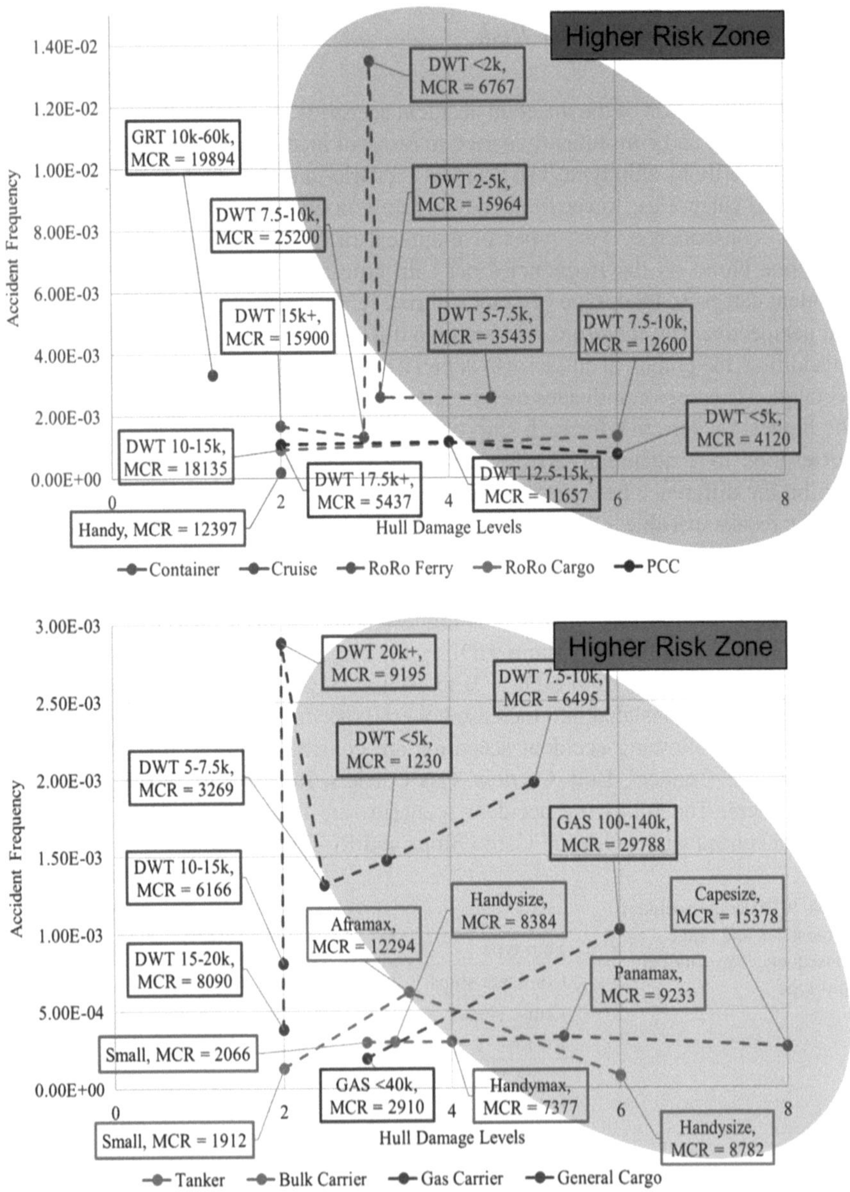


**Fig. 3** Graphical risk-related curves per ship type and size class

average MCR is the average value for the specific size class in the N = 100 risk sample.

What stands out in this graph is the spike in accident frequency for small Ro-Ro Ferries, i.e. with DWT <2000 t, which clearly places this size class in the higher risk zone, even though the consequence severity is relatively low (HDI = 3). This result is dominated by greater accident frequency and, therefore, indicates that small Ro-Ro Ferries are more susceptible to power-manoeuvrability problems in adverse weather conditions due to their relatively small power rating. Furthermore, Capesize Bulk Carriers have the highest consequence value (HDI = 8) but coupled with relatively low accident frequency. Therefore, even though large Bulk Carriers do not face such issues in adverse weather frequently, they may sustain hull damage severe enough to result in LOWI and total loss of the ship. It should be noted that these are examples of informative results that may form the basis for further research into the details of the underlying cause mechanisms.

Table 10 lists the accident scenarios in the higher risk region, per ship type and size class, and classifies them into lower frequency-higher consequence and higher frequency-lower consequence categories.

Figure 4 shows the graphical risk-related curves that were produced by ship type and accident category. It is noted that each point represents a different accident category for each ship type, which corresponds to an average accident frequency and an average hull damage level.

It is observed that contacts and groundings have the highest risk values across all ship types. This is due to the fact that they are the most frequent types of accidents in the compiled database (which consists of 46% groundings and 30% contacts) but also because they have higher consequence values.

Table 11 lists the accident scenarios in the higher risk region, per ship type and size class, and classifies them into lower frequency-higher consequence and higher frequency-lower consequence categories.

**Table 10** Classification of accident scenarios per ship type and size class

| Ship type | Lower frequency-higher consequence | Higher frequency-lower consequence |
|---|---|---|
| Pure Car Carriers | DWT <5 k | |
| Ro-Ro Cargo | DWT 7.5–10 k | |
| Tankers | Handysize | |
| Bulk Carriers | Panamax Capesize | |
| Gas Carriers | Cap. 100–140 k $m^3$ | |
| Ro-Ro Ferries | | DWT <2 k<br>DWT 5–7.5 k |
| General Cargo | | DWT <5 k<br>DWT 7.5–10 k |

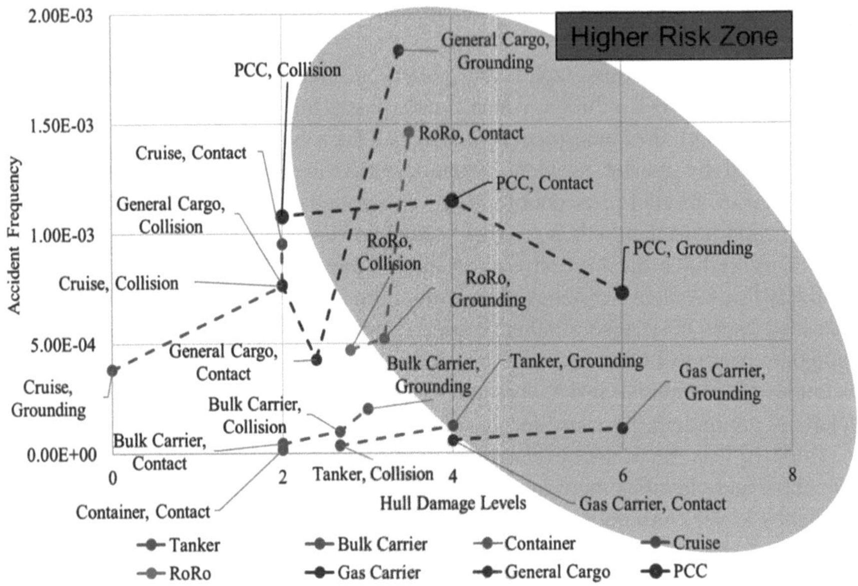


**Fig. 4** Graphical risk-related curves per ship type and accident category

**Table 11** Classification of accident scenarios per ship type and accident category

| Ship type | Lower frequency-higher consequence | Higher frequency-lower consequence |
|---|---|---|
| Gas Carriers | Grounding<br>Contact | |
| Pure Car Carriers | Grounding<br>Contact | |
| Ro-Ro | | Contact<br>Grounding |
| General Cargo | | Grounding |

## 5 Discussion

As mentioned in the introduction section, the maritime industry has been debating the effectiveness of the IMO EEDI regulation in ensuring the sufficiency of propulsive power for safe navigation in adverse weather conditions since its introduction in 2011. An interesting question that arose from the findings of the presented analysis was whether ships that are involved more frequently and with greater consequences (higher risk) in accidents in adverse weather conditions, and may be related to insufficient propulsion power, comply with the IMO 2013 Interim Guidelines. To address this question, the minimum propulsion power requirement is compared to the average installed propulsion power of the ships in the accident sample. The

**Table 12** Comparison of minimum propulsion power requirement (IMO 2013 Interim Guidelines) with the average installed power in the accident sample

| Ship type | Size class | DWT range | Minimum propulsion power requirement range | Average installed propulsion power in the accident sample |
|---|---|---|---|---|
| Bulk Carriers | Panamax | 65–100 k | 7390–9794 kW | 9233 kW |
| | Capesize | >100 k | >9794 kW | 15,378 kW |
| Tankers | Handysize | 10–60 k | 3942–7387 kW | 8782 kW |

results for the size classes of Tankers and Bulk Carriers that have the highest calculated risk are listed in Table 12, where the minimum propulsion power requirement is calculated according to the following formula (IMO 2013a):

$$\text{Minimum Power Line Value} = a\ (\text{DWT}) + b$$

It is observed that Tankers and Bulk Carriers that were involved in accidents in adverse weather conditions that may be related to insufficient propulsion power comply with the minimum propulsion requirements, as defined in the IMO 2013 Interim Guidelines. In fact, the average installed power for Capesize Bulk Carriers in the accident sample is about 60% over the minimum requirement, while Handysize Tankers are about 19% over the requirement. These results may indicate a need for revising the minimum propulsion power requirements in adverse weather conditions, while also reconsidering the limits for the adversity of the weather conditions that a ship faces during its operation. Future work to address this issue may be directed towards more detailed analysis and further examination of the causes of this type of accident, to provide a solid link to insufficient propulsion power.

## 6 Conclusions

The implementation of the concept of risk triplets, which are vector quantities that determine the risk level for specific accident scenarios, enabled the identification of ship types, size classes, and accident categories that face high risk when involved in accidents in adverse weather conditions that may be related to insufficient propulsion power. Informative visualizations of the constructed risk triplets, in the form of graphical risk-related curves, were used to provide insights into the relative risk between ship types, size classes, and accident categories.

The analysis presented herein was based on the calculation of relative accident frequencies from an accident sample and the specification of the consequence severity. The calculated accident frequencies are in the range of $10^{-5}$ to $10^{-3}$ accidents per shipyear. When compared to corresponding accident frequencies from published FSAs, they are found to be about one order of magnitude lower. This is reasonable considering that the frequencies from FSAs do not consider the prevailing weather conditions, while this work examined accidents that occurred only in adverse weather conditions. Due to the limited available information regarding the consequences of individual accidents, the analysis focussed only on hull damage which was mostly available (44% of the examined cases), compared to other consequence data. In addition, due to the qualitative nature of the available hull damage information, a semi-quantitative approach was employed where increasing levels of severity were assigned a dummy index value.

The comparison of relative risk levels between the different ship types yielded the following results: Ro-Ro Ferries and General Cargo ships have high risk values due to high accident frequency and medium level of hull damage consequences; Gas Carriers, Tankers, and Bulk Carriers have high risk values due to the observed high level of hull damage consequences; and Cruise ships and Container ships have relatively lower risk values that do not place them in the higher-risk zones.

The comparison of relative risk levels between the different accident categories showed that groundings and contacts in adverse weather conditions present the highest risk values across all ship types.

The data that supported the analysis indicates that ships with seemingly adequate installed propulsion power, as defined by the IMO 2013 Interim Guidelines, may, in fact, be involved in navigational accidents in adverse weather conditions with possible adverse consequences, ranging from minor hull damage up to and including total loss of the ship. As was the case for Bulk Carriers (about 60% over the minimum requirement) and Tankers (about 19% over the minimum requirement) in the accident sample, ships may be involved in this type of accident even if they exceed the minimum requirement by a relatively large margin. In addition, considering that Bulk Carriers and Tankers were found to be in the higher risk zones, compliance with the current minimum propulsion requirements in adverse weather conditions does not seem to ensure navigational safety in these conditions. The results are also indicative of the fact that ships may operate in a far more hostile environment that may surpass the worst-case design specifications. Therefore, because maintaining manoeuvrability in adverse weather conditions is paramount for safety, a revision of what may be considered adequate propulsion power for safe navigation in the full range of weather conditions that a ship may face during its operation may be necessary. The insights from the present analysis regarding potential safety issues of ships in adverse weather conditions may form a solid basis for focusing further research on the underlying cause mechanisms for specific ship types and accident categories. Finally, the fact that such informative results may be used in establishing and fine-tuning new regulatory requirements for efficient reduction of maritime emissions is a solid argument for supporting and broadening science-based policy making in the maritime industry.

**Acknowledgements** The work presented in this paper was partly supported by the Collaborative Project (Grant Agreement number 605221) SHOPERA (Energy Efficient Safe SHip OPERAtion) co-funded by DG Research of the European Commission.

## References

ATSB. (2008). *Independent investigation into the grounding of the Panamanian registered bulk carrier Pasha Bulker on Nobbys Beach, Newcastle, New South Wales, 8 June 2007*. Canberra: ACT.

Bitner-Gregerse, E. M., Soares, C. G., & Vantorre, M. (2016). Adverse weather conditions for ship manoeuvrability. *Transportation Research Procedia, 14*, 1631–1640. https://doi.org/10.1016/j.trpro.2016.05.128

Devanney, J. (2011). EEDI absurdities. *Cent Tanksh Excell*, 1–6.

Eliopoulou, E., Papanikolaou, A., & Voulgarellis, M. (2016). Statistical analysis of ship accidents and review of safety level. *Safety Science, 85*, 282–292. https://doi.org/10.1016/j.ssci.2016.02.001

Hassel, M., Asbjornslett, B. E., & Hole, L. P. (2011). Underreporting of maritime accidents to vessel accident databases. *Accident Analysis and Prevention, 43*, 2053–2063. https://doi.org/10.1016/j.aap.2011.05.027

IACS. (2010). EE-WG 1/4 Minimum required speed to ensure safe navigation in adverse conditions.

IHS. (2013). Sea-Web Casualties-Comprehensive Vessel Casualty Database. http://maritime.ihs.com.

IMO. (2011). Resolution MEPC.203(62), Amendments to the Annex of the Protocol of 1997 to amend the International Convention for the Prevention of Pollution from Ships, 1973, as modified by the Protocol of 1978 relating thereto.

IMO. (2013a). Resolution MEPC.232(65), Interim guidelines for determining minimum propulsion power to maintain the manoeuvrability in adverse conditions.

IMO. (2013b). Global Integrated Shipping Information System (GISIS). https://gisis.imo.org.

IMO. (2014). Resolution MEPC.255(67), Amendments to the 2013 Interim Guidelines for Determining Minimum Propulsion Power to Maintain the Manoeuvrability of Ships in Adverse Conditions.

IMO. (2015). Resolution MEPC.262(68), Amendments to the 2013 Interim Guidelines for Determining Minimum Propulsion Power to maintain the Manoeuvrability of ships in Adverse Conditions.

Kaplan, S. (1997). The words of risk analysis. *Risk Analysis, 17*, 407–417. https://doi.org/10.1111/j.1539-6924.1997.tb00881.x

Kaplan, S., Garrick, B. J., Kaplin, S., & Garrick, G. J. (1981). On the quantitative definition of risk. *Risk Analysis, 1*, 11–27. https://doi.org/10.1111/j.1539-6924.1981.tb01350.x

Kristiansen, S. (2005). *Maritime transportation: safety management and risk analysis*. Routledge.

MAIB. (2013). *Report on the investigation of the grounding of the cargo ship Carrier at Raynes Jetty in Llanddulas, North Wales on 3 April 2012*. Southampton, UK.

Papanikolaou, A., Zaraphonitis, G., Bitner-Gregersen, E., et al. (2015). Minimum propulsion and steering requirements for efficient and safe operation (SHOPERA). In *Proceedings of the 37th Motorship Propulsion and Emissions Conference* (pp. 1–13). Hamburg.

Papanikolaou, A., Zaraphonitis, G., Bitner-Gregersen, E., Shigunov, V., El Moctar, O., Guedes Soares, C., et al. (2016). Energy efficient safe ship operation (SHOPERA). *Transportation Research Procedia, 14*, 820–829. https://doi.org/10.1016/j.trpro.2016.05.030

Psarros, G., Skjong, R., & Eide, M. S. (2010). Under-reporting of maritime accidents. *Accident Analysis & Prevention, 42*, 619–625. https://doi.org/10.1016/j.aap.2009.10.008

Shigunov, V., & Papanikolaou, A. (2014). Criteria for minimum powering and manoeuvrability in adverse weather conditions. *Ship Technology Research, 62*, 140–147. https://doi.org/10.1080/09377255.2015.1104090

Uluşçu, Ö. S., Ozbaş, B., Altıok, T., & Or, İ. (2009). Risk analysis of the transit vessel traffic in the strait of Istanbul. *Risk Analysis, 29*, 1454–1472. https://doi.org/10.1111/j.1539-6924.2009.01287.x

van Dorp, J. R., & Merrick, J. R. W. (2011). On a risk management analysis of oil spill risk using maritime transportation system simulation. *Annals of Operations Research, 187*, 249–277. https://doi.org/10.1007/s10479-009-0678-1

Ventikos, N. P., Koimtzoglou, A., Louzis, K., & Eliopoulou, E. (2015). Statistics for marine accidents in adverse weather conditions. In *Maritime Technology and Engineering - Proceedings of MARTECH 2014: 2nd International Conference on Maritime Technology and Engineering*.

# Simulation-Based Support to Minimize Emissions and Improve Energy Efficiency of Ship Operations

**Michael Baldauf, Syed Raza Ali Mehdi, Michele Schaub, Knud Benedict, Gerd Milbradt, Georg Finger, and Sandro Fischer**

## 1 Introduction

For sailing/manoeuvring, ships need energy. Nowadays, the vast majority of ships still burn fossil fuel to produce the energy needed. Burning fuel causes emission of greenhouse gases (GHG), which have a negative impact on the environment and on humans (see for example Oeder et al. 2015, or BTRE 2005). Saving energy and reduction of GHG emissions is an urgent compelling need for shipping.

So far the mandatory ship's voyage planning process addresses energy efficiency aspects especially through weather routing on a macro level, covering the passage of oceans only. In these open seas, ships operate in 'sea mode' in stable conditions, with almost no change to the engine parameters and, except when performing for example a course change, limited use of controls. However, when ships are approaching coastal areas and entering or leaving ports, they operate in 'manoeuvring mode' to safely pass fairways to harbour basins for berthing. In these coastal areas, close to ports and in harbours, ships operate in transient conditions, characterized by frequent use of controls and substantial changes to engine

M. Baldauf (✉)
World Maritime University, Malmö, Sweden

Hochschule Wismar, Department of Maritime Studies, University of Applied Sciences: Technology, Business and Design, Institute for Innovative Ship Simulation and Maritime Systems (ISSIMS), Rostock, Germany
e-mail: mbf@wmu.se

S. R. A. Mehdi
World Maritime University, Malmö, Sweden

M. Schaub · K. Benedict · G. Milbradt · G. Finger · S. Fischer
Hochschule Wismar, Department of Maritime Studies, University of Applied Sciences: Technology, Business and Design, Institute for Innovative Ship Simulation and Maritime Systems (ISSIMS), Rostock, Germany

 
A. I. Ölçer et al. (eds.), *Trends and Challenges in Maritime Energy Management*, WMU Studies in Maritime Affairs 6, https://doi.org/10.1007/978-3-319-74576-3_17

status. Energy consumption and emissions are comparably high in this phase of the voyage and research on the specific processes promises very substantial improvements for energy efficiency and reduction of emissions.

Empirical studies into manoeuvring processes testify that manoeuvring begins with the navigator's thinking ahead of a ship's path and following this virtually planned path by adapting the control handles of the ship (rudder angle, thruster power, and engine revolutions) in order to match the real ship's positions and track with the plan in the navigator's mind (Baldauf et al. 2012). This practice implies a certain kind of trial and error strategy to find the suitable combination of control settings to realize the mental plan of the navigator. Both energy efficiency and the issue of emissions are more or less neglected in this type of planning; the mental plan focuses on safety only. Moreover, quite often greater rudder angles are ordered and used; thruster and engine power are, if at all, considered marginal during planning because their specific effects are not precisely known. It is, therefore, assumed that pre-planning of manoeuvring regimes have potential to decrease negative effects and may potentially contribute to reducing fuel consumption and emissions.

The e-Navigation initiative of the International Maritime Organization (IMO) and the International Association of Marine Aids to Navigation and Lighthouse Authorities (IALA) addresses the safety of navigation from 'berth to berth', but at the same time addresses the protection of the marine environment (IMO 2006). This, although not specifically mentioned, also implies the reduction of greenhouse gas (GHG) emissions because avoiding or at least reducing air pollution substantially contributes to the protection of the environment. New electronic means, such as decision support systems, will make use of the latest information and communication technologies and support the human operator on the ship's navigational bridge to ensure sustainable operation under varying environmental conditions and for different statuses of the ship in terms of draught, trim etc., and shall support environmentally-friendly operation of the ship by the crews.

The analysis at hand focuses on operational measures that may contribute to energy-efficiency and reduction of emissions. The research is related to the development of tools and functions that will support decision making for manoeuvring. The tools under development are aimed at real-time manoeuvre support and have already proven their potential for integration into maritime education and training programs. The authors are specifically researching the impact of manoeuvring assistance with a prototype decision support system basing on Fast-Time-Simulation (FTS) technology (ISSIMS 2016). The core functions allow prediction of path and ship status in almost real-time, as well as planning of different steering sequences to achieve a certain manoeuvring objective well in advance of approaching the next harbour. This prototype, so far, allows simulation of different strategies and will be further extended by motor process models. The ultimate aim is to enable ship operators to plan manoeuvres, as well as to provide immediate feedback on the consequences of any manoeuvring action in real-time not only in terms of safe manoeuvring but also in relation to energy consumption and emissions. Covering those aspects will enable detailed pre-planning to identify and decide on the manoeuvre plan with the minimum need for fuel/energy and the steering sequence with the lowest GHG emissions. The bridge team then becomes aware of the consequences of an intended action, can improve its understanding and learn which steering sequences are convenient and beneficial and which are rather adverse and harmful (Benedict et al. 2015).

# 2 Aspects of Manoeuvring and Manoeuvring Support

## *2.1 Principal Background*

The three main external factors influencing the manoeuvring of a ship are, firstly, safety requirements, secondly, time pressure resulting from tight scheduling of ship operations, and thirdly, the economic constraints resulting in demands for efficient use of fuel. Finally, under the umbrella of all the aforementioned aspects, the requirements for environmentally-friendly manoeuvring with minimal emissions of carbon, nitrogen-sulphur oxides and particles apply as a fourth factor of rising importance. All these aspects need to be taken into account when navigating a ship from berth to berth.

Different phases of a ship's voyage have different priorities and require different measures with respect to the principle manoeuvring strategies. Overall safety has to be ensured, but it usually competes with demands for fuel savings while keeping the time schedule of arrival/departure. Minimizing air pollution by exhaust gases of the engine is usually considered of minor importance in the immediate decision making process for manoeuvring in fairways, harbour areas and basins.

As already mentioned, during sea trials in open sea, the ship's engine is operating in constant and stable conditions. There are few manoeuvres required. More frequent manoeuvres are necessary in coastal areas and, finally, when approaching ports in river fairways. Morover in harbour basins, there is a high frequency of engine and rudder and thruster manoeuvres. Berthing and un-berthing are usually supported by the use of thrusters. In those final phases of a ship's voyage, engine status, rudder angles and power for thrusters are constantly changing. Nowadays, ships can also be equipped with electric podded azimuth thruster systems (Azipod) providing better manoeuvrability. However, in respect to energy consumption the same principals apply. When revolutions, pitch and angle of the thrusters are varied according to the specific needs of a nautical manoeuvre, a higher engine load occurs, as is depicted in the graphs of Fig. 1.

In such phases a ship engine operates in transient conditions, which is the main cause of higher emission rates as has been proven by measurements in test-bed installations in laboratories and on-board ships in real operational conditions (Schaub et al. 2015).

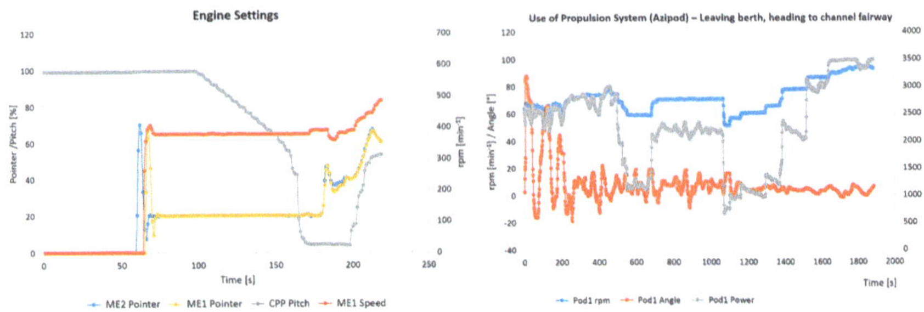


**Fig. 1** Sample of time history of engine parameters during manoeuvring to leave the berth and harbour for a ferry equipped with Azipod propulsion system (source: real time measurement)

## 2.2 *Manoeuvring Strategies*

A manoeuvring strategy focussing on safety may usually (but not necessarily) take more time, while manoeuvring strategies aimed at keeping a certain berthing time are usually characterized by more intensive use of steering controls and, consequently, require more energy, consume more fuel and, as a result, produce more GHG emissions.

Safe manoeuvring is dedicated to always keeping a safe distance from fairway borders or shallow water areas, meaning trying to keep to the centre of a fairway or to a reasonably safe distance from obstacles. The manoeuvres are oriented in always having a "reserve"; this translates into not using the full range of the control settings to be able to react to sudden changes or events.

A time efficient manoeuvring strategy is, on the other hand, aimed at the quickest completion of a manoeuvre using almost the full range of the available control settings at a relative high speed.

An economic manoeuvring strategy's focus would have to take into account manoeuvring with the lowest possible use of resources like fuel or electrical power. This may result in a "soft" and rather cautious use of rudder, thruster and engine. However, this cannot always be considered as safe manoeuvring because low speed and little use of propulsion leads to less rudder forces and, consequently, to lower turning rates and greater stopping distances. Finally, an environmentally-friendly manoeuvring strategy should look at the sequence of manoeuvres that have minimum production of exhaust gases and emissions.

However, there is a great number of combinations of manoeuvring commands for engine, rudder and thruster to safely berth a ship, but the most energy-efficient, economic sequence of manoeuvring commands is not exactly (or not at all) known by the bridge team during the decision making process for manoeuvring. That is why a related decision support systems seems to be a compelling need.

## 2.3 Enhanced e-Navigation-based Manoeuvring Support

With respect to global trends in research and technological development, the current situation is characterized by increasing automation of all on-board processes (inter alia by means of digitalisation and big data), the growing direct connection between shore-based and on-board systems (inter alia through satellite, broadband and other connections, networking) and the tendency to use all technological developments to realize unmanned shipping and autonomous navigation (among others Rødseth and Burmeister 2012; Rødseth et al. 2013).

However, e-Navigation as one of the most important present activities is not aimed at taking the human operator out of the loop, but at supporting the operator by better, more effective provision of related information through electronic means. Developments under the e-Navigation initiative are to address the human operator on-board and ashore. That is why the development of enhanced manoeuvring support is aimed at 'better' presentation of information to improve the situational awareness of the bridge team (the captain, officer of the watch (OOW) and the pilot) and support their decision making. For this specific purpose, predictions that take into account actual settings of manoeuvring controls as well as the immediate visualisation of the reaction of the ship to these changes are needed. A basic concept has been developed and the principal structure of such enhanced manoeuvre support is shown in Fig. 2.

The enhanced comprehensive manoeuvring decision support system comprises of a data collection module (Fig. 2 left), a data storage and processing module (Fig. 2 centre light blue) and a display module (Fig. 2 right). It will consider not only the actual and future navigational states of the ship but also the technical states of the main engine and its auxiliary systems. The first module (Fig. 2, left hand side) is to provide the parameters describing the actual status of the ship (incl., among others, moving and steering parameters). At the same time, the actual environmental conditions including the sea area with its hydrographical and hydro-meteorological data but also other targets and objects that constrain the navigable space will be gathered.

The core element of the system, the data storage and processing module, enables taking into account aspects of environmental-friendly manoeuvring. It provides a data base and a simulation module. The database is used to collect and analyse data related to the manoeuvring performance and will also provide the corresponding engine data when the ship manoeuvres in shallow water areas, like port approaches or even in harbour basins when berthing. The FTS module calculates and predicts the future ship states according to the prevailing circumstances of the area and weather conditions as well as the actual settings of the steering equipment (e.g. engine revolutions, rudder angles, thruster power). FTS allows calculation of the ship's motion for more than 24 min ahead of time within less than a second of a computer's calculation loop. The following basic equations are used to model a ship's hydrodynamics:

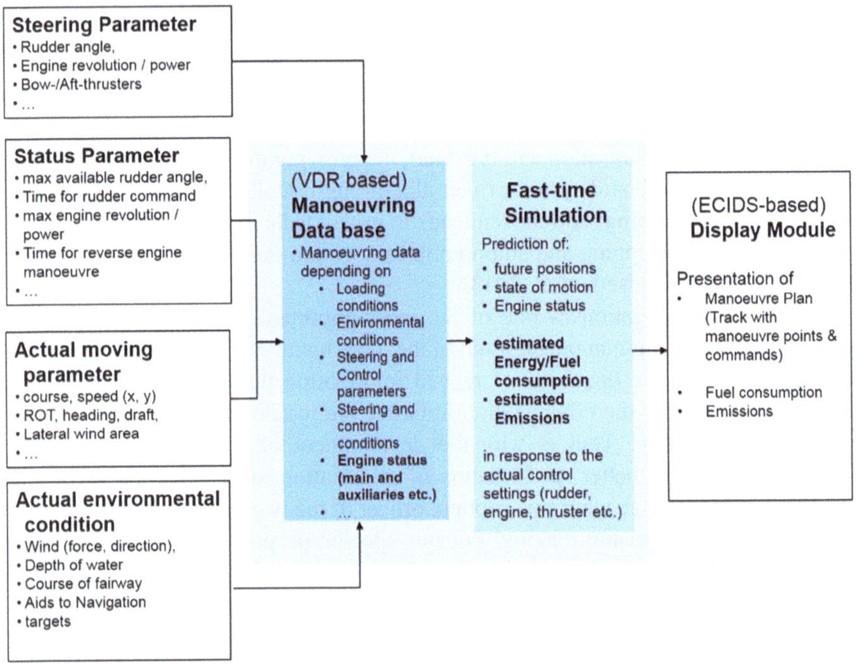


**Fig. 2** Structure of an enhanced e-Navigation-based manoeuvring assistance system

$$X = m(u - rv - x_G r^2)$$
$$Y = m(v + ru + x_G r)$$
$$N = I_z r + m x_G (v + ru)$$

The right hand side describes the effects of inertia, where u and v are the speed components in longitudinal and transverse direction, while r represents the ship's turning rate. The mass of the ship is given by m, $x_G$ is the distance from origin to re of gravity and $I_Z$ is the moment of inertia. The forces X and Y and the turning moment N are on the left hand side. The complete details of the calculations are described in Benedict et al. (2012).

Finally, the display module provides visualisation of the future states and can be, for instance, the mandatory ECDIS or conning display of an Integrated Navigation System (INS). A sample is provided in Fig. 3.

Figure 3 demonstrates the advantages of FTS-based predictions. Future states of the ship are presented in an ECDIS-environment for the next 5 min (the time horizon is freely configurable); each ship state is given in one minute time steps and the turning of the ferry using actual port rudder, reverse engine and full thruster is visualized. In order to highlight the innovation of the simulation-based prediction, the conventional curved heading vector is shown (magenta) in addition. The advantage of the FTS based prediction is obvious, as the curved heading vector indicates a

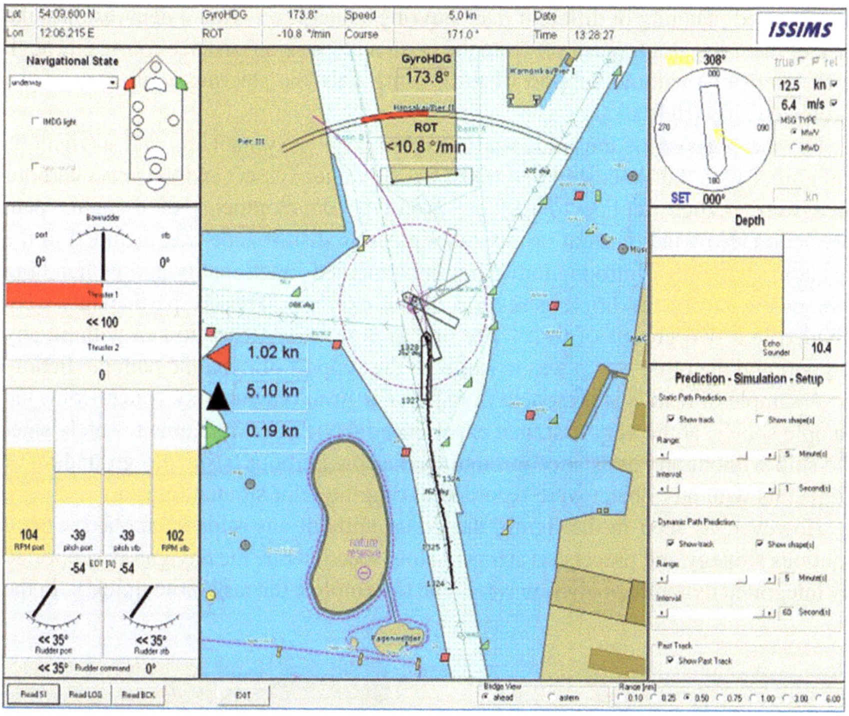


**Fig. 3** Integrated prediction of future ship states (position, heading in response to a combined rudder, thruster and engine manoeuvre) adapted from Baldauf et al. (2012)

further longitudinal movement straight ahead with tendency to port-side on the basis of historical data and, especially, not taking into account the actual control settings.

# 3 Potential Impact of Enhanced Manoeuvring Assistance

## *3.1 Pilot Study to Investigate Potential Effects of Dynamic Predictions*

One of the fundamental questions in respect to the development of tools for manoeuvring support and decision making is the users' level of acceptance and the potential impact of the use of such tools on the manoeuvring performance of navigators.

Initial preliminary studies have been conducted in the full-mission ship-handling simulator at the Maritime Simulation Centre Warnemuende (Germany) in order to investigate the potential impact of the developed and integrated functions for dynamic prediction of the future ship's path and status and to study potential effects of advanced planning of different manoeuvring strategies. Figure 4 demonstrates the task of the navigator to push off from the berth and leave on easterly courses through a channel (left) without support of tugs, using only the steering equipment of the simulated ship (right).

For the purpose of the pilot study, experienced ferry captains and navigators, unfamiliar with the exact ship (a 4.000 TEU container vessel) and the exact harbour area were invited ad hoc to test the prototype developments of dynamic path prediction. No wind, current or any other external disturbance was included in the scenario. A detailed introduction to the implemented functionality and its handling was given during the briefing session of the exercise. The six participants were divided into two groups of three. The first group was assigned to sail without any support and the other group was to sail with the support of dynamic path prediction.

A sample of results is presented in Fig. 5. Far from drawing any conclusions, the sample shows, at the very first, that experienced navigators were able to safely steer the ship without any assistance through the narrow harbour basin. No grounding or collisions with any object were recorded during the pilot simulation runs.

However, as seen in the figure, the group without any support applied a more cautious strategy and proceeded at a very slow speed, while the navigators supported by integrated dynamic prediction were able to complete the task more quickly. In the

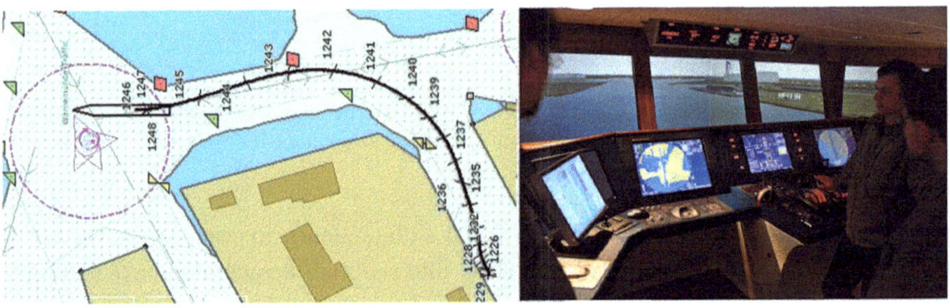


**Fig. 4** Pilot study on the impact of enhanced manoeuvring assistance. Required task (left) and Simulator Bridge with dynamic prediction integrated in the INS

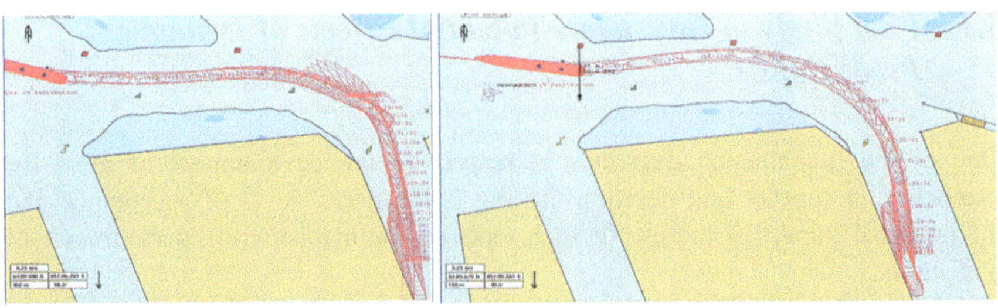

**Fig. 5** Recorded tracks of a simulated manoeuvre to leave the berth from a harbour basin; left: without prediction for manoeuvring support; right: with dynamic prediction)

samples presented in Fig. 5, the navigator without support took around 6 min more than the navigator who was supported by prediction. Furthermore, also very obvious is the smaller space that is covered by the navigator with the prediction. It was observed that he adapted the rudder angle and thruster power in a way that the predicted path was near to the centre of the fairway. Unlike the navigator without any support who made numerous "correcting" rudder and thruster manoeuvres to carefully turn into the fairway, the curve made with the support of the prediction looks smoother and covers less space.

The simulation studies will, of course, need to be elaborated further and detailed quantitative analysis is on-going. However, the tendency that enhanced manoeuvring support contributes to smoother manoeuvring with fewer overall steering commands was already observed in earlier investigations.

## 3.2 *On-going Developments*

Further experimental studies are under preparation to investigate the potential effects of preparing detailed manoeuvre plans well in advance (before arriving at the port areas).

A similar experimental set-up is chosen and implementations are on-going. The scenario will require the participants, again, to steer a ship coming from open sea, entering a channel fairway up to the turning area where the ship has to be turned and then go astern to the dedicated berth.

As in the first experiment, it is planned to split the participants into three groups, where one group will perform the task without any support, the second with the support of a manoeuvre plan and the third group will use the prepared manoeuvre plan and dynamic prediction for the execution of the harbour manoeuvres and the berthing operation.

The experimental set-up is prepared again in a full-mission ship-handling simulator providing the equipment and the full hydrodynamic model of a combined RoRo-Passenger ferry of 200 m length and 30 m breadth, operating in the simulated harbour area. The provided steering equipment consists of two propellers and rudders and bow thruster. The steering sequence for the complete complex manoeuvre was planned together with the candidates, using the newly developed manoeuvre planning module. A sample of a complete manoeuvring plan is given in Fig. 6.

Figure 7 shows a screenshot of a potential version of the prototype's display module. It provides an ECDIS-based graphical (left) and an alphanumerical presentation (table bottom) of the manoeuvring plan integrated into the modified conning display interface.

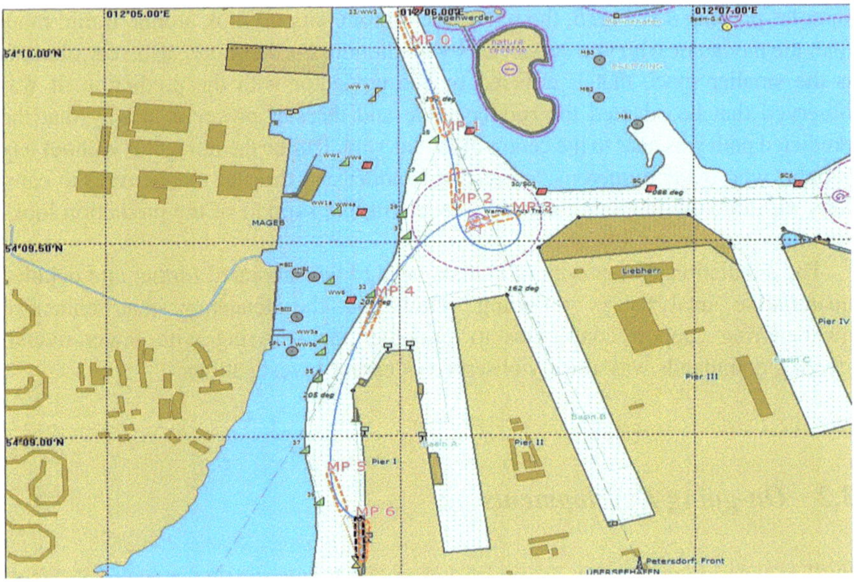


**Fig. 6** Enhanced Manoeuvring Assistance through pre-planning of the steering sequence for a complex manoeuvre coming from open sea, passing a channel fairway turning the ship in the turning basin and proceed backwards to the berth

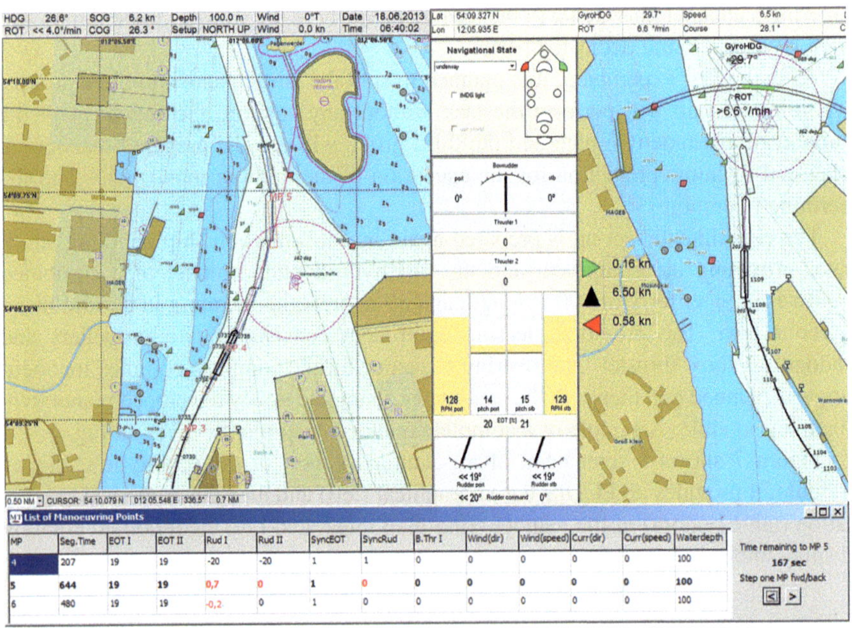


**Fig. 7** Enhanced Manoeuvring Assistance through monitoring the execution of manoeuvres in comparison to the prepared manoeuvre plan

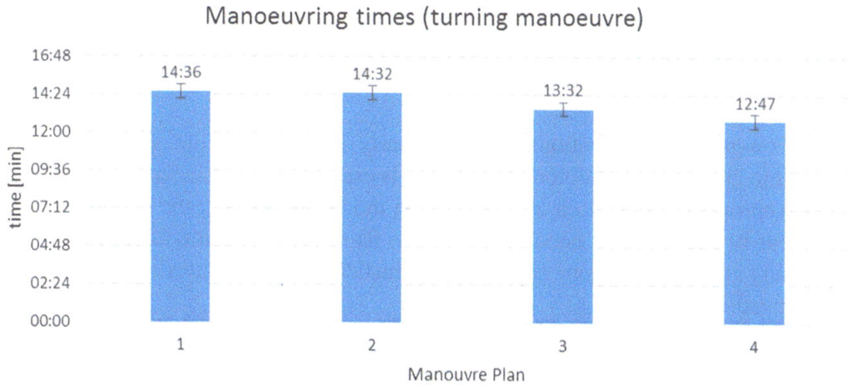


**Fig. 8** Estimated amount of time for four different manoeuvre plans with different steering sequences to reach the same manoeuvring goal

The second experiment is also designed to research different manoeuvring strategies by testing several rather unusual combinations of the control settings to widen knowledge and improve the manoeuvring database in general.

Figure 8 presents the amount of time for four different pre-planned manoeuvring sequences to reach the same manoeuvring aim for the scenario depicted in Fig. 6. The different times indicate that there is obviously potential for detailed pre-planning of steering sequences. This is especially evident when taking into account varying environmental conditions such as wind speeds/directions that will influence the turning rates and stopping distances, the greater the lateral area of the exact ship is.

Even though the advanced motor model is not yet available, the variety of potential solutions to manoeuvre for berthing in a harbour shows that there is potential for time saving, fuel saving and for reducing emissions, accordingly.

# 4 Summary, Conclusions and Outlook

Investigations into the applicability and the extension of existing simulation models for a manoeuvre decision support system are on-going. The overall aim is the provision of information on fuel and energy consumption, as well as estimations of the related emissions in order to allow a comparison of complex sequences of manoeuvres and selection of the most suitable one. The information can be provided on-board a ship e.g. through suitable displays in order to adapt the steering commands for engine, rudder, thrusts etc. to a more energy-efficient and environmental-friendly manoeuvring regime. In case of pre-planning well in advance of a port approach, different manoeuvring strategies can be tested and compared in order to find the sequence with minimal fuel consumption and emissions. The prototype under development, in the same manner, supports simulation based manoeuvre training and makes navigators aware of the concept and the potential of green manoeuvring.

The on-going investigations already clearly demonstrate the significance of taking into consideration motor/engine models especially for transient engine operation. Appropriate integration into simulation models describing the hydrodynamic behaviour is a compelling need. Strategically, as a long term aim, the potential for integrating information from additional sophisticated sensors needs to be further investigated.

Until now, a significant database has been produced serving as an input source for further investigations, especially for the development and improvement of user interfaces for real-time manoeuvre decision support and for existing training regimes and courses.

Implementing information on fuel and energy consumption as well as the estimated amount of GHG emissions for a planned or commanded manoeuvre will allow navigators to consider these aspects and choose steering handles with lower consumption of fuel and, consequently, with less emissions.

The results, even though still preliminary, have clear potential to also improve modules developed and tested in earlier research for the purpose of training bridge teams and making them aware of how they can contribute to energy-efficiency and reduction of air pollution (Baldauf et al. 2012, 2013; Baumler et al. 2014).

The objectives of the next steps in the on-going research activities are to find a level of detail of engine models that is sufficiently accurate and at the same time ensures acceptable calculation times. The ultimate aim is to provide the bridge team with reliable information to assess and influence ship operation to maintain the "green mode". Enhanced engine simulation models integrated into the existing FTS-Software modules will serve this aim and, at a later stage, will allow the application of multi-criteria optimization strategies to consider safety, economic and ecological aspects.

**Acknowledgements** The research work presented in this analysis is partly performed within the frame of the research and development projects MEmBran (Modelling of Emissions and Fuel Consumption during Manoeuvring Operation of ships" and "MTCAS - electronic maritime collision avoidance", which are funded by the Federal Ministry for Economic Affairs and Energy (Germany). Some parts of the work relevant for application are performed in the frame of the project "Research Port Rostock" financially supported by the Ministry of Education and Research of Mecklenburg Pomerania. Further on-going work that contributes to this study belongs partly to WMU's project on further development and implementation of the e-Navigation concept funded by Korea Research Institute Ships & Ocean Engineering (KRISO) as well as the European project for research and technological development "OpenRisk", co-financed by the EU – Civil Protection Financial Instrument as project 2016/PREV/26.

Furthermore, the shipping company Stena Line and its crew on the RoPax ferry "Mecklenburg-Vorpommern" kindly supported the studies through their invaluable collaboration for which the authors thank the crew very much.

Finally, it has to be mentioned that the professional version of the SAMMON software tools has been further developed by the start-up company Innovative Ship Simulation and Maritime Systems GmbH (see ISSIMS (2016)) and provided without costs for research studies.

## References

Baldauf, M., Baumler, R., Ölçer, A., Nakazawa, T., Benedict, K., Fischer, S., et al. (2013). Energy-efficient ship operation – training requirements and challenges. *International Journal on Marine Navigation and Safety of Sea Transportation, 7*(2), 283–290. https://doi.org/10.12716/1001.07.02.16

Baldauf, M., Klaes, S., Benedict, K., Fischer, S., Gluch, M., Kirchhoff, M., et al. (2012). Application of e-Navigation for ship operation support in emergency and routine situations. *European Journal of Navigation, 10*(2), 4–13.

Baldauf, M., Pourzanjani M., Brooks, B., de Melo, G., & Benedict, K. (2012). *Simulation-based training module to promote green energy-efficient ship operation, Part I: Basics.* Tokyo: International Association of Maritime Universities Tokyo.

Baumler, R., Ölçer, A., Pazaver, A., Nakazawa, T., Baldauf, M., Moon, D., et al. (2014). Train-the-trainer course on energy efficient operation of ships. *American Journal of Climate Change, 3*(4), 404–412. https://doi.org/10.4236/ajcc.2014.34035

Benedict, K., Gluch, M., Kirchhoff, M., Schaub, M., Fischer, S., & Baldauf, M. (2015). Innovative simulation tools for learning & teaching ship's dynamic and investigation of manoeuvring capabilities. In *Proceedings of MARSIM conference 2015*, Newcastle, United Kingdom, September 8–11, 2015.

Benedict, K., Kirchhoff, M., Gluch, M., Fischer, S., Baldauf, M., & Klaes, S. (2012). Support for design and operation of ships by combining fast time and full mission manoeuvring simulation technology. In *IMDC – International Marine Design Conference*, Glasgow, UK, June 2012.

BTRE. (2005). *Health impacts of transport emissions in Australia: Economic costs* (Working Paper 63). Canberra: Bureau of Transport and Regional Economics, www.bitre.gov.au/publications/2005/wp_063.aspx.

International Maritime Organization. (2006). *Report of the Maritime Safety Committee on its eighty-fifth session* (MSC85/26/Add.1). Annex 20. Strategy for the development of an e-Navigation. London: Author.

ISSIMS. (2016). SAMMON, SIMOPT & SIMDAT, online: http://www.issims-gmbh.com/joomla/index.php/software-products

Oeder, S., Kanashova, T., Sippula, O., Sapcariu, S. C., Streibel, T., Arteaga-Salas, J. M., et al. (2015). Particulate matter from both heavy fuel oil and diesel fuel shipping emissions show strong biological effects on human lung cells at realistic and comparable in vitro exposure conditions. *PLoS ONE, 10*(6), e0126536. https://doi.org/10.1371/journal.pone.0126536

Rødseth, Ø. J., & Burmeister, H.-C. (2012). Developments toward the unmanned ship. In *Proceedings of International Symposium Information on Ships • ISIS 2012*, Hamburg, Germany, August 30–31, 2012.

Rødseth, Ø. J., Kvamstad, B., Porathe, T., & Burmeister, H.-C. (2013). *Communication Architecture for an Unmanned Merchant Ship.* OCEANS - Bergen, 2013 MTS/IEEE, https://doi.org/10.1109/OCEANS-Bergen.2013.6608075, last accessed September 2013.

Schaub, M., Benedict, K., Klaes, S., & Baldauf, M. (2015). Fast-time simulation for prediction of fuel consumption and emissions in ship manoeuvres. In O. Turan (Ed.), *International conference on shipping in changing climates. SCC 2015 Proceedings* (Vol. 1, pp. 171–183). Glasgow, UK: University of Strathclyde.

## Further Reading

MEPC.177(58): Draft Model Course "Energy efficient operation of Ships" London, November 2011 Online in Internet: URL: http://www.imo.org/blast/blastDataHelper.asp?data_id=23761&filename=177%2858%29.pdf [date 2014-01-10].

Woerner, K. (2014). *COLREGS-compliant autonomous collision avoidance using multi-objective optimization with interval programming*. PhD Thesis, Massachusetts Institute of Technology, Cambridge (MA), US.

# Fuel Saving in Coastal Areas: A Case Study of the Oslo Fjord

Karina Hjelmervik, Halvor Schøyen, Hao Wang, and Ottar Osen

## 1 Introduction

Today, 70–90% of goods are transported by sea (Stopford 2009; IMO 2014; UNCTAD 2016). Even though maritime transportation is generally considered environmentally friendly compared with other transportation types (Dalsøren et al. 2009), fuel is an essential cost in shipping operations and fuel combustion results in air emissions that have a significant impact on the environment, particularly in coastal areas (Corbett et al. 2007; Oeder et al. 2015). In addition, the impact of human activities such as the burning of fossil fuels further contributes to climate change; phenomena like increasing sea-levels and more powerful storms are indicative of the new environmental paradigm (Watson et al. 1996; Bode et al. 2002; IMO 2014).

Fuel consumption accounts for up to 50–60% of the overall shipping operating costs, depending on the type of ship and service (Word Shipping Council 2008). Reduced fuel consumption can therefore bring those costs down significantly and decrease marine pollution levels in relation to air emissions.

Air emissions from ships represent a significant contribution to the global human-made $CO_2$ equivalent greenhouse gas (GHG) emissions, combining $CO_2$, $N_2O$ and $CH_4$ (IMO 2014). According to the International Maritime Organisation (IMO), maritime transport emits around 1000 million tonnes of $CO_2$ annually and is responsible for 2.5% (IMO 2014) to 4–5% (Harrould-Kolieb 2008) of total GHG emissions. By 2050, the EU aims to reduce GHG emissions by 80% and shift 50% of road freight over 300 km in Europe to other transportation modes such as waterborne

K. Hjelmervik (✉) · H. Schøyen
University College of Southeast Norway, Borre, Norway
e-mail: karina.hjelmervik@usn.no

H. Wang · O. Osen
Norwegian University of Science and Technology, Trondheim, Norway



A. I. Ölçer et al. (eds.), *Trends and Challenges in Maritime Energy Management*,
WMU Studies in Maritime Affairs 6, https://doi.org/10.1007/978-3-319-74576-3_18

transport in order to achieve a more competitive, greener and resource-efficient transport system (European Commission 2011). Moreover, the local environment is especially impacted by air emissions of particulate matter (PM), $NO_x$ and $SO_x$ (Endresen et al. 2003; Corbett and Köhler 2003; Eyring et al. 2005; Oeder et al. 2015). IMO has established certain emission control areas (ECAs) with more stringent controls on $SO_x$ and $NO_x$ emissions (IMO 2014). In the Baltic and North Seas the contribution of $NO_2$ from shipping contributes up to 83% of the total concentration of $NO_2$ near the surface. The corresponding percentage for $SO_2$ is up to 88% (Haglund et al. 2016).

Although previous studies state that GHG emissions could be reduced by 25–75% with known measures (Buhaug et al. 2009 and references therein), there is a gap between research and implementation of energy efficient shipping (Styhre and Winnes 2013). Ports operate on a "first come, first served" basis, which, in turn, normally leads to ships going full speed ahead and then laying at anchor while waiting to be served (Porathe et al. 2014). Studies conducted of short sea dry bulk shipping in the North and Baltic seas reveal that some ships spent more than 40% of their time in ports and that half of this lay time was not productive (Johnson and Styhre 2015). If the ships are to adjust their speed according to the currents and other weather conditions, the ports have to change their practice accordingly and operate on a "just in time" basis (Bichou 2013). The increased complexity and volume of ship traffic are other challenging factors, especially in coastal areas. The North Sea is a region of high dense traffic with about 3000 ships sailing its waterways at any time (Aulinger et al. 2016) and the English Channel is the world's most congested sea passage (Porathe et al. 2014).

Emitted pollutants are regarded as proportional to fuel consumption, but depend on, inter alia, actual hull shape and surface roughness, loading condition and engine condition (Endresen et al. 2003). In any case, fuel consumption per time travelled is typically a cubic function of speed (Ronen 1982; Fagerholt et al. 2010). One response to rising fuel costs is to reduce the vessel speed (Notteboom and Vernimmen 2009).

Weather routeing has a high potential for fuel savings on specific routes (IMO 2012). Relevant information, such as current and tidal atlases, and climatological and oceanographic forecasts, should be taken into account when planning a voyage (IMO 1999). By exploiting the weather and current conditions, speed made good (SMG) can be gained and thereby reduce fuel consumption, transport costs, and air emissions (Fagerholt et al. 2010). Systems for route optimization and weather routeing are therefore becoming increasingly sought after for the planning and execution of voyages. These systems exploit the favourable currents and weather for both ocean-going and coastal ships (i.e. Padhy et al. 2008; Takashima et al. 2009; Delitala et al. 2010; Tsou 2010; Lin et al. 2013).

The effect on fuel saving using weather routeing can be considerable on longer passages (Fagerholt et al. 2010). One of the many variables that affect ship motion is currents. In the case of following currents, speed through water (STW) can be reduced accordingly and the ship will still maintain the same speed made good (SMG) and thereby reduce fuel consumption and cost and, consequently, air emissions. However, in coastal areas, navigation is constrained by rules, including the existing Traffic Separation Scheme (TSS), restricted and protected areas and fishing areas, in addition to numerous pleasure crafts, especially during summer.

This study investigates whether fuel consumption can be reduced on a relatively short voyage in the Oslo Fjord, Norway. Previous studies based on Automatic Identification System (AIS) reveal that most vessels in the Oslo Fjord can benefit more from exploiting the currents to a larger extent (Hjelmervik and Schøyen 2015). As in most fjords, both the currents and the ship traffic are headed either in or out of the fjord. The Oslo Fjord can therefore be regarded as a representative fjord in terms of navigating currents in fjords.

In order to have full control of the currents and other factors influencing the voyage, this study was conducted on an advanced ship bridge simulator in combination with an engine room simulator. High-resolution current data from a numerical model were implemented on the bridge simulator. Bridge simulators have previously been applied for testing voyage plan systems for the Baltic Sea (Porathe et al. 2014), analysis of vessel collisions (Gralak and Juszkiewicz 2010), and are becoming increasingly popular in maritime human factor research (i.e. Porathe et al. 2015; van Leeuwen et al. 2013).

## 2 The Oslo Fjord

The Oslo Fjord, in the south-eastern part of Norway, is selected as the area of interest, since it is has the largest density of vessel traffic in Norway; previous studies also imply that the currents can be exploited more in this area (Hjelmervik and Schøyen 2015). This area has the country's highest traffic density of ferries, cargo ships, charter boats, and pleasure craft. More than 40% of the Norwegian population lives less than 1 h by road from the fjord. The majority of vessels operating in the fjord head either North or South.

The Oslo Fjord is approximately 150 km long and can be considered as restricted water. Its width varies from about 25 km at the mouth (~59°N) to about 1–2 km in the inner areas. The currents in the fjords are affected not only by tides, but also by fresh water input from rivers, atmospheric pressure, local water depths, the coastline, and winds. Seasonal and daily changes in addition to variations in depth- result in a time dependent flow pattern with vertical layers. Sills, numerous skerries, islets, and several basins at different depths contribute to a complex flow pattern. The dominant tidal constant in this fjord is $M_2$ with a period of 12.4 h, which indicates that the direction of the currents changes every 6.2 h (Norwegian Mapping Authorities 2016). The direction is northwards during rising tide and southwards during ebb tide. The strength of the tidal currents in the fjord typically ranges from 0.5 to 2 knots (Hjelmervik et al. 2017).

Recently, a high resolution, three-dimensional model covering the fjord has been developed (Røed et al. 2016; Hjelmervik et al. 2017). The numerical model calculates the currents based on tidal forcing at the mouth of the fjord, fresh water inputs from the rivers, bottom topography, and atmospheric pressure, winds, air temperature, and humidity. The results generated from the numerical model are validated against observations (Hjelmervik et al. 2017).

This study focuses on a selected area of the fjord; details are presented in Fig. 1. As the Port of Oslo is located in the innermost part of the fjord, all vessels to and from the capital city travel through the selected area. All vessels longer than 24 m have to apply to the Traffic Separation Scheme (TSS) established in this area. The TSS is approximately one nautical mile (1.85 km) wide and divided into northbound

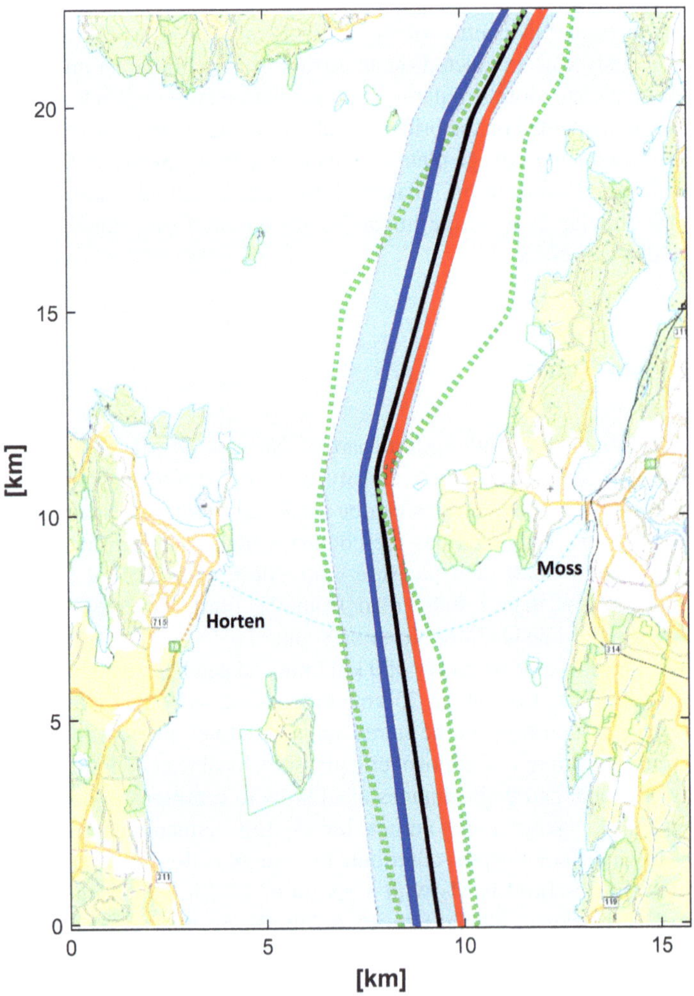


**Fig. 1** The routes in the selected part of the Oslo Fjord. The TSS is indicated by the shaded area. The blue, black, and red lines indicate the western, middle, and eastern routes inside the northbound lane respectively. The TSS does not bind the routes marked by green lines. The routes are drawn on a map supplied by the Norwegian Coastal Administration

and southbound lanes. In this study, only northbound vessels are considered (thus only the northbound lane is of interest). Due to complex flow patterns, the currents in this area vary over short distances, also inside the TSS. Different routes inside the northbound lane are therefore examined.

## 3 Experimental Setup

In order to have full control over the currents and other factors influencing fuel consumption, the experiments were conducted by a full bridge team using the advanccd K-Sim Navigation bridgc simulator from Kongsbcrg Digital (Fig. 2). The K-Sim navigation system is found to comply with Class A- Standard for Certification of Maritime Simulators based on the requirements of the STCW Convention, Regulation 1/12 (DNV 2007). The simulator thereby fulfils international standards for maritime training and allows vessels, objects, and equipment to operate and interact realistically using hydrodynamic models.

Three current fields from a numerical model (Røed et al. 2016; Hjelmervik et al. 2017) have been chosen and implemented into the navigation simulator. The surface current fields at the beginning of each run are shown in Fig. 3. The fields implemented into the simulator are three-dimensional and change with time. The first and third fields are dominated by north going currents, while the currents are south going in the second field. Note that the currents change over relatively short distances, which is typical for currents in coastal areas. In the southern part of the second field, there is a strong current shear inside the northbound lane with stronger south going currents toward the western side of the lane. Calm weather is assumed

**Fig. 2** The K-Sim Navigator Bridge applied in the experiments. Photo: Svend Nordby

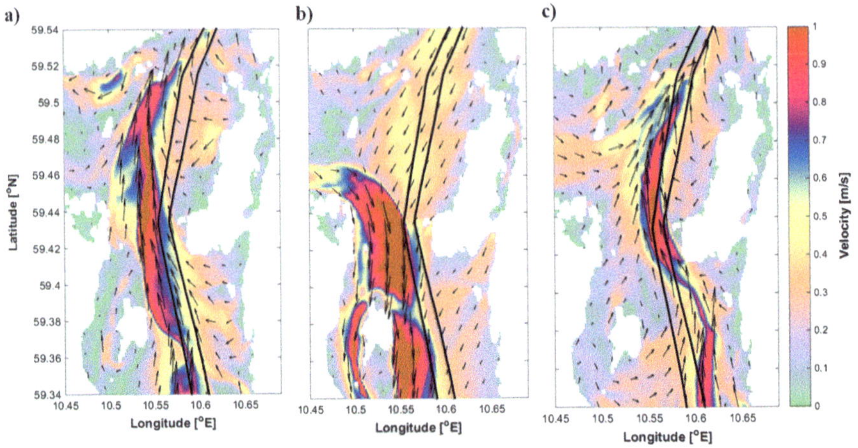


**Fig. 3** The surface currents in the beginning of the three selected current fields. The colours indicate the magnitude and the arrows the direction of the currents. The solid lines mark the northbound traffic lane. The currents are retrieved from a numerical ocean model (Røed et al. 2016). (**a**) Field 1, following current. (**b**) Field 2, counter current. (**c**) Field 3, following current

**Table 1** Experimental setup consists of 10 runs

| ID | Current field | STW [kn] | Choice of route |
|---|---|---|---|
| 1 | No current | 16.7 | Middle route |
| 2 | Field 1, following current | 16.7 | Western route |
| 3 | Field 1, following current | 16.7 | Eastern route |
| 4 | Field 1, following current | 16.7 | No TSS restrictions |
| 5 | Field 1, following current | 15.7 | Eastern route |
| 6 | Field 2, countercurrent | 16.7 | Western route |
| 7 | Field 2, countercurrent | 16.7 | Eastern route |
| 8 | Field 2, countercurrent | 16.7 | No TSS restrictions |
| 9 | Field 3, following current | 16.7 | Western route |
| 10 | Field 3, following current | 16.7 | Eastern route |

Note the reduced speed in run 5

with sun, no wind, and no waves. No other ships or activities are interrupting the passage. Consequently, only the effect of the currents is considered.

The experiment consists of ten different setups with northbound routes; one on no current, four on the first current field, three on the second current field, and two on the third current field (Table 1). The selected routes are approximately 25 km long. Both the eastern and western routes inside the northbound lane were defined in addition to two routes with no TSS restrictions. The two routes with no TSS restrictions have been chosen in order to follow the most favourable currents and avoid stronger counter currents. An Electronic Chart Display and Information System (ECDIS) was used to standardize the routes. Each setup was run twice for

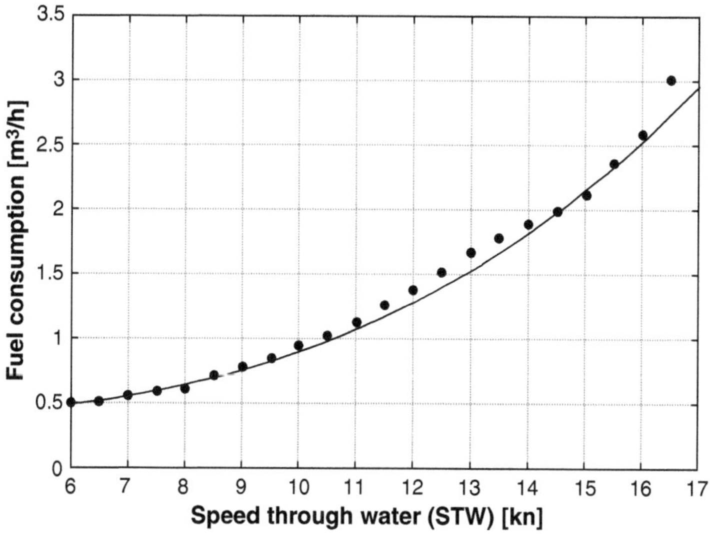


**Fig. 4** Calculated fuel consumption (dots) for the selected vessel using the machine room simulator together with a curve which is proportional to cubic speed

the sake of consistency. If the two runs diverged, more runs were performed with the same setup, and the diverging runs were deleted. Every 10 s, the position, course made good (CMG), speed made good (SMG), heading (HDG), and speed through water (STW) were recorded.

An anchor handling vessel of 79 m length overall, 5.6 m draft, 8991 tonnes displacement and a loading capacity of 4743 dwt was chosen for the experiments. The vessel was fitted with a flap rudder, or Becker rudder, with a max. Angle of 45°. The propulsion schema was a diesel electric one, with an output of 86,109 kW (117,075 hp). It was fitted with a controllable pitch propeller (CPP) with a constant rpm of 130. Before the experiments were conducted, fuel consumption for the selected vessel was calculated using the Kongsberg Digital K-Sim Engine room simulator (see Fig. 4). As highlighted before, in accordance with Ronen (1982) and Fagerholt et al. (2010), the consumption per time travelled can be estimated by a cubic function of speed through water.

## 4 Discussion and Results

The crossing time varied from run to run (see Fig. 5). The crossing time on following currents was approximately 6% less than in the reference runs with no currents and 12–13% less than with the counter currents. It is therefore advised to wait for a favourable tide if possible. Instead of reducing the crossing time, the speed can be

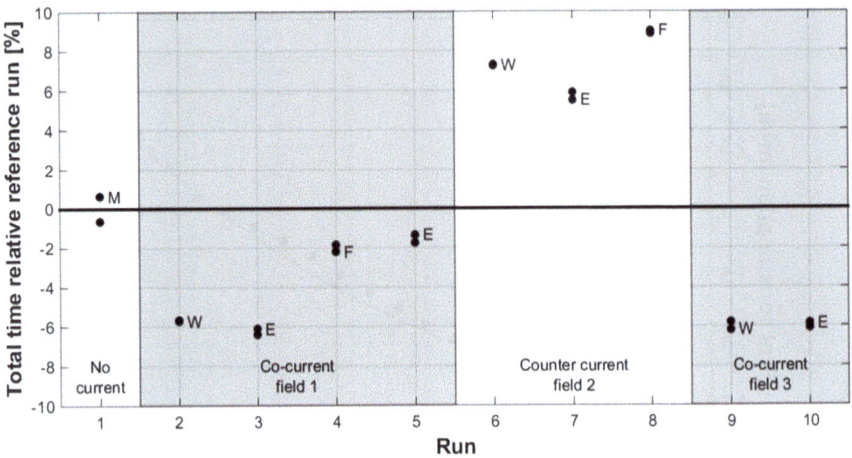


**Fig. 5** Total crossing time for the 10 setups of the study. Each setup was run twice. The letter M indicates middle of lane, W and E indicate western and eastern routes inside the lane respectively, and F indicates free route with no TSS restrictions. NB. Run 4 is the one with a lower STW than the others

reduced. In run 5, the speed of the vessel was 1 knot less than in the others, but due to the following currents, the crossing time was slightly shorter than in the reference runs. According to the calculated fuel consumption (Fig. 2), this implies approximately a 15% reduction in fuel consumption for the vessel tested.

There is a noticeable time difference between the eastern and the western routes inside the northbound lane on the counter-currents (see Fig. 5), which is caused by differences in SMG during the first part of the passage (see Fig. 6). As can be seen from the counter current field (Fig. 3b), the currents in the first part of the passage are

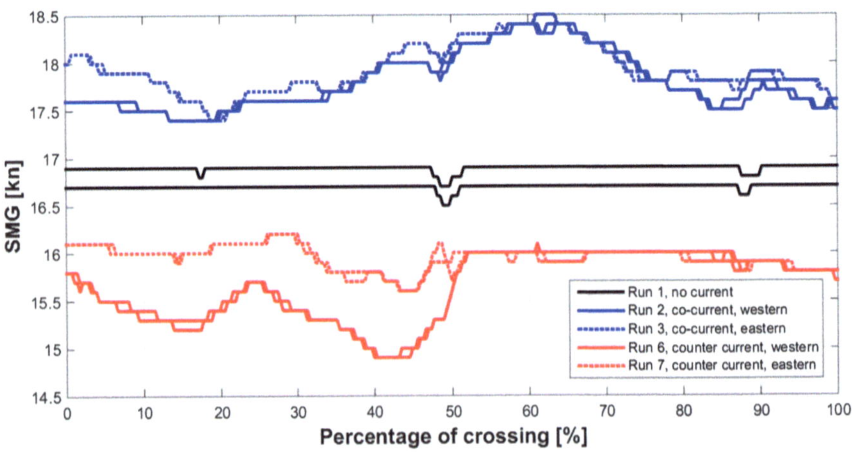


**Fig. 6** Speed made good (SMG) during the crossing

approximately 0.5 m/s (1 knot) stronger in the western part than in the eastern part of the northbound lane. The SMG will therefore be more reduced if the vessel lies west instead of east in the lane. In the last part of the route, the currents are more uniform across the lane.

In addition to the eastern and western routes inside the northbound lane, two routes not bound by TSS restrictions were tested (green lines in Fig. 1). These "free routes" were chosen in order to follow the most favourable currents and avoid the less favourable ones. For field 1, the tested route followed the strongest following currents west of the TSS. For field 2, the tested route avoided the strongest counter current by going east of the TSS. In both cases, the length of the route was increased and thereby the crossing time did not decrease as expected when the SMG increased. Favourable currents are not always beneficial if the length of the route has to be increased in order to exploit the currents.

This study focused on northbound vessels. Similar results are expected for southbound vessels as the tidal currents change direction every 6.2 h. In the spring, the fresh water inputs from the rivers increase due to melt water from the mountains flowing into the rivers. The lower density fresh water will accumulate near the water surface and this often results in stronger southbound and weaker northbound surface currents. During the autumn storms, the pressure in the Skagerrak outside the fjord forces water into the fjord and drags water out of the fjord. This causes sea level changes and increased current strengths. The position of the most efficient route, the favourable timing and potential fuel savings depend therefore on both daily and seasonal changes. Detailed current forecasts are therefore crucial if currents in coastal areas are to be exploited.

In addition to currents, other weather conditions such as wind, waves and ice have a direct impact on manoeuvring the vessels. When the waves and the currents propagate in the same direction, the wave lengths increase and the wave heights decrease. The opposite would be the case when the waves and the currents propagate in opposite directions. This phenomenon could be applied to locate the currents, but can also represent a safety risk especially at the entrance of fjords when outgoing currents encounter the ocean waves or along the coastline when the coastal current encounters opposing waves (Norwegian Hydrographic Service 2017).

# 5 Conclusion

Fuel consumption accounts for 50–60% of overall maritime transport operating costs; emissions from internal combustion engines represent a significant contribution to the degradation of the environment. Reduced fuel consumption can therefore positively impact those operating costs and significantly reduce the associated air pollution. Weather routeing has a high potential for efficiency savings, both on short and longer passages.

This study tested whether accurate and detailed current fields could be used to select the most energy efficient routes between ports in coastal areas. As the experiments were conducted using a bridge simulator, it was possible to have full control over the currents; the Oslo Fjord was selected for the conduct of these experiments. Due to the character of the currents and the movements of the vessels in the Oslo Fjord, this area is regarded as representative relative to navigation on currents in fjords. It is expected that the results of this study will be relevant for fjords in other areas with similar geographic characteristics.

The results of the analysis in hand are promising because they demonstrate that significant fuel and air emission savings may be achieved. An anchor handling vessel of 79 m length overall can save up to 12–13% time if a favourable current rather than a counter current is present in the Oslo Fjord. If one knot of favourable current is present, the ship's captain may voluntarily reduce the speed from 16.7 knots to 15.7 knots, resulting in an SOG of 16.7 knots, arriving at the Port of Oslo on schedule and saving approximately 15% fuel. Alternatively, a speed through water of 16.7 knots, combined with one knot favourable current instead of opposing current, would decrease the time of arrival by approximately 19 min. If possible, vessels are therefore advised to time their voyages with a favourable tide and avoid counter currents. Fuel reduction of up to 15% may generate considerable economic and environmental savings.

Shipping operations in coastal areas are geographically limited. The results reveal that even the route of the vessel inside the traffic lanes has an impact on fuel consumption since currents in coastal areas vary over short distances. Ferries and other vessels operating with very precise timetables could thereby save fuel without changing the time of arrival.

The track of the most efficient route, the favourable timing and potential fuel savings depend on both daily and seasonal changes. Detailed current forecasts are therefore crucial if currents are to be exploited in coastal areas. Actual weather can differ from predicted conditions during the voyage. The fuel savings may therefore differ from the predictions. In addition to currents, other weather conditions such as wind, waves and ice may have an impact on manoeuvring vessels, as well as upon their fuel consumption. When the waves and the currents propagate in an opposite direction wave height increases and thus represents a safety risk, at least for small ships. This is typically the case at the entrance of fjords and along the coastline. Wind and waves should therefore be included in further studies.

Since the routes examined in this study are relatively short (approximately 25 km), further experiments have to be conducted in order to test the extent to which the results can be generalized for weather routeing in other coastal areas. Tests involving different types of vessels should also be conducted in order to estimate potential savings for the all the various vessels navigating the specific area.

In addition to the possible fuel reduction during voyages, detailed current forecasts may benefit various ship operations in connection with safe ship handling, i.e. ship-to ship rescue and recovery, dredging, and helicopter handling (Perera and Soares 2017). Quantification of possible fuel consumption effects during specified ship operations could be included in further studies.

**Acknowledgments** The experiments were part of a collaborative research project between Kongsberg Digital and the University College of Southeast Norway. The project was partly funded by the Regional Research Fund Oslofjordfondet (Research grant no. 248723). The project group would like to thank Theis Arvesen, Jens Ferdinand Karvel, Aksel Stuve, and Severin Bøe Synnevåg for performing the experiments.

## References

Aulinger, A., Matthias, V., Zeretzke, M., Bieser, J., Quante, M., & Backes, A. (2016). The impact of shipping emissions on air pollution in the greater North Sea region–part 1: Current emissions and concentrations. *Atmospheric Chemistry and Physics, 16*, 739–758.

Bichou, K. (2013). *Port operations, planning and logistics*. Boca Raton: CRC Press.

Bode, S., Isensee, J., Krause, K., & Michaelowa, A. (2002). Climate policy: Analysis of ecological, technical and economic implications for international maritime transport. *International Journal of Maritime Economics, 4*(2), 164–184.

Buhaug, Ø., Corbett, J. J., Eyring, V., Endresen, Ø., Faber, J., Hanayama, S., et al. (2009). *Prevention of air pollution from ships: Second IMO GHG study*. London: International Maritime Organization. http://www.imo.org/en/OurWork/Environment/PollutionPrevention/AirPollution/Documents/SecondIMOGHGStudy2009.pdf

Corbett, J. J., & Köhler, H. W. (2003). Updated emissions from ocean shipping. *Journal of Geophysical Research: Atmospheres, 108*(D20), 4650.

Corbett, J. J., Winebrake, J. J., Green, E. H., Kasibhatla, P., Eyring, V., & Lauer, A. (2007). Mortality from ship emissions: A global assessment. *Environmental Science & Technology, 41* (24), 8512–8518.

Dalsøren, S. B., Eide, M. S., Endresen, Ø., Mjelde, A., Gravir, G., & Isaksen, I. S. (2009). Update on emissions and environmental impacts from the international fleet of ships: The contribution from major ship types and ports. *Atmospheric Chemistry and Physics, 9*(6), 2171–2194.

Delitala, A. M. S., Gallino, S., Villa, L., Lagouvardos, K., & Drago, A. (2010). Weather routing in long-distance Mediterranean routes. *Theoretical and Applied Climatology, 102*(1–2), 125–137.

DNV. (2007). *Statement of compliance regarding the K-Sim navigation, bridge operation simulator*. Norway: Det Norske Veritas.

Endresen, Ø., Sørgård, E., Sundet, J. K., Dalsøren, S. B., Isaksen, I. S., Berglen, T. F., et al. (2003). Emission from international sea transportation and environmental impact. *Journal of Geophysical Research: Atmospheres, 108*(D17), 4560.

European Commission. (2011). Roadmap for moving to a competitive low carbon economy in 2050. White Paper, COM 112. 8. March 2011.

Eyring, V., Köhler, H. W., Van Aardenne, J., & Lauer, A. (2005). Emissions from international shipping: 1. The last 50 years. *Journal of Geophysical Research: Atmospheres, 110*(D17).

Fagerholt, K., Laporte, G., & Norstad, I. (2010). Reducing fuel emissions by optimizing speed on shipping routes. *Journal of the Operational Research Society, 61*(3), 523–529.

Gralak, R., & Juszkiewicz, W. (2010). Analysis of Vessels Collision "m/v Ziemia Łódzka" and "m/v Vertigo" with Fullmission Bridge Simulator. In *Proceedings of the 8th International Probabilistic Workshop* (p. 101). Dirk Proske Verlag.

Haglund, K., Claremar, B., & Rutgersson, A. (2016). Deposition of sulfur, nitrogen and particles originating from shipping activities in the Baltic and North Seas. In *Multiple drivers for Earth system changes in the Baltic Sea region*, 197.

Harrould-Kolieb, E. (2008). *Shipping impact on climate: A source with solutions*. Washington, DC: Oceania.

Hjelmervik, K., Kristensen, N. M., Staalstrøm, A., & Røed, L. P. (2017). A simple approach to adjust tidal forcing in fjord models. *Ocean Dynamics, 67*, 949–958.

Hjelmervik, K., & Schøyen, H. (2015). Correlation between coastal tides and the speed of commercial vessels. In *OCEANS 2015-Genova* (pp. 1–5). IEEE.

Hjelmervik, K., Staalstrøm, A., Kristensen, N. M., & Røed, L. P. (2017). *Evaluation of the FjordOs-model* (MET Report 11/2017). Norway: Norwegian Meteorological Institute.

IMO. (1999). Resolution A.893 (21). Guidelines for voyage planning, A 2/Res. 893.

IMO. (2012). Resolution MEPC.213 (63). 2012 Guidelines for the development of a ship energy efficiency management plan (SEEMP).

IMO. (2014). *Third International Maritime Organization (IMO) greenhouse gas study 2014*. London, UK: IMO.

Johnson, H., & Styhre, L. (2015). Increased energy efficiency in short sea shipping through decreased time in port. *Transportation Research Part A: Policy and Practice, 71*, 167–178.

Lin, Y. H., Fang, M. C., & Yeung, R. W. (2013). The optimization of ship weather-routing algorithm based on the composite influence of multi-dynamic elements. *Applied Ocean Research, 43*, 184–194.

Norwegian Hydrographic Service. (2017). *Norwegian pilot guide – Sailing direction, Vol. 1, general information* (8th ed.). Stavanger, Norway: Norwegian Hydrographic Service.

Norwegian Mapping Authority. (2016). *Tide tables from the Norwegian coast and Svalbard*. Stavanger, Norway: Norwegian Mapping Authority.

Notteboom, T. E., & Vernimmen, B. (2009). The effect of high fuel costs on liner service configuration in container shipping. *Journal of Transport Geography, 17*(5), 325–337.

Oeder, S., Kanashova, T., Sippula, O., Sapcariu, S. C., Streibel, T., Arteaga-Salas, J. M., et al. (2015). Particulate matter from both heavy fuel oil and diesel fuel shipping emissions show strong biological effects on human lung cells at realistic and comparable in vitro exposure conditions. *PLoS One, 10*(6), e0126536.

Padhy, C. P., Sen, D., & Bhaskaran, P. K. (2008). Application of wave model for weather routing of ships in the North Indian Ocean. *Natural Hazards, 44*(3), 373–385.

Perera, L. P., & Soares, C. G. (2017). Weather routing and safe ship handling in the future of shipping. *Ocean Engineering, 130*, 684–695.

Porathe, T., Brodje, A., Weber, R., Camre, D., & Borup, O. (2015). Supporting situation awareness on the bridge: Testing route exchange in a practical e-Navigation study. *Information, Communication and Environment: Marine Navigation and Safety of Sea Transportation*, 85.

Porathe, T., de Vries, L., & Prison, J. (2014). Ship voyage plan coordination in the MONALISA project: user tests of a prototype ship traffic management system. In *Proceedings of the Human Factors and Ergonomics Society Europe Chapter 2013 Annual Conference* (pp. 1–11).

Røed, L. P., Kristensen, N. M., Staalstrøm, A., & Hjelmervik, K. B. (2016). *A high-resolution, curvilinear ROMS model for the Oslofjord. MET Report 4/2016*. Norway: Norwegian Meteorological Institute.

Ronen, D. (1982). The effect of oil price on the optimal speed of ships. *Journal of the Operational Research Society, 33*(11), 1035–1040.

Stopford, M. (2009). *Maritime economics* (3rd ed.). Oxon: Routledge.

Styhre, L., & Winnes, H. (2013, July). Energy efficient shipping–between research and implementation. In *Proceedings of the IAME2013 Conference* (pp. 3–5).

Takashima, K., Mezaoui, B., & Shoji, R. (2009). On the fuel saving operation for coastal merchant ships using weather routing. In *Proceedings of International Symposium TransNav* (Vol. 9, pp. 431–436).

Tsou, M. C. (2010). Integration of a geographic information system and evolutionary computation for automatic routing in coastal navigation. *Journal of Navigation, 63*(02), 323–341.

UNCTAD. (2016). Review of Maritime Transport 2016. No. UNCTAD/RMT/2016. United Nations Publication.

van Leeuwen, W. M., Kircher, A., Dahlgren, A., Lützhöft, M., Barnett, M., Kecklund, G., et al. (2013). Sleep, sleepiness, and neurobehavioral performance while on watch in a simulated 4 hours on/8 hours off maritime watch system. *Chronobiology International, 30*(9), 1108–1115.

Watson, R. T., Zinyowera, M. C., & Moss, R. H. (1996). *Climate change 1995 impacts, adaptations and mitigation of climate change: Scientific-technical analysis*. Cambridge: Cambridge University Press.

World Shipping Council. (2008). Record fuel prices place on ocean. Annual Report, 2008.

# A Bayesian Belief Network Model for Integrated Energy Efficiency of Shipping

**Onder Canbulat, Murat Aymelek, Osman Turan, and Evangelos Boulougouris**

## 1 Introduction

Shipping of industrial products and raw materials within the system of the global logistics chain is vital for every industry and the economy. The shipping business is growing continuously and becoming more sophisticated, efficient and effective (Cullinane 2014). Shipping is a global scale activity linking the supply and demand of global commodities as well as international manufacturers and consumers. On the other hand, minimisation of energy consumption allows shipping to reduce transportation costs, minimise negative influences on climate change, and prevent environmental pollution. One of the primary goals of the industry during the last decade and for the near future is the decrease of emissions generated by shipping and port operations, while expanding global marine transport activities (Helfre and Boot 2013).

Shipping is a highly energy efficient transport mode (Johnson 2014) when compared to other modes of transport. However, fuel costs represent the highest component of total transportation cost. For instance, more than 65% of the total operating expenditures (OPEX) in the tanker shipping industry are attributed to bunker cost (Buxton 1985; Nandakumar 2012; Poulsen and Sornn-Friese 2015). In order to minimise the total operating cost in multimodal transportation, the inte-

O. Canbulat (✉) · O. Turan · E. Boulougouris
Department of Naval Architecture, Ocean and Marine Engineering, Faculty of Engineering, University of Strathclyde, Glasgow, UK
e-mail: onder.canbulat@strath.ac.uk

M. Aymelek
Department of Marine Engineering, Faculty of Naval Architecture and Ocean Engineering, Izmir Katip Celebi University, Izmir, Turkey



A. I. Ölçer et al. (eds.), *Trends and Challenges in Maritime Energy Management*, WMU Studies in Maritime Affairs 6, https://doi.org/10.1007/978-3-319-74576-3_19

grated operational energy efficiency of ships, ports and their terminals is becoming increasingly significant. There are several important issues which should be examined in order to ensure optimal energy utilisation, such as proper selection of ship size, hull form design and optimal speed, optimisation of short term and long term engine performance, efficient cargo handling on board and port operation performance, and arrangement of communication (Plessas et al. 2013; Johnson 2014; McKinnon 2014; Sharma 2014).

This research focuses on all the variables mentioned earlier with an integrated approach in order to maximise energy efficiency, provide cost efficiency and reduce greenhouse gas emissions generated from shipping and port operations. The study suggests an integrated system inter-operability approach, including both port and ship operations by utilisation of the probabilistic Bayesian Belief Networks. The outcome of the study develops a conceptual framework for quantitative analysis applications, which may help to measure energy efficiency more holistically. The structure of the paper is developed in a systematic manner to generate sufficient methodological and practical knowledge regarding the application of Bayesian Belief Networks in ship-port interface. In Sect. 2, extensive literature on the subject is reviewed and the research gap is demonstrated based on the review of the literature. In Sect. 3, the research aim and objectives of the study are clearly indicated. In Sect. 4, the methodology of the study is explained in depth and an adaptation of the BBN methodology to the given problem is provided. In Sect. 5, a selected case study is described and data collection steps are identified and clarified. In addition, the results of the case study are presented and discussed. In Sect. 6, a brief conclusion and future research direction of the study are given.

## 2 Literature Review

The international response to climate change evolved after the United Nations Framework Convention on Climate Change (UNFCCC) as shown in Fig. 1, which provides details of the developments for the period of 1979–2015 (UNFCCC 2016). Recently, the outcomes of the United Nations climate change conference in Paris

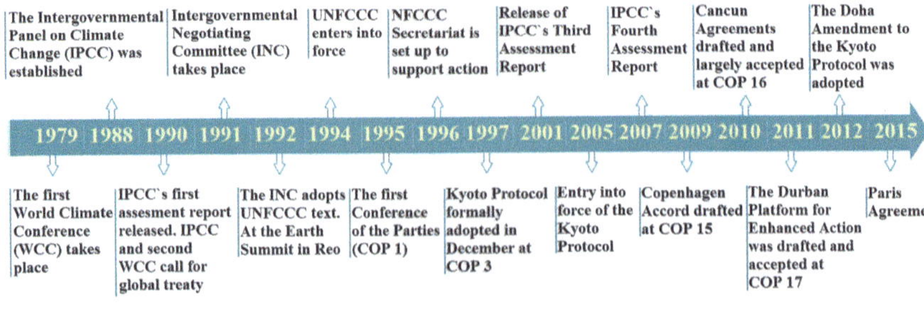


**Fig. 1** Timeline of the international response to climate change between 1979 and 2015

were declared for further fundamental new courses. These agreed courses have also confirmed the goal of limiting global temperature increase to less than 2 °C. The outcomes of the conference did not mention shipping directly because international transport, including shipping, has been removed from the final agreed text (Wessels 2016). However, one of the actions to reach this aim was to make more investment in clean energy research and development within a five-year period. Therefore, energy efficiency improvement in the shipping industry should reduce the $CO_2$ emissions substantially and keep its influence on climate change as low as possible.

The Ship Energy Efficiency Management Plan (SEEMP) is established as a mechanism to improve the energy efficiency of ship operation in line with cost efficiency. The SEEMP is linked to wider corporate energy management policy (IMO 2011). Compared to the Energy Efficiency Design Index (EEDI), the SEEMP improves operational efficiency of ship and unit operation (Wang and Lutsey 2013). Managing ship and fleet efficiency performance over a period by using operational indicators is remarkably vital for shipping companies. The Energy Efficiency Operational Indicator (EEOI) is a ratio of the mass of $CO_2$ (M) emitted per unit of transport work done by the ship. The EEOI estimation is affected by various parameters. Before performing the EEOI calculations of a particular voyage, all possible uncertainties must be clarified (Acomi and Acomi 2014) to attaining a better outcome (Coraddu et al. 2014). Recent studies indicate that the type of charter contract and size of a ship have an impact on the EEOI ratings (Parker et al. 2015). For instance, owner-operated ships achieved more efficient EEOI ratings than time-chartered ships due to pressure for timely delivery of the contract. Arslan et al. (2014) considered that the EEOI value could be improved by depreciating fuel utilisation for the same voyages, or by expanding the measures of payload conveyed and use of the ship (i.e. decreased time in ballast and port).

The concept behind potential Market Based Measures (MBMs) is focus on the financial incentives, in detail to serve two primary purposes. The first one is to ensure financial encouragement for the reduction of fuel consumption by investing resources in energy-efficient ships. This would also assist in promoting innovation and operating procedure enhancement. The second purpose is offsetting different sectors of ship emission reduction.

Integrated Energy Efficiency (IEE) does not have any standards for quantification, analysis or ship performance monitoring, and the analysis of operational data was not widely established (Banks 2015). Armstrong and Banks (2015) argued that the range of possible measures for improving energy efficiency is highly necessary and available in the sector. These measures include improvement of fleet planning, such as weather routing for just-in-time arrival, speed and power optimisation, optimised ship handling such as ballast, trim, use of rudder and autopilot, strategic fleet management, efficient cargo handling, and well-structured energy management policy.

Lu et al. (2014) developed an accurate and practical prediction model for ship operation performance, which is beneficial for selecting the optimal route. They utilised weather forecasts and multi-objective weightings in order to determine the most energy efficient route options. Their model also assisted in building a more efficient system to manage a fleet. However, the only optimal route was not capable of solving the problem of the scheme. Complex maritime logistics require more developments to manage shipping as an integrated system. Individual elements can be developed to maximise the utilisation (days of sailing as laden) of the ships.

Voyage optimisation is the practice of selecting the optimum route for ship operators to increase energy efficiency and reduce Green House Gas emissions in the shipping industry. Varelas et al. (2013) developed an innovative toolkit to optimise ship routing. In addition, Lu et al. (2015) utilised a tool that can accurately predict operational performance for a particular commercial ship under various drafts, at changing speeds and in changing encounter angles. It enables the user to examine the relationship between fuel consumption and different sea conditions. It also clarifies the technical issues which the ship may face during its journey. Based on the outcomes from the operational performance forecast model and real time weather data, different alternatives for the ship's navigation plan can be evaluated depending on a number of aims, including maximisation of safety and minimisation of fuel consumption, as well as voyage time.

Ports and fleet operations can be integrated to maximise the combined efficiency of shipping (Windischhofer and Lepistö n.d.). Moreover, Intertanko and OCIMF (2010) reported fuel wastage by ships which steam ahead at full speed in order to berth at harbours where there are known delays in cargo handling. The ships may avoid wasting time at anchor waiting for port cargo handling by adopting intelligent slow steaming. Thus, emissions can be reduced, congestion can be prevented and safety can be improved in port areas. For instance, where an inevitable inefficiency is observed (such as a port delay) then proper communication and management can allow for alternative operational energy efficient measures to be implemented, such as just in time arrival (Intertanko and OCIMF 2010). Thus, it is possible to say that communication may be considered an essential element of integrated energy efficiency. Starcrest Consulting Group (SCG 2009) investigated the energy demand of a transpacific transit operation from Shanghai to Los Angeles ports. Their results demonstrate that a significant amount of a vessel's energy is consumed during the trans-Pacific part without the local inventories. Because the propulsion engines were performing at their highest power level, the time spent at the port is much longer than the time spent at sea.

According to Banks et al. (2013), the time at sea of an Aframax vessel generally constitutes the longest period of operation time. Based on their case study, the time in port is approximately 42% of the total operation time; the voyage time in ballast is approximately 27% of the total operation time, and the voyage time in loaded condition is 31% of the total operation time. Moreover, Banks (2015) proved that the larger the tanker, the less amount of time is spent in a port per year. Banks' case study shows that approximately 54% of the total operation time for Handysize, 42% of the total operation time for Aframax and 32% of the total operation time for Suezmax tankers are port operation times. Such data underlines the importance of fleet and port integration in terms of time management. Therefore, the analysis of the ship's operating profile and the port's efficiency can be valuable inputs for the study of the port-ship interface and interoperability.

## 3 Research Aims and Objectives

According to existing literature, there are very few examples of holistic energy efficiency efforts in the shipping industry. Integrated efficiency management between port and fleet requires detailed analytical research. The research gap in existing literature leads this study to develop a holistic approach towards integrated energy efficiency of shipping. In other words, this paper aims to specifically focus on the inter-operability of different management systems of the ship-port interface in order to increase energy efficiency more holistically. The primary objectives of this study are as follows:

- Applying Bayesian Belief Network modelling to identify nodes of the integrated system to understand port and fleet operation inter-operability in a more coherent way.
- Obtaining analytical solutions and creating an interdependency framework in order to increase energy efficiency as a whole in the marine transport system. This framework may improve inter-operability among ship and port services utilising a probabilistic approach. The software application of that approach may also help to create an accurate and detailed visual illustration of the Bayesian Belief Networks. "GeNIe" software (Tungkasthan et al. 2010) is an ideal tool to be utilised in Bayesian Belief Networks.
- Improving the resource efficiency of integrated coordination between port and fleet management, and optimising the energy consumption of the integrated marine transport system by utilisation of the Bayesian Belief Networks.

## 4 Methodology

The dependability of marine transport systems is a growing interest in contemporary research in the maritime industry. The recent research mainly aims to improve the dependability/inter-operability of these systems in order to make them more efficient and sustainable. Many different standards and models generated are trying to address the dependability challenges (Neil et al. 1996). Some research (Cao et al. 2013; Neil et al. 1996; Yuqing and Tong 2008) recommended that the most intelligent approach to support this kind of argumentation is Bayesian Belief Networks (BBNs) (invented by Thomas Bayes in 1763). A Bayesian Network (also called Bayesian belief network, belief network, Bayesian net, BBN, BN, graphical probability model, graphical probability networks, causal probabilistic networks, causal nets and probabilistic influence diagrams) is a model for reasoning about uncertainty. Applications of this centuries-old probability theory have been increasing rapidly in recent years due to advances in algorithms and philosophy (Neil et al. 1996; Yuqing and Tong 2008; Sutrisnowati et al. 2014). The inter-operability/dependability analysis of marine transport is a sophisticated approach assisting commercial shipping in becoming more efficient. In marine transport, there are many operational details.

Therefore, each functional relationship should be examined and described, such as vessel speed, port performance, communication challenges, engine performance, weather conditions, and ship types or sizes. This research applies BBNs to marine transport systems. This approach may provide both details on the availability of information during operation, as well as the possible operational influence of any determined deficiencies.

## 4.1 Bayesian Belief Networks (BBNs)

Bayes' Theorem (BT) provides the background knowledge to reversing conditional probability statements from the prior and the probability functions. Bedford and Cooke (2001) briefly explained the BT theorem as follows:

*A* partition is a collection of events $A_1, \ldots, A_n \subset S$ such that $A_j \cap A_j = \emptyset$ whenever $i \neq j$, and $A_1 \cup, \ldots, \cup A_n \subset S$. Suppose that B is an event and if it has $A_1, \ldots, A_n$ a partition, in other word $P(B) \neq 0$ *with* $P(B) > 0$. Then,

$$P(Ai|B) = P(B|Ai)P(Ai)/P(B) \tag{1}$$

Proof of above definition of conditional probability can be written as follows:

$$P(B) = \sum\nolimits_{j=1}^{n} P(B \cap A_j) \tag{2}$$

$$= \sum\nolimits_{j=1}^{n} P(B|A_j)P(A_j), \tag{3}$$

It remains to expand *the* $P(B)$ term. Law of Total Probability states as follows:

$$P(B) = \sum\nolimits_{j=1}^{n} P(B|A_j)P(A_j), \tag{4}$$

The initial probability $P(A_i)$ is called the prior probability, and the updated probability $P(A_i| B)$ is the posterior probability. The term $P(B| A_i)$ is known as the likelihood. The probability of *B* given *A* is indicated as follows:

$$P(B|A) = \frac{P(B|A)P(A)}{P(B)} = \frac{P(B|A)P(A)}{\sum\limits_{j=1}^{n} P(B|A_j)P(A_j).} \tag{5}$$

Yuan and Druzdzel (2006) define a Bayesian Belief Network as a directed acyclic graph (DAG) in which the nodes represent stochastic variables. Also the arcs used for directed probabilistic dependences between the nodes. A BBN model encodes the joint probability distribution across a set of variables $\{A_i, \ldots, A_n\}$ where n is finite and disintegrates it into an output of dependent probability distributions across each variable assigned its parents in the graph. In the case of nodes with parents, the prior probability is practiced. The joint likelihood distribution across $\{A_i, \ldots, A_n\}$ can be achieved by using the output of each of these earlier where;

$$Pr(A_i, \ldots, A_n) = \prod_{i=1}^{n} Pr\left(A_i | PA(A_i)\right). \tag{6}$$

$PA(A_i)$ is indication of the parent nodes of $A_i$.

## 4.2 Stages to Apply a BBN Model

The stages of this method were developed by Cao et al. (2013). A BBN model may be abstracted to surveillance and reconnaissance which also help to build a new methodology based on BBNs. This may require following key actions which are illustrated in the Fig. 2.

This research applies the following three general steps to set up a BBN model on the aforementioned issue:

- Specifying the structure (nodes and connections) and developing Casual Concept Map (CCM);
- Parameterisation of the structure (utilisation of conditional probability tables (CPTs), conditions, or potentially gaining from data cases); and
- Running alternatives and situations including tests on the structure, sensitivity examinations, and so forth (Smith et al. 2015; Cao et al. 2013). These stages are distributed with some stakeholder discussions, as described by the flow diagram illustrated in Fig. 3.

The BBNs are dependent on straightforward layouts which include processes and components; meanwhile, connections indicating how the systems are associated are commonly inferred in utilisation of expert knowledge. Uusitalo (2007) believes that combining different sources of knowledge and small or incomplete data sets is applicable for the BBNs. Although the Bayesian model is a valuable approach to

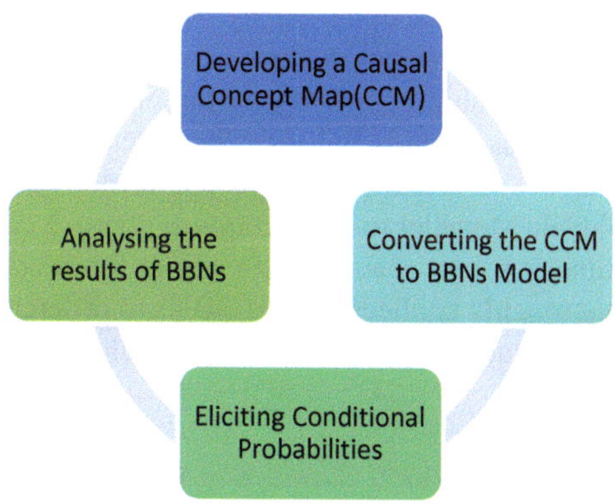


**Fig. 2** A BBN requires four actions (adapted from Cao et al. 2013)

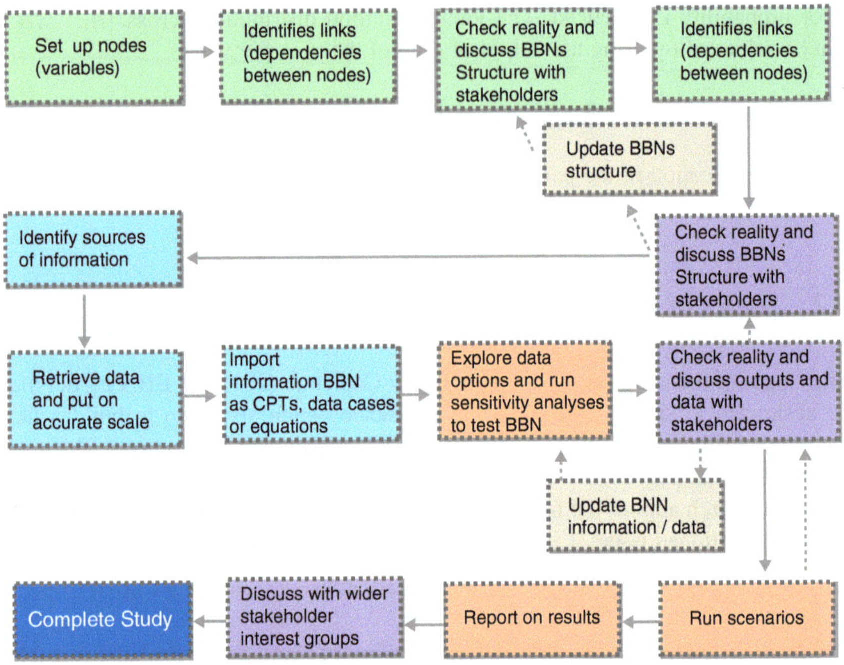


**Fig. 3** Flow diagram showing the steps required to develop and apply a BBN (adapted from Smith et al. 2015)

designing a model based on expert knowledge, Bayesian nodes may change to extract information depending on the opinions of the specialists in the steps shown in the above diagram. Therefore, extracted information may change depending on probability distributions. On the other hand, BBN allows us to estimate probabilities based on all of the accessible information. Moreover, it is not only a recognised and established approach but also easy and quick to use once some experience has been gained (Smith et al. 2015).

With the development of computer technology, inter-operability is increasing its actuality in academia. Firstly, it focused on inter-operability on the Operating System level; but now, it emphasises more on other issues. The utilisation of BNN in inter-operability is shown in the previous figure. BNN will assist us in analysing inter-relationships and inter-dependencies between integrated shipping system elements and creating a unique application by applying a software-based tool called GeNIe Modeller to take actions to improve energy efficiency as a whole in the maritime transport chain. GeNIe allows for interactive model building, and it has been thoroughly applied to build BBNs in different disciplines like business, health, and agriculture since 1998 (GeNIe Modeler 2017).

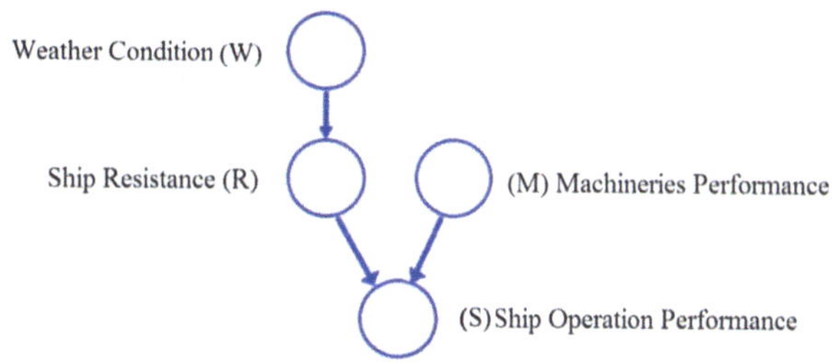


**Fig. 4** An example of the application of BBNs on ship operation

### 4.3 BBNs Model of Integrated Ship-Port Energy Efficiency

Figure 4 shows a highly-simplified example of BBNs which models the causes of equipment performance and handling delay at a port. The variables in this model are Weather Condition (W), Ship Resistance (S), Machineries Performance (M), and Ship Operation Performance (S). For the sake of simplicity, it is assumed that each of these variables has 2 or 3 states. For instance, W has three outcomes, calm, moderate and heavy, representing "Beaufort wind scale 0-3", "Beaufort wind scale 4-6", and "Beaufort wind scale 7 or over", respectively.

A directed arc between W and R accounts for the fact that heavy, moderate or calm weather conditions will impact the likelihood of them affecting Ship Resistance. Similarly, an arc from R to S represents that "Ship Resistance" influences the probability of decreasing "Ship Operation Performance". Lack of directed arcs is similarly a way of expressing information, particularly statements of (conditional) independence. For example, lack of directed arcs between node W and node S means that the state of the weather can influence the chance of decreasing the Ship Operation Performance, S, only indirectly through Ship Resistance, R. These causal statements can be converted to accounts of conditional independence: S is independent of W given R. There is no arc between W and M, which means that the state of the weather will not affect the Machinery's Performance. The absence of any links between Machinery Performance, M and the rest of them expects that M is independent of others. Based on the graph, the joint probability distribution nodes can affect the production of the conditional probabilities of each node (Yuan and Druzdzel 2006).

## 5 Case Study

### 5.1 Case Description and Data Collection

In this case study, real data of an Aframax oil tanker is taken to apply a simplified model developed with BBNs. The departure and destination ports are respectively selected as the Port of Los Angeles (US LAX) and the Port of Chiba (JP CHB) which is the largest seaport in Japan, shown in Fig. 5. Departure day from US LAX is 13.09.2011 and arrival day in JP CHB is 30.09.2011.

The Bayesian network helps to represent the quantitative relationships among the modelled variables of the case study. Each variable is represented as a node in Fig. 6

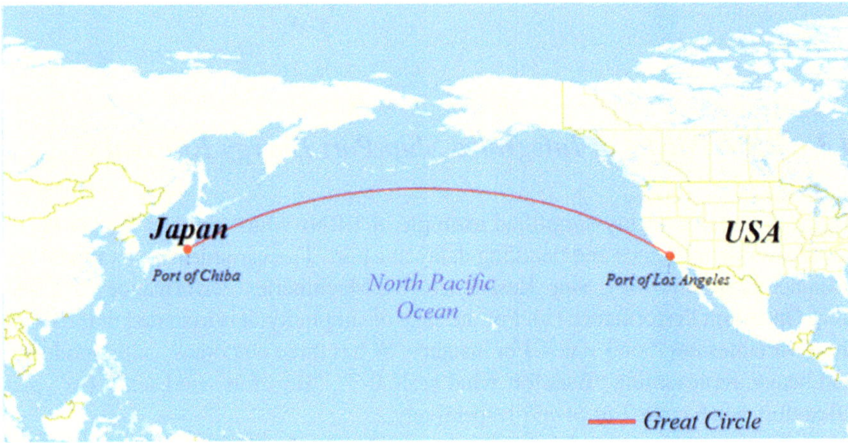


**Fig. 5** The origin and destination ports in the case study

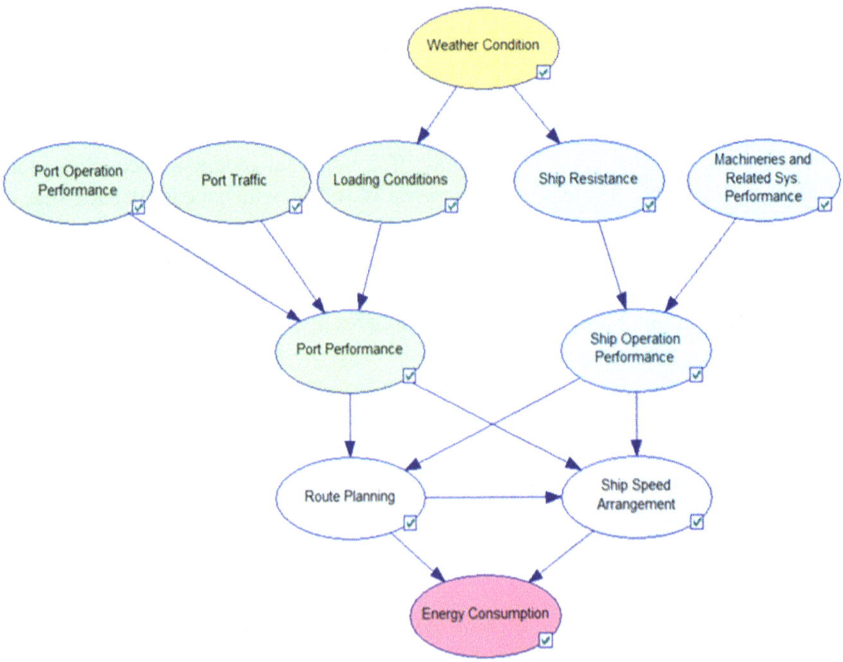


**Fig. 6** The BBN model of port and ship integration created in GeNIe Modeller

which is generated by applying the steps of BBNs application fully explained earlier in Sect. 4.2. The figure shows application of a BBN model on the port and ship interface. The specific structure (nodes and connections) of the model is developed based on the literature and expert decision (Hansen 2012). Numerically, the BBN model represents the joint probability distribution of these variables. This distribution is indicated efficiently by examining the probabilistic independencies amongst the modelled nodes based on the data illustrated in Table 1. Each node is described by a probability distribution conditional on its primary predecessors. For example, nodes in Fig. 6 will be described by the prior probability distributions over its outcomes shown in Table 1. Node 'Loading Conditions' was described by a probability distribution over its outcomes (Low/Moderate Traffic and Heavy Traffic), conditional on the outcomes of its predecessor (node Weather Condition, outcomes 0-3, 4 6, and 7-12). Both the structure and the numerical parameters of a Bayesian network were elicited based on expert discussions and some of them were learned from measurements, weather data, ship and port data, and secondary sources as illustrated in Table 1. For this case study, Optimal Route data with minimum fuel consumption was used for Route Planning and Energy Consumption nodes to calculate their CPTs, based on the academic research of Cui et al. (2016). Data of port weather conditions were used from publicly available web data based on Wunder Ground (n.d.).

## 5.2 *Results and Discussion*

After full implementation of the case study according to a model based on the Bayes theorem by utilisation of the GeNIe version 2.1, the main results were generated. Figure 7 illustrates the value of each node's outcome. As a result of this BBN model application, the interface system is highly likely to be energy efficient by 68%. However, there are still possibilities of lower than average energy consumption value. On the other hand, when we know that the route planning is 'not_fuel_efficient' and also the average ship speed (arrangement) is 'slow_steaming_speed', the probability of having lower than average consumption of energy decreases from 68% to 40%. It certainly clarifies the importance of slow steaming and weather routing on energy consumption. When we know that the port performance is lower than mean but our ship speed arrangement has the average slow steaming rate, the ship is 22% more likely to have lower energy consumption compared to the original result which is 68%. Moreover, if we assume that the ship operation performance is greater than mean, it causes a 25% increase on the probability of having higher energy consumption of more than average value. The result basically shows that port related positive changes in terms of performance affect energy consumption less than ship operation related parts. However, when both port performance and ship operation are chosen as the greater than mean value, this integration increases the possibility of being efficient in terms of energy consumption value more than average value from 68% to 95% as shown in Fig. 8. It

**Table 1** Input variables of each nodes of the BBN model

| Input node | Description | States | Data type |
|---|---|---|---|
| Weather Condition | Beaufort Numbers obtained from recorded weather data. Frequency of the weather condition in terms of sea state | 0-3<br>4-6<br>7-12 | Noon Data and Real Weather Data |
| Port Operation Performance | Capacity of discharging/loading cargo type based on port specification and operation. It is assumed based on Expert Views. 'Mean Value' refers to the main of the port operation performance. Due to the data limitation assumptions are used | "Greater_Mean" or "Lower_Mean" | Secondary Resources, and Expert Views |
| Port Traffic | Existence of queuing based on marine traffic data. 'Busy' mean that there is a queue in the port. It is calculated based on Marine Traffic Statistic for the selected port cases | "Busy" or "Not Busy" | Marine Traffic Data |
| Loading Conditions | Difficulties of the loading conditions based on port operation research and weather data. Calculated based on how often operation stops due to the operational problems such as heavy weather. Ports reports and expert views are used for estimation | "Moderate_Lower_ Difficulties" or "Heavy_Difficulties" | Secondary Resources, and Expert Views |
| Ship Resistance | Total resistance of ship. Weather effect is calculated based on Y. J. Kwon and R. L. Townsin theory cited in Kwon (2008) | "Greater_Mean" or "Lower_Mean" | Real Results Of Sea Trial, Secondary Resources and Noon Data |
| Machineries and Related System Performance | Performance of all machinery parts of ship based on actual ship data. The 'Mean Value' is calculated as the main of the calculated sea trail performance based on the studies of Shao (2013) and Cui et al. (2016). Case study data are used in calculation | "Greater_Mean" or "Lower_Mean" | Real Results Of Sea Trial And Noon Data, Secondary Resources |
| Port Performance | Performance of whole port operation based on actual port research. Due to the data limitation an expert's views are used as well as partly available port data and Marine Traffic Data | "Greater_Mean" or "Lower_Mean" | Secondary Resources, and Expert Views |

(continued)

**Table 1** (continued)

| Input node | Description | States | Data type |
|---|---|---|---|
| Ship Operation Performance | Only performance of ship based on energy consumption from actual ship data. The 'Mean Value' is calculated as the main of the calculated sea trail energy consumption based on Shao (2013) and real operation data are used to calculate how often the operation performance gets greater value then Mean Value | "Greater_Mean" or "Lower_Mean" | Real Results Of Sea Trial and Noon Data, Secondary Resources |
| Route Planning | Possible efficient route options based on the research of Cui et al. (2016) | "Fuel_Efficient_ Route" or "Not_Fuel_Efficient_ Route" | Secondary Resources, and Expert Views |
| Ship Speed Arrangement | Based on actual recorded speed and optimum speed relation. It is calculated according to analysis of Cui et al. (2016) | "Slow_Steaming_Speed" or "Other_speed" | Noon Data |
| Energy Consumption | Total energy consumption of ship based on actual ship data and study of Cui et al. (2016) | "Lower_Average" or "Greater_Average" | Noon data, Secondary Resources, and Expert Views |

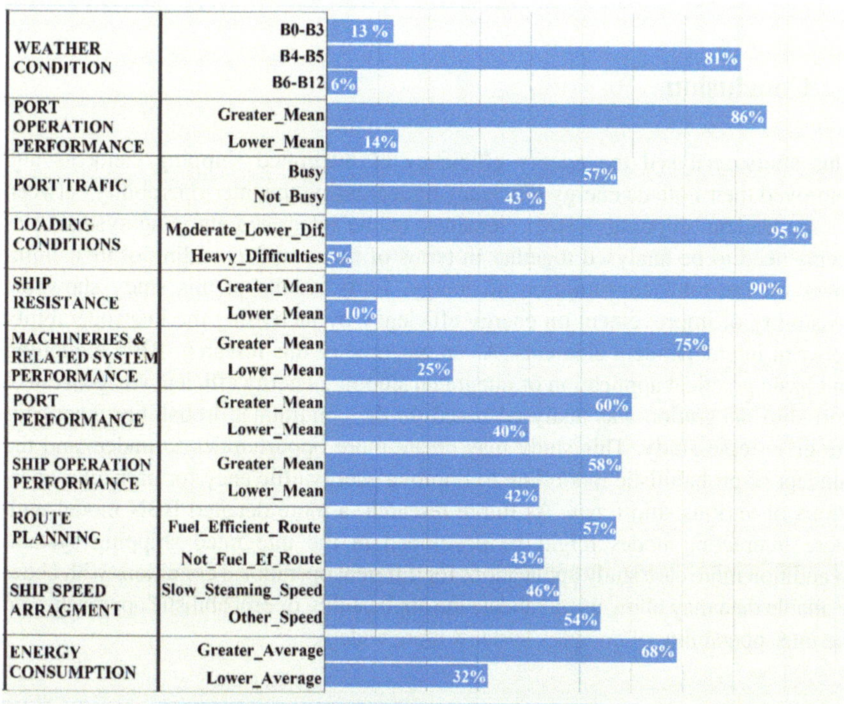


**Fig. 7** Calculated probabilistic value of each nodes' outcome

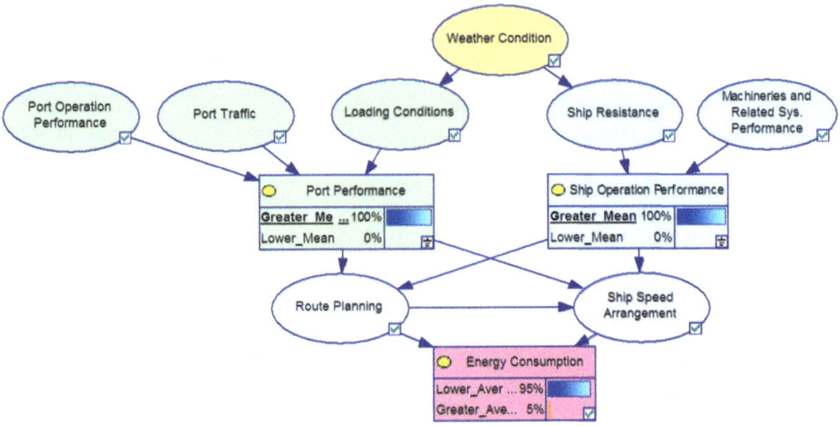


**Fig. 8** Results when the Port Performance and Ship Operation performance are known as greater than mean value

proves that the integration works on any other combinations of mentioned cases. The main reason behind that might be the chosen route of the Aframax tanker which is a long ocean passing journey that usually requires substantially more time for sailing compared to the port time.

## 6 Conclusion

This study analysed the energy efficiency of integrated shipping elements and improved their holistic energy efficiency by increasing the inter-operability between interdependent shipping system elements. In the integrated shipping system, elements need to be analysed together in terms of better understanding of their influences on the total consumption of energy. Early results of this study show the possibility of improvement on energy efficiency by increasing the inter-operability between interdependent elements. As an outcome of this research, BBN modelling and basic practical application of integrated shipping energy efficiency is generated. Port-ship integration was analysed based on the conditional probability theory for the given case study. This study may create more opportunities to understand the concept of probabilistic modelling to improve energy efficiency for particular operations of various ship types. As future research, a more detailed BBN model with more interacting nodes might be developed in the integrated shipping system. In addition more case study applications for different operational conditions with better available data may allow us to validate the applicability of probabilistic approaches on the inter-operability of complex systems more widely.

**Acknowledgements** This research is funded by Turkish Government. Authors specially thank to National Education Ministry of Turkey and Bursa Technical University, University of Strathclyde, and Izmir Katip Celebi University for their support.

## References

Acomi, N., & Acomi, O. C. (2014). Improving the voyage energy efficiency by using EEOI. *Procedia-Social and Behavioral Sciences, 138*, 531–536.

Armstrong, V. N., & Banks, C. (2015). Integrated approach to vessel energy efficiency. *Ocean Engineering Journal, 110*, 39–48.

Arslan, O., Besikci, E., & Olcer, A. (2014). Improving energy efficiency of ships through optimisation of ship operations. No. FY2014-3 IAMU.

Banks, C. (2015). *Operational practices to improve ship energy efficiency*. Doctoral dissertation, University of Strathclyde.

Banks, C., Turan, O., Incecik, A., et al. (2013). Understanding ship operating profiles with an aim to improve energy efficient ship operations. In *Proceedings of the Low Carbon Shipping Conference*, London, 9 Sep 2013.

Bedford, T., & Cooke, R. (2001). *Probabilistic risk analysis: Foundations and methods*. Cambridge, UK: Cambridge University Press.

Buxton, I. (1985). Fuel costs and their relationships with capital and operating costs. *Maritime Policy and Management Journal, 12*(1), 47–54.

Cao, T., Coutts, A., & Lui, F. A. (2013). *Combined Bayesian belief network analysis and systems architectural approach to analyse an amphibious C4ISR system*. Paper presented at the 22nd National Conference of the Australian Society for Operations Research, Adelaide, Australia, 1–6 Dec 2013.

Coraddu, A., Figari, M., & Savio, S. (2014). Numerical investigation on ship energy efficiency by Monte Carlo simulation. *Proceedings of the Institution of Mechanical Engineers, Part M: Journal of Engineering for the Maritime Environment, 228*(3), 220–234.

Cui, T., Turan, O., & Boulougouris, E. (2016). *Development of a ship weather routing system for energy efficient shipping*. Paper presented at the Annual Conference of the International Association of Maritime Economists, 26 Aug 2016.

Cullinane, K. (2014). Targeting the environmental sustainability of european shipping: The need for innovation in policy and technology.

GeNIe Modeler. (2017). *GeNIe modeler user manual*. http://support.bayesfusion.com/docs/genie.pdf. Accessed 17 May 2017.

Hansen, S. V. (2012). *Performance monitoring of ships*. Doctoral dissertation, Technical University of Denmark.

Helfre, J. F., & Boot, P. A. C. (2013). Emission reduction in the shipping industry: Regulations, exposure and solutions.

IMO. (2011). Final SBSTA EEDI SEEMP COP17. http://www.imo.org/en/OurWork/Environment/PollutionPrevention/AirPollution/Documents/COP%2017/Submissions/Final%20SBSTA%20EEDI%20SEEMP%20COP17.pdf. Accessed 05 January 2017.

Intertanko and OCIMF. (2010). *Virtual arrival, optimising voyage management and reducing vessel*. London, UK: Oil Companies International Marine Forum.

Johnson, H. (2014). GHG emissions and the energy efficiency gap in shipping, European panel of sustainable development.

Kwon, Y. J. (2008). Speed loss due to added resistance in wind and waves. The Naval Architect, March, RINA, UK (pp. 14–16).

Lu, R., Turan, O., Boulougouris, E., Banks, C., et al. (2014). *A semi-empirical ship operation performance prediction model for voyage optimization towards*. Paper presented at 2nd International Conference on Maritime Technology (ICMT), Glasgow, 23 January 2014.
Lu, R., Turan, O., Boulougouris, E., Banks, C., et al. (2015). A semi-empirical ship operational performance prediction model for voyage optimization towards energy efficient shipping. *Ocean Engineering, 1*(110), 18–28.
McKinnon, A. C. (2014). *Options for reducing logistics-related emissions from global value chains*. Robert Schuman Centre for Advanced Studies Research Paper No. RSCAS, 31.
Nandakumar, C. G. (2012). *Environmental impact of non metallic hull ships*. Paper presented at Green Technologies (ICGT), 2012 International Conference IEEE, 18 Dec 2012.
Neil, M., Littlewood, B., & Fenton, N. (1996). Applying Bayesian belief networks to system dependability assessment. In *Safety-critical systems: The convergence of high tech and human factors* (pp. 71–94). London: Springer.
Parker, S., Raucci, C., Smith, T. W. P., et al. (2015). *Understanding the energy efficiency operational indicator: An empirical analysis of ships from the Royal Belgian Shipowners' Association*. Energy Institute. s.l. Royal Belgian Shipowners' Association.
Plessas, T., Papanikolaou, A., Pytharoulis, M., Boulougouris, E., & Adamopoulos, N. (2013). Simulation of loading/discharging procedure of tankers, 7 Oct 2013. In C. Guedes Soares & F. Lopez Pena (Eds.), *Developments in maritime transportation and exploitation of sea resources: IMAM 2013* (pp. 501–510). London: CRC Press.
SCG. (2009). *2006 Expanded Greenhouse Gas Inventory*. Available at: https://www.portoflosangeles.org/DOC/REPORT_GHG_Inventory_2006.pdf. Accessed: 17 January 2016.
Shao, W. (2013). *Development of an intelligent tool for energy efficient and low environment impact shipping*. Doctoral dissertation, University of Strathclyde.
Sharma, S. (2014). *Global Voyage Centre: Energy efficiency and route optimisation*. http://maersklinesocial.com/global-voyage-centre-energy-efficiency-and-route-optimisation/#sthash.Qz0BkOzJ.dpuf. Accessed 18 November 2014.
Smith, R., Madsen A. L., & Barton, D. N. (2015). *Bayesian Belief Networks, Preliminary guidelines describing the set of methods for mapping and modelling ecosystem service supply and their application in the WP5 case studies*. http://www.esmeralda-project.eu/getatt.php?filename=OpenNESS_D3.2_FINAL_13634.pdf. Accessed 05 January 2017.
Sutrisnowati, R. A., Bae, H., & Park, J. (2014). Bayesian network learning for port-logistics-process knowledge discovery. *International Journal of Industrial Engineering, 21*(3), 141–152.
Tungkasthan, A., Jongsawat, N., Poompuang, P., et al. (2010). *Automatically building diagnostic Bayesian networks from on-line data sources and the smile web-based interface*. Rijeka: INTECH Open Access Publisher.
UNFCCC. (2016). *Background on the UNFCCC: The international response to climate change*. http://unfccc.int/essential_background/items/6031.php. Accessed 15 April 2015.
Uusitalo, L. (2007). Advantages and challenges of Bayesian networks in environmental modelling. *Ecological Modelling, 203*(3), 312–318.
Varelas, T., Archontaki, S., Dimotikalis, J., Turan, O., Lazakis, I., & Varelas, O. (2013). Optimizing ship routing to maximize fleet revenue at Danaos. *Interfaces, 43*(1), 37–47.
Wang, H., & Lutsey, N. (2013). *Long-term potential for increased shipping efficiency through the adoption of industry-leading practices*. The International Council on Clean Transportation, Retrieved July. 30 Sep 2013.
Wessels, G. (2016). *Actual market and regulation change/input* [Seminar of department of NAOME]. University of Strathclyde, 31 March.
Windischhofer, R., & Lepistö, M. (n.d.) *Integrated operations: ABB's digital business transformation for the maritime industry*. http://new.abb.com/marine/generations/technical-insight/integrated-operations. Accessed 17 February 2017.
Wunder Ground. (n.d.) *Historical weather*. https://www.wunderground.com/history/. Accessed 17 January 2017.

Yuan, C., & Druzdzel, M. J. (2006). Importance sampling algorithms for Bayesian networks: Principles and performance. *Mathematical and Computer Modelling, 43*(9), 1189–1207.

Yuqing, L., & Tong, Z. (2008). Bayesian network to construct interoperability model of open source software. In *Computer Science and Software Engineering, Presented at the 2008 International Conference on 12 Dec 2008* (Vol. 3, pp. 758–761). IEEE.

# Smart Micro-Grid: An Effective Tool for Energy Management in Ports

**Monica Canepa, Giampaolo Frugone, and Riccardo Bozzo**

## 1 Introduction

Energy generation and distribution in ports is like an INTERNET of energy flows to be optimally managed, it is a target that requires terminal operators and port authorities to be aware of how energy is used, which and how operations are carried out and from where energy comes or is sold. A well-coordinated approach results in energy cost savings and in a possible source of revenue. It is a matter of fact that ports around the world are trying to reduce the use of fossil generated energy in favor of renewable green energy generated either within the port border or outside (off-shore wind farms). Ports are intensive work areas that require many tens of "reliable" megawatts for their operations. The life and operations of any port have their "lymph" on a capillary electrical grid interacting with almost any services, machines, equipment, ship, building, warehouses, lighting, and communications. All these services and the size and quantity of equipment grow and must be served efficiently by the grid that is generally supplied with energy from an external utility, which is rather rigid in configuration and offers limited opportunities for improving efficiency. However, things are changing with the introduction of smartness into the grid, moving it to a smart micro-grid capable of integrating safely distributed generation, interacting efficiently with users and controlling flow of energy.

The increasing introduction of renewable energies (offshore wind, photovoltaic panels, cogeneration and trigeneration) and of energy storage, with its consequent benefit for the environment and the operational cost. This is a strong trend for a port infrastructure; it is an evolution that influences the technical, social and economic aspects of port life and of surrounding areas.

M. Canepa (✉) · G. Frugone · R. Bozzo
Department of Naval, Electrical, Electronic and Telecommunication Engineering (DITEN), University of Genoa, Genoa, Italy
e-mail: moc@wmu.se; monica.canepa@unige.it; riccardo.bozzo@unige.it



A. I. Ölçer et al. (eds.), *Trends and Challenges in Maritime Energy Management*, WMU Studies in Maritime Affairs 6, https://doi.org/10.1007/978-3-319-74576-3_20

Port energy distribution still suffers from some rigidity since it has inherited the concepts of traditional configuration and operations of a traditional grid, characterized by unidirectional flow of energy, and it lacks the flexibility to exploit the opportunities stemming from internal generation, controllable loads, local energy policy and supply paid services. Actions that bring efficiency, reliability, revenues, and flexibility are:

- Upgrade existing grid and build a flexible port micro-grid able to face out easily expansions and to foster new services
- Balance integration of renewable energy sources under minimal environmental impact,
- Minimize ownership and operational cost by taking advantage of internal generation, trading surplus of energy, allowing load curtailment, selling paid ancillary services, and minimizing cost of maintenance, while keeping target reliability levels
- Capture user cooperation via friendly and intuitive approaches, with voluntary suasion as the main governing criteria
- Manage energy production and consumption efficiently
- Comply with emission reduction directives

It is a set of objectives that can be pursued in different ways, each having pros and cons. For example, a systematic review of energy consumption vs cost associated with energy procurement helps to identify inefficiencies, but it is often difficult to work out, which are the causes in relaxation to the results. Getting all actors genuinely involved is an rational approach that may be difficult to maintain for a consistent time.

Purely localized solutions are not the right way for efficient energy management, instead it is better to use a connected smart micro-grid that is well accepted by decision makers and operators, and technically suited to accomplish different kind of operations.

A micro-grid controlled on the basis of multiple optimality conditions provides the necessary flexibility to fit the changing environment for many years. A micro smart grid, is an effective solution since it is a pervasive infrastructure that extends its coverage to connection points, to controller of equipment, to buildings of different types, to local generation, and to future energy transaction systems.

There are several challenges to be take into consideration, such as:

- Which is the best design of the grid taking into account the legacy structure?
- Which is the best control structure for the specific environment and targets?
- What is the impact of price policy on course of operations and future development?
- What is the expected impact of foreseeable changing regulations?

Under the umbrella of a smart micro-grid, all energy consuming systems take advantage and may cooperate amongst themselves for a common target. Dispatchable systems take an active role in the operations (for example smoothing, deferral of loads, etc.) while non-dispatchable ones get the advantage of improved quality of service. The main reason for these advantages lies in the coordination of operations from the broader perspective of micro-grid control joined with the search for optimal use of energy at all influencing levels of utilization. The result is reliability, quality, resiliency, cheaper rates, capability to serve more loads and revenues streaming from selling back energy or services to utilities or customers.

### *1.1 General Aspects*

All the aspects of modern industrial society and its economy require cheap, suitable and reliable electric power infrastructures; the US National Academy of Engineering cited the power grid as the "supreme" achievement of engineering of the twentieth century. University of Glasgow states "When Smart Grid implementation becomes reality, everyone wins – and what were once our risks become our strengths."

A smart micro grid is capable of providing electrical energy efficiently and reliably in a port scenario characterized by fast changing operations, growing presence of distributed intermittent generation, and varying loads. Since users such as buildings, parking areas, plant owners and warehouses are often owners of PV generators, they are included in the same smart flexible scheme and are expected to use dynamic incentives and disincentives as guidelines for their operations.

A port is representative of an energy intensive industrial area; it contains civil and technical structures, generation plants, ship terminals, cranes, transport systems, electrical vehicle fleet, warehouses, extensive lighting, and the ship themselves. Efficiency of its operations depends on infrastructure and on behavior of port operators that are exposed to international and national competition.

Designing the grid and managing operations is guided by criteria of availability and quality of delivered energy, level of emissions and by the fact that any port has "hidden" resources that are often unutilized, for example biomass that might be used for power generation.

Utilization of renewable generators, even small ones, opens new technical and economic possibilities, provided that requirements are well analyzed and set so that a correct design and control policy is pursued.

## 2 Scenario

Energy demand in a port consists primarily of operations and heat/cooling. The main energy consumers of a port are terminals including ship to shore, refrigerated containers, cranes, electrical rubber tire gantries and warehouses. They represent a large amount (between 60 and 80%) of the port total energy demand. The remaining is accounted for by lighting, communications, buildings, tower and other equipment.

Electricity is largely procured from external grid utilities with the only exception of emergency contributions that come from UPS, but now more and more comes from internal generation (PV, CHP, Biomass) or sea wind farms, a possibility that is gaining increasing importance. Often smaller size generation (mainly PV) is associated with buildings and large and small local storage systems.

Full benefit can be achieved if all systems' work is correlated. Assuming a reasonable level of internal distributed power generation and of controllable loads, the micro smart grid infrastructure has the distributed intelligence needed for management to guarantee cost effective delivery of energy in spite of randomly changing requirements. Appropriate use of storage facilities reduces the need to continuously change the level of energy import from outside utilities, generally necessary to compensate the random nature of internal renewable generation, and provides the possibility of revenues from selling back energy and services. Integration of distributed power generation, power demand forecasts and plant monitoring is a key point for optimal operations.

Control policy of the smart grid is set to ensure continuity and efficient operations, stability, power quality, energy efficiency, compliance with environmental regulations, and an optimal mix of import of energy and internal generation. It guarantees quality of delivery of energy, contributes to reduction of operational cost and maximization of potential profits that derive from selling energy to external and internal customers, including ships under a cold ironing scheme, and provide support services to assist operations and trading ancillary services.

## 2.1 Reference Micro-Grid

The port area micro-grid grid includes:

- Point of common coupling with an external grid.
- Local generation (PV, CHP, Diesel generators, biomass, off-shore wind plants, UPS etc.)
- Large and small energy storage systems
- Cogeneration and trigeneration plants
- Available electrical batteries used by local transportation fleet
- EV infrastructure and EV charging points
- Refrigerated warehouses
- Docks
- Heavy equipment (cranes for example)
- Buildings
- Lighting for public areas

It is useful to split the micro smart grid into a number of nanogrids each one with its own a nano-controller. A nano-grid includes loads or generation plus storage, but mixed ones can exist depending on the benefit. The aggregation of the nano-grid constitutes the overall micro-grid and its own controller. Nano-grids are coordinated from micro-grid controllers under some coordinating principles inside a hierarchical system that takes into account the autonomous capability of each nano-grid.

## 2.2 *Reference Data*

Typical characteristics for a port are described through interviews of the authors to different Italian ports authorities (Table 1).

## 2.3 *Energy Management and Infrastructure*

The Energy management system supervises, controls and optimizes the operations of the smart micro-grid and of nano-grids; it is organized as a high-level controller and a collection of lower level controllers connected by a communication system. The high level defines policies and targets for local controllers; it balances between electric power imported from external grids and energy generated on site, distinguishing between in-facility generation and external purchasing.

The multiplicity of requirements, often uncorrelated and conflicting, requires a high degree of control flexibility in order to:

- Pursue the global objective (technical and economic) and manage the grid accordingly after having split it into many sub-objectives pursued locally
- Allow proper design of configuration of the grid infrastructure; key factors are use of smart equipment, distribution of intelligence, fast solution methods, and efficient communication
- Split the micro-grid into many nano-grids or even pico-grids each covering homogeneous sub-areas and energy locality. (Nano-grid is a collection of equipment, devices, loads and policies, characterized by similar profiles, logic of control and mode of control that is something unifying even if physical equipment is not closed)
- Privilege a soft-control approach via an appropriate policy, price signals, rewards and penalties.

# 3 Objectives

Two general objectives are defined, their balance depending on time-varying situations.

- TECHNICAL: provide a stable, resilient and reliable micro-grid capable of delivering high quality energy. It implies optimization, state and load estimation, load and power flow, optimal economic and emission dispatch, AGC, P/Q

**Table 1** General characteristics of port area (average values)[a]

| Type | Quantity, value or description |
|---|---|
| Total energy utilization | Up to 250,000 MWh |
| Typical power rating | 10 MW (cold ironing not included) |
| Ship to Shore Crane | 30 kWh (per operating hour)<br>3.91 kWh (energy use per container) |
| Full reefer box | 61 kWh (per storage day) |
| Tower lights | Up to 10 kW, around 50 for a typical port area |
| Commercial building | Two commercial buildings 300 kW installed on average |
| Generic commercial area | 2 MW |
| Transtainers | 3 |
| Portainer | 8 |
| Refer park (for refrigerated containers) | 1 |
| Generic area lighting | High pressure Sodium, LEDs, some hundreds of kW, 0.5–1 MW |
| Electrical substations (MV) +MVA | Five on average |
| Frequency | 50–60–400 Hz available to ships |
| Electrical substations (LV) +MVA | Depends on configuration, many tenths |
| Frequency converters | One for each shore to ship connection |
| Local generators (voltage, power, type) | Photovoltaic (some MW), Diesel (some hundreds of kW), μCHP (some tents of kW) |
| Local storage (voltage, power, type) | Lithium batteries, up to some hundreds of kW |
| Number of load points | Many tenths, of different size |
| Seasonality | 30% influence (plus-minus) on the average value |
| Availability of a local control center | Generally yes |
| Electrical structure: feeders/rings | Main redundant ring plus feeders deriving from bus in substations |
| μCHP | Some units |
| Emission control | Monitoring |
| Market operator | In perspective |
| Operational areas | 100 ha typical |
| Container ship | Voltage from 380 V to 6.6 kV @50 or 60 Hz<br>Average power 2 MW, peak 8 MW |
| Cruise ship | Average power 5.8 MW, peak 11 MW |
| Average duration of docking | 24 h for commercial ship, some days for cruise ships |

[a]Interviews of the authors to different Italian port authorities

control, V/f control, inadvertent energy exchange control, contingency evaluation, security assessment, load and price forecast, outage detection, islanding detection, network restoration, reconnection and re-synchronization, operation in islanded mode, etc.

- ECONOMIC: provide cost effective energy and services to a variety of customers; generate revenues from management and trading of energy; minimization of energy procurement (bilateral contracts, capability services, hedging), of operational costs, of penalties incurred due to non-compliance with contractual terms.

## 4 Structure of Decentralized Control

The organization of a port with very different profiles of utilization and generation of energy suggests that a decentralized hierarchical scheme is well suited; it is a system where

- high level defines a general policy on the basis of more complete knowledge of the status and of the overall system target and decomposes it into many policies and transmits down along the hierarchy
- lower level smart components look for their own optimization objective elaborated on the basis of information, targets entered by owner and operational profile.

A decentralized structure works well when a proper set of policies is identified and a set of "coordination principles" is used. The coordination principle reflects the fact that each local controller operates on its own, without taking into account influence exerted on the grid by other controllers (for simplicity, no peer-to-peer connection is considered) or influencing other controllers.

It is therefore assumed that the peripheral controllers are able to perform, by themselves, a control sequence for a local optimization based on the state they are aware of, a family of objective functions/constraints and the sub-policy received by the high-level controller. See Fig. 1.

The decentralized scheme implements a distribution of the intelligence over different levels (two in our case, but they may be more) where the higher level operates according to a global policy that is mapped into local policies constrained by specific conditions and corrected on the basis of coordination principles.

To make the complete model manageable enough and computable in real time, the low-level controllers are considered decoupled while the high-level controller selects and delivers conditions and policies, but does not directly control the operations of the low-level controllers.

Local decoupling permits each controller to operate in conflict with other controllers in the same layer or in a higher-level layer whereby the conflict is worked out by the controller at the highest level.

An important point is reliability of the operations. In case of failure of the local control system, the decentralized scheme, differently from the case of a centralize one, offers more possibilities to quickly reassign functions and restart operations. In case of a fault, the control system recognizes it and, at the high level, recalculates a new strategy to minimize technical disturbances or economic loss.

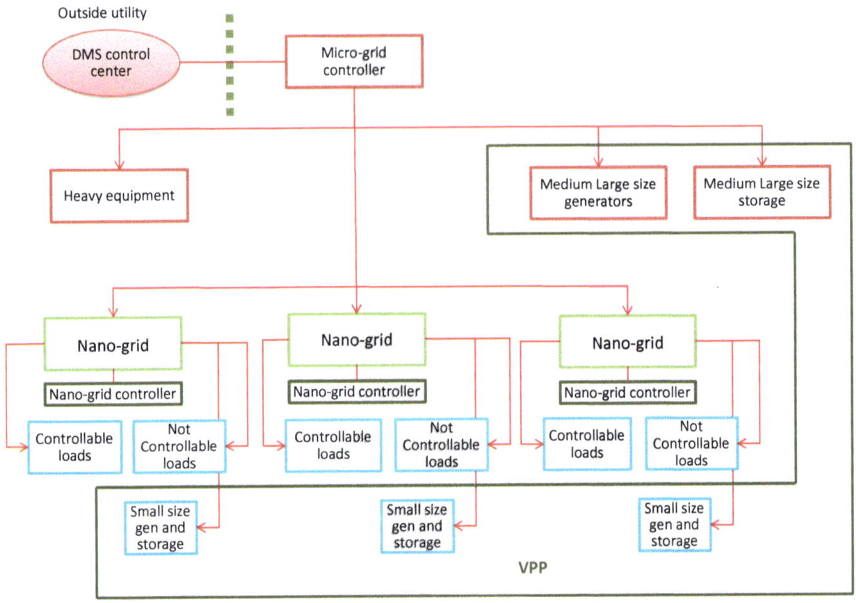


**Fig. 1** Implementation (conceptual)

# 5 Implementation Criteria

The following criteria guide a possible implementation for a port infrastructure:

- Splitting the micro-grid into a number of nano-grids. The nano-grid approach offers increased flexibility in management due to the capability to fit the behavior of a small set of energy related customers. It is easier to manage each nano-grid since optimization is performed at a level that is simpler to optimize.
- Concurrent exploitation of renewable generation and of battery storage.
- Coordinate generation and loads aggregated as a Virtual power plant (VPP). Control of generators and loads forming the VPP is performed either through dedicated communication links or through INTERNET connections.

## 5.1 *Micro-Grid General*

General functionalities are:

- Grid wide energy management performed at various levels of the control hierarchy
- Islanding, reconnection and synchronization
- Islanded emergency operations
- Trading of services and sell back of excess of energy

## 5.2 Nano-Grid General

The nano-grid functionality is a substantial replica of the micro-grid functionalities but with less information and shorter intervention time.

## 5.3 Micro-Grid Advanced

- Optimization of energy flow, voltage and frequency regulation, synchronization, real and reactive power control, load forecasting and scheduling, micro-grid monitoring, outage detection and restoring protection and black start.
- Local control at low level of the control hierarchy.
- Primary voltage and frequency regulation, primary real and reactive power control for each local generation and energy storage unit
- Emulation of effects of kinetic energy.

## 5.4 Generation

- Support the reduction of any sudden large load/generation imbalance and maintain a state of frequency.
- Provide energy to serve critical customer demand should a power facility fail
- Support the prevention of inadvertent unplanned transfer of energy between one utility and another.
- Store energy imported or generated at low cost to use during peak demand periods

## 5.5 Storage

- Support frequency and voltage control
- Provide kinetic energy support necessary for quick stabilization of the grid, especially in case of disconnected operations (for whatever reason)
- Supervise "State of Charge" and manage charge and discharge cycles
- Calculate best charge and discharge cycles also in view of economic optimization

## 5.6 Distribution

- Support maintenance of components in synchronous operation
- Reduce power oscillations by injection and useless absorption of real power.

- Manage voltage profiles via reactive power
- Support storage operations

## 5.7 *Smart Metering*

- Electrical coordination with shipyards, transportation systems and the like
- Coordination of BMS inside port borders
- Enable Demand response and Demand Side Management
- Support billing for EV charging points of electrical internal transportation services

# 6 Micro Grid Structure

This section highlights a general structure of a micro-grid in a reference port area; first the main components are listed and then the organization is introduced.

## 6.1 *Configuration*

A loop configuration (See Fig. 2 as an example) that connects nano-grids and customer areas and then returns to the original point is employed to increase flexibility and reliability. The loop structure copes well with nano-grid control organization since it is organized as:

- a main loop encompassing primary MV distribution in all port areas, including energy intensive equipment and the VPP;
- secondary loop-islands of local MV/LV distribution systems, that is the nano-grid.

## 6.2 *Smart Subsystems*

### 6.2.1 SCADA & EMS

The micro-grid is supervised and controlled by a SCADA & EMS that is connected to:

- DMS dispatching centre of the external utility
- RTUs and PLCs directly controlling heavy equipment not belonging to a nano-grid
- Nano-grid controllers

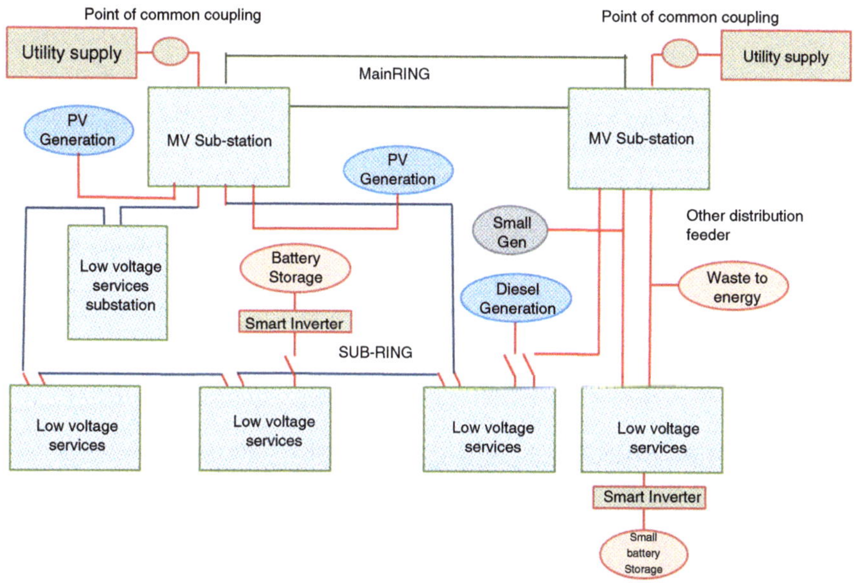


**Fig. 2** Hierarchical organization

### 6.2.2 VPP

The concept of VPP is introduced to integrate controllable and uncontrollable energy production systems and is managed to operate as one unified and flexible physical resource (like a physical power plant). A VPP makes efficient use of renewable, but less predictable, resources in the electricity grid. In the port environment, it includes:

- different sources of generation (mainly PV, but also waste to energy plants),
- diesel generators, μCHP and UPS
- large storage and local storage
- dispatchable load (a reduction in load is equivalent to an increase in generation) such as ships, large mechanical equipment, and HVAC in buildings

Some advantages:

- Operated and dispatched as a single power plant
- Facilitates the best use of port generation and reduces the need for energy procurement
- Responds effectively to fast change patterns of loads and copes with un-dispatched change of PV generation

### 6.2.3 Port Cranes

Port cranes are heavy energy consumers and are often not connected to the port electricity grid as they are equipped with their own electrical generators that can supply excess of energy to the micro-grid.

The same concept can be applied to ships, in addition to shore to ship. It is possible to foresee, under some circumstances, a ship to shore mode of operation (provided that ship engines comply with pollution regulations), allowing also an effective use of ship batteries.

### 6.2.4 Storage

A Battery Energy Storage System (BESS) is deployed as farm (generally on the side of large renewable generators, i.e. HV or MV bus) or as local battery on the side of loads and small generators (i.e. on LV bus). Charging and discharging capabilities permit energy to be stored and used when technically necessary or economically convenient in this order of priority. BESS is used to:

- suppress generation fluctuations due to renewable sources
- support power quality, voltage regulation, reactive power, and operating reserves.
- permit load levelling, valley filling and peak shifting

Remark: The fleet of electrical vehicles used in the port area is a good reserve of energy (via their batteries) and can be efficiently used as an important resource to balance exchange of energy.

## *6.3 Micro-Grid Components*

### 6.3.1 Equipment

- Point of common coupling (PCC)
- Breakers, switches, isolators
- PV generation plants
- Smart inverters and Source Voltage Converters
- Cogeneration and trigeneration
- EV charging points (part of EV infrastructure)
- Docks including points for "shore to ship powering (High voltage shore connection systems (HVSC), including frequency conversion)
- Frequency converters
- Static Var Compensator

The micro-grid is built around one or more substations for interconnections with the external grid (usually a redundant connection is foreseen) that feed a number of subrings or feeders with proper disconnecting devices.

### 6.3.2 Metering

Metering infrastructure has three important functions:

- Support data collection for processing load profiling (profiles are used by central high level controller).
- Enable bidirectional reading (to enable rewarding of local generation)
- Provide proper billing to enabling different tariffs and dynamic pricing

The metering system consists of:

- AMI (Automated metering system)
- Bidirectional smart meters
- Communication support

# 7 Control Organization

The distributed electrical organization of the system fosters a hierarchical distributed control organization spanning on the nano-grids. Control is organized according to the following schema, which reflects a hierarchical organization (Fig. 3).

The micro-grid spans over the entire network and sees the nano-grids as a collection of subsystems. In this way, the hierarchical structure is established where:

- The micro-grid controller supervises, obtains coordination information, organizes and devises policies (suggested, not mandatory) to low level micro-nano controllers
- Local operations are carried out by nano-grid controllers, all connected through a communication infrastructure to a Master Controller (i.e. the high level controller)
- Each nano controller pursues its optimum on the basis of its own policies
- Nano controllers can start with their optimization, and transmit feedback to the micro-grid controller, which in turn verifies how general optimality is achieved and modifies the policies of nano controllers.

## *7.1 Layer Representation*

The control and supervision structure is graphically depicted in Fig. 4. The control wherein our representation economic and optimization and policies are collapsed.

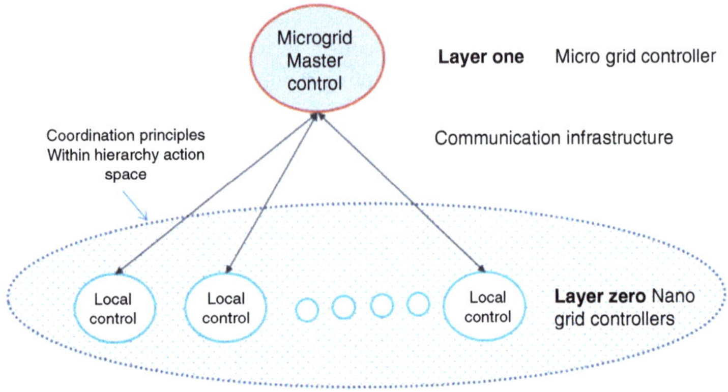


**Fig. 3** Hierarchical organization

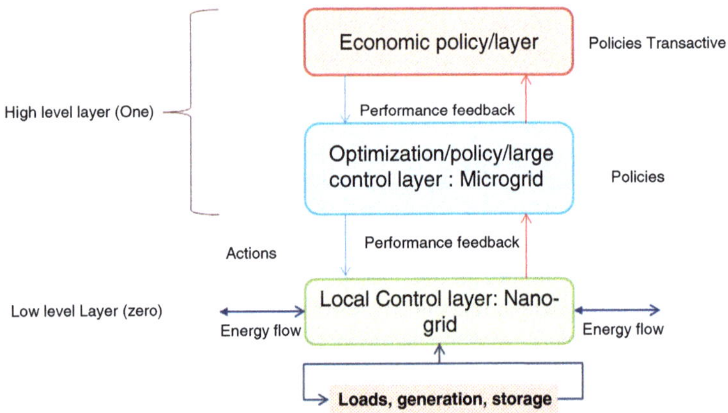


**Fig. 4** Control and supervision structure

The policy elaborated by the central controller stems from:

- The optimization rules (functions) defined for the whole infrastructure
- The "coordination principle", which is the set of relations that describes the fact that each controller operates on its own without exchange of information with other peer controllers

In general, it is possible to say that the performances of a decentralized structure are not worse than the ones of a centralized structure. Indeed, assigning a cost $\tau_c$ to a centralized structure, it is possible to define an expected cost of the decentralized structure as

$$C = \tau_c - \tau_D$$

C is the maximum cost that is possible to pay to assign control responsibilities to decentralized (distributed) local controllers.

The use of local controllers permits parallelized control operations that are localized to low level layers (through problem decoupling, there is less information to be processed) and specialized for the particular functions to be carried out.

Figure 5 represents both the organization (the infrastructure) and the control itself and is substantially representative of the real control network.

It is noted that noted that nodes on the same hierarchical level work asynchronously when they exchange data and commands, while the process between two layers is synchronous, wherein the synchronous period is set by the control algorithm that takes the longest time to be executed.

## 7.2 *Optimization Problem*

Let us consider a structure consisting of members $i \in i = \{1,2,3,\ldots,n\}$. Each member collects information $z_i$ and controls the decision variable $u_i$. The objective function J is defined as:

$J = J(\gamma_1, \gamma_2, \ldots, \gamma_n)$ Where $\gamma_i$ is a control law defined as $U_i = \gamma_i(z_i) \cap \gamma_i \in \Gamma$ and where $\Gamma_i$ is the set of control laws that are admissible for the $i_{th}$ member. The optimization problem is written as

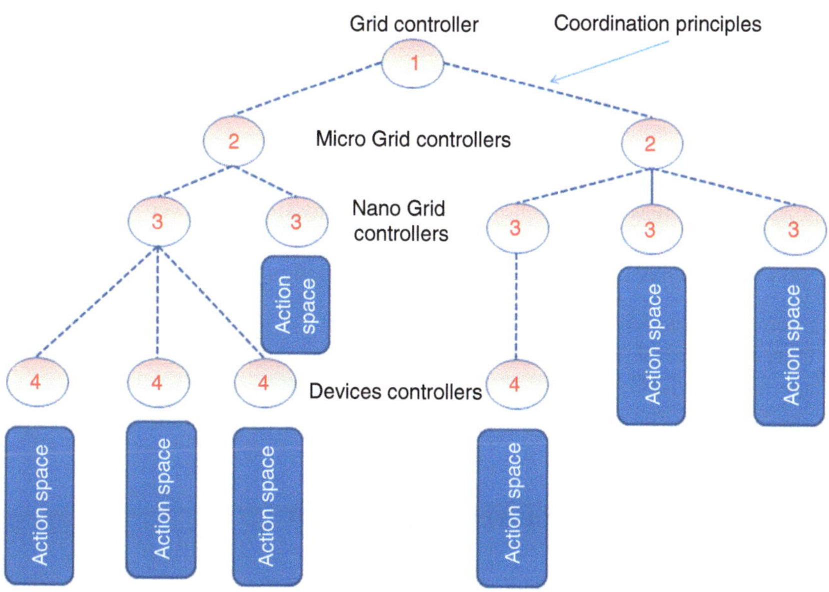


**Fig. 5** Graph of the real control network

$$\gamma^* \epsilon \Gamma i \forall / \mathrm{J} = \min \mathrm{J}(\gamma_1, \gamma_2, \cdots \gamma_n)$$

This represents a problem of optimum search within an environment where the structure evolves dynamically and the local decisions have an influence on the object function and on the type and quantity of collected information.

### 7.3 Generation Control

The first level of generation control of the controller in the nano-grid utilizes fast feedback controls that control real and reactive power changes to inverters associated with local generation included in the micro-grid or nano-grid.

The second level controller, still at low level of hierarchy, is a fast feedback decoupled real and reactive power controller. This control loop supervises the voltage and voltage angle set points of the "regulation" controller and will be automatically reverted to another controller if the current regulation controller saturates.

The third level controller, operating instead at the high level of the hierarchical organization, optimizes the micro-grid by determining the policy for optimal state variables in the system (The state of the system is defined as the voltage magnitude and angle at each branch of the micro-grid). It receives generation and consumption plans and data from local controllers (nano-grids). As part of policy, targets are sent to VPP and nano-grids that are responsible for maximizing their objective function.

### 7.4 Load Management

Load management in electrical distribution networks serving port installations is important for increasing efficiency and sustainability of port power systems. Even more interesting is active load management, based on the idea of Active Networks (ANs) (Van Overbeeke and Roberts 2002). The solution is based on:

- interconnection among networks,
- local load control areas in nano-grids
- ancillary services

## 8 The Rationale of Control Organization

The typical port environment include users, loads and generator that are characterized by different size, exigencies, time of utilization and requirements.

In principle, the general objective is agreed upon by all participants, but in practice each has its own local horizon, profile and constraints that can lead to a conflict with the general objective. These characteristics make the optimization more complex; it is a value to leave a certain degree of freedom to users and set up a general policy so that what is not performed in accordance with the main target of one user is recovered by a rescheduled action of other users. This means splitting operations and communications to a certain level of granularity. Operational logic is based on the following steps:

1. The high-level controller gets the status of the entire micro-grid and of external connections (tie lines) and inside and outside costs
2. It elaborates an optimum control policy (a set of Pareto optimal policy), one set for each other controller, belonging to equipment/subsystem of the micro-grid or of nano-grids
3. Each local controller receives the policy and operates accordingly; a set of local constraints and objectives is set up by its owner.
4. The high-level controller processes all feedback and elaborates alternate appropriate politics (adjustment process) until an optimum policy and low level agreement is reached within a certain period of time
5. If an optimum is not achieved, a set of optimum/suboptimum operations is executed and a cost/reward recorded to be used during the next computational period
6. Under high level supervision, low level subsystems compensate mutually.

# 9 Pro and Cons

## *9.1 PROS*

Deployment of a smart micro-grid provides a common framework for working out somewhat conflicting objectives, namely general reduction of operational cost, while improving performance, reliability, flexibility and resilience. Delocalization of intelligence as well as decoupled control across layers permits flexibility of operation and inclusion of new loads, generation, storage, and services without losing efficiency. Transactive control contributes to stream revenues from operation and particularly in cooperation with external utilities.

Some key advantages are summarized as:

- Allowing industrial and civil users to adjust their usage to take advantage of renewable generation sources.
- Tracking energy use online.
- Supporting efficient use of generators and of the grid infrastructure
- Reducing carbon footprint
- Managing an electric infrastructure for efficient use of port electric vehicles
- Improving quality of energy services (KPI)

- Enabling consumers to tailor their energy consumption based on individual preferences, such as price and/or environmental directives
- Reducing problems during grid peak conditions through Demand-Response
- Allowing deferral of capital investments as future peak loads are reduced
- Minimizing spinning reserve requirements

### 9.2 CONS

Clearly the deployment of a smart micro-grid is not only a scenario of advantages. Shortcomings exist, especially in a short time horizon. Problems may arise from the legacy infrastructure that may cause a cost overrun to achieve the required upgrade level. Controllers should be added and sometimes devices are inherently uncontrollable. Change in users attitudes is advisable, but might not be easy to achieve, at least in the beginning. Certain habits should be changed and at the same time it has to evaluated how not to force behavior and to avoid refusals. Payback might not be immediate. Some points to be taken into consideration are:

- Completely rebuilding the existing electrical may be time-consuming and expensive. One alternative to completely rebuilding the existing grid is to add smaller nano-grids to which to reassign equipment, users, building, etc.
- Need for updated communication infrastructure to communicate in both directions
- If internet is used to provide real time grid data then the risk of privacy and security concerns is increased
- Concerns about cyber-security and privacy
- Complex rate structures proposed might not incentivize efficiency if customers do not understand their structure.
- Independence, ownership and control: Are customers allowed to add assets to their nano-grids? Would customers relinquish any level of control of their assets in the master controller of the micro grid?

## 10 Conclusions

Energy service management for electrical distribution systems in ports must be carried out in a safe way, exploiting all the technological developments, reducing cost and emissions, and increasing revenues for the port authority.

Even if the electrical infrastructure of port areas was or is a (partially) uncoordinated system, it needs to be operated in an integrated way with a common goal while leaving a certain degree of flexibility and freedom to each customer (either generator or load point). This helps acceptance and cooperation to maintain a constant level of optimized operations.

A micro smart grid together with a cluster of nano-grids is able to work out all the requirements and provide early and ongoing technical and economic benefits.

Furthermore, a hierarchical decentralized control structure such as the one suggested is well suited to efficiently perform the various tasks, maintaining a level of independence and best fitting the user profile preferences.

Energy service management of port areas can be carried out as an opportunity for economic business for the port authorities and for customers.

## Further Reading


Adams, J. S. (1965). Inequity in social exchange. In L. Berkowitz (Ed.), *Advances in experimental social psychology* (Vol. 2, pp. 267–299). New York: Academic Press.

ESPO. (2013) *Green Guide*, Annex 1, Good practice examples inline with the 5 Es. European Sea Port Organization.

International Universities Power Engineering Conference UPEC 2018, scheduled on 4th – 7th September 2018 Glasgow Caledonian University, Glasgow, Scotland

Li, Z., & Yao, T. (2010). Renewable energy basing on smart grid. In *Proceedings of the 6th international conference on wireless communications networking and mobile computing (WiCOM)*. Chengdu, China: Institute of Electrical and Electronics Engineers.

Parise, G. LNG and Electric shore power infrastructure in the ports of future, Trieste, 9-10 March 2016, Wise Port & Business Energy Management: Port Facilities, Electrical Power Distribution. Italy: Sapienza University of Roma.

Parise, G. *Wise port & business energy management: port facilities, electrical power distribution* (Life Fellow IEEE). Italy: Sapienza University of Roma.

Reid, R., Roberts, B., & Tran, T. (2009). Management of Dynamic Power Demands for Large Container Port Loads IEEE Xplore. doi: https://doi.org/10.1109/UPEC.2008.4651655.

The Modern Grid Initiative. Version 2.0, conducted by the National Energy Technology Laboratory for the US Department of Energy Office of Electricity Delivery and Energy Reliability; January 2007. http://www.netl.doe.gov/moderngrid/resources.html.

Van Overbeeke, F. and Roberts, V. (2002). Active networks as facilitators for embedded generation. *Cogeneration and On-Site Power Production.*

Wilmsmeier, G., Froese, J., Zotz, A., & Meyer, A. (2014). *Energy consumption and efficiency: Emerging challenges from Reefer Trade in South American Container Terminals* (FAL Bulletin N° 32). Digital Repository Economic Commission for Latin America and the Caribbean Publishing Web. http://repositorio.cepal.org/handle/11362/37283. Accessed 01 June 2017.

Wilmsmeier, G., & Spengler, T. (2016a). *Energy consumption and container terminal efficiency FAL* (Bulletin N° 6). Digital Repository Economic Commission for Latin America and the Caribbean Publishing Web. http://repositorio.cepal.org/bitstream/handle/11362/40928/1/S1601301_en.pdf Accessed 01 June 2017.

Wilmsmeier, G., & Spengler, T. (2016b). Energy Consumption and Energy Efficiency Indicators in Container Terminals - a national inventory.

Wulf, W. A. (2000). Great achievements and grand challenges. *The Bridge*, *30*, 5–10, Fall/Winter.

# Energy Manager Role in Ports

**Fabio Ballini and Aykut I. Ölçer**

## 1 Introduction

With the world's population growing to nine billion by 2050 (UNDESA 2015), and with the world moving beyond planetary boundaries (Rockström et al. 2009), sustainable resource management has become a top priority among policymakers worldwide.

According to the U.S Energy Information Administration (EIA), world energy consumption will grow by 28% between 2015 and 2040, especially in countries where demand is driven by strong economic growth, particularly in Asia (IEO 2017).

Human influence on the climate system is increasingly supported by evidence, and recent anthropogenic emissions of greenhouse gases (GHG) are furthermore the highest in human history (IPCC 2014). Energy conservation has become imperative to curb the growing global trend in energy consumption and the related rise of emissions causing global warming. Recent studies show that the major developing economies will play a significant role in global energy transition and climate change mitigation over the next decades (Garnaut 2014; Barbi et al. 2016). Increasingly, decision-makers are looking toward technological solutions, management and operational plans to reduce energy consumption, costs and externalities. Once fuel consumption is decreased, externalities and the negative consequences of air pollutants resulting from shipping will be reduced (Ballini and Bozzo 2015).

As transportation is one of the main pillars of globalization in our era, shipping is forecast to become one of the fastest growing sectors in terms of GHG emissions, along with the aviation sector (Winnes et al. 2015). Around 90% of world trade in terms of volume is transported by sea, which is why port cities can be considered the

F. Ballini (✉) · A. I. Ölçer
World Maritime University, Malmö, Sweden
e-mail: fb@wmu.se



A. I. Ölçer et al. (eds.), *Trends and Challenges in Maritime Energy Management*,
WMU Studies in Maritime Affairs 6, https://doi.org/10.1007/978-3-319-74576-3_21

gateways to globalisation (Jung 2011). Of the world's 20 economically leading and most competitive cities, 14 are port cities. Additionally, 65% of cities in the world with populations above 1.3 million are coastal cities (Vallega 2001). Port cities are, therefore, predominantly integrated with urban industrial complexes, supply chains, utility services and transport systems.

Sea terminals around the world are currently adopting and implementing plans to shift from fossil fuel to cost-effective sustainable alternatives, increasingly from renewable sources located within the port perimeter (Acciaro et al. 2013). Port Energy Managers have, therefore, become the key to achieving the overall goals of higher, more cost-effective energy supply.

This paper aims to describe the position of the Port Energy Manager as a preliminary attempt to suggest a trajectory for further scientific analysis in such research domains as sustainable economic development and certification processes in the context of maritime transport.

## 2 Energy Management Programme Overview

The primary objectives of an energy management programme are to minimise energy consumption and its cost while maximising profit for the corporate entity. This will include several sub-objectives, such as reducing energy consumption and increasing energy efficiency to comply with overall regulations and enhance competitiveness. This is secured by putting in place monitoring, reporting and management strategy mechanisms (Ölçer et al. 2017).

Energy management covers the organised and systematic coordination of procurement as well as the convention, distribution, and use of energy, aiming to enhance cost-efficiency (VDI 4602, 2016).

The overall objectives of an energy management system can be summarised as follows (Ölçer and Ballini 2018; Boile et al. 2015; Hippinen and Federley 2014):

- Reduce energy consumption for a fixed output of products or services
- Enhance competitiveness through lower consumption and greater efficiency
- Increase energy supply from renewable sources
- Provide better planning and management
- Introduce production systems with low environmental impact
- Deliver on relevant Corporate Social Responsibilities (CSR) targets
- Promote recycling
- Support the transition from a linear to circular economy

The ISO 50001 certification system defines an Energy Management system (EnMS) as a "set of interrelated or interactive elements to establish an energy policy, energy objectives, and processes and procedures to achieve those objectives" (ISO 50001 2011). In compliance with the ISO 50001 standard, the structure of an EnMS can be divided into seven phases (see Fig. 1) and their related elements.

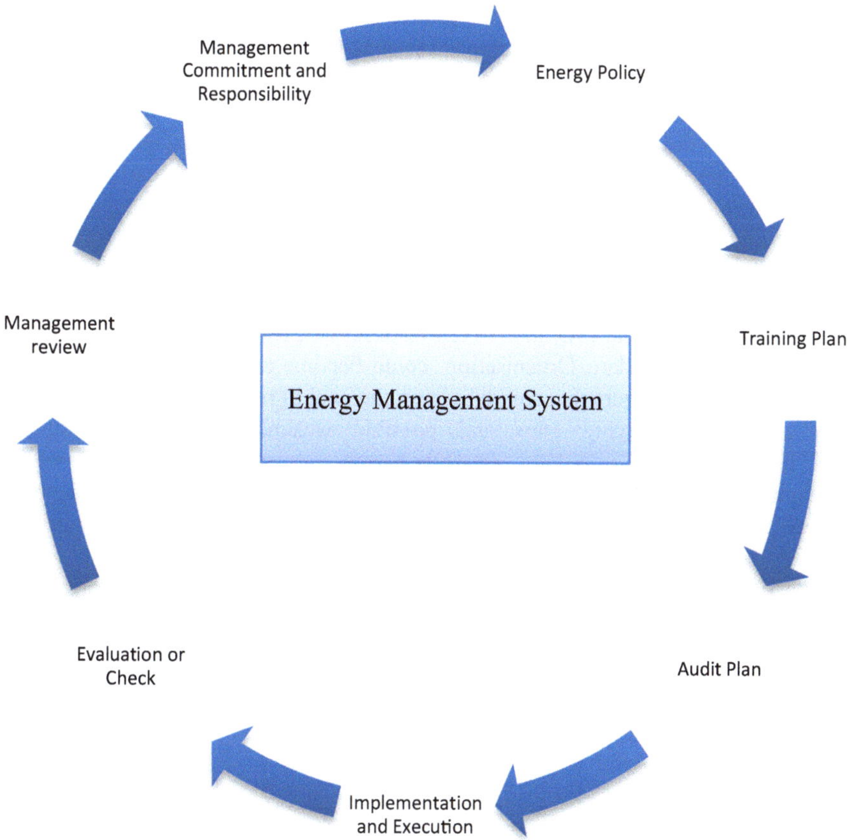


**Fig. 1** Energy Management System overview

*Management commitment and corporate responsibility* are one of the first and most important elements. It defines the general requirements and roles of the top management in relation to energy management.

*An energy policy,* approved by the top management, provides guidance and direction for the organisation in terms of energy reduction, convention, and other energy actions.

*The planning process* involves different elements, such as understanding the legal and other requirements, conducting energy audits, defining the energy baseline, defining Energy Performance Indicators (EnPI), developing objectives, setting out targets with a reasonable payback time, and developing an action plan for delivery.

*Implementation* is a phase in which the organisation focuses on the methods and operational and technical measures in order to achieve an improvement of energy performance in line with the approved targets.

*Evaluation and check*. During this phase, the organisation will invest effort in assessing and evaluating the effectiveness and compliance of the implemented agreed measures.

*Management review* is a very crucial step where the Energy Manager evaluates and establishes the changes required to achieve the targets set. Management review also contributes to ensuring EnMS' suitability, adequacy, effectiveness and continual improvement in energy performance.

To mitigate the effect of climate change, governments are stepping forward to introduce legislation, voluntary procedures, and standard certification (ISO 50001, ISO 14001 as well as Eco-Management and Audit Schemes (EMAS)). Focused on energy management, the ISO 50001 certification, which was introduced in 2011 by the International Standard Organisation, could become an important and efficient tool for Energy Managers that can deliver on the goal of reducing energy consumption e.g. by identifying gaps and possible weaknesses and energysaving opportunities.

Figure 2 describes the different phases of the EnMS, which aims to achieve its targets through developing and implementing a plan and subsequently monitoring and reporting on its progress.

The EnMS includes a PDCA (Plan-Do-Check-Act) cycle to ensure that energy activities are systematically identified, controlled, monitored and recorded. Through this approach, an organisation can ensure that the performance of the EnMS is continuously improved.

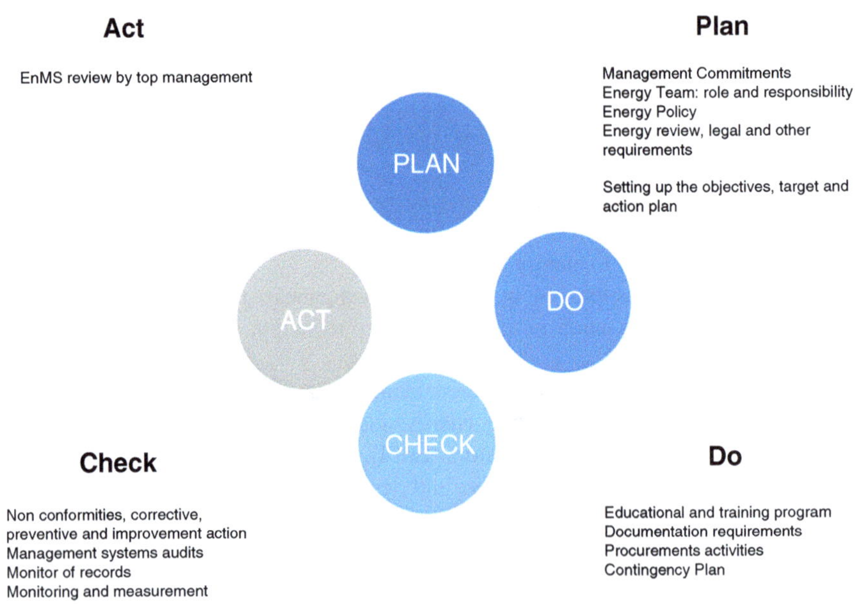


**Fig. 2** Plan-Do-Check-Act elements and activities of the Energy Management System Systems (based on ISO 50001)

## 3 Energy Manager

The top management of an organisation will first establish, communicate, implement, and maintain an official energy policy. The energy policy will provide a guide for the organisation with which to develop, implement, maintain and continually improve energy management (Howell 2014).

Once an Energy Manager (EnMr) has been appointed, the EnMr is to play a pivotal role in setting up and structuring the Energy Team that will support the development, implementation and subsequent evaluation of the energy management plan. This takes place in liaison with the energy committee. The positions of EnMr and the members of the energy team, therefore, require multidisciplinary skills from economic, legal, technological and scientific backgrounds. Figure 3 describes the main activities and responsibilities of an EM in an organisation.

It should be highlighted here that an EnMr needs more management skills and related competencies rather than technical or engineering ones since EM deals with the management of energy. These skills should include personal and interpersonal management skills and their pertinent competencies (Oung 2013).

To be able to achieve the targets set by the management, the EnMr is to formally design a specific operation strategy in line with the Energy Policy and define the Energy Key Performance Indicator (EnPI). A new set of key factors is designed for

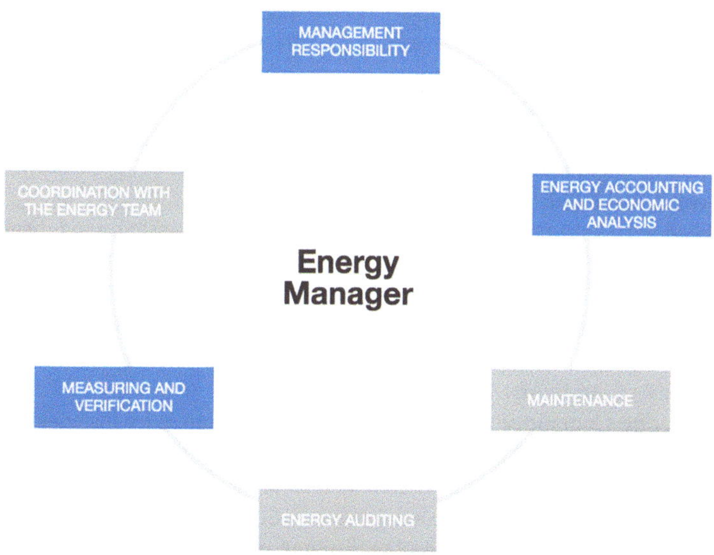


**Fig. 3** Overview of Energy Manager responsibilities

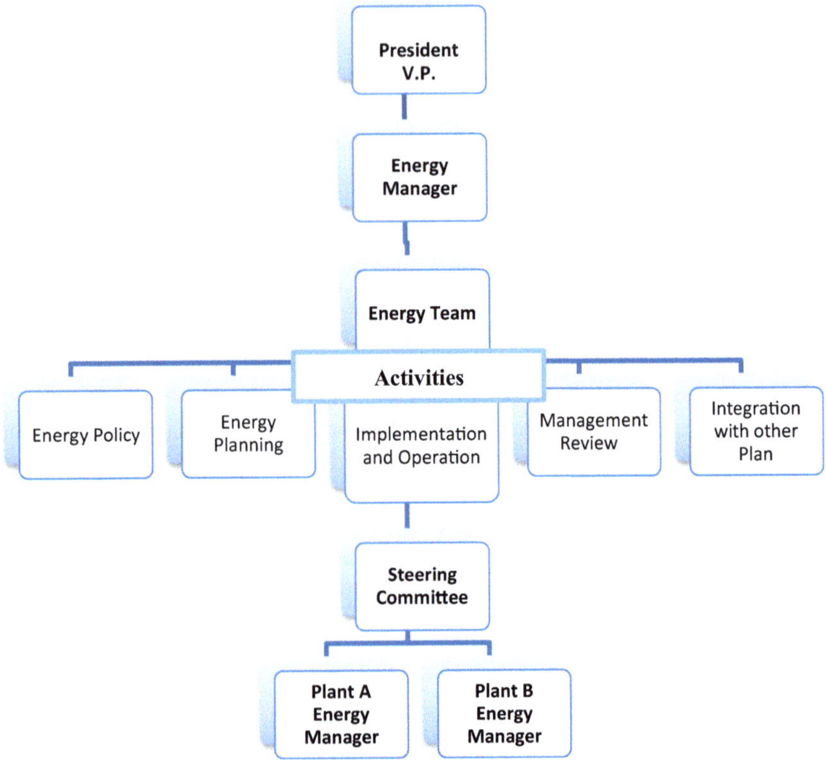


**Fig. 4** Typical structure of Energy Management Programme in an organisation chart

each different operations strategy (Costa et al. 2012). Figure 4 provides a general framework of the EnMr's responsibilities and duties.

To develop and structure an EMP it is crucial for the organisation to hire an EnMr who has the responsibility to manage the Energy Management Programme. Figure 4 shows a typical scenario of multidimensional organisations when the primary EnMr coordinates the different activities, while the Energy Team is an operational and technical body that executes the different activities under the coordination and responsibility of the EnMr.

The Steering Committee is a supporting body represented by the major division of the organisation and its main function is to support the Energy Management Programme in terms of communications and awareness among all the staff of the organisation. Finally, a multidivisional organisation may need several EnMr's in order to coordinate the different tasks at different organisational levels.

# 4 Port Energy Management Plan

Energy management has, in the recent decade, become a top priority for ports. The European Sea Ports Organization (ESPO) has monitored the top-10 environmental priorities of the European port sector over time. The issue of energy management was ranked, in 2016, as the second-most important concern (European Sea Ports Organization 2016). Ports in the cities of Genova, Rotterdam, Hamburg, Los Angeles and others have already adopted strategies, measures and energy management plans to systematically track, analyse and improve energy efficiency and energy saving measures. This will help port authorities become more cost-effective and increase environmental benefits and awareness.

A Port Energy Management Plan (PEMP) is a strategic tool that allows port authorities to address energy saving and environmental objectives by obtaining a structured and detailed analysis of the current energy consumption in the port area and subsequently propose potential energy-saving solutions (Ölçer et al. 2017).

A PEPM consists of a structured set of analytical documents, technical papers, policy documents, strategies and communication tools that provide the EM with the necessary tools with which to ensure energy optimisation. A PEMP includes a timeline and energy monitoring methods. It also defines the responsibility of each involved port actor as well as defining short-term management targets, etc.

## *4.1 Port Energy Manager and Port Energy Team: A Case Study in Genova*

The Port Energy Manager (PEM) oversees the development, implementation, and revision of the Environmental Port Energy Plan in accordance with the objectives and targets of the port.

Roles and responsibilities of the PEM:

- Develop, implement and maintain the PEMP
- Review and approve the relevant reports and decisions made by the energy team
- Communicate periodically with the top management (such as Head of the Port Environmental Division, Legal and Policy Departments, General Secretariat, President)
- Conduct periodical energy reviews
- Ensure that energy performance indicators are in place and in line with the targets selected
- Ensure and coordinate internal and external Auditing
- Monitor the activities of the energy team and provide assistance and training activities
- Collect records, meeting minutes and management review documents

- Encourage the use of renewable energy resources within an organisation or community
- Develop carbon management solutions
- Highlight the importance of energy conservation
- Increase awareness and participatory involvement among employees

The Port Energy Team is responsible for the day-to-day energy management activities. Under the management of the EnMr, the Port Energy Management Plan is monitored periodically, taking into consideration the targets and key performance indicators.

The multidisciplinary role of the EnMr and the Port Energy Team means that emphasis must be given to training in the complexity of port activities, including interaction with public and private stakeholders.

As a case study, the Port Authority Port System of Western Ligure Sea Port of Genova, in 2011 developed a planning tool, namely Port Energy Environmental Plan (PEEP). The aims of PEEP are the promotion and use of renewable energy sources and the increase of energy efficiency in the Port area in compliance with EU objectives concerning GHG reduction.

The main activities of the PEEP include evaluation in terms of energy saving, including the potential energy production from renewable sources, short-term targets for investment, the implementation of an operational tool for the management, and day-by-day monitoring of the activities. The Port system of Western Liguria Sea Port of Genoa has exclusive competences and authority in terms of developing and implementing PEEP.

Since the port system is a complex mix that includes different concession companies (i.e. port terminals, port operators, etc.) this makes it difficult to adopt its own PEM because each concessioner makes its own autonomous decisions in terms of management and policies, including energy policy. Therefore, the port system of Western Liguria Sea Port of Genoa can be defined as an energy management promoter in relation to an / the increased focus on energy flow rationalisation within the port and proactive approaches to energy conservation (Acciaro et al. 2013).

Generally, the EnMr in ports manages the development, implementation, and updating of the Environmental Port Energy Plan in relation to the objectives and targets of the Port system.

In recent years, the issues of energy saving have become increasingly important both from environmental and economic points of view. Aware of this, the Municipality of Genoa (Italy) in April 2015 established, within its Technical Area, the specific position of EnMr to deal with all the activities related to energy efficiency.

The structure, as defined by Municipal Decree DGC-2014-254, consists of four offices within the Technical Area that collaborate transversely with the various Directorates of the entity in order to implement the energy sustainability policies envisaged by current legislation as well as favouring the development of a green economy, energy saving and obtaining funding in the energy field. Figure 5 describes the organisational structure of the EM position under the Municipality of Genova.

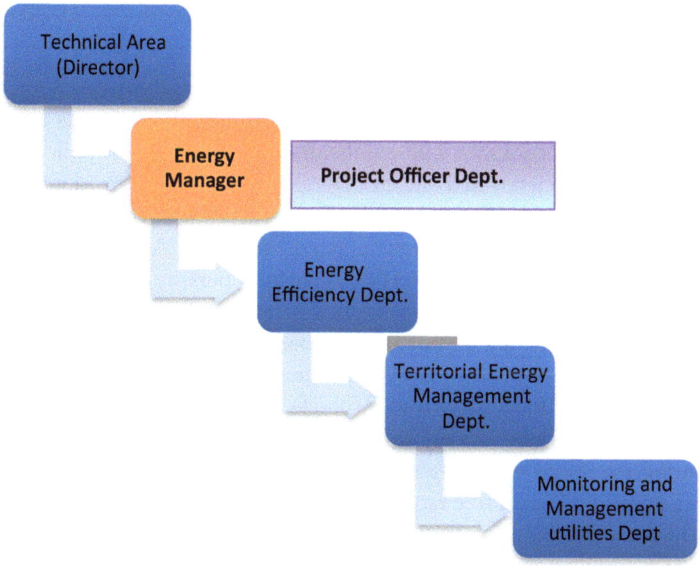


**Fig. 5** Energy Manager structures, Municipality of Genova

The EnMr also collaborates with local authorities (Province and Municipality) on the preparation of projects and plans, both national and for the EU. The EnMr maintains a database of energy consumption of ships calling at the port. In addition, the EnMr supports and guides the Energy Committee by gathering all the data required to evaluate projects submitted by stakeholders. With the support of the energy team, the EnMr performs the energy audits and the energy assessment of the port infrastructure. With the support of the energy team, the EM develops a strategy for the use of renewable energies in compliance with the energy standards of the Kyoto Protocol (electrical mobility in the harbour, wave energy, solar and geothermal energy, etc.). The EnMr collaborates with public and private stakeholders for the promotion and development of national and European smart city projects.

The EnMr at the Municipality of Genova refers to the Environmental Department, specifically the Port Energy Environmental Plan Office, which is supervised by the Regional Planning Department.

The EnMr, therefore, collaborates with the different managers' departments for the definition of operating programs and responds to the actions and objectives assigned in accordance with the guidelines and directives agreed, including also the identification of actions, interventions and procedures needed to promote rational use of energy, energy savings and the development of renewable sources.

## 5 Concluding Remarks

Global warming and GHG emissions have recently received increased attention in the global shipping sector. Ports have started to introduce programmes and policies to curb emissions in accordance with regulations. In addition, these policies also aim to increase energy efficiency and enhance competitiveness through adopting energy-saving measures and new environmentally-friendly technologies.

Energy consumption is a major operative cost for ports. A Port Energy Management System (in compliance with ISO 50001) helps a port organisation to internalise policies, procedures and tools to systematically track, analyse and improve energy efficiency. Energy audits are part of such an energy management system.

In this perspective, the role of the PEM is crucial to the implementation of a Port Energy Management Plan based on corporate energy policies. This undertaking requires skills in working closely with different public and private stakeholders and ensuring the success of different operational and technical measures adopted by the organisation. Such a role requires a multidisciplinary background not only in relation to the PEM but also to the individual members of the corporate Energy Team, which is why training of such personnel is an important component that port authorities and port communities need to bring more into focus if the transition to a low-emission and energy-saving economy is to succeed.

## References


Acciaro, M., Hoffmann, P. N., & Eide, M. S. (2013). The energy efficiency gap in maritime transport. *Journal of Shipping and Ocean Engineering, 3*, 1–10.

Ballini, F., & Bozzo, R. (2015). Air pollution from ships in ports: The socio-economic benefit of cold-ironing technology. *Research in Transportation Business & Management.*

Barbi, F., da Costa Ferreira, L., & Guo, S. (2016). Climate change challenges and China's response: Mitigation and governance. *Journal of Chinese Governance, 1*(2), 324–339.

Boile, M., Theofanis, S., Sdoukopoulo, E., & Plytas, N. (2015). *Developing a port energy management plan: Issues, challenges and prospects*. Transportation Research on Board of the National Academic (GREENBERTH project).

Costa, A., Keane, M. M., Rafttery, P., & O'Donnell, J. (2012). Key factors methodology—A novel support to the decision-making process of the building energy manager in defining optimal operation strategies. *Energy and Buildings, 49*, 158–163.

ESPO/EcoPorts Port Environmental Review 2016, Insight on port environmental performance and its evolution over time, April 2016.

Garnaut, R. (2014). China's role in global climate change mitigation. *China & World Economy, 22*(5), 2–18.

Hippinen, I., & Federley, J. (2014). *Fact-finding study on opportunities to enhance the energy efficiency and environmental impacts of ports in the Baltic Sea Region*. Motiva Services Ltd.

Howell, M. T. (2014). Effective implementation of an ISO 50001 Energy Management System (EnMS). ISBN: 978-0-87389-872-0.

ISO 50001:2011, Energy management systems — Requirements with guidance for use. International Standard Organization (ISO).

Jung, B. (2011). Economic contribution of ports to the local economies in Korea. *The Asian Journal of Shipping and Logistics, 27*(1), 1–30.
Municipal Decree DGC-2014-254, 2014, Genova, Italy.
Ölçer, A., & Ballini, F. (2018). Energy management in the maritime industry. In L. L. Froholdt (Ed.), *Corporate social responsibility in the maritime industry*, Book Chapter 8. Springs International Publishing.
Ölçer, A., Baumler, R., Ballini, F., & Kitada, M. (2017). Maritime energy management. In I. D. Visvikis & P. M. Panayides (Eds.), *Shipping operations management*. Heidelberg: Springer.
Oung, K. (2013). *Energy management in business the manager's guide to maximising and sustaining energy reduction*, Chapter 15. ISBN: 9781409452454.
Rockström, J., Steffen, W., Noone, K., Persson, Å., Stuart Chapin, F., Lambin, E. F., et al. (2009). Planetary boundaries: Exploring the safe operating space for humanity. *Ecology and Society, 14*(2), 32.
UNITED NATIONS, Department of Economic and Social Affairs, Population Division, 2015, World Population Prospects: The 2015 Revision, Key Findings and Advance Tables. Working Paper No. ESA/P/WP.241.
Vallega, A. (2001). Urban waterfront facing integrated coastal management. *Ocean Coastal Management, 44*, 379–410.
Winnes, H., Styhre, L., & Fridell, E. (2015). Reducing GHG emissions from ships in port areas. *Research in Transportation Business & Management, 17*, 73–82.

## Further Reading

Capehart, B. L., Turner, W. C., & Kennedy, W. J. (2008). *Guide to energy management*. Lilburn (GA): The Fairmont Press.
Ciriminna, R., Meneguzzo, F., Pecoraino, M., & Pagliaro, M. (2016). Reshaping the education of energy managers. *Energy Research & Social Science, 21*, 44–48.
ISO 50002:2014, Energy audits — Requirements with guidance for use. International Standard Organization (ISO).
Kanneganti, H., Gopalakrishnana, B., Crowe, E., Al-Shebeeba, O., Yelamanchi, T., Nimbarte, A., et al. (2017). Specification of energy assessment methodologies to satisfy ISO 50001 energy management standard. *Sustainable Energy Technologies and Assessments, 23*, 121–135.
Rogelj, J., Chen, C., Nabel, J., Macey, K., Hare, W., Schaeffer, M., et al. (2010). Analysis of the Copenhagen Accord pledges and its global climatic impacts—A snapshot of dissonant ambitions. *Environmental Research Letters, 5*(3).
The Association of German Engineers. (2007). *VereinDeutscherIngenieure-Guideline VDI 4602* (pp. 3–5). Berlin: BeuthVerlag.
U.S. Energy Information Administration. International Energy Outlook 2013. EIA. www.eia.gov/forecasts/ieo/.

# Part IV
# Economics and Social Dimensions of Maritime Energy Management

# The Impact of SECA Regulations on Clean Shipping in the Baltic Sea Region

**Eunice O. Olaniyi, Gunnar Prause, and Jan Boyesen**

## 1 Introduction

The shipping industry is responsible for the transport of approximately 90% of world trade and is one of the most environmentally benign forms of transport if the transport of goods on a tonne mile basis is considered (Jiang et al. 2014). Ships are also major producers of pollutants such as carbon oxides ($CO_2$), sulphur oxides (SOx), ozone depleting substances (ODS), volatile organic compounds (VOC) and nitrogen oxide (NOx) emissions (Wiśnicki et al. 2014). Although much of the pollution emitted from ships is deposited over the sea, it does not undermine the fact that it is the largest single source of acidification and eutrophication in many countries in Europe. The output of SOx and NOx by shipping surpasses that of land transport (Panagakos et al. 2014). Sulphur especially burns in fuels to form SOx, which is an environmental pollutant, especially in the formation of acid rain (Notteboom 2010).

It also contributes significantly to increased levels of health-damaging particulate matter and ozone (Airclim 2011). Air pollution is the cause of about 50,000 premature deaths every year in Europe, resulting in a social cost of about €58 billion annually. After reacting with air, these emissions are converted to dangerous compounds that cause premature death by poisoning the blood through infiltration from the lungs, leading to inflammation and heart and lung failures after continuous exposure over a long period of time (Alcamo et al. 1987).

Additionally, much deep-sea shipping occurs near coastlines, polluting densely populated coastal areas and causing significant health damage (Holmgren et al.

E. O. Olaniyi (✉) · G. Prause
Department of Business Administration/TSEBA, Tallinn University of Technology, Tallinn, Estonia
e-mail: eunice.olaniyi@ttu.ee

J. Boyesen
Maritime Development Center of Europe, Copenhagen, Denmark

 
A. I. Ölçer et al. (eds.), *Trends and Challenges in Maritime Energy Management*, WMU Studies in Maritime Affairs 6, https://doi.org/10.1007/978-3-319-74576-3_22

2014). Emissions at ports during manoeuvring and berthing account for only a small fraction of ship emissions. Even though this is the case, it is important not to undermine their harmful health effects on the population living around the coastal area. Emissions during free sailing account for a larger percentage of the overall ship emissions, but have less damaging effects on human health because of the sparse population (Jiang et al. 2014).

Maritime transport is currently far less regulated than land-bound transport. One reason is that while land transport is not as international as sea transport, most countries are often logistically, organisationally, and legally involved in shipping and often tend to obstruct the legislation process (Holmgren et al. 2014). Another explanation would be that most governments are careful not to discourage export trade by increasing its cost and, finally, popular views assume that most sea transport occurs on international waters and its pollution does not reach land—an assumption that is far from the truth.

The clean shipping vision was set to make maritime transport greener, and is being achieved through new technologies and changed behaviour on board across all maritime stakeholders in concerted and integrated efforts of multiple measures (Stipa 2013). The Integrated Clean Shipping approach focuses on three fronts: *Ships*—the ship that is welcome in every port, and causes no negative impact on the environment with a zero target emission; *Ports*—highly efficient with excellent environmental services and strong incentives to facilitate and encourage Clean Shipping; and *Cargo*—excellent corporate footprint and cargo owners include environmental issues in their decision-making process when contracting carriers (NSF 2008).

The International Maritime Organization (IMO) is employing a global clean shipping approach to improving the maritime carbon footprint, emissions of SOx and particulate matter from ships, especially through technical and operational reduction measures (IMO 2015). The International Convention for the Prevention of Pollution from Ships (MARPOL 73/78) had a three-tier regulatory scheme to reduce NOx and SOx emissions from ships, Tier I in 1997 and Tier II/III in 2008 (IMO 2008, 2009), all of which were formed to ensure harmonisation with international law and better enforcement. The basic legislation for regulating sulphur emissions from ships was Directive 1999/32/EC. This was amended by Directive 2005/33/EC, adopted at the Marine Environment Protection Committee (MEPC) 53rd session in July 2005, which designated the Baltic Sea, the North Sea and the English Channel as sulphur emission control areas (SECAs) and limited the maximum sulphur content of the fuels used by ships operating in these sea areas to 1.5% (IMO 2008). The Baltic Sea was named as the first SECA from 19 May 2006, followed by the North Sea Area and the English Channel in 2007 (IMO 2009).

It is noteworthy to point out that SECA is not solely applicable to SOx emissions as this criterion was also amended at the 58th MEPC session in October 2008 on Appendix III of Annex VI. The new criteria stated that SOx, NOx, PM and Noxious Liquid Substances, Sewage, and garbage are all restricted in this area from 2010. By 1 July 2010, MARPOL in Annex VI reduced the Sulphur content in marine fuel from 1.5% to 1.0 % (IMO 2013). This was further reduced to 0.1 % from 1 January 2015 (IMO 2015). Another addition to SECA came through China and the regulation states that from 2 December, 2016 there will be three emission control areas in the China Sea—the Pearl River Delta, the Yangtze River Delta and Bohai Sea (North 2016).

The global sulphur cap was set not to exceed 4.5% by the amended Annex VI law. This limit was further dropped to 3.5% in 2011, effective from January 2012. Currently, the global limit has reduced to 0.5% from 2020. This was confirmed at the MEPC 70th session in October 2016 (IMO 2016). There will have been twenty-three preparatory years (1997–2020) for maritime stakeholders to comply with the strict regulation on global sulphur reduction by 2020.

Strengthened regulations and environmental awareness are of vital importance to stimulate Clean Shipping but it has a number of consequences for shipping business (Panagakos et al. 2014). Going by Panagakos premise, the implications of the Sulphur emission Directive, maritime sector stakeholders, such as shipping companies, will be directly or indirectly linked to the economic decisions that will ensue in their efforts to comply. One of the consequences will be increased freight costs. In other words, the high costs of the clean shipping regulations could cause a modal backshift from the sea to the road and could result in more road pollution and congestion, which will be contrary to other land regulations (Hämäläinen et al. 2016). Furthermore, export companies like heavy industrial companies that have maritime transport at the core of their supply chain activities are worried about how this will affect their profit (AirClim 2011).

At the end of 2015, the European Union (EU) approved, in the frame of the Baltic Sea Region (BSR) Interreg Programme, the "EnviSuM (Environmental Impact of Low Emission Shipping Measurements and Modelling Strategies) Project" aiming to study and assess the technical efficiency and the socio-economic impact of clean shipping solutions in BSR. The project seeks to provide tested and analysed results on the efficiency of the different clean shipping solutions and proffer recommendations that will benefit not only the environment and the general health of the BSR populace, but also maritime businesses by promoting their economic growth.

This work as part of the ongoing EnviSuM project studies the economic impacts of SECA regulations on maritime stakeholders in the BSR. The objective of this work is to study the various measures taken by stakeholders towards Sulphur emission reduction and maritime stakeholders' assessments of the economic impact of SECA regulations on their businesses and on BSR. The research questions are, specifically, (1) How has the SECA regulation impacted on the maritime business activities in BSR?; (2) What are the stakeholders' reactions to these impacts?; and (3) How will SECA impact on blue growth and innovation activities in the BSR?

The discussions that are related to SECA compliance in BSR were mostly obtained from expert interviews, learning café and focus groups, while the economic impact of SECA regulation on the maritime industry and BSR were from a BSR-wide survey, targeted at top management executives in the maritime business. This contribution will improve the integration of new knowledge that will provide new systems for cleaner and more cost-effective shipping for the maritime sector.

This work is organised as follows: Sect. 2 discusses the sulphur emissions regulations, shipping activities on the Baltic Sea and SECA compliance options.

Section 3 describes the system of methods used for this research. Section 4 highlights SECA regulation compliance in BSR as well as the economic impact of SECA regulations on maritime businesses and BSR. Section 5 discusses the implications of the results and provides conclusions.

## 2 Sulphur Emissions Regulations Compliant Shipping on Baltic Sea

According to Annex VI of MARPOL 73/78, the Baltic Water means "the Baltic Sea with the Gulf of Bothnia, the Gulf of Finland and the entrance to Baltic Sea bounded by the parallel of the Skaw in the Skagerrak at 57°44'.8 N" (Nugraha 2009). Measuring about 415,000 $km^2$, and considered the largest brackish water in the world, the Baltic Sea, despite being shallow (53 m mean depth) with a narrow canal that makes it difficult for ships to navigate, is subject to intense shipping and covers a significant share of the world's seaborne trade (Karl-Heinz et al. 2013). It is one of the most heavily plied seas in the world and accommodates up to 15% of the world's cargo freight. Over 2000 ships operate on the Baltic Sea at a time and carry as much as 166 million tonnes of oil. The shipping industry can be considered important for all Baltic Sea countries because significant shares of foreign trade are routed through the Baltic Sea Ports. These countries are Finland, Sweden, Denmark, Poland, Russia, Germany, Estonia, Latvia and Lithuania (Helcom 2012). Shipping activities have been predicted to increase in the coming years due to the economic growth of the BSR. For example, Russian oil exports through the Baltic alone have been predicted to reach up to 180 million tonnes by 2020. These activities make the Baltic Sea very susceptible to heavy pollution, whose cumulative effect comes with heavy negative consequences to all living things.

The SECA regulation mostly affects the cost of ship operations, especially those which operate more or solely in the SECAs. For the benefits of Sulphur regulation to outweigh the costs, the reduction measure chosen by the ship owners is critical (Jiang et al. 2014). A substantial amount of studies have already discussed the different abatement or compliance methods for the SECA regulations (Notteboom 2010; Wiśnicki et al. 2014; Jiang et al. 2014; Nugraha 2009; Brynolf et al. 2014; Bergqvist et al. 2015; OECD/ITF 2016). Some of the different options available to reduce sulphur emissions in the SECAs are changing the source of fuel from the usual Heavy Fuel Oil (HFO) to other possible alternative sources of ship fuel, such as Marine Gas Oil (MGO), Marine Diesel Oil (MDO), Liquefied Natural Gas (LNG), biogas, biodiesel, ethanol, methanol, coal, nuclear power, wind, solar panels, and hydrogen cells. The other is the use of abatement technologies such as scrubbers. The scope of this work will, however, be limited to switching to low sulphur fuel, use of LNG and the installation of scrubbers.

A complete switch to 0.1% low-sulphur content fuel seems like the easiest option and is expected to become the most common solution for shipping lines. However, one of the salient challenges of this option is the difficulty in predicting trends in fuel prices due to the volatility of market forces (Jiang et al. 2014; Hämäläinen et al. 2016). Hämäläinen et al. (2016) gave examples of several studies that have attempted to forecast fuel prices in their bid to analyse the sulphur directive effects and have recurrently come-short.

The use of LNG has been argued to be the most attractive alternative fuel to meet the sulphur directive. Apart from its political backing, LNG is naturally low in sulphur and, therefore, meets the new restrictions without any problems during combustion (Jiang et al. 2014). When the gas is in liquid form, it contains more energy and is easier to process in a combustion engine. However, the disadvantage is that while it may be possible to convert existing vessels, it is very costly to convert and requires twice as much storage capacity as the usual marine fuel to provide the same amount of energy; thus, it is most likely that LNG will be used only in new LNG vessels (Brynolf et al. 2014).

The extension of the MARPOL Convention also allows the use of technical equipment that cleans emissions from sulphur and allows the continuous use of cheaper high-sulphur fuel, while still meeting the stricter sulphur restrictions. One such abatement technology is the installation of scrubbers on existing vessels or new vessels (Bergqvist et al. 2015). There are principally two types of scrubbers, namely dry and wet scrubbers. The wet scrubber is further classified into open loop, closed loop, and hybrid. The disadvantages of scrubbers, especially wet scrubbers, range from the dilemma of discharging the acidic wash water, which requires a marine environment with high salinity and high alkalinity to neutralise, to the fact that the scrubber itself occupies a considerable amount of space on the ship. The cost of installation of a scrubber on a vessel, especially on an old vessel, is also high (about €2–€8 million, depending on the age and the size of the ship) and requires about 10 years before a return on investment (Brynolf et al. 2014).

# 3 Method

This research employs multiple and cross-sectional methods. The empirical data collection was executed in the frame of the EnviSuM project between September and December in 2016.

Maritime stakeholders were identified and divided into categories from 1 to 5. Each category was also index based on its perceived direct impact from SECA regulations. As shown in Table 1, 3-stars signifies high priority, 2-stars medium priority and 1-star low priority for data collation. Thus, categories 1, 2 & 3, with 3 and 2 stars, were the major focus for data gathering. The target group consisted of top management level stakeholders (MD/CEOs, Directors, top-level managers), mostly because experts' opinions and decisions, as well as the knowledge of what activities are being embarked on were the core information needed for the research.

First, to explore the activities of the maritime stakeholders towards SECA regulation compliance, interviews (structured, face-to-face, telephone, and Skype)

**Table 1** SECA maritime stakeholders priority index (Source: Compiled by the authors)

| Category | Maritime stakeholders | Order of impart |
|---|---|---|
| Category 1 | Shipping companies, ship operators, ship owners | 3 stars |
| Category 2 | Ports | 3 stars |
| Category 3 | Ship suppliers (maritime equipment and services) | 2 stars |
| Category 4 | Port authority, Non-governmental organisation (NGO) Public sector (Ministry, Municipal, Government agency, education, Research), Police/Customs | 1 star |
| Category 5 | Shippers (Logistics service providers, Production and Trading, Real Estates/Facilities managers) | 1 star |

were conducted. This was guided by the exploratory nature of this objective, which was to produce, according to Mason (1996), "... a rounded understanding of rich, contextual, and detail data that can also be generalised in some way." Furthermore, Mason pointed out that interviews are suitable for a research position that suggests people's knowledge, opinions, understanding, interpretations, experiences and interactions. Each interview lasted between 1 and 2 hours. Notes were taken, and voice recordings were made. Further follow-up questions were asked for clarification and validation. Descriptive analysis of the interview data was conducted to provide an account of SECA-related activities around BSR, using a case study method (Yin 1989).

Second, a BSR-wide online survey study was conducted with a selected number of maritime experts focusing on the experts' perception and conclusions concerning the financial impact of SECA on their businesses and then on BSR. A pilot of the survey instrument was executed to establish content validity from EnviSuM partners from Germany, Estonia, Finland, Denmark, Norway and Poland. The focus of the pilot was to provide clarity, value, and the importance of the survey items. Sixty respondents, representing a broad cross-section of maritime experts at management levels from Estonia, Denmark, Germany, Finland, Norway, and Poland completed the survey.

Other sources of information were from focus groups and EnviSuM learning cafés organised in maritime fairs and exhibitions where workshops were organised by the EnviSuM project. Letters of invitation were sent to participating companies and confirmations were received beforehand for the organisation of such events.

# 4 Results

## *4.1 SECA Regulations Compliance in BSR*

Results from the experts' interviews and focus group meetings revealed the following:

**The Use of Low Sulphur Content Fuel** Because of the availability of different types of marine fuel as well as SECA regulations, most shipping companies, in the interim, tend towards fuel distillates such as MGO and MDO. Further, ultra-low sulphur fuel oil (ULSFO) has been gaining popularity as a cheaper alternative to other distillates since 2015. There is already a 180,000 mt demand, which is expected to grow beyond 300,000 mt by the end of 2017. Another reason for its rise in demand is the heavy investment that is needed for scrubbers or for LNG retrofit.

Some ship operators have, however, expressed their concern over the long-term effect of using this fuel and the effect it would have on ship engines originally designed for HFO. Others are optimistic that its usage can flush out the sludge that accumulates in the fuel tank. There is also a concern about the availability of low sulphur fuel with the 0.5% global sulphur cap coming up in 2020. Most cruise ships use low sulphur fuel, which seems to be the interim strategy at present until they can be sure about the fuel market trend.

So far, because of intense competition, no shipping line has increased its transportation cost. Many of them are waiting for the fuel price to increase. In addition, the new global sulphur cap for 2020 seems to be a game changer. Nobody seems to be sure of the full impact of this regulation on fuel price until then. There are anticipations that the global sulphur cap will be reviewed and pushed back to 2025.

**LNG as an Alternative Fuel** The use of LNG as a marine fuel has especially attracted the high attention of the shipping industry in the BSR. Some ship-owners have stated their preference for the use of the alternative fuel path for their sulphur compliance requirements, especially the use of LNG. One of the major reasons for this is that the use of LNG ensures compliance with not only SECA, but also the anticipated NOx emissions regulation. LNG is seen as the total package for emission regulation compliance, ranging from SOx, NOx, and PM to $CO_2$ and black carbon. LNG is not a popular option when compared to MGO/MDO/ULSFO and scrubbers.

The major reason is the cost implication for retrofitting or building a new ship, but its demand seems to have soared. For example, by the end of 2014, the only LNG-fuelled ships in the BSR were Viking Grace, the dual-fuel ferry owned by Viking, and two LNG-powered (gas-only engine) cruise ferries of the Fjord Line operator. However, by the end of 2015, there were 24 new builds and 3 LNG retrofitted ships and as of 2016, there are over 40 LNG-propelled ships ordered from different shipping lines with delivery dates up to 2018.

**The Use of Scrubbers** In the BSR, ferries, ro-ro and general cargo (feeder container ships) account for the majority of ships that are equipped with scrubbers. Even though fuel price has been on a downward trend since 2014 and the price spread between HFO and MGO has greatly reduced, some ship owners still invest in scrubber technology.

According to the Baltic Port Organization (BPO), by the end of 2015, 5.4% of the total short sea shipping vessels were equipped with scrubbers, which amounts to 73 vessels in the BSR. More so, more ships are already scheduled to be retrofitted before the end of 2017. By the end of 2020, about 200 ships are expected to be retrofitted with the scrubbers. Most port operators have brought up arguments concerning the ecological safety of the open loop scrubber since its wash water is being discharged back into the sea together with an unclear rule as to whether it has been banned or is still being allowed in the BSR.

There was a lot of favour towards the hybrid model because the closed loop scrubber will be useful in the shallow Baltic waters and, at the same time, its open loop part will be useful while in non-SECA waters. Dry scrubbers are not popular and only one ship is equipped with a dry scrubber in the BSR. Tests in Poland showed that using a scrubber requires up to 13 MWh of additional electricity per day. In the case of an electricity production price on board the ship of around 80 € per MWh, this mean that it would face additional costs of €100,000/year.

**Compliance Cases from Shipping Companies** Three cases are presented to highlight the efforts made by the shipping companies in terms of their compliance to SECA regulations as follows:

**Case 1: Tallink, Estonia**. Tallink has a total of 17 vessels: 2 cargo, 6 Ro/Ro, and 9 cruises. Currently, three of these vessels are chartered out. At the moment, Tallink's strategy is to use MGO while in SECA and HFO in non-SECA. However, one of the latest additions to its fleet because of SECA is the Megastar, which is equipped with dual fuel engines and will be propelled by LNG, and it was built by the Meyer wharf in Turku, Finland. The propulsion is equipped with 5 engines, which run around 6.000 h per year, 3 Wärtsilä 12V50DF (11.4 MW/machine) and 2 Wärtsilä 6L50DF (5.7 MW/machine). This means that the engine can be switched automatically from fuel oil to LNG operation at loads below 80% of the full load. In order to switch, the operator's command can be an automatic initiate transfer that takes only about a minute allowing the fuel oil to slowly change to LNG.

**Case 2: Viking Line, Finland**. Viking Line operates 7 vessels. Viking's strategy for SECA is to use fleets propelled by LNG. An already existing LNG-driven ship is the Viking Grace. The Viking Grace started its service in 2013 and is equipped with 4 Wärtsilä 8L50DF engines (7.7 MW/machine). It sails on the route between Stockholm, Åland and Turku. Viking line has already ordered one more LNG-driven ship that will arrive by 2020. There is no mention on any scrubber retrofit.

**Case 3: DFDS Seaways Denmark**. DFDS owns 53 vessels, five are on charter while one is laid up. Leveraging on the reduced price spread between oil fuel and gas oil, the biggest investment in scrubber technology in the BSR is by DFDS Seaways. DFDS has 17 scrubber retrofitted ships that cost the company about DKK 568M (Approx. €80M). The company's target is to install scrubbers on 21 of its ships by 2017. At the end of the 2015 financial year, the company reported that it had achieved the budgeted return on investment for the scrubber strategy. DFDS is one of the scrubber grant recipients from the EU. It has received DKK 35M (Approx. €5M) on this premise to date.

**Compliance Activities in the BSR Ports** This section presents four key SECA compliance activities in the BSR ports: (a) LNG infrastructure; (b) Onshore Power Supply; (c) Incentives for shipping companies; and (d) Compliance Monitoring and Control.

(a) **LNG Infrastructure**. The development of LNG bunkering infrastructure within the BSR was initiated by the Baltic Port Organization "LNG in Baltic Sea Ports" and LNG in Baltic Sea Ports II projects for the harmonisation of pre-investment activities of Ports networks to standardise the LNG system. This initiative was due to the EU Directive 2014/94/EU that TEN-T core seaports must be equipped with accessible LNG bunkering points by 2025. So far, there has been a steady increase in the development of LNG bunkering infrastructure within the Baltic ports. There are terminal feeds for vessels with LNG going through bunkering upgrades. Some of them currently use ship-to-ship, and tank truck to ship bunkering. The port of Stockholm was the first port in the BSR to take the LNG initiative and it was the first port in the world to offer LNG bunkering solutions through a bunkering boat for LNG supply to Viking Grace, a cruise ferry. Other LNG terminal infrastructures are in Świnoujście, Poland; Port of Klaipeda, Lithuanian; and Lysekil, Sweden. There are other small-scale LNG terminals in Nynäshamn and Gothenburg, Sweden. There is an LNG gas bunker station used for Fjord Line in the Port of Hirtshals, Denmark.

(b) **Onshore Power Supply**. Onshore Power Supply (OPS), also known as shore connection, shore-side electricity and cold ironing are solutions that allows ships in port to turn off their auxiliary engines in order to stop the use of fuel. The use of OPS allows a substantial reduction in emissions, noise and vibrations generated by ships berthing at ports. This technology is particularly important for crews, passengers and for residential areas close to the ports as obtainable in Stockholm, Ystad, Gdynia, and Tallinn. The ports of Gothenburg, Lübeck, Helsinki, Ystad, and Stockholm currently provide OPS. Helsinki introduced shore power in 2012 for Viking Line's ferries. The Port of Ystad has a large OPS system for both 50 and 60 Hz. Stockholm port was planning to have all of its five new quays equipped with the OPS system by the end of 2016, but these are not ready yet.

(c) **Incentives for Shipping Companies**. Currently, most incentive activities are geared towards ships that use LNG, generate low NOx emissions or use OPS. The Port of Gothenburg, for example, uses the Environmental Ship Index (ESI) and the Clean Shipping Index (CSI) for its reward system, and awards ships fuelled by LNG a 20% discount, saving the ships SEK 0.05 per gross tonnage (GT). This will make a vessel like Viking Grace, calling at Stockholm daily, earn a rebate of around SEK 1 million (approx. €105,400) every year until 2019, when the port plans to stop. In addition, ships with ESI score of 30 points or above are categorised as "green" by CSI yardsticks and are awarded a 10% discount off port dues according to their GT. The Port of Gothenburg also offers a rebate for the use of OPS. The Port of Rostock offers vessels with ESI of 40 points and above a range of 5–10% discounts. In the Port of Riga, a 10%

rebate on port charges is given to crude oil tankers with a Green Award Certificate.

(d) **Compliance Monitoring and Control**. Compliance and monitoring are done through inspections and fuel samplings. The common practices are the inspections of the IAPP6 and IOPP7 certificates, Oil Record Book, Bunker Delivery Notes (BDN), logbooks and records relating to fuel switchover before entering the SECA and navigational activities. Inspections are made mostly on ships without installed emission abatement equipment. In some countries like Poland, fuel samples are collected and analysed for inspection, but in other countries, samples are only analysed if there were indications of abnormalities noticed or perceived during document inspections. Samples are taken from the service tank, or return pipe or fuel filter and analysed by accredited laboratories. Analysis usually takes between 1 and 7 days. Sometimes the inspectors make use of handy equipment that can immediately detect the sulphur content of the fuel used on board, although the results provided by this method are only indicative and usually need further confirmation from laboratory analyses. According to Mellqvist (2016) during the EnviSuM project partner meeting, 6801 inspections were carried out from 2015 up until March 2016. Surveillance aircraft are also strategically stationed at the water belt of Helsinki and Denmark to monitor emissions based on optical and sniffer measurement. An optical ultraviolet spectroscopic sensor analyses reflected solar light that passes through the smoke (also called DOAS (Differential Optical Absorption Spectroscopy)) to measure the path concentration of $SO_2$ and $NO_2$ from the ship. Airborne measurement of about 100 ships is carried out monthly. The preliminary report from the EnviSuM technical team reported a 96.5% compliance in the BSR ports, which has been consistent (Mellqvist 2016).

## 4.2 *Impact of SECA Regulations on Maritime Business Activities in BSR*

Tallinn University of Technology together with the Maritime Development Centre started an ongoing survey on the impact of SECA regulations on maritime business in BSR from October 2016. The underlying questionnaire included ten questions assessing the impact of SECA regulations from the view of different maritime stakeholders. The ten questions covered the overall economic impact of SECA as well as the impact on blue growth, costs, pricing, company development, innovation, FDI, cargo flows, modal split and branding of BSR as shown in Table 2. Five-point scaling from $-2$ = very negative to 2 = very positive was used to measure the maritime stakeholders' views for all questions. Until now, 60 questionnaires from all BSR were collected showing a first trend on how SECA regulations affect or are assumed to impact maritime business activities.

Based on the completed questionnaires, statistical analysis was carried out in order to find core patterns in the maritime community. The first observation shows

**Table 2** Description of economic impact of SECA regulations

| *What is the impact of SECA regulations on the maritime businesses and the BSR?* |
|---|
| Overall economic impact on BSR |
| Impact on blue growth in BSR |
| Impact on your product/service costs |
| Impact on your pricing |
| Economic impact on your company development |
| Impact on innovation of maritime sector in BSR |
| Impact on attractiveness of BSR for foreign direct investments (FDI) |
| Change of cargo flows within Europe |
| Change on transport modal split within the BSR |
| Reputation/Branding of BSR |

Note: N of data = 60

that nearly all Mean (M) results for all the questions were close to zero, except the M for innovation and branding, which were around 1. Statistical tests confirmed the significance of this result and also revealed that the overall Ms are not significantly different for the high impact priority groups of maritime stakeholders (i.e. ship owners, ports, maritime suppliers), which suggests that the assessment of SECA impact is relatively homogenous among the maritime sectors. The interpretation of these first results indicates that from 2 years after the 0.1% SECA in BSR until now, the impact on economic parameters such as costs, pricing, FDI, cargo flows and modal splits are considered negligible by the maritime stakeholders. Significant SECA impacts are, however, attributed to innovation and reputation of BSR.

Additional regression analysis was used to determine which of the remaining nine variables affects the variable "overall economic impact of SECA". The F-test showed a significant fit for the resulting model despite the fact that the R-square was 52%. The regression analysis coefficients and the related t-test revealed that variables "blue growth" and "FDI" are statistically significant, both with a positive leading sign. The variable "pricing" appeared to be "nearly significant" with a p-value of 7% but with a negative sign. The Beta-coefficients percentage shows that three variables "blue growth", "FDI" and "pricing" are accounted for in about 60% of the impact in "overall economic impact of SECA". The negative sign of the "nearly significant" variable "pricing" suggests that increasing pricing will have a negative impact on the overall perception of SECA.

Finally, the ANOVA analysis of the responses of the ship owners and the Ports concerning the SECA regulation impact on the modal split revealed a significant difference. Whereas the Mean of ship owners on the modal split was slightly positive with a value of about 0.4, the corresponding value for the ports was about −0.4.

## 5 Implications and Conclusions

SECA regulations have already been in force in BSR for 2 years now and have no doubt enhanced clean shipping in BSR with significant changes seen in vessels that operate on the Baltic Sea as well as the Ports. The EnviSuM project aims to reveal, empirically, the impact of SECA regulations in BSR, and the first measurements, which show a compliance rate of 96.5% in most BSR ports, confirm significant health benefits for the BSR. Furthermore, ongoing expert interviews and surveys suggest that the maritime stakeholders in BSR expect a very small economic impact of SECA regulations on the shipping sector, especially on pricing, costs, FDI, cargo flows or modal split. Only innovation, branding and reputation of BSR are expected to improve significantly through SECA regulations. Surprisingly, this assessment does not differ significantly among all the maritime stakeholder groups in BSR. Some of the expert interviews and studies of secondary data also revealed that most of the ship owners took advantage of the low oil price to comply with SECA regulations in BSR by using MGO. By doing so, most of the ship owners have tried to postpone expensive abatement investments. Economic implications and unreliability of fuel prices are the major reason for this investment stalling. Other reasons are fear of revenue loss during installations, time spent on retrofitting, further and post-installation and modification activities as well as the constant need for intervention from manufacturers, especially regarding maintenance.

There were many views that the implementation of SECA should have been better prepared for and many ship owners were confused regarding the expected costs and benefits related to the different abatement and emission control technologies, which has led to sub-optimal solutions. Most ship owners would have preferred to have access to reliable information to help them make decisions. However, with the current situation in the shipping market, where there is an excess capacity for demand, there is very little willingness to invest in LNG vessels and scrubber technology. The interesting information from the expert interviews indicated that the trend for the decision on abatement strategy is that ship owners would prefer a scrubber installation for used vessels and LNG installations for new builds, which can be confirmed by the fact that there are only 3 LNG retrofitted vessels running in BSR. However, the BSR can still be considered a bedrock for the implementation of LNG infrastructure, but its specialised infrastructure for bunkering purposes is necessary to allow the growing number of LNG-fuelled ships to operate. It is, therefore, necessary that ports be involved in LNG bunkering facilities synergies with ship owners and other stakeholders like the LNG suppliers. It was generally accepted that the use of LNG as ship fuel would be more popular in the coming years.

Another interesting matter that kept cropping up was the use of OPS, which has considerably improved the air quality and the noise level in ports. However, the OPS requires adaptation and suitable equipment (e.g., frequency converters) not only in ports but also on-board ships. This is another investment inclined action which also requires the cooperation of both ports and ship owners.

A thought provoking question that emerged from the different interactions is, since MGO and other compliant fuel are distillates, what are the necessary actions to prevent the residual fuel from becoming technical waste? Scrubber technology seems to provide a much-needed technology push effect for European technology, but how many ship owners can afford the costs?

Prior to the strengthened requirements for sulphur reductions, there was a great fear that the SECA regulations would have a negative effect on the maritime industry. However, there have been fewer effects than initially anticipated, especially by ship owners and ports. There are probably several reasons for this. The first is that there has been a strong drop in oil prices, which has absorbed a lot of the feared cost increase. Since fuel costs are the main cost component for ship owners, this has led to substantial savings. It will, therefore, be interesting to see what will happen if the oil prices go up again. Another reason for the low impact might also be an over-capacity witnessed in the shipping market. Some ship owners have closed down routes, or changed routes. Some have changed to larger vessels in order to adopt to SECA and reduce negative economic impacts. Most ship owners have used the over-capacity opportunity to slow steam, achieving significant financial savings. It seems like this will continue for many years to come.

The SECA research in BSR will be providing a very useful case study, especially when the global sulphur cap comes into full effect in 2020. Although a cause-effect situation might be difficult to extrapolate without a higher population size, this work contributes to the body of knowledge in the following ways. It postulates the economic consequences and the market potential for abatement technologies beneficial for knowledge integration in the maritime sector in BSR and Europe. It also provides a good prospect for using the BSR SECA as a larger test bed and demonstration platform for solutions that can be used globally. Furthermore, since parameters such as fuel prices and market demand are also important to capture the fuller impact of the different SECA compliance options, further research is needed to measure the full impact of the SECA regulations, especially towards the creation of an "Investment Decision Tool" accessible to all maritime stakeholders.

**Acknowledgement** The authors thank to the European Regional Development Fund sponsors who has supported the EnviSuM project.

## References

AirClim. (2011). Air pollution from ship. Air Pollution & Climate Secretariat Seas at Risk Bellona Foundation North Sea Foundation Transport & Environment European Environmental Bureau. https://www.transportenvironment.org/publications/air-pollution-ships-0.

Alcamo, J., Amann, M., Hettelingh, J. P., Holmberg, M., Hordijk, L., Kämäri, J., Kauppi, L., Kauppi, P., Kornai, G., & Mäkelä, A. (1987). Acidification in Europe: A simulation model for evaluating control strategies. *Ambio, 17*, 232–245.

Bergqvist, R., Turesson, M., & Weddmark, A. (2015). Sulphur emission control areas and transport strategies - The case of Sweden and the forest industry. *European Transport Research Review, 7* (2), 10.

Brynolf, S., Magnusson, M., Fridell, E., & Andersson, K. (2014). Compliance possibilities for the future ECA regulations through the use of abatement technologies or change of fuels. *Transportation Research Part D: Transport and Environment, 28*, 6–18.

Hämäläinen, E., Hilmola, O. P., Prause, G., & Tolli, A. (2016). Forecasting maritime logistics costs – Will the low oil price remain? In A. Serry & L. Lévêque (Eds.), *Le transport maritime à courte distance (Short-Sea Shipping): Devport 2016 Conference (Le Havre, France)* (pp. 19–34). Le Havre, France: EMS Geodif.

Helcom. (2012). *Maritime activities in the Baltic Sea.* Baltic Sea Environment Proceedings No. 123. Accessed December 20, 2016, from https://www.google.ee/search?q=helocme+report+&ie=utf

Holmgren, J., Nikopoulou, Z., Ramstedt, L., & Woxenius, J. (2014). Modelling modal choice effects of regulation on low-sulphur marine fuels in Northern Europe. *Transportation Research Part D: Transport and Environment, 28*, 62–73.

IMO. (2008). *Revised MARPOL Annex VI: Regulations for the prevention of air pollution from ships and NOx technical code.* London: IMO Marine Environmental Protection Committee (MEPC).

IMO. (2009). Prevention of air pollution from ships Second IMO GHG Study 2009 Update of the 2000 IMO GHG Study Final report covering Phase 1 and Phase 2: noted by Secretariat (MEPC 59/INF.10).

IMO. (2013). Sulphur oxides (SOx)—Regulation 14. http://www.imo.org/OurWork/Environment/PollutionPrevention/AirPollution/Pages/Sulphur-oxides-(SOx)-Regulation-14.aspx. Accessed 2016-09-20.

IMO. (2015). *Guidance on the application of regulation 13 of MARPOL annex VI tier III requirements to dual fuel and gas-fuelled engines.* MEPC.1/Circ.854. London.

IMO. (2016). IMO sets 2020 date for ships to comply with low sulphur fuel oil requirement. Press briefing release 28/10/2016. http://www.imo.org/en/MediaCentre/PressBriefings/Pages/MEPC-70-2020sulphur.aspx. Accessed 30/11/2016.

Jiang, L., Kronbak, J., & Christensen, L. P. (2014). The costs and benefits of sulphur reduction measures: Sulphur scrubbers versus marine gas oil. *Transportation Research Part D, 28*, 19–27.

Karl-Heinz, B., Möller, M., & Wenske, C. (2013). Baltic maritime transport, its structure, competitive situation and economic weight. In K. Breitzmann, & M. Hytti (Eds.), *Pan-Baltic manual of best practices on clean shipping and port operations.* Turku, Finland: UBC Environment and Sustainable Development Secretariat, Vanha Suurtori 7, FIN-20500.

Mason, J. (1996). *Qualitative researching.* London: SAGE.

Mellqvist, J. (2016). SOx compliance monitoring, EnviSuM partner meeting, 26.10.2016, Copenhagen.

North. (2016). China: Emission Control Areas. http://www.nepia.com/news/industry-news/china-emission-control-areas-starupdatestar/. Retrieved on 22.12.2016.

Notteboom, T. (2010). The impact of low sulphur fuel requirements in shipping on the competitiveness of roro shipping in Northern Europe. *WMU Journal of Maritime Affairs, 10*(1), 63–95.

NSF - North Sea Foundation. (2008). *Clean shipping: Towards an integrated approach of sustainable shipping.* Utrecht, The Netherlands: North Sea Foundation.

Nugraha, F. (2009). Effective implementation of emission control area towards cleaner shipping operations: Focusing on sulphur oxides (SOx) emission reduction. *World Maritime University Dissertations*, 186.

OECD/ITF. (2016). Reducing Sulphur Emissions from Ships: The Impact of International Regulation. http://www.itf-oecd.org/reducing-sulphur-emissions-ships-impact-international-regulation. Retrieved on June 10, 2016.

Panagakos, G. P., Stamatopoulou, E. V., & Psaraftis, H. N. (2014). The possible designation of the Mediterranean Sea as a SECA: A case study. *Transportation Research Part D: Transport and Environment, 28*, 74–90.

Stipa, T. (2013). Clean shipping current. In K. Breitzmann, & M. Hytti (Eds.), *Pan-Baltic manual of best practices on clean shipping and port operations*. Turku, Finland: UBC Environment and Sustainable Development Secretariat, Vanha Suurtori 7, FIN-20500.

Wiśnicki, B., Czermański, E., Droździecki, S., Matczak, M., & Spangenberg, E. (2014). Sulphur regulation–technology solutions and economic consequences for the Baltic Sea Region shipping market. In E. Czermański (Ed.), *Institute of maritime transport and seaborne trade*. Gdańsk: University of Gdańsk.

Yin, R. K. (1989). Case study research: Design and methods, revised edition. Applied Social Research Methods Series, 5.

# Life Cycle Assessment of Marine Coatings Applied to Ship Hulls

**Yigit Kemal Demirel, Dogancan Uzun, Yansheng Zhang, and Osman Turan**

## 1 Introduction

It is a well-known truth that hull resistance is a significant parameter for ships with regard to fuel consumption, increment in power requirement for desired operation capability and greenhouse gases (GHG) emissions. Basically, ship resistance is divided into two parts, frictional resistance which directly relates to roughness of surface and residuary resistance that occurs because of waves created by ship actions at sea (Demirel 2015).

For low-speed merchant ships, frictional resistance accounts for 80–90% of ship total resistance (van Manen and van Oossanen 1988). Since the fouling on ships strongly increases the roughness of ship hull surface, fouling is a quite detrimental phenomenon for ship frictional resistance.

In the study done by the Wood Hole Oceanographic Institution, it was observed that the ships in the British Navy have a frictional drag increase of about 0.25% and 0.5% per day for temporal and tropical waters, respectively. This led to 35–50% extra fuel consumption for these ships (Hole 1952).

Fouling organisms' effects on ship resistance is pointed out in respect to their type and coverage rate. According to Schultz (2007), ship resistance is increasing because of fouling in a range between 2% and 80% compared to ship resistance while the ship's surface is accepted as clean. In a recent study, artificial barnacles printed via 3D printers and attached to plates were towed in a towing tank. As a result of the experiment for 5% and 10% coverage rates, 36.8–97.5% and 22.5–59.7% increases were observed in respect to bare plates for frictional resistance and effective power, respectively (Turan et al. 2016).

Y. K. Demirel (✉) · D. Uzun · Y. Zhang · O. Turan
University of Strathclyde, Glasgow, UK
e-mail: yigit.demirel@strath.ac.uk



A. I. Ölçer et al. (eds.), *Trends and Challenges in Maritime Energy Management*,
WMU Studies in Maritime Affairs 6, https://doi.org/10.1007/978-3-319-74576-3_23

To mitigate these detrimental effects of fouling organisms, antifouling paints are widely used for ships all over the world. Thanks to advances in chemistry and material science today, the shipping industry has the opportunity to use a vast range of antifouling paints. State of the art antifouling paints are basically divided into two groups: biocide contended Self-Polishing Co-Polymers (SPC); and Foul-Release (FR) antifouling coatings.

The selection of best antifouling paint for any considered ship is not an easy task and is related to various parameters such as performance of antifouling paints, operation profile of ship, and route. An antifouling coating system can be claimed to be better than another coating system only if the total life-cycle costs and environmental impacts of the coating are less than those of another one. To decide which paint is better than the others, paints should be evaluated in all their aspects, which can be done with the life-cycle assessment (LCA) method. The aim of using the LCA method is to quantify the impact of such technology based on improvement on existing ships, as calculated for the remaining life of the ship. This can be extended to fleet-level and global-fleet figures. Actual cost-savings as well as benefit can be established by considering the following points:

- Life cycle costs.
- Life cycle energy consumption and savings in terms of fuel and cost.
- Potential life cycle environmental impacts in terms of carbon footprint reduction and impact on the marine life.
- Energy and raw material used in each process as well as the waste and emission generated.
- Initial paint cost in terms of material and application.
- Durability of the paint and maintenance cost in terms of frequency of hull cleaning and painting (extended service life).
- Duration of dry-docking/hull cleaning and effect of the paint on the availability of the ship for hire (more income).

The evaluation will consider impacts related to all life cycle stages, by simplifying some of the stages (especially production of antifouling) and focusing on the effects of antifouling coatings on fuel consumption of ships during operation, and hence emission and maintenance of ships.

To the best of the authors' knowledge, no specific life cycle assessment model exists to predict the impact of antifouling coatings applied to ship hulls. The aim of this paper is, therefore, to fill this gap by developing an LCA model consisting of several prediction methods, and to show how to use the proposed LCA model by investigating the effect of the application of two existing marine coatings on the fuel consumption, cost and GHG emissions of a specific real ship over a 30 year life cycle.

This paper is organized as follows: In Sect. 2, general LCA method is introduced along with LCA of antifouling coatings. The proposed LCA model is explained in detail by giving mathematical relations that are used in the model in Sect. 3. A case study was carried out to show how to use the developed LCA model in Sect. 4. Finally, the results of the study are discussed in Sect. 5, along with recommendations for future avenues of research.

**Table 1** Major stages of LCA of an antifouling (AF) coating

| Stage | Activities |
|---|---|
| 1. Production of AF coatings | Extraction of natural resources, mining non-renewable material and transporting these materials to processing facilities |
| 2. Application | Application of antifouling coatings on ship hulls |
| 3. Operation of ships with AF coatings | Extra fuel is consumed due to the effects of antifouling coatings/fouling |
| 4. Maintenance of ships (Hull cleaning and recoating) | System maintenance (dry-dock and in water hull cleaning) |
| 5. End of life | Dismantling of ship |

# 2 Methodology

Since the performance and efficiency of an antifouling coating can be assessed by its effect on ship fuel consumption due to hull fouling and on maintenance costs and emissions of a ship, the analysis focuses on the life cycle of paints on a ship hull rather than the life-cycle of a coating itself. In other words, the antifouling coating is taken as a system used in the whole life-cycle of a ship and hence the life-cycle of an antifouling coating system is taken as the whole life-cycle of a ship. Table 1 depicts these five stages and details the activities and processes partitioned among each stage.

Given that a new LCA methodology and model is developed for the assessment of an antifouling coating, a tailor-made methodology was used to highlight the important parameters of an antifouling coating. For this reason, the LCA results provided in this paper are specific to antifouling coatings and the methodology does not follow the formal LCA.

## *2.1 Data Requirements*

The required data for complete LCA of antifouling coatings cover an extensive range of information. Therefore, it would be beneficial to split the required information into two parts in order to have a better understanding.

### 2.1.1 Ship Operations

The information required in this part is used to estimate the total costs and emissions from fuel consumption of a ship due to the use of different antifouling coatings. The required data can be listed as below:

- Type of the ship
- Main dimensions of the ship including the wetted surface area
- Sea trial data
- Ship operational data (speed, draft, operational regions and durations of port calls or being stationary)
- Type of the coating on ship hull

### 2.1.2 Antifouling Coating Applications

The information required in this part is used to estimate the total costs and emissions from the activities in the initial and dry-dock paint applications to ships due to the use of different antifouling coatings. The required data can be listed as below:

- Dry-dock intervals
- Paint related information
- Details of the surface preparation methods

## 3 Foul-X-Spel Life Cycle Assessment Model

The life-cycle model in question is a computer program that aggregates the financial and environmental costs of ship operation and maintenance in relation to fouling. The model captures the aspects of operations and processes that influence and are directly impacted by marine fouling and, hence, time-dependent drag performances of antifouling coatings. Figure 1 shows a diagram representing the workflow of FOUL-X-SPEL (Environmentally Friendly Antifouling Technology to Optimize the Energy Efficiency of Ships, Project number 285552, FP7-SST-2011-RTD-1) LCA model. The essential structure of the workflow is that the relevant aspects of the model start from the shipbuilding stage all the way to the dismantling of the ship is simulated.

Fouling primarily accumulates while the ship is static, and rate of growth of fouling depends predominantly on the water temperature. Warmer water is more teeming with life than colder water, and the longer a submerged body stays stagnant in water, the more likely it is to be colonized by marine flora and fauna. These facts have been well-known qualitatively since the beginning of marine transport. However, in order to make fine decisions about modern anti-fouling measures, we need a quantitative model of these factors. The following are the key ingredients of the model:

1. A representation of voyages and anchorages of the ship in question. The representation should describe the time and location of the ship over its lifetime. The developed model can generate such representations based on probability distributions derived from the operating life of real ships, and can import fully specified voyage data from noon-reports.

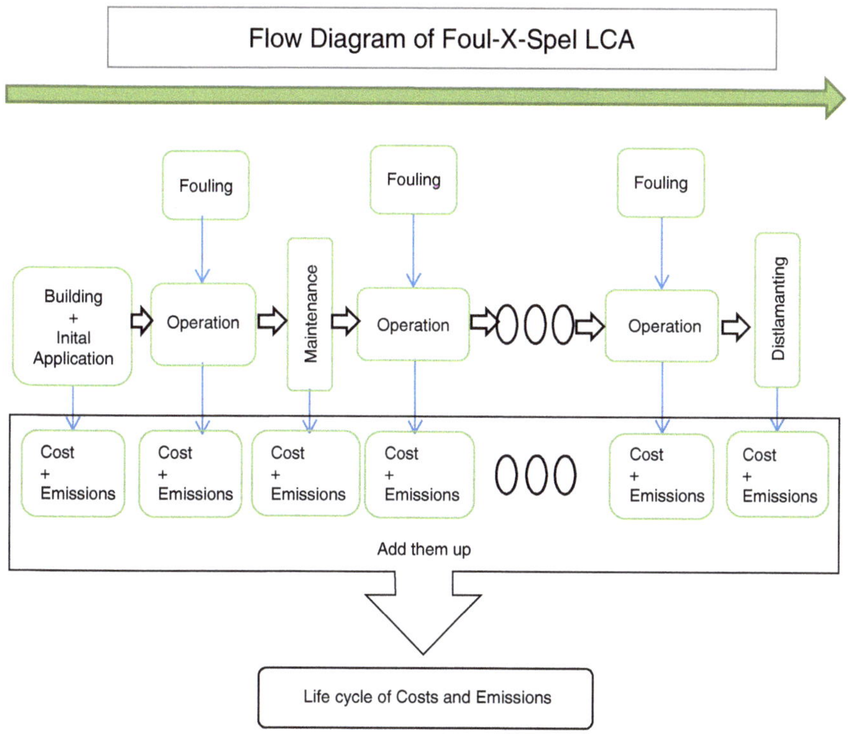


**Fig. 1** Schematic diagram of the simulator in FOUL-X-SPEL LCA methodology

2. Model of temperature-dependent and time-dependent growth of fouling.
3. Model of variation of sea-temperature with location.
4. Model of the costs and effects of hull maintenance activities.
5. Model of fuel-consumption behavior of the ship.

These elements come together in a computer program and allow us to evaluate alternative strategies. It must be noted that the drag-coefficient is re-baselined at the beginning of each maintenance cycle.

The change in drag coefficient is calculated starting from that point. Thus, if we identify the j-th voyage of the i-th maintenance cycle using the double index "ij", we could write the overall cost and emissions using the following expression:

$$Cost_{total} = \sum_{i=1}^{N} \left( \mathrm{Cost}_{\mathrm{dry-dock(i)}} + \sum_{j=1}^{M_i} \left( Cost_{voyage(i,j)} \right) \right) \quad (1)$$

where $\mathrm{Cost}_{\mathrm{total}}$ is the total cost over the life cycle. The index "i" runs over the maintenance cycles. $\mathrm{Cost}_{\mathrm{dry-dock(i)}}$ is the cost of the i-th dry dock. $\mathrm{Cost}_{\mathrm{voyage(i,j)}}$ is the cost of the j-th voyage of the i-th cycle. The cost of the voyages may be computed as follows:

$$Cost_{voyage(i,j)} = F_i P T_{ij} \left(1 + \sum_{k=1}^{j-1} \left(\frac{\Delta CF_{ik}}{CD_i}\right)\right) \tag{2}$$

where P is the fuel price, $F_i$ is the base-line fuel consumption per day at the beginning of the $i^{th}$ maintenance cycle, $CD_i$ is the drag coefficient at the beginning of the $i^{th}$ maintenance cycle. $\Delta CF_{ik}$ is the change in the frictional resistance coefficient in the $k^{th}$ anchorage in the $i^{th}$ maintenance cycle. Finally, $T_{ij}$ is the sailing duration of the $j^{th}$ voyage in the $i^{th}$ maintenance cycle.

Similarly, we can express the total emission as follows:

$$Emission_{total} = \sum_{i=1}^{N} \left(\mathrm{Emission}_{\mathrm{dry-dock(i)}} + \sum_{j=1}^{M_i} \left(Emission_{voyage(i,j)}\right)\right) \tag{3}$$

where $Emission_{total}$ is the total emission over the entire life cycle. $Emission_{dry-dock(i)}$ is the emissions from the $i^{th}$ dry dock. $Emission_{voyage(i,j)}$ is the emissions from the $j^{th}$ voyage of the $i^{th}$ cycle. The emissions from individual voyages may be computed as follows:

$$Emission_{voyage(i,j)} = F_i M T_{ij} \left(1 + \sum_{k=1}^{j-1} \left(\frac{\Delta CF_{ik}}{CD_i}\right)\right) \tag{4}$$

Most of the terms in the $Emission_{voyage}$ expressions are common with the corresponding expression for cost. The only different term in the expression is M which replaces P. M is the mass conversion factor between the fuel and $CO_2$. Please note that the inner summation runs between 1 and (j−1), representing that the fouling accumulation for the $j^{th}$ voyage in the cycle is done until the previous anchorage.

### 3.1 Fuel Consumption Model

Fuel consumption over time is strongly dependent on the time-dependent drag performance of antifouling coatings. Once the time-dependent drag performance of the antifouling coating and the ship's sea trial information, which includes the fuel consumption corresponding to specific speed and draft, are known, it is possible to predict the fuel consumption over time.

The fuel consumption depends on several factors. Engine condition, transmission efficiencies and sea-states are significant factors, determining the fuel consumption. However, in the context of this study, we do not have any control over these factors. If the objective was to carry out weather-based routing, the sea-state would be an indispensable part of the model on which to perform the analysis. Likewise, if the objective was to plan engine maintenance, the engine's detailed behavior would have to be an integral part of the model. In the FOUL-X-SPEL LCA model. However, the focus is on the variation of hull resistance due to the fouling and the surface finish, which allows us to ignore variations in the other aforementioned factors. While ignoring the variations in those factors, some baseline values of those factors are used as scaling parameters. Thus, we use the baseline values of the ship's drag coefficient and fuel consumption, and linearize the dependence of fuel consumption in the neighborhood of the baseline. The linearized model which is shown at below will allow us to compute the incremental changes caused by the changes in surface fouling.

$$\Delta FC = FC_{baseline} * \frac{\Delta C_F}{C_{D\_baseline}} \tag{5}$$

It is of note that the initial baseline values are obtained using the ship details, including main dimensions, wetted areas, ship sea trial results, etc.

Since fuel consumption is proportional to the drag coefficient, if there is a time-dependent model of fouling growth, the linearization is used in conjunction with that model to compute the changes in fuel consumption rates. Therefore, a crucial component of the simulation required for FOUL-X-SPEL LCA is the expression of fouling growth as a function of time, given the antifouling coating type. It is well understood that fouling accumulates more aggressively when the ship is stagnant for an extended length of time (Tupper and Rawson 2001).

The simulation model must capture this phenomenon in terms of equations or algorithms, but since fouling growth is a complex biological process, the model has to be based upon empirical evidence. In this study, semi-empirical models for fouling were developed based on two substantial investigations. The first investigation was a numerical study of hull surface roughness, which led to the expression of a ship's frictional resistance as a function of the surface roughness. This is a theoretical model which correlates the different hull fouling conditions into frictional resistance (Demirel 2015). In this study, the roughness function values of Schultz and Flack (2007) were used to predict the effect of a range of representative coating and fouling conditions on the frictional resistance of ships, based on the description of surface conditions given by Schultz (2007), using Granville's similarity law scaling procedure (Granville 1978).

The present predictions were made based on the assumptions that the given fouling conditions can be represented by these roughness functions and roughness length scales. Schultz (2007) validated these assumptions and this method by comparing his results with other studies such as Hundley and Tate (1980) and Haslbeck and Bohlander (1992), documenting the effects of coatings and biofouling on ship powering through full-scale trials. Recently, Demirel et al. (2017) used the roughness function values of Schultz and Flack (2007) to predict the effect of biofouling on a ship's resistance using Computational Fluid Dynamics (CFD).

The second investigation was that of long-term immersion of several coated plates in two different oceanic environments accompanied by intermittent measurement of fouling growth on these plates over more than 2 years. The data from this experiment and the first theoretical method together gave rise to time-dependent formulae for fouling growth for three different coatings and two locations. These formulae culminate in a calculation procedure for obtaining the added resistance coefficient due to fouling. Currently only two locations and three coatings are supported by the time-dependent model but that is only due to the limited nature of the immersion test dataset. The same methodology can be applied to new paints and locations once we plug in experimental results as new data points.

Although the semi-empirical model developed in this study provides users with a very useful indication of the increase in resistance of a ship due to the use of a specific coating, it may not reflect the resistance increase of ships under real speed-activity conditions since it was developed based on the static immersion data. Hence, this model was calibrated using the real operational data derived from the noon data database of Strathclyde University and using the data provided by a paint company.

Developed time-dependent formulation of fouling was derived from immersion experiments. These formulations are based on static immersion experiments done in port waters very close to the land. Our assumption is that such experiments preserve the relative behaviours across different paints of the same type, but the absolute figures of growth rate are not really representative of the fouling growth rate for ships that frequently undergo motion. Thus, we used 5-year fouling growth rates of two ships under typical operating conditions in order to calibrate the fouling growth equations derived from the immersion experiments.

### 3.1.1 Operational Behavior of Ships

There is no doubt that the time-dependent drag performances of antifouling coatings depend on so many operational parameters of a ship along with coatings' own particulars. The most dominant operational parameter is the operational route of the ship, which determines geography and the temperature of the waters and the anchorage behavior of a ship. Hence, these parameters must be well defined to be able to accurately calculate the life cycle values.

For this purpose, an operational behavior prediction sub-model was also developed within the LCA model. The model's approach is for the defined operational behavior of a ship. If the operational behaviors of the ship are defined day-by-day, the time-dependent drag performances are calculated based on the given behavior. This approach is also able to generate the whole life-cycle behavior of a ship, using the existing operational data. In other words, for instance, if the operational behavior of a ship in the first 2 years is known, the rest of the life cycle behavior of the ship may be assumed to be the same as the first 2 years. It means that the sub-model is educated using the existing data and it assumes that the behavior of the ship in question is kept the same and the fouling accumulation is estimated based on this assumption.

### 3.1.2 Initial and Dry-Dock Paint Application

The first application of the antifouling coatings and the renewal of the coatings in dry-dock are very important aspects of the life-cycle of an antifouling coating, because the life-cycle costs and emissions of an antifouling coating system are consequences of the paint application and ship operations. Therefore, the first coating application activities and all the coating related activities in dry-docks are taken into consideration in the model. The model requires the details of such activities as input and then it calculates the total costs of the first application and renewal of the antifouling coatings.

The costs may be classified under individual headings, and may be refined to arbitrary level of detail. It must be noted that the broken-up costs and emissions are not a necessary condition for the LCA to work. The break-ups are supported because it may be interesting for the stake-holders to understand the individual drivers of cost.

# 4 Case Study

A case study was carried out to show how to use the LCA model, using a tanker which is assumed to be coated with two different types of coatings, namely a fouling release coating (FoulXSpel 1) and a tin free self-polishing antifouling paint (FoulXSpel 2). A real operating tanker of around 110,000 DWT was selected for this case study. Firstly, the model was validated against the real noon data of the tanker and then the total costs and emissions due to the use of different types of antifouling coatings were calculated for the whole life-cycle of the ship.

## 4.1 Validation

The validation process involved simulating the life-cycle of a tanker using its 6.6 year trail of noon-reports and comparing the behavior against the simulation. The simulated result was within 1.65% of the actual value. Figure 2 is a plot showing the actual fuel consumed against simulated values.

## 4.2 Results

The production of the paint generates emissions indirectly because of the energy consumption and refining the raw materials to the atmosphere due to the use of energy and raw materials. However, only the emissions due to electricity consumption were taken into consideration as output for this LCA. It was assumed that the paints were produced using the purchased electricity and the conversion factor of 0.53936 $kgCO_2/kWh$ was assumed according to Defra and DECC (2010). Besides, the selling rate to ship owner was taken into consideration since the life cycle costs were also to be evaluated.

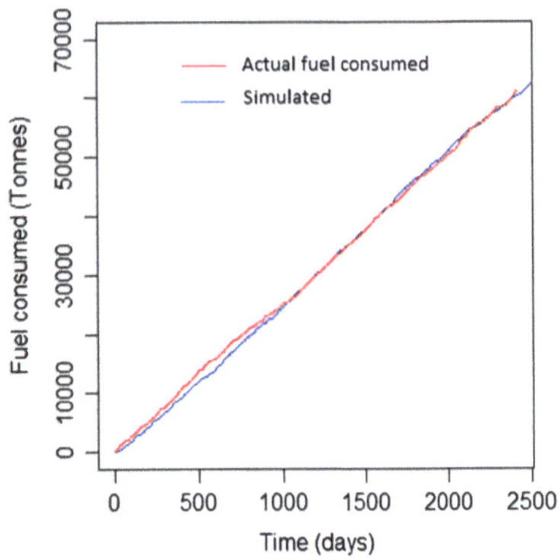


**Fig. 2** Actual fuel consumed against simulated values

The activities in initial and dry-dock paint application require energy inputs and have corresponding emissions that impact the environment and human health. The emissions occurring due to these energy inputs were also ignored due to the lack of the data. On the other hand, the costs of each action of the initial and dry-dock paint application stage as well as the paint costs were considered. Major differences between the maintenance actions of the ship coated with FoulXSpel 2 and that of the ship coated with FoulXSpel 1 were taken into consideration.

Another input was the heavy fuel oil (HFO) used during the sailing of the ship and it was the most important parameter if the amount of the HFO consumed was taken into account for the operation of the ship. It is of note that the time-dependent drag performances of different coatings may differ significantly due to the ship type and operational region and it directly affects the amount of HFO consumed.

Figures 3 and 4 compare the overall cost and emission categories of ships coated with FoulXSpel 1 and FoulXSpel 2 over 30 years of life cycle. Figure 5 clearly shows that the initial and maintenance costs of FoulXSpel 1 are much higher when compared with those of FoulXSpel 2. However, the use of fouling release coatings, e.g., FoulXSpel 1, compensates its high investment costs in terms of fuel savings with respect to the self-polishing antifouling, e.g., FoulXSpel 2, as shown in Fig. 3. Besides, the total emissions due to the use of FoulXSpel 1 are much less than those of FoulXSpel 2 as depicted in Fig. 4. It can be translated into a harmful effect of self-polishing antifauling in terms of $CO_2$ emissions along with the other harmful effects on the marine environment such as the release of biocides.

Figures 6 and 7 compare the overall cost and emission categories of ships coated with FoulXSpel 1 and FoulXSpel 2 over 30 years of life cycle. It is seen from Figs. 6 and 7 that initial and dry-dock paint application (maintenance) costs and emission are much less than the cost and emission from HFO. The assessment reveals the differences between these coatings and highlights the advantages of fouling release coatings against self-polishing coatings. The use of FoulXSpel 1 provides a 2.5%

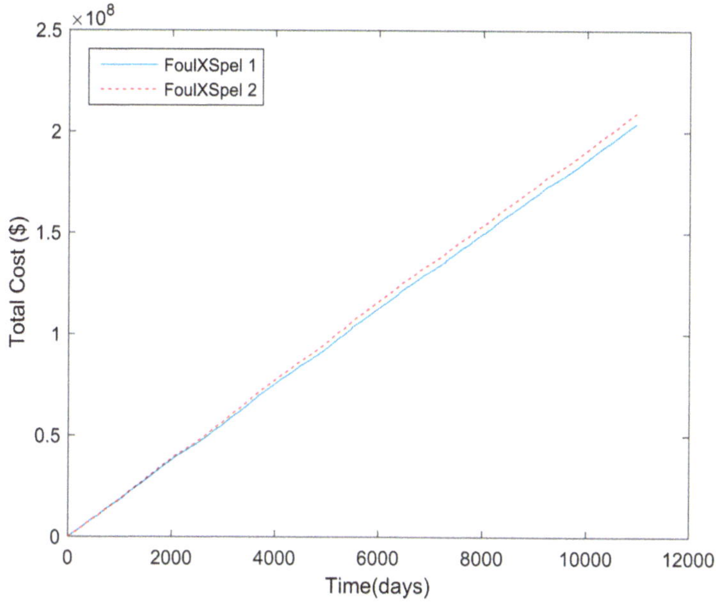


**Fig. 3** Comparison of total costs of FoulXSpel 1 and FoulXSpel 2

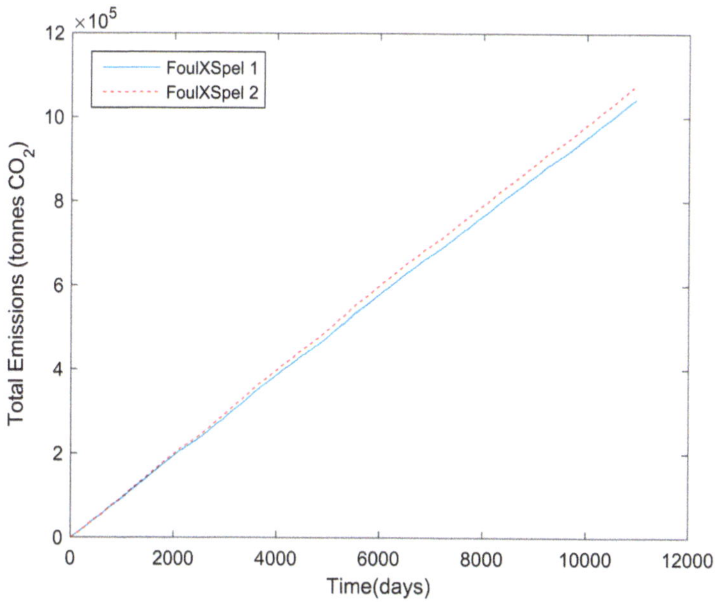


**Fig. 4** Comparison of total emissions of FoulXSpel 1 and FoulXSpel 2

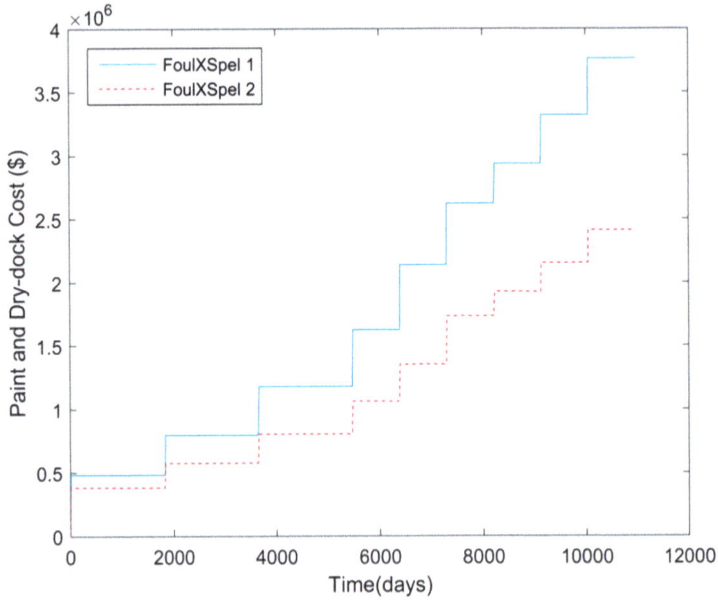


**Fig. 5** Comparison of total paint and maintenance costs of FoulXSpel 1 and FoulXSpel 2

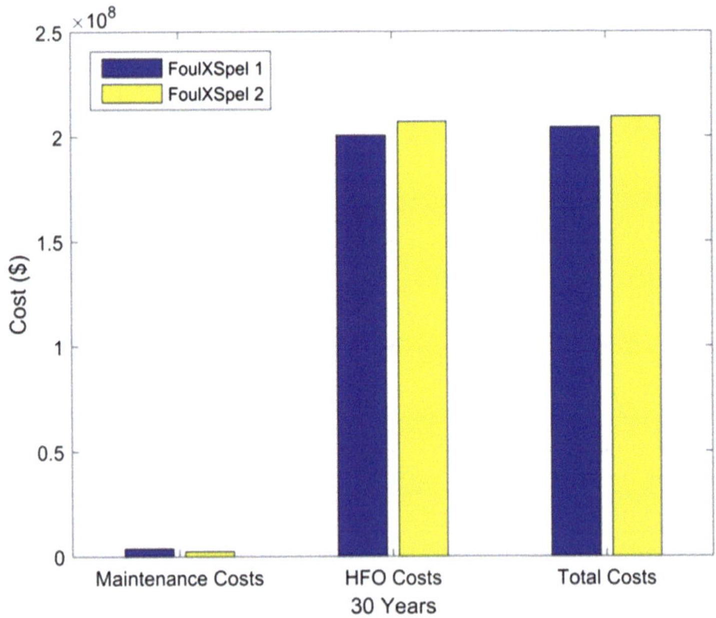


**Fig. 6** Overall cost categories over 30 years

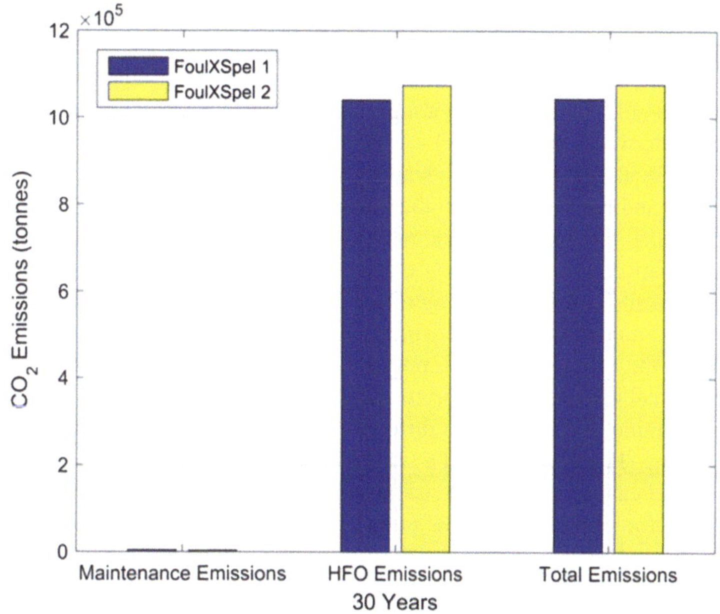


**Fig. 7** Overall emission categories over 30 years

saving on costs and a 3% saving of $CO_2$ emissions with respect to the use of FoulXSpel 2, in total over 30 years for this particular ship.

## 5 Conclusion and Discussion

A new Life Cycle Costs and Environmental Impact Assessment model was developed for the assessment of antifouling coatings with regard to the life-cycle of a ship. For this reason, a new methodology is also proposed within the model. The most important contribution which makes this model novel is that all existing LCA models rely on fuel consumption data that is recorded daily in noon reports, whereas the proposed model makes a prediction of fuel consumption using static and dynamic paint experiments before the ship launches to sea. It should be kept in mind that noon reports include many effects that affect the fuel consumption of a ship such as weather conditions, main engine performance and transmission system. Therefore, creating an LCA system by introducing these non-target effects would not provide a biofouling prediction based LCA model.

In this study, the parameters affected by the use of antifouling coatings were investigated and all the relevant items were selected. The required data to assess the antifouling coatings were defined. Since the assessment of an antifouling coating over a life-cycle is not a straightforward procedure, various modelling techniques were used to predict the time-dependent drag performances of different antifouling coatings. The semi-empirical formulae developed in this study were calibrated using the real operational data of various ships. In this way, the real-time dependent-drag performances of antifouling coatings under speed-activity conditions were modelled. Afterwards, a sub-model was developed to predict the operational behavior of a ship in its life-cycle, using real operational data of a ship.

The costs and emissions due to the initial and dry-dock paint applications were modelled using the real dry-dock reports. This data clearly reveals the major differences between the application methods of fouling release and self-polishing antifouling coatings. All the differences were also considered in the model. Finally, a generic LCA model for the assessment of antifouling coatings was developed, gathering all the sub-models and input data. The model was then validated against 6.6 years of real operational data from a ship. Then, a case study was carried out to show an example of LCA of two different antifouling coatings, namely a fouling release coating (FoulXSpel 1) and a self-polishing antifouling paint (FoulXSpel 2).

This model can be used as a decision-making tool which determines the suitable coating type for particular types of ships. It may also be used to decide the best maintenance and/or hull cleaning activities and/or intervals. A techno-economic feasibility study of a newly developed antifouling coating can be carried out using this LCA model.

## References

Defra & DECC. (2010). *Guidelines to Defra/DECC's GHG conversion factors for company reporting: Methodology paper for emission factors*. http://archive.defra.gov.uk/environment/business/reporting/pdf/101006-guidelines-ghg-conversion-factors-method-paper.pdf. Accessed 30 July 2014.

Demirel, Y. K. (2015). *Modelling the roughness effects of marine coatings and bio-fouling on ship frictional resistance*. Dissertation, University of Strathclyde.

Demirel, Y. K., Turan, O., & Incecik, A. (2017). Predicting the effect of biofouling on ship resistance using CFD. *Applied Ocean Research, 62*, 100–118. https://doi.org/10.1016/j.apor.2016.12.003

FOUL-X-SPEL. (2011). *FOUL-X-SPEL: Environmentally friendly antifouling technology to optimize the energy efficiency of ships*. http://www.foulxspel-antifouling.com/. Accessed 01 February 2017.

Granville, P. S. (1978). *Similarity-law characterization methods for arbitrary hydrodynamic roughnesses*. Final Report Naval Ship Research and Development Center, Bethesda, MD Ship Performance Dept. 1978;1.

Haslbeck, E. G., & Bohlander, G. (1992). *Microbial biofilm effects on drag-lab and field*. 1992 Ship Production Symposium Proceedings, SNAME1992.

Hole, W. (1952). *Marine fouling and its prevention bureau of ships*. Navy Department (580).

Hundley, L., & Tate, C. (1980). *Hull-fouling studies and ship powering trial results on seven FF 1052 class ships*. D W Taylor Naval Ship Research and Development Center Report # DTNSRDC-80/027 111 p. 1980.

Schultz, M. P. (2007). Effects of coating roughness and biofouling on ship resistance and powering. *Biofouling, 23*(5–6), 331–341. Available at: http://www.ncbi.nlm.nih.gov/pubmed/17852068.

Schultz, M. P., & Flack, K. (2007). The rough-wall turbulent boundary layer from the hydraulically smooth to the fully rough regime. *Journal of Fluid Mechanics, 580*, 381–405.
Tupper, E. C., & Rawson, K. J. (2001). *Basic ship theory, combined volume*. Butterworth Heinemann.
Turan, O., Demirel, Y. K., Day, S., & Tezdogan, T. (2016). Experimental determination of added hydrodynamic resistance caused by marine biofouling on ships. *Transportation Research Procedia, 14*, 1649–1658. https://doi.org/10.1016/j.trpro.2016.05.130.
Van Manen, J. D., & Van Oossanen, P. (1988). Resistance. In E. V. Lewis (Ed.), *Principles of naval architecture second revision: Volume II: Resistance, propulsion and vibration*. Jersey City, NJ: The Society of Naval Architects and Marine Engineers.

## Further Reading

Naval Ships' Technical Manual. (2002). *Waterborne underwater hull cleaning of navy ships*. S9086-CQ-STM-010/CH-081R5. Naval Sea Systems Command.

# The Human and Social Dimension of Energy Efficient Ship Operation

**Martin Viktorelius**

## 1 Introduction

Between 2007 and 2012 the shipping industry was responsible for annually emitting 1016 million tonnes of carbon dioxide ($CO_2$) on average. This accounted for 3.1% of global $CO_2$ emissions (Smith et al. 2014). The European Commission (EC) has set the target to reduce $CO_2$ emissions from ships by 40–50% compared to 2005 levels before 2050 (EC 2011). The urgency to reduce emissions from ships is enhanced by the fact that projections into the future predict an increase in $CO_2$ from ships of up to 250% in the period up to 2050 (Buhaug et al. 2009).

The emission of $CO_2$ is directly related to the fuel consumption of ships. Improving the energy efficiency of new and existing ships can, therefore, be a means of reducing ship emissions.[1] This can be achieved by technical solutions and operational measures (Buhaug et al. 2009). The assessed potential for improving energy efficiency (defined as the emitted $CO_2$ per tonne-mile) in operations of existing ships through voyage optimization has been estimated to be up to 10% (Wang et al. 2010). In its efforts to create regulations that facilitate improvements in energy efficiency, the International Maritime Organization (IMO) has made it mandatory for all ships to carry a Ship Energy Efficiency Management Plan (SEEMP) specifying how the work for improvement is organized on board.

The success of such an organizational and regulatory effort to increase energy efficiency in the shipping industry, and in particular the implementation of operational measures, depends on the ships' crews and their capabilities of developing

[1] It should be noted, however, that even if measures that increase energy efficiency are needed, they are not likely to reduce the total amount of emission from ships due to the expected growth of the maritime sector (Bazari and Longva 2011).

M. Viktorelius (✉)
Chalmers University of Technology, Göteborg, Sweden
e-mail: martin.viktorelius@chalmers.se



A. I. Ölçer et al. (eds.), *Trends and Challenges in Maritime Energy Management*,
WMU Studies in Maritime Affairs 6, https://doi.org/10.1007/978-3-319-74576-3_24

new practices, skills and knowledge. This has been emphasized by researchers (Banks et al. 2014; Beşikçi et al. 2016; Bännstrand et al. 2016; Kitada and Ölçer 2015) and classification societies (DNV-GL 2014, 2015). In a study purported to identify fuel saving strategies, the "development of crew understanding, motivation, cooperation, and participation" was ranked as having the highest potential and "to be the most important single ingredient of a successful fuel savings program." (Bertram et al. 1983, p. 162). Baldauf et al. (2013) argued that, in order for the IMO conventions and guidelines on energy efficiency to change the shipping industry, it is vital to integrate the environmental standards into the daily work practices on board: "Only mariners and crews who have background knowledge and who know how they can contribute in the best way to energy efficient and emission reduced ship operation will be able to contribute to the ambitious aims (p. 283)".

In spite of the consensus about the importance of crew motivation and competence, few empirical studies have addressed the issue explicitly. The study presented in this paper aimed at understanding the nature of the skill and knowledge utilized in actual efforts on board ships to save fuel as well as how these capabilities are developed in practice. Thus the following research questions are addressed in this paper: (1) What are the characteristics of knowledge related to the practical accomplishment of energy efficient ship operation?; and (2) How can such knowledge be developed on board ships?

## 2 Knowing in Action and Practice: Embodied, Local and Social

In order to analyze the nature of knowledge in energy efficient ship operation it is useful to first introduce some recent theory in organizational knowledge and learning. Three central themes emerge from the literature on knowledge in organizations. These can be summarized as describing knowledge as embodied, situated/local and social.

The assertion that knowledge is embodied emphasizes the inherently action-oriented nature of practitioners' knowing. From this perspective knowledge is not seen as a static entity inside isolated individuals' heads but rather as an active process taking place among practitioners embedded in socio-technical systems (Cook and Brown 1999; Orlikowski 2002). This type of knowing is part of practice and the actions carried out. It also emphasizes that knowing is "ordinarily tacit, implicit in our patterns of action and in our feel for the stuff with which we are dealing" (Schön 1984, p. 49). This means that the professional practitioner is not always able to state adequate criteria or refer to explicit rules or procedures for the innumerable qualitative judgments he or she makes in everyday work practice. The know-how of the practitioner is *in* his or her action, rather than stemming from some prior intellectual operation. Or as expressed by Ryle (1949, p. 32) "thinking what I am doing does not connote both thinking what to do and doing it. When I do something intelligently. . .I am doing one thing not two". Knowing-in-action is, according to Schön (1984) expressed as the ability to spontaneously carry out actions, recognitions and judgments without having to think about them prior to their performance.

The second theme highlights that the knowledge that is part of organizational practices often has a local and situated character (Nicolini 2011; Gherardi 2008). This means that a substantial part of what practitioners know, has been developed in the context where they work, that is, through "interaction among people with the programs, operations or objects (physical artefacts) that are specific to a local context" (Yanow 2004, p. 10). Since the effective carrying out of work tasks is often dependent on contingent and contextual circumstances, the knowledge required to handle particular situations needs to be adapted to the context of the required action. One implication of the local nature of knowledge is that when experts change their work context they are not always able to directly apply their knowledge before they have acquired an understanding of the new setting (Sole and Edmondson 2002).

The third, and final theme found in the literature on knowing in practice recognizes that what practitioners know is highly associated with the social interactions taking place during work (Wenger 1998; Brown and Duguid 1991). It has been demonstrated that learning in organizations is a process where novices gradually become members of the community of practice giving meaning and structure to the activities and work performed (Lave and Wenger 1991). Becoming a legitimate member of a community of practice is, moreover, a process involving the development of a professional identity and a sense of what to do and how to do it well (Gherardi et al. 1998).

Given this background in previous research on knowledge, skill and expertise in organizational settings it is now possible to explore the knowledge processes on board ships in order to develop accurate accounts of the present practices of energy efficiency and energy management as well as formulate grounded frameworks for the improvement of energy efficiency in the shipping industry.

## 3 Method and Research Setting

In order to investigate the unquantifiable dimension of knowledge described above and its relation to energy efficient ship operation, a qualitative field study (Czarniawska 2014) was performed on board five RoPax vessels (carrying passengers, freight, cars) operated by an international transport and travel service business with the most comprehensive route network in Europe. The purpose and value of qualitative, or interpretative, research methods is their capacity to examine the social context and subjective perspectives of humans. It can be argued that without an understanding of peoples' contexts of action and perspective on things it is impossible to understand the reasons for their actions.

The data collection method included observations of the work performed during voyages and interviews with crew members. In qualitative research, observation and interviews are considered to be two of the most crucial methods for collecting data (Flick 2014). The combination of observation and interviews is particularly useful when studying work practices because it reflects both practitioners' beliefs and actions (Schatzki 1996).

The observations in this study were conducted during 28 voyages spread over twelve months, in order to allow intermittent and iterative analysis. Each voyage took between 4 and 14 h. Approximately 195 h of observation was spent onboard the ships during the data collection phase. The observations were primarily made on the bridge and in the engine control room.

In addition to the observations, semi-structured interviews were held with 40 crew members (Master Mariners, Officers, Chief Engineers, Engineers). The interviewees ranged from novices (3rd officers with only a few months of experience and seniors and Masters, with several years of experience). The interviews lasted between 30 and 120 min and covered questions related to the crew members' work with energy efficiency, the technology they used and the challenges they perceived in improving energy efficiency.

Field notes from the observations and recordings from the interviews were transcribed. The analysis of the material (collected data) followed the grounded theory approach to data analysis where conceptual categories (themes) are identified in the collected data and later compared to each other and to the emerging theoretical constructs (Glaser and Strauss 1967). The structured analytical process is ended when the conceptual categories (themes) are saturated by the data, that is, when new dimensions of the themes are no longer discovered by examining the data (Corbin and Strauss 2015). It is widely agreed that the concept of validity in qualitative research has less to do with capturing the objective reality and more with grounding the scientific constructions in the constructions of the participants of the study (Flick 2014). In order to assure this, the research procedure of this study involved several re-visits to the research field (the five ships) in order to affirm the interpretation of the data collected on previous visits.

The observations and interviews of this study covered many different practices regarding energy efficiency, such as trim optimization, ballast water use, engine load, cargo handling, autopilot use, and performance monitoring. However, due to limitations of space, this paper will focus on voyage execution and speed regulation, in particular. Ship speed and just-in-time arrival have an important role in energy efficiency and is therefore relevant to focus on (Buhaug et al. 2009).

## 4 Results

The results illustrate a number of important aspects of the epistemic dimension of voyage optimization found in the present case-study. The findings described below represent common beliefs, ideas and opinions expressed by the crew members.

## 4.1 The Skills Required for Energy Efficient Ship Navigation

A highly salient theme in the data, expressed by all interviewees, was that sailing with a constant speed during a voyage was not energy efficient and there was a consensus on board the five ships that a lot of energy could be saved by dynamically adjusting the speed in relation to the context of the operation, that is, to the locally changing conditions during each particular journey. Since the resistance due to winds, currents, waves and water depth varied during each voyage, all Masters, officers, and engineers believed that a constant speed would result in unnecessary peaks in fuel consumption. Avoiding the peaks, by accounting for the changing external conditions while dynamically adjusting the speed, without causing a delay, was explained to require a lot of skill and experience regarding the ship and the route. The proficient execution of this task was considered, by all crew members, to be based on taking several situational factors into account, such as the varying strength and direction of the wind and sea current, the varying water depth along the route, the distance and duration of the voyage, waves, traffic, the amount and type of cargo and time to load the cargo in the next port, safety, the characteristics and behaviour of the ship and engines, and passenger comfort. Since fuel consumption varied, according to the crew members, from ship to ship, voyage to voyage and from minute to minute the context of energy efficient navigation and speed regulation was highly volatile and required constant situated judgments.

However, the ability to account for all the situational variables was based on judgements that could not easily be converted into explicit reasoning but depended, rather, on officers' perception of the situation. Most officers with several years of experience of the ship and the route were, for example, able to use their perception of the ship's interaction with the sea as a possible indication of the need to change the course and slightly deviate from the route in order for the ship to sail more smoothly through the water with less resistance created by the waves and wind. The experience from previous voyages on the same ship and route was considered important and gave officers a sense of how the ship normally reacted to varying water depths and strengths and directions of the winds and currents, which made situated judgments during navigation easier.

Moreover, the situated factors reflected in the officers' decisions on the course and speed were perceived to sometimes be in contradiction with each other. High speed in shallow waters, e.g., in the archipelago, resulted in decreased energy efficiency because of the so-called squat effect (causing the vessel closer to the seabed). However, lower speed in shallow waters made it necessary to compensate with a relatively higher speed on the open and deeper sea. But higher speed and stronger winds on the open sea reduced the energy efficiency. Officers thus evaluated the trade-offs, affected by the contingent conditions on every particular voyage, and reflected on the different alternative actions affecting the total energy efficiency of the voyage.

### 4.2 The Difference Between Novices and Experienced Officers

A majority of the Masters, officers, and engineers agreed that skills for energy efficient navigation do not come immediately to new officers, even if they have experience from working on other ships. New officers who did not yet have the "right feel" for the ship were considered less energy efficient operators. According to several more experienced officers and engineers, before an officer learns the local practices and characteristic of the ship and route, he or she was not able to navigate energy efficiently. As explained by one Chief Engineer: *we cannot demand of an officer when he comes here that he should know this route after one day. It takes time. A new officer costs. It takes a few months to get into the other people's work practices and understand the ship.*

New officers were described as anxious to be on time and thus often maintained a somewhat higher speed without being able to consider the context and the changing conditions along the route. Novices were, moreover, perceived to be less proficient in handling dense traffic situations in energy efficient ways. Another beginner's mistake, according to many experienced officers, was to unnecessarily increase or decrease the speed as soon as the estimated arrival time differed from the scheduled arrival time. More experienced officers, however, were considered better at judging whether a particular speed was too high or low in relation to the conditions of the remaining voyage and scheduled arrival time, which required a higher or lower speed. Experienced officers were more proactive, rather than reactive, and could remain calm even if the estimated time of arrival indicated a delay or a premature arrival. If, for example, bad weather further down the route was expected, a higher speed (than the required average speed over ground indicated by the estimated time of arrival) could be used in order to avoid a peak in fuel consumption later.

### 4.3 The Acquisition of Energy Efficiency Skills

Many crew members believed that operating energy efficiently was difficult to learn by formal education and training or through information campaigns purporting to raise the environmental awareness of seafarers. The shipping company had held an internal course about a newly installed fuel management system and a web-learning course about green shipping. However, a majority of the interviewed crew members found both courses to be too general and superficial to actually help them operate more energy efficiently. Instead the process of learning was thought to be embedded in the interactions between the crew members and the ship in actual operations and by testing different tactics and trying out how the energy system as a whole reacted to different ways of operating. A Chief Engineer explained that it was not possible to derive the most efficient way of operating the ship by calculation only. The idiosyncrasies of the ship could only be learned by experiencing it functioning under actual conditions in practice.

The role of social interaction and knowledge-sharing among crew members was explained to be important for the acquisition of the collective skills of energy efficient ship operation. Mutual engagement was described as crucial both for the learning process of novices but also for the development of the senior officers' skills. An experienced officer explained, for example, how he helps novices become familiar with the route and its different characteristics, such as the depth, frequent weather conditions (e.g., winds, currents) and traffic situations and how to handle them and plan for them ahead of time. The officer explained how he shows novices where to allow deviations from the route (lessen the accuracy of the autopilot) and where staying on the route is more important because of deep water, where there are possibilities to take short cuts and where short cuts are less efficient because of shallow waters.

# 5 Discussion

This study addressed the research gap pertaining to the nature of crew capabilities and knowledge characterizing energy efficient voyage execution. Previous research has developed few conceptions of how to understand the behavioural and organizational dimension of energy efficiency. As demonstrated in this study, knowledge, skill and expertise in energy efficient ship handling was fundamentally situated (i.e., knowing how to navigate efficiently given a specific ship in locally changing circumstances), embodied (i.e., manifested in collective tacit experience and action) and social (i.e., grounded in mutual efforts to develop energy efficient speed regulation and navigational practices).

The execution of an energy efficient voyage was seen to involve many situated decisions. Simply reducing the speed was not possible due to the fixed time table. However, crew members did not consider having a constant speed to be the most efficient way of operating the ship, although this would have been the easiest way of executing the voyage. Instead officers were keen to save fuel by learning how to adapt the speed to the contingent circumstances during the voyage in order to avoid unnecessary peaks in consumption.

The crew members' strategies had been developed on board through years of discussion, negotiation and learning. The motivation to learn and try to save fuel was more the result of a positive energy culture rather than the enforcement of regulation, policy or instruction from shore-based ship management. Crew members' practices were thus formed primarily by the local communities and social interactions in work task environments rather than by the overarching formal system of the whole organization or the shore office managers' monitoring strategies and policies.

This has implications for the management of energy knowledge. The literature on energy management in the shipping industry has discussed the possibility of enhancing knowledge and competence through the application of various so-called "best practices" such as IMO SEEMP and ISO 50001 (Johnson and Andersson 2014; Johnson et al. 2014; Poulsen and Johnson 2016; Poulsen and Sornn-Friese 2015). This study has illustrated that the skill and knowledge of energy efficient voyage execution cannot be reduced to the application of some general formula but exists rather in an ability for situated adjustments developed through local interactions in the physical and social context. This implies that energy management should be more focused on creating the proper social conditions for workplace or organizational learning than it has been before. Wenger (2000) argues that organizational success depends on the ability to develop social learning systems in organizations and to participate in broader learning systems in the industry. The progress of learning systems in communities of practice is determined by:

- *The level of learning energy* (initiative, leadership, spirit of inquiry, recognizing and addressing knowledge gaps, being open for new directions and opportunities)
- *The depth of social capital* (trust, mutual engagement between practitioners)
- *The degree of self-awareness* (reflection on the repertoire of tools, actions and practices)

This could be achieved by the development of training programs that are true to the actual, rather than the canonical, work practices on board, which increases the likelihood of crew members understanding and adopting the material (Brown and Duguid 1991). This involves facilitating discussions between ship operational departments, creating boundary processes where practitioners from different communities of practice meet, for example, by creating cross-disciplinary learning projects where shore managers, Masters, officers and engineers collaborate and experiment together, supporting reflection on current practices, developing leadership on board ships that focuses on the facilitation of learning among crew members (Engeström 2005).

Practices of energy efficient ship operation can be developed locally, in real work settings, by promoting and supporting collective knowing in action. It is therefore crucial that the methods employed by organizations to facilitate crews' knowledge, skills and expertise in energy efficiency are focused on the organizational and technological contexts in which crew members work.

# 6 Conclusion

This paper is an attempt to understand the human and social dimension of energy efficient ship handling. Acknowledging the relationship between energy efficiency and the knowledge, skills and expertise of the crews is an important step towards building the capacity to mitigate emissions from ships.

This study explored the nature of the knowledge needed for energy efficient ship operation and has emphasized the importance of understanding the social conditions under which organizational learning takes place. The findings illustrate the importance of organizational knowing as emerging from the contextually situated and embodied interactions of the crew members. It was shown that energy efficient practices are intimately connected to situated forms of knowledge. It is proposed that an important dimension of these practices is the practical, localized, tacit and collective know-how of the crew members on board.

Future research should investigate social mechanisms for developing situated knowledge on energy efficiency in the shipping industry, understand how the boundaries between different communities within organizations and across the sectors can be bridged, understand how situated curricula can be developed and what kind of technology can facilitate the learning and development of new practices.

**Acknowledgements** This paper is written within the framework of a PhD project financed by the Swedish Energy Agency. The author wishes to express sincere gratitude to the interviewed crew members on board the vessels.

## References

Baldauf, M., Baumler, R., Ölçer, A., Nakazawa, T., Benedict, K., Fischer, S., et al. (2013). Energy-efficient ship operation—Training requirements and challenges. *TransNav: International Journal on Marine Navigation and Safety of Sea Transportation, 7*(2), 283–290.

Banks, C., Turan, O., Incecik, A., Lazakis, I., & Lu, R. (2014). Seafarers' current awareness, knowledge, motivation and ideas towards low carbon-energy efficient operations. *Journal of Shipping and Ocean Engineering, 4*(3–4).

Bännstrand, M., Jönsson, A., Johnson, H., & Karlsson, R. (2016). *Study on the optimization of energy consumption as part of implementation of a ship energy efficiency management plan (SEEMP)*. IMO.

Bazari, Z., & Longva, T. (2011). *Assessment of IMO mandated energy efficiency measures for international shipping*. London, UK: International Maritime Organization.

Bertram, K. M., Saricks, C. L., & Gregory, E. W., II. (1983). Developing a comprehensive programme for improving shipboard energy efficiency. *Transportation Planning and Technology, 8*(3), 161–175. https://doi.org/10.1080/03081068308717249

Beşikçi, E. B., Kececi, T., Arslan, O., & Turan, O. (2016). An application of fuzzy-AHP to ship operational energy efficiency measures. *Ocean Engineering, 121*, 392–402.

Brown, J. S., & Duguid, P. (1991). Organizational learning and communities of practice: Toward a unified view of working, learning and bureaucratization. *Organization Science, 2*(1), 40–57.

Buhaug, O., Corbett, J. J., Eyring, V., Endresen, O., Faber, J., Hanayama, S., et al. (2009). *Prevention of air pollution from ships—Second IMO GHG study*. London: International Maritime Organization.

Cook, S. N., & Brown, J. S. (1999). Bridging epistemologies: The generative dance between organizational knowledge and organizational knowing. *Organization Science, 10*(4), 381–400.

Corbin, J., & Strauss, A. (2015). *Basics of qualitative research: Techniques and procedures for developing grounded theory*.

Czarniawska, B. (2014). *Social science research: From field to desk*. Sage.

DNV-GL. (2014). *Energy management study 2014*.

DNV-GL. (2015). *Energy management study 2015*.

EC. (2011). *White paper: Roadmap to a Single European Transport Area – Towards a competitive and resource efficient transport system*. Luxembourg: European Commission.

Engeström, Y. (2005). *Developmental work research: Expanding activity theory in practice*. Berlin: Lehmanns Media.
Flick, U. (2014). *An introduction to qualitative research*. Sage.
Gherardi, S. (2008). Situated knowledge and situated action: What do practice-based studies promise? In D. Barry & H. Hansen (Eds.), *Sage handbook of new approaches in management and organization* (pp. 516–525). London: Sage.
Gherardi, S., Nicolini, D., & Odella, F. (1998). Toward a social understanding of how people learn in organizations: The notion of situated curriculum. *Management Learning, 29*(3), 273–298.
Glaser, B., & Strauss, A. (1967). *The discovery of grounded theory: Strategies of qualitative research*. London: Wiedenfeld and Nicholson.
Johnson, H., & Andersson, K. (2014). Barriers to energy efficiency in shipping. *WMU Journal of Maritime Affairs*, 1–18.
Johnson, H., Johansson, M., & Andersson, K. (2014). Barriers to improving energy efficiency in short sea shipping: An action research case study. *Journal of Cleaner Production, 66*, 317–327.
Kitada, M., & Ölçer, A. (2015). *A contemporary outlook on human element in energy efficient seaborne transportation*. Paper accepted to the International Association of Maritime Universities Conference, Opatija.
Lave, J., & Wenger, E. (1991). *Situated learning. Legitimate peripheral participation*. Cambridge: Cambridge University Press.
Nicolini, D. (2011). Practice as the site of knowing: Insights from the field of telemedicine. *Organization Science, 22*(3), 602–620.
Orlikowski, W. J. (2002). Knowing in practice: Enacting a collective capability in distributed organizing. *Organization Science, 13*(3), 249–273.
Poulsen, R. T., & Johnson, H. (2016). The logic of business vs. the logic of energy management practice: Understanding the choices and effects of energy consumption monitoring systems in shipping companies. *Journal of Cleaner Production, 112*, 3785–3797.
Poulsen, R. T., & Sornn-Friese, H. (2015). Achieving energy efficient ship operations under third party management: How do ship management models influence energy efficiency? *Research in Transportation Business & Management, 17*, 41–52.
Ryle, G. (1949). *The concept of mind*. London, UK: Hutcheson.
Schatzki, T. R. (1996). *Social practices: A Wittgensteinian approach to human activity and the social*. Cambridge: Cambridge University Press.
Schön, D. A. (1984). *The reflective practitioner*. New York: Basic Books.
Smith, T. W. P., Jalkanen, J. P., Anderson, B. A., Corbett, J. J., Faber, J., Hanayama, S., et al. (2014). *Third IMO GHG Study 2014*. London, UK: International Maritime Organization (IMO).
Sole, D., & Edmondson, A. (2002). Situated knowledge and learning in dispersed teams. *British Journal of Management, 13*(S2), 17–34.
Wang, H., Faber, J., Nelissen, D., Russell, B., & Amand, D. (2010). Marginal abatement costs and cost effectiveness of energy-efficiency measures, IMO Technical rapport: MEPC 61/INF.18.
Wenger, E. (1998). *Communities of practice: Learning, meaning and identity*. New York: Cambridge University Press.
Wenger, E. (2000). Communities of practice and social learning systems. *Organization, 7*(2), 225–246.
Yanow, D. (2004). Translating local knowledge at organizational peripheries. *British Journal of Management, 15*(S1), 9–25.

# The Need for Education and Training in Maritime Energy Management in Myanmar

**Thet Hlaing Swe, Momoko Kitada, and Aykut I. Ölçer**

## 1 Introduction

Current technological developments in the maritime industry have been largely led by so-called developed countries in terms of the quantity and quality of maritime transport. Containerisation, faster and bigger ships and ports are good examples of technological developments in the twentieth century. In the last decade, the idea of development has evolved from the mere implication of economic development to the more comprehensive implication of sustainable development. Over 150 world leaders adopted the 2030 Agenda for Sustainable Development, including the Sustainable Development Goals (SDGs) at the United Nations (UN) Sustainable Development Summit on 25 September 2015. In this new direction of a sustainable future, the role of developing countries is more emphasised in the realisation of ideological sustainable development.

One of the key tasks in the maritime sector is to respond to its contribution to the accomplishment of SDGs. In particular, SDG 7 'Ensure Access to Affordable, Reliable, Sustainable and Modern Energy for All' along with SDG 12 'Ensure sustainable consumption and production patterns', and SDG13 'Take urgent action to combat climate change and its impacts' are the goals that the maritime industry can contribute to through the implementation of energy efficiency. This global challenge has been addressed by various stakeholders, such as policy makers, regulators, shipowners, manufacturers, port operators, ship builders, educators, and NGOs.

T. H. Swe (✉)
Myanmar Maritime University, Yangon, Myanmar
e-mail: s14011@alumni.wmu.se

M. Kitada · A. I. Ölçer
World Maritime University, Malmö, Sweden



A. I. Ölçer et al. (eds.), *Trends and Challenges in Maritime Energy Management*, WMU Studies in Maritime Affairs 6, https://doi.org/10.1007/978-3-319-74576-3_25

However, when SDG 7 is localised in developing maritime countries, like Myanmar, there are a number of challenges and restrictions in terms of available resources to achieve the Goal by 2030. Considering the fact that Myanmar is producing seafarers to work in the high seas, it is reasonable to assume that Myanmar's role in advancing SDG 7 in the maritime sector will be to educate future seafarers and equip them with relevant knowledge and, thereby, build competence for energy efficient ship operations. Therefore, this paper discusses how Myanmar, as a developing country, can respond to energy efficient shipping through education and training.

To begin with, the paper reviews the situation of Myanmar's maritime industry and examines both strengths and weaknesses of the industry. Then, it looks into the measures developed by the International Maritime Organization (IMO) regarding energy efficiency on ships.

## 2 Review of the Maritime Industry in Myanmar

In the discourse of energy efficient shipping within the future landscape, it is important to consider the role that developing countries to play in achieving the global goal, such as SDG 7. This section presents a review of Myanmar's maritime industry in terms of both its strengths and weaknesses in its national context.

### *2.1 Strengths of Myanmar's Maritime Industry*

Among many strengths that Myanmar's maritime industry has, it is notable that Myanmar is geographically advantageous in maritime operations and services. Myanmar has great potential in its location to be a strategic hub for South-East Asia in terms of maritime-related facilities, such as ports, ship building and repair.

In Myanmar, there are a number of deep-sea ports situated along the coastline (see Fig. 1). The benefits of deep seaports are that large international seagoing vessels can berth and goods can be transported by connecting the inland route to China and other East-Asian countries. Therefore, the effective use of deep seaports in Myanmar would bring the huge merit of shorter and more economical transportation routes than the current seaborne lane, which passes the Malacca strait. With this potential, Myanmar ports are expected to play a vital role in the intermodal transportation of East Asian and South-East Asian trades in the near future (Embassy of the Kingdom of The Netherlands in Myanmar 2016).

In addition, Myanmar has another advantage in inland water transport. Four major rivers, which are Ayeyarwaddy, Chindwin, Thanlwin, and Sittaung, flow down through the country from north to south. These rivers are a lifeline for Myanmar people who rely on agriculture as well as inland water transportation. In particular, the Ayeyarwaddy River is commercially navigable and its navigational

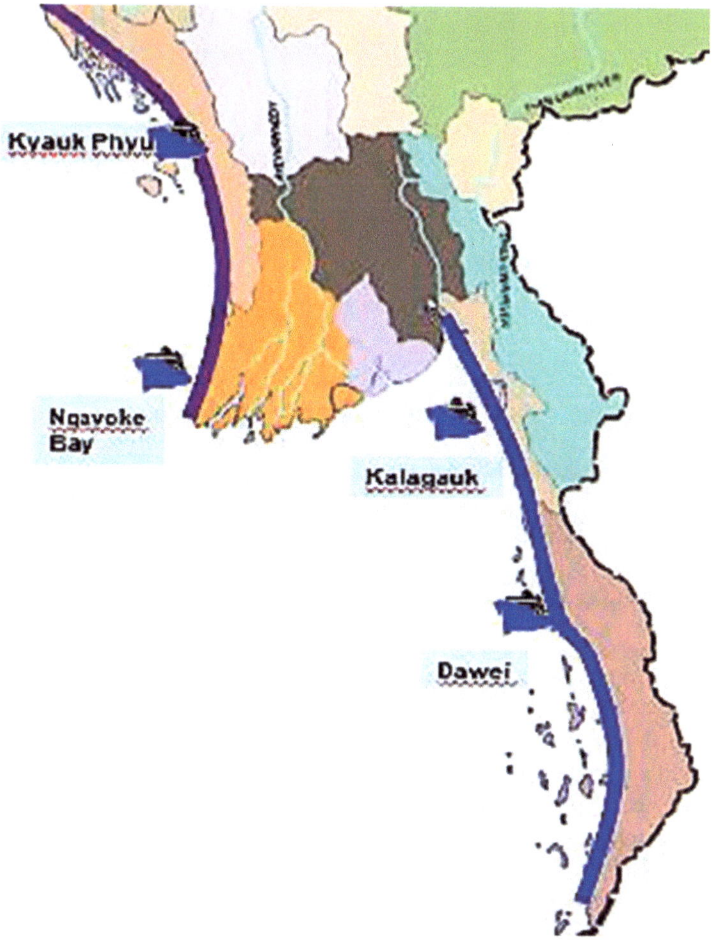


**Fig. 1** Ports in Coastal region of Myanmar (originally published by Embassy of the Kingdom of The Netherlands© Ministry of Foreign Affairs (2016); published with kind permission of © Ministry of Foreign Affairs. All rights reserved)

length is 2170 km out of the total of 6951 km. Myanmar has a long history of maritime transportation that dates back to the ancient era when it is recorded that maritime transportation was widely used.

It is also important to note that Myanmar is rich in natural resources, such as gems, oil, natural gas, and minerals. According to the Myanmar Oil and Gas Enterprise (MOGE), there are 51 oil and gas fields in total; of which 26 are in shallow water and 25 are in deep water (see Fig. 2). Some of these sites have already been located but the majority is still unexploited (Embassy of the Kingdom of The Netherlands in Myanmar 2016). In terms of the implementation of energy efficiency measures, Myanmar is suitable for investment in renewable energy because of its

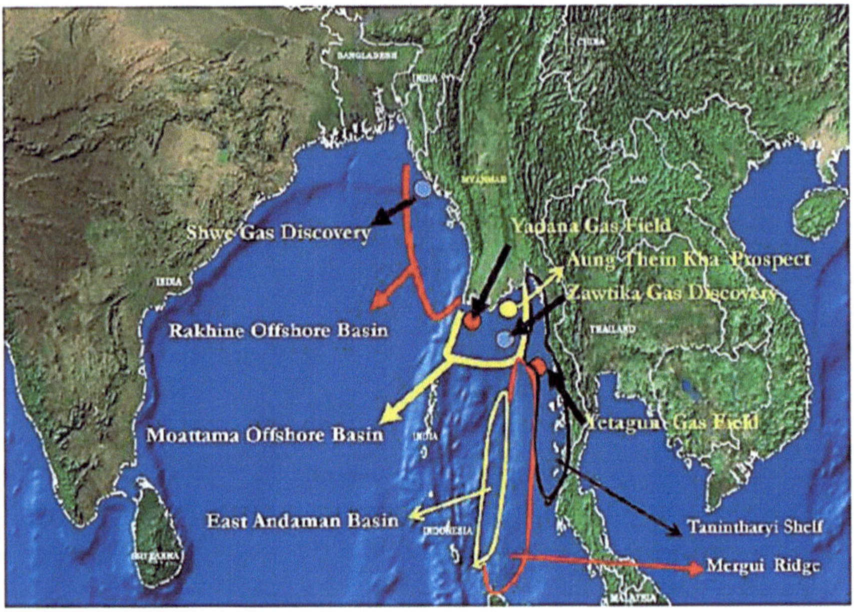


**Fig. 2** Myanmar Offshore Oil and Gas fields (originally published by Embassy of the Kingdom of The Netherlands© Ministry of Foreign Affairs (2016); published with kind permission of © Ministry of Foreign Affairs. All rights reserved)

climate and geographical features in favour of sunlight, wind, rain, tides and geothermal heat. This would create a great opportunity for land-based industries that want to employ renewable energy, such as hydraulic energy or bio-diesel gas in producing electricity for domestic use.

From a human resource perspective, the characteristics of people in Myanmar are generally well-received, such as obedience, humbleness and quick uptake of new things, including technology. This is known as one of the valuable assets of the country.

In summary, maritime and energy industries have great potential in Myanmar and South-East Asia; hence, they are recognised as inevitable and important in the future development of Myanmar. The country shares long international borders as well as a long coastline, which makes Myanmar a strategic maritime country. Regional co-operation in various aspects of the maritime industry has become essential not only for Myanmar but also for the region as a whole.

## 2.2 *Weaknesses of Myanmar's Maritime Industry*

Inland Water Transport (IWT) under the Ministry of Transport has a fleet of about 100,000 tons along with 500,000 tons owned by private owners. These vessels are mainly barges, and mixed cargo/passenger ships. IWT is currently facing financial issues due to the emergence of private fleet operators in inland waters. Consequently, most vessels owned by IWT are becoming old and, thus, they are producing various types of pollution in rivers, such as oil pollution, air pollution, garbage pollution and sewage pollution. Most of the pollutions are due to either the age of vessels or poor operational and maintenance practices resulting from the lack of knowledge and skills of seafarers on board.

Regarding offshore regions where a number of oil and natural gas rigs are located in Myanmar's territorial waters, there are no effective regulations or policies for pollution prevention and contingency plans.

Shipbuilding in Myanmar is a small-scale industry, which emphasises the construction and repair of domestic inland cargo/passenger vessels. In spite of relatively cheap labour costs, Myanmar's shipbuilding sector is neither mature nor competitive at the international level due to the lack of skills, facilities, and technology as well as weak shipbuilding rules and regulations and an unfavourable tax regime. The majority of shipbuilders in Myanmar does not have sufficient knowledge of the Energy Efficiency Design Index (EEDI) technology. Moreover, there are no effective national rules or regulations for Energy Efficiency Implementation Plans for ships. These are the challenges that Myanmar shipowners face when building their ships. Environmental conservation does not receive enough attention although fuel consumption is the main economic driver of operational costs.

## 2.3 *Education and Training*

To maximise Myanmar's strengths while minimising the effects of the above-mentioned weaknesses, it appears that education and training of people is the key to transforming Myanmar into a more energy efficient country in South-East Asia. The main income of Myanmar relies on exporting raw materials such as natural resources. The over-dependence on natural resources is anticipated to be unsustainable for the future of Myanmar. A more balanced approach is necessary to support sustainable development in Myanmar, which is only possible by investing in people through education and training for human resource development. With acquired knowledge and skills, Myanmar is highly likely to develop in the long term.

For example, Myanmar already has natural gas in its ocean basin. If the country has facilities, infrastructure and experts in LNG bunkering, it will enable to build LNG bunkering stations within their deep-sea ports. At the same time, a new transportation lane can be explored if the country launches LNG bunkering facilities in its deep-sea ports. Such a technology-driven vision will help the advancement of Myanmar's maritime industry. To make it happen, it is vital to cultivate human resources in Myanmar through education and training.

# 3 Current Status of Energy Efficiency Education in Myanmar

It is evident that Myanmar's maritime industry is growing but is still immature in many areas where modern technologies might be required. With a number of potentials highlighted earlier, Myanmar's maritime industry would be able to develop in a sustainable manner if more experts in maritime technology are produced. A paradigm shift is needed from natural resource-oriented to human-resource oriented thinking to strengthen Myanmar's maritime industry.

Currently, two government-owned training institutions, Myanmar Maritime University (MMU) and Myanmar Mercantile Marine College (MMMC), are the main players contributing to the advancement of education and training in maritime technology for maritime professionals. There are also private training centers under the framework of the Department of Marine Administration providing short training courses to meet the requirements of seafarers' competences. The following sections present how each training institution in Myanmar provides maritime energy efficiency related education and training in order to analyse the gap between the current and future programmes.

## *3.1 Current Activities of Training Institutes*

### 3.1.1 Myanmar Maritime University (MMU)

The Human Development Function of MMU is to train naval architects, marine engineers, marine mechanical engineers, port and harbour engineers, river and coastal engineers, electrical and electronic engineers, and deck officers. In addition, MMU offers a course in Shipping Management and Port Management to maritime experts.

Myanmar also needs to respond to the requirement in Annex VI of the International Convention for the Prevention of Pollution from Ships (MARPOL), which requires all new ships to implement EEDI. In Myanmar, where only inland and coastal vessels are built, shipbuilders do not need to comply with the EEDI measures. However, such updated knowledge, which has been kept among the faculty of Naval Architecture and Ocean Engineering (NAOE) of MMU, can be transferred to shipbuilders and other maritime professionals to design and produce more energy efficient inland water ships. The latest information on maritime rules and regulations as well as maritime technologies are important to be disseminated among the relevant stakeholders in order to prepare them for new trends and changes in the global maritime industry.

The NAOE faculty members have been working on upgrading the naval architecture courses to support EEDI through research since 2014 (see Table 1).

**Table 1** An overview of energy efficiency research in MMU

| Title | Type of activities | Brief description | Outcome |
|---|---|---|---|
| 1. Study of Trim Dependence for Ship performance in the Actual Sea Conditions by Computational Fluid Dynamics (CFD) Simulations | Joint project (cooperation with Weather News Inc., Japan) | The research purposal is to build up a foundation for employing CFD techniques for ship trim optimisation in actual sea conditions. | The University established the CFD application for hull form optimization practice in the students' final year projects and shipbuilding industry of Myanmar. |
| 2. Study on Hull Form Optimisation for Minimum Wave Resistance Based on Rankine Source Method | Master Thesis | The main purpose is to get optimum hull form inland vessels in Myanmar with minimum wave making resistance. Optimisation is by using a nonlinear programming method based on Rankine source method. | In Progress |
| 3. Analysis of the Passenger Ship with Energy Efficiency Specification: Using Light Material for Superstructure | Bachelor Thesis | The key target of the paper is to obtain an Energy Efficient Ship by Renovation Technology. Therefore, the paper analyses reforming the superstructure of an existing ship from steel to aluminium to check the effectiveness of lightweight material by verifying whether the newly renovated ship attained EEDI with required EEDI. | – We could see whether using the lightweight material in inland and coastal Myanmar vessels is a possible resolution in saving fuel consumption and cutting $CO_2$ emission.<br>– We could know how to recover the advantages of aluminium instead of steel in shipbuilding.<br>– Naval Architecture students could understand how to apply and verify EEDI measures in design state. |
| 4. Trim Optimisation of container ship by test tank and numerical simulation | Master Thesis | The main purpose is to develop and maintain a strategy under the mandatory Ship Energy Efficiency Management Plan (SEEMP) as well as achieving real savings to remain competitive and optimise the draft for enhanced fuel savings. To obtain the optimal result, the model ship is tested in towing tank. | In process |

These research activities are monitored by experienced faculty members who are well-trained by overseas maritime universities such as the World Maritime University (Sweden), Osaka University (Japan), Korea Maritime University (South Korea), and Shanghai Maritime University (China).

### 3.1.2 Myanmar Mercantile Marine College (MMMC)

The MMMC was established in 1972. It offers four-year bachelor's degree programmes in merchant marine science and merchant marine technology as well as a post-graduate diploma course in maritime transportation and maritime technology. The training courses are based on the International Convention on Standards of Training, Certification and Watchkeeping for Seafarers (STCW), 1978 as Amended.

Though MMU and MMMC seem to have some overlap, they actually have different approaches. MMU focuses on training both competent seafarers and experienced maritime engineers whilst MMMC is specialised in training competent seafarers in deck and engine departments.

### 3.1.3 Private Training Centers Authorised by the Department of Marine Administration (DMA)

The courses relating to the Certificate of Competence (CoC) are also taught in private training centers, authorised by the Department of Marine Administration (DMA) under the Ministry of Transport and Communications. Table 2 shows

**Table 2** Training courses, relevant to energy efficiency, of private training centers and MMMC (originally published by Lin (2015); modified with kind permission of ©Lin (2015))

| Course titles | Related topics for SEEMP | Covered topics in the courses |
|---|---|---|
| 1. Certificate of competency course based on the IMO model course 7.01 (this course is for Master and Chief Officer) | Plan a voyage and conduct navigation | Voyage planning |
| | Forecast weather and Oceanographic conditions | Weather routing |
| | Plan and ensure safe loading, stowage, securing, and care during the voyage and unloading of cargoes<br>Control Trim, Stability and Stress | Trim optimisation |
| 2. Certificate of competency course based on the IMO model course 7.02 (this course is for Chief Engineer officer and Second Engineer officer) | Physical and chemical properties of Fuels and lubrications | Low sulphur fuel content |
| | Operation and maintenance of Marine Engine and Auxiliary Engine<br>Waste heat utilisation | Waste heat recovery system |

training courses relevant to energy efficiency conducted at private training centers and MMMC.

Furthermore, workshops, seminars and conferences are occasionally organised by DMA and Myanmar Engineering Society. For example, 'Workshop on prevention of air pollution from ships and action by IMO to address GHG emissions' was held in Yangon City between 18 and 19 March 2013. With such events, the professionals in the Myanmar maritime industry are kept updated on specific subjects like energy efficiency.

### 3.2 *Review of Academic Activities in Myanmar Towards Energy Efficiency*

Both Tables 1 and 2 summarise ongoing academic and research activities within Myanmar's maritime industry in relation to energy efficiency. Some areas, such as resistance minimisation, hull form optimisation, and EEDI calculation and verification, have been addressed. However there are still gaps in which some missing elements need to be fed into the academic activities. For example, a towing tank experiment is needed at the EEDI preliminary verification stage. At present, MMU's towing tank is used for making calculations for resistance and powering of domestic vessels. For high-speed craft models and complex hull forms, the power and resistant data can be obtained by the application of a CFD simulation. The infrastructure in MMU and the experience gained for domestic vessels provide great opportunities for the maritime industry to take the first step of the EEDI verification stage for complex hull forms.

On the other hand, MMMC and private training centers are making their way to upgrade seafarers' competences and skills relevant to energy efficient ship operations. They offer training courses to cover onboard Ship Energy Efficiency Management Plan (SEEMP) and its evaluation tool, Energy Efficiency Operational Indicator (EEOI). Though the current situation in Myanmar does not fully reflect the best scenario for effective implementation of maritime energy management, various efforts can be initiated to promote energy efficiency as one of the main streams of future maritime education and training in Myanmar.

## 4 Future Training Model for Maritime Energy Efficiency in Myanmar

The previous section analysed the current situation of Myanmar's maritime industry in terms of maritime energy management. This section proposes four key areas to improve the quality of Myanmar's education and training on energy efficiency.

These four areas, which need to be focused, are Legislation; Research collaboration and dissemination; IMO model course; and Regional cooperation.

## 4.1 Legislation

The first area to support Myanmar's education and training is legislation. Myanmar has already signed MARPOL Annex I, III, and V. However, Annex VI has not been ratified by Myanmar as of November 2017. This hinders the development of national regulation and policy in Myanmar to prevent air pollution resulting from the maritime industry.

Moreover, most new ships currently being built are to be used for inland water transportation and coastal liners. According to Regulation 19 of MARPOL Annex VI, EEDI does not apply to ships sailing entirely within flag state waters.

If MARPOL Annex VI is ratified by Myanmar and strict national laws are set up, it will encourage various stakeholders to work towards a low carbon and energy efficient future for the maritime industry; for example, classification societies will come to follow those regulations; shipowners and shipbuilders must comply with compulsory rules in their ship construction process; port and terminal operators need to provide adequate reception facilities for exhaust gas cleaning residues in Myanmar Ports.

## 4.2 Research Collaboration and Dissemination

The Myanmar Maritime University has, itself, been conducting research to build capacity in maritime energy management, for example, topics related to technical measures and EEDI. These research activities are kept rather in-house and the dissemination of research is not strong. Thus, promoting research in maritime energy management at both national and international levels is crucial. To do this, exploring collaborative opportunities with local and overseas universities would be helpful. For example, the Myanmar Maritime University has been a member of the International Association of Maritime Universities (IAMU) since February 2015. IAMU has 62 universities of the world's maritime education and training universities/faculties as of November 2017. Such a global network of maritime education and training will provide a forum to exchange research findings and collaborations and benefit the maritime energy management studies in Myanmar.

Another issue is the lack of connection between academia and industry. The research conducted by academics does not go beyond; therefore, in order to make effective use of the research outcomes, academia and industry need to cooperate on applying energy efficiency measures. This will also trigger future research topics through understanding of the industry's needs and may create funding opportunities for research.

## 4.3 IMO Model Courses

There are many ways to facilitate education and training on energy efficiency in the maritime industry. One of the most well-known methods is to follow the IMO model courses. This method is effective especially in developing countries like Myanmar where resources are limited and some technical guidance such as IMO Model Courses is a good start to design a suitable curriculum. The other useful guide is a handbook called *"Training course on Energy Efficient Operation of Ships"* developed by the World Maritime University (Baumler et al. 2014) and recently updated by IMO (IMO 2016).

It is also worth noting that training is important not only for trainees but also for trainers. In this philosophy, IMO and WMU developed a train-the-trainer (TTT) package in 2013. The primary goal of the TTT course is to support developing countries as part of the IMO's capacity building activities. In 2015, TTT was updated by adding a new module on the regulatory framework related to energy efficiency of ships, an EEDI Calculator for training purposes, and relevant training resources, such as posters, assignments, exercises and presentations (IMO 2016). The modules in this course and their potential benefits are given in Table 3.

**Table 3** IMO train the trainer courses on energy efficiency ship operation

| Module materials | Benefits for trainers in developing countries |
|---|---|
| M1: Climate Change and the Shipping Response | – Logically explaining air pollution and climate change issues<br>– Could know the current reactions of International Organizations and IMO |
| M2: Ship Energy Efficiency Regulations and Related Guidelines and EEDI Calculator and Relevant Document | – Clear instruction to distribute to the trainees on how to verify EEDI<br>– Could clearly understand the EEDI's expectations<br>– Easily to understand Attained EEDI and EEOI calculation |
| M3: From Management to Operation | – Comprehensive information of advantages and disadvantages of Best Practices of SEEMP<br>– Clear explanation how to apply the best practices |
| M4: Ship Board Energy Management | Good guide to the responsibilities of seafarers in their respective ranks |
| M5: Ship Port Interface for Energy Efficiency | Knowledge on technical compatibility aspects of ship-port interface and communication mechanism between port and ship |
| M6: Energy Management Plans and Systems | Good Steps to hold energy efficiency implementation by monitoring, reporting from the shipping company sector |

## 4.4 Regional Cooperation

In addition to national and international scopes of interventions, it is also important to highlight the need for regional cooperation in maritime energy management. Myanmar cannot maintain itself without regional cooperation. Regional unity and amity provide a good foundation for Myanmar and other neighbouring countries to move toward a zero emissions and energy efficient future of the region (UNDP 2013).

Myanmar is a member of regional associations, such as the Association of Southeast Asian Nations (ASEAN), the Bay of Bengal Initiative for Multi-Sectoral Technical and Economic Cooperation (BIMSTEC), and the Colombo plan. These regional associations strengthen the opportunities for their Member States to initiate energy efficiency in the maritime sector and in other transport sectors in the region. In fact, regional countries have already been active in establishing environmental conservation, protection and contamination plans in the current decade.

### 4.4.1 Energy Efficiency Initiatives in ASEAN

Myanmar has been actively participating in the regional initiatives for energy efficiency within the ASEAN countries. Although the preventative measures against air pollution in the maritime industry is still at a very early stage in the regional countries, ASEAN fully supports global goals, such as the Paris Agreement, and is keen to contribute to those measures to mitigate the effect of climate change (Table 4).

Even though maritime-related energy efficiency is not yet very clear from each national goal, the IMO's leadership and its Member States in the region are expected to facilitate the implementation of energy efficiency in the maritime industry. In

**Table 4** Setting goal for energy efficiency regionally (adopted from © OECD/IEA 2017 Southeast Asia Energy Outlook, IEA Publishing. Licence: www.iea.org/t&c; as modified by WMU)

| Member States | Energy efficiency saving goal(s) |
|---|---|
| Brunei | Reduce total energy consumption by 63% from BAU levels by 2035 |
| Indonesia | Reduce energy intensity by 1% per year to 2025 |
| Lao PDR | Reduce final energy consumption from BAU level by 10% |
| Malaysia | Promote energy efficiency in the industry, buildings and domestic sectors with methods of standard setting, labelling, energy audits and building design |
| Myanmar | Reduce primary energy demand by 8% by 2030 from 2005 level |
| Philippines | Reduce energy intensity by 40% by 2030 from 2010 level<br>Decrease energy consumption by 1.6% per year against baseline forecasts by 2030 |
| Singapore | Improve energy intensity by 36% by 2030 from 2005 levels |
| Thailand | Reduce energy intensity by 30% by 2036 from 2010 level |
| Vietnam | Increase commercial electricity savings to more than 10% of total power consumption by 2020 relative to BAU |

*BAU* business-as-usual scenario

order to achieve this globally concerning issue, it is essential to promote regional cooperation to create a regional dialogue and response to the global concern. Smooth flow of information and technology in a relatively limited time is necessary to accelerate the process of implementation toward the sustainable future. By learning best practices from the European Union (EU), a large regional block, South-East Asian countries can also develop an institutional framework for the implementation of energy efficiency measures across all the industries, including the maritime industry.

## 5 Discussion

Among many strengths and weaknesses in maritime energy management in Myanmar, the paper has identified key areas to build capacity to support the process toward a sustainable future. So, how can we achieve this goal? To apply energy efficiency technology in Myanmar, this paper proposes a hierarchical model shown in Fig. 3. The government's leadership and the awareness of energy efficiency amongst local people and maritime professionals become a foundation. This leadership of the government also helps to take actions, such as developing an environmental conservation plan, sharing best practices of energy efficiency measures among both developed and developing countries, and cooperation with international organisations, regional organisations, and national and overseas universities.

For effective implementation of such action plans, cross-ministerial cooperation within the Myanmar government is necessary, for example, the Department of Marine Administration, under the Ministry of Transport and Communications as well as the Ministry of Natural Resources and Environmental Conservation.

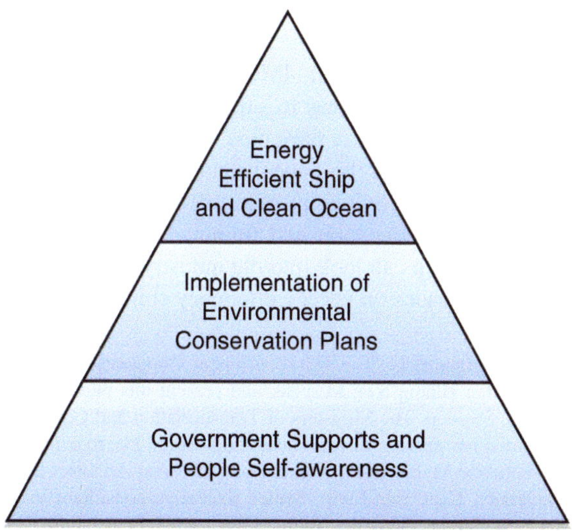


**Fig. 3** Hierarchies for implementations of marine energy efficiency measures

That is why, to establish a green future of the maritime industry, the authorised bodies can coordinate national-level actions to promote best practice of energy efficiency by:

1. Cooperating at an international level to embed regulations and standards which promote best practices and reward innovative solutions, and translating them into the national context;
2. Promoting and creating awareness amongst the recognised classification societies; and
3. Creating incentives for shipowners and builders.

# 6 Conclusion

Maritime energy management has become a hot topic in the global maritime industry. The majority of discussions and decisions are currently made by the leadership of developed countries. It is not well-known how the maritime sector in developing countries is reacting and taking actions to a global concern, such as climate change. The principle of the UN's SDGs is not to leave anyone behind. In this philosophy, it is important to address the perspective of developing countries in the context of maritime energy management.

This paper focused on the role of education and training to implement energy efficiency measures in Myanmar's maritime industry. Several notable strengths and weaknesses of Myanmar with regard to energy efficient shipping were highlighted. As a long-term strategy, Myanmar needs to build capacity in human resources to support maritime energy management in the country. The role of Myanmar does not end at the national level, but can be extended to the regional level of Southeast Asia by mobilising its resources within existing regional cooperation networks, such as ASEAN, BIMSTEC, and Colombo plan. Finally, the paper identifies four areas in which to focus on maritime energy management: Legislation; Research collaboration and dissemination; IMO model course; and Regional cooperation. To operationalise these areas to support maritime energy management in Myanmar, a hierarchical model was presented.

The limitation of this paper is that it is more like a snapshot of the current situation of Myanmar's activities in maritime energy management. Since the focus of this paper was on education and training, there are few implications to the industry. Future research can look into the industry's perspectives as well as the national law and policy issues on energy efficiency shipping.

**Acknowledgements** The IMO Technical Cooperation Division provided generous funding to enable Thet Hlaing Swe to speak and present this paper at the MARENER 2017 conference in Malmö, Sweden. The Ministers of Transportation and Communication Ministry in Myanmar also supported her mission travel. The authors would like to express appreciation to the guidance made by Professor Myint Myint Khine, Head of Naval Architecture Department of Myanmar Maritime University, Capt. Soe Lwin, Senior Executive consultant in Myanmar Maritime Education and Training Sector, Mr. Ye Myint, Director of Legislation Department, Department of Marine Administration.

## References

Baumler, R., Ölçer, A. I., Pazaver, A., Nakazawa, T., Baldauf, M., Moon, D., et al. (2014). Train-the-trainer course on energy efficient operation of ships. *American Journal of Climate Change, 3*(4), 404–412.
IMO. (2016). *IMO Train the Trainer Course on Energy Efficiency Ship Operation*. London: IMO.
IEA. (2017). *Southeast Asia Energy Outlook 2017*. Paris: International Energy Agency.
Lin, M. T. (2015). *Listing training institutes in Myanmar*. Marine Engineering Department. (H. o. Department, Ed.) Yangon, Yangon, Myanmar.
Embassy of the Kingdom of The Netherlands in Myanmar. (2016). *Maritime in Myanmar*. Retrieved October 2016, from myanmar.nlambassade.org : http://myanmar.nlembassy.org
UNDP. (2013). *Accelerating Energy Access for All in Myanmar*. Yangon: UNDP Myanmar.

# The Role of Maritime Transport from the Perspective of Energy and Gender: The Case of the Pacific Islands

**Momoko Kitada, Kuini Rabo, Ore O. Toua, and Thierry Nervale**

## 1 Introduction

The link between energy and gender has been increasingly recognised in the last decade, with particular attention to economic growth and sustainable development. All human beings need energy, but the fact that women and men in many countries have different expectations of their roles, responsibilities, and voices within households, markets, and communities can lead to differences in their access and choice of use of energy, and the impact of energy services on their lives. Such gender dimensions of access to services and benefits as well as exposure to risks and benefits are considered as important for effective policy making and project design in the energy sector (World Bank 2010). For example, the differences in gender roles may result in different energy need, use and choice; men may prioritise fuel for fishing boats whereas women may prioritise fuel for cooking.

Whilst such discussions are a good starting point to create awareness in the relationship between energy and gender, the scope is often limited within their countries or even communities. Recently, discourse on energy and gender has evolved to another level because of its link to a global concern, climate change. Energy choice ultimately influences climate change, and when displacement occurs as a result of climate change, it affects women and men differently in terms of, for example, their coping strategies (Karlsson 2007). This global link between energy and gender is helpful to understand the impact of every action that we take in everyday life.

M. Kitada (✉)
World Maritime University, Malmö, Sweden
e-mail: mk@wmu.se

K. Rabo · O. O. Toua · T. Nervale
Pacific Community, Suva, Fiji



A. I. Ölçer et al. (eds.), *Trends and Challenges in Maritime Energy Management*,
WMU Studies in Maritime Affairs 6, https://doi.org/10.1007/978-3-319-74576-3_26

In the meantime, the implication of energy and gender at a global level is not limited to the ways in which women can learn how to mitigate the effect of climate change, but also includes opportunities to actively participate in economies operated within a global supply chain. However, there is little research on how maritime transportation can support women's contribution to sustainable energy use and economic growth. This paper, therefore, explores the nexus of maritime energy and gender by examining the case of the Pacific Islands and contributes to building a new socio-economic dimension of maritime energy management research.

In order to explore how maritime transportation is relevant to energy and gender, this paper looks into the case of the Pacific Islands where the economy largely depends on maritime transportation. Another specific reason why the Pacific Islands was chosen as a case study is because the idea of this paper originated from a discussion held during the Regional Conference for Pacific Women in Maritime, organised by the International Maritime Organization (IMO) and the Pacific Community (SPC), in Tonga, 11–15 April 2016.[1] Such a regional forum on women in the maritime sector feeds local needs, such as energy and gender, to experts. It exemplifies how successful networking helps to produce new knowledge and research agendas.

The paper begins with the discourse on energy and gender in general, followed by more specific implications on maritime transportation, energy, and gender. The paper also refers to new business opportunities for women entrepreneurs to take part in the "circular economy". To set the context of energy and gender discussions, the features and characteristics of the Pacific Islands are presented. Two cases from the Solomon Islands and the Federated States of Micronesia are used to illustrate how gender is relevant to maritime energy. It discusses the role of maritime transport and energy from a gender perspective and concludes that the gendered nature of maritime transport is limiting access of minority users, including rural women in the Pacific Islands.

## 2 Energy and Gender: Broader Perspectives

The link between energy and gender is often undermined but important in terms of political, legal, economic, social, and environmental perspectives. The United Nations (UN) Fourth World Conference on Women, in Beijing, in 1995 called for action in gender mainstreaming 'in all policies and programmes so that, before decisions are taken, an analysis is made of the effects on women and men, respectively' in Objective G of the Platform for Action. This view was further reinforced at the ninth session of the UN Commission on Sustainable Development (CSD-9) in

[1]The conference outcomes were reported to the IMO Technical Cooperation Committee through the information paper, TC 66/INF.2 (Partnership and capacity building: Strengthening the impact of Pacific women in the maritime sector).

2001, urging governments to 'support equal access for women to sustainable and affordable energy technologies through needs assessments, energy planning and policy formulation at the local and national level' (Oparaocha and Dutta 2011).

With these efforts, it has been increasingly recognised in the energy sector that gender dimensions of access to services, exposure to risks, and access to benefits are important elements for effective policy making and project design (ESMAP 2013). For example, women and girls in developing countries are often responsible for cooking for the family, using solid fuels, such as wood fuels and charcoal, in poorly ventilated kitchens. This indoor air pollution is known to be the cause of major health hazards (e.g., cataracts, causing blindness) to women and girls (ESMAP 2013).

An important step for the integration of gender into energy initiatives is to acknowledge that not everybody has the same opportunities and same power within their households, communities and societies. Economic inequalities create disharmony in access to energy usage, priorities, and decision-making. Thus, understanding and responding to these inequalities in programme design and implementations will make a difference and is anticipated to narrow the missing link between energy and gender.

Within such discourses relating to energy and gender, most discussions seem to happen at the national or even community level, but not much further beyond. What is concerning here is that there is little debate on energy and gender at the global level of economic activities. In many cases, women tend to engage in activities in informal economies. Hence, it is not difficult to understand why most of the discussions about energy and gender are local rather than global.

To scale up women's participation in formal economies at the global level, women should be encouraged to be part of a global supply chain and market. According to CDP (2017), an international NGO working on climate change, their supply chain report emphasises that there is a missing link between gender and global supply chains. For example, when a company understands where their supply chain issues are, they can set up a goal for supply chain sustainability by mitigating their operational externalities and any negative impacts they may have. Such actions may be motivated by their stakeholders, including customers and stockholders, and can be linked to the global sustainability agenda, including women's economic participation. Women's empowerment and entrepreneurship cannot be achieved without the emancipation of women from "private patriarchy", which tends to extend to public spheres (Walby 1990, 1997). To encourage and support women's participation in a global supply chain helps to scale up economic growth in the country and region.

Here, it is also important to understand how maritime transport can support women's entrepreneurship and economic empowerment and provide access to the global supply chain. In this respect, the role of maritime transport in empowering women needs to be explored.

## 3 Circular Economy and Women's Entrepreneurship

The European Commission (EC) recognises that energy and gender are key drivers for development and economic growth. Economic empowerment of women is considered to be one of the four main pillars of the EU Gender Action Plan: 'Fighting violence of any kind against women and girls'; 'Economic and social empowerment'; 'Strengthening voice and participation'; and 'Shifting institutional culture'. It is also noted that women's entrepreneurship is an important source of growth, prosperity and poverty reduction. At the national level, it will have a positive impact on families and children, health and nutrition, communities and economy (EC 2015).

Maritime transport has contributed significantly to the process of industrialisation by shipping raw materials to one country and manufactured products to another country. International trade was, therefore, expanded, but humans have focused on the consumption of materials in this economic model. A linear economy repeats a pattern of take-make-dispose as far as resources are available. However, it is evident that such an economic model is not sustainable. In the majority of so-called "development" that dominated the twentieth century, serious environmental harm was created by linear economy. Linear economy is also known as risky in business because of its dependence on cheap and stable resource supply, which is not always predictable in global supply markets (World Economic Forum 2014).

Circular economy, on the other hand, emphasises 'product, component and material reuse, remanufacturing, refurbishment, repair, cascading and upgrading as well as solar, wind, biomass and waste-derived energy utilization throughout the product value chain and cradle-to-cradle life cycle' (Korhonen et al. 2018). Growing concern about a sustainable way of living to save the planet is a motivation to transform our mind-set from a linear to a circular economy. Indeed, this new shift towards a sustainable economy is opening up new business opportunities. Resource efficiency is a keyword here and can be translated into new types of services, products, public relations, and technologies. Considering the general fact that women have less access to formal economy (Gokulsing and Tandrayen-Ragoobur 2014), the question is how women can take part in a circular economy.

It is important to address the impact of a circular economy on women's lives and the ways of supporting this process from the maritime transport sector. It is also possible to generalise the absence of gender inquiries in the global maritime transport sector except for the limited advocacy of women seafarers (see, for example, IMO 2013). When it comes to maritime transport, energy and gender, there is very little knowledge in the industry. This area undoubtedly needs to be explored in future research. Hence, this paper addresses the linkage between maritime transport, energy and gender as a solid economic case. In particular, the paper pays attention to women entrepreneurs in the circular economic model and recognises the role of maritime transport in supporting sustainable development by empowering rural women in the Pacific Islands.

## 4 Exploring the Relevance to Maritime Transport, Energy, and Gender

Though the nexus between energy and gender is being more recognised at national and regional levels, it is rarely discussed how gender relates to maritime energy management. Maritime energy management itself is a relatively new research area, covering multidisciplinary subjects, for example, laws and regulations, technology, economics, and sociology (Kitada et al. 2017). Technical aspects of maritime energy management tend to be the focus; however, the importance of the human element in maritime energy management research is slowly being recognised (Kitada and Ölçer 2015). Among those human element studies in maritime energy management, the implication of gender in maritime energy has not yet been developed. However, this does not mean that gender is not relevant to maritime energy management. In fact, even before discussing maritime energy management and gender, maritime transport itself has been behind in terms of the integration of women in the industry (Kitada et al. 2015).

Nevertheless, it is important to include gender considerations from the beginning or as early as possible, in order to ensure that gender is considered at all levels. Gender and maritime energy management cannot be discussed without the implication of economic impacts. This paper addresses this gap from the viewpoint of global maritime transport and discusses how it affects women's economic participation and entrepreneurship, especially in relation to circular economy.

In this paper, we look into the case of the Pacific Islands to highlight the link between maritime energy management and gender. The reason why the Pacific Islands were chosen as the case is because of the inspiring discussion on maritime energy management and gender at the Regional Conference for Pacific Women in Maritime, organised by IMO and SPC, held in Tonga, 11–15 April 2016. It was probably the first conference in the world that dealt with the topic of energy and gender in the maritime sector as a formal session agenda. This is significant, in terms of the voices from the Pacific Islands, to advance the knowledge of maritime energy management and gender on which many decisions (e.g., rules and regulations) are being made in the North. From a feminist perspective, it is very important to listen to "the others" who are far from decision-making. The case of the Pacific Islands is, therefore, of particular interest in terms of social inclusion and political science. The next section focuses on the characteristics of Pacific Island communities from economic, environmental and social perspectives.

## 5 Pacific Islands: Their Isolation and Limited Capacities

The Pacific Islands geographically stretch over 10,000 square kilometers and are so remote, not just from the nearest continent, but also from neighbouring countries (Becker 2012). The Pacific Islands region contains 16 small island developing states (SIDS): the Cook Islands; the Federated States of Micronesia; Fiji; French Polynesia;

**Fig. 1** Isolation has significant economic, environmental and social impacts on Pacific Islands' communities (Modified by the authors based on the United Nations Map No. 3974)

Kiribati; Nauru; New Caledonia; Niue; Palau; Papua New Guinea; the Republic of the Marshall Islands; Samoa; Solomon Islands; Tonga; Tuvalu; and Vanuatu. In addition to these countries, 10 more countries[2] are the members of the Pacific Community.

Isolation has significant economic, environmental and social impacts on the Pacific Islands in terms of their overall sustainable development. To be more specific, the challenges facing the Pacific Islands can be summarised as below (see also Fig. 1):

(a) Large distances, high fuel costs and low economies of scale make the cost of developing and maintaining infrastructure, such as transport and communications, prohibitively high;
(b) A small population base tends toward a narrow range of resources and skills, increasing the costs of public administration and limiting institutional capacity;
(c) Narrow markets for local products and dependence on international trade create vulnerability to global developments as well as fewer employment and livelihood opportunities;
(d) Trade in remote locations is limited by high freight costs and marketing difficulties leading to increasing urbanisation;

[2]They are American Samoa; Australia; France; Guam; New Zealand; Northern Mariana Islands; Pitcairn Islands; Tokelau; United States of America; and Wallis and Futuna.

(e) The resulting rise in population density is placing pressure on resources and infrastructure, leading to problems such as overfishing, freshwater depletion, and pollution; and
(f) The vulnerability to natural disasters is also high. Cyclones and tsunamis often hit the Pacific Islands, and lifelines such as electricity and water are affected for a period of time.

These challenges that are commonly shared by the Pacific Islands and their remoteness have been acknowledged as "disadvantages" of SIDS (Briguglio 1995). In 2005, the Mauritius Strategy for Implementation (MSI) was adopted to recognise the unique vulnerabilities of SIDS and to identify the actions needed at national, regional and international levels for sustainable development. However, is this vulnerability of SIDS equally applied to both women and men in the Pacific Islands?

In fact, Pacific Island women are often found to be more vulnerable in terms of their socio-economic status. The 5-year review of MSI in the Pacific Islands reports that 'despite gender equality in lower levels of education in most Pacific SIDS, women remain under-represented in technical and professional education and are over-represented in low-paid informal sectors (ESCAP 2010)'. Couper et al. (2015) discuss women's socio-economic problems in Papua New Guinea where fishery is an important income source as its export value is USD760 million per year. Transnational corporations based in the Philippines and Taiwan exploit Pacific women in the workforce (the majority of some 10,000 workers) by engaging them in long-hour, low-pay, low-skilled work in their six factories, remotely built from their villages and families. Nevertheless, women's domestic workload still remains the same and women face a huge challenge in work-life balance (Couper et al. 2015).

Women generally have less access to formal economies, though a small number of Pacific women seafarers exist. In Kiribati, many young men serve on international ocean-going vessels as seafarers. The International Labour Office (ILO) reports that the number of Kiribati seafarers on board international merchant vessels was 1452 in 2006, falling to 750 in 2015 due to the impact of the global financial crisis; however, the number is expected to slowly rise again. In addition, there are also an estimated 325 I-Kiribati (native Kiribati) crew working on 100–200 fishing vessels of the Republic of Korea, Taiwan and China (ILO 2015). According to the 2006 Kiribati national household survey, single female-headed households in South Tarawa accounted for 28 per cent of all households due to many seafaring husbands being absent from home (AusAID 2012). On the other hand, the number of women seafarers in Kiribati was reported as two in April 2016: one certified as an Engineer Class 4 and another as a Master Class 5[3]; however, both women seafarers left sea in the latest figure (see Table 1). In Tonga, there used to be a few women seafarers but they have already quitted seafaring in order to look after their families. There are also

[3]Presentations made by the delegations of Kiribati during the Regional Conference for Pacific Women in Maritime, organised by the International Maritime Organization (IMO) and the Pacific Community (SPC), in Tonga, 11–15 April 2016.

**Table 1** The number of female seafarers and marine pilots in the Pacific Island states as of November 2017 (Source: Database from SPC, 2017)

| Number of female seafarers and marine pilots in the Pacific Island states | Cook Islands | Federate State of Micronesia | Fiji | French Polynesia | Kiribati | Marshall Islands | Nauru | New Caledonia | Niue | Palau | Papua New Guinea | Samoa | Solomon Islands | Tonga | Tuvalu | Vanuatu | Total |
|---|---|---|---|---|---|---|---|---|---|---|---|---|---|---|---|---|---|
| Employed in domestic ships | 0 | 0 | 30 | 0 | 0 | 0 | 0 | 0 | 0 | 0 | 22 | 19 | 5 | 0 | 1 | 15 | 92 |
| Employed in foreign ships | 0 | 0 | 7 | 0 | 0 | 0 | 0 | 0 | 0 | 0 | 0 | 0 | 0 | 0 | 0 | 0 | 7 |
| Certified or under-training marine pilots | 0 | 0 | 0 | 0 | 0 | 0 | 0 | 0 | 0 | 0 | 0 | 0 | 2 | 0 | 0 | 0 | 2 |
| Total | 0 | 0 | 37 | 0 | 0 | 0 | 0 | 0 | 0 | 0 | 22 | 19 | 7 | 0 | 1 | 15 | 101 |

a small number of women seafarers in Fiji, Papua New Guinea, and the Solomon Islands and the majority is employed on domestic vessels. Many other Pacific Island States do not have women seafarers.

This example suggests that women's participation in the formal economy tend to be limited and will hinder the process of sustainable development of the region. In order to break the barriers, energy and gender may be the keywords to examine the current socio-economic mechanisms and see whether the maritime sector can play a role to support women's entrepreneurship and energy access. The next section looks into the cases of the Pacific Islands in terms of energy, gender, and economy.

## 6 Examining Energy-Gender-Economy Issues in the Pacific Islands

The Pacific Islands, consisting of 16 remote island states, have been facing energy security issues, which are commonly observed in SIDS but even more acute in their extremely remote environment. As stated earlier, in the Pacific Islands, resources are limited; thus, maritime transport and fisheries are vital for supplying necessary foods and other materials to those islanders. It is not clear how many women engage in fishery in the Pacific Islands, but the research on fishery in Samoa shows that 18% of all village fishers are women (Lambeth et al. 2002).

By considering these facts, this section examines the issues related to energy, gender, and economy in the Pacific Islands by looking into two cases from the Solomon Islands and Federated States of Micronesia.

### *6.1 A Case in the Solomon Islands*

Similar to many other Pacific Island States, women in the Solomon Islands play a vital role in marine food gathering at low tide, shallow waters, or inside barrier reefs (Couper et al. 2015). In addition, women also transport fish to sell at local markets.

In the Choiseul Province of the Solomon Islands, only small boats are used for access to the main market (i.e., Taro Island) due to irregular shipping services. Those small boats operate in various activities, such as transport of goods, passengers, fish, and anything needed for living. Figure 2 shows how women in the Solomon Islands use traditional canoes as transportation in order to sell fish to the main market centre. Often, safety is compromised. Even though fuels remain expensive, it is the only transport mode in most of the coastal islands in the Solomon Islands. Indeed, the cost of transporting goods in the outer islands is high and, thus, unaffordable to most local communities.

Limited energy sources and budget for fuels are likely to be prioritised for running cargo ships and fishing boats rather than for lighting to improve household

**Fig. 2** Women use traditional canoes for transport to the main market centre to sell fish (esky ice box store fish) [Source: Koin Etuati, Geoscience, Energy and Maritime (GEM), SPC; Pacific Community (SPC); reprinted with kind permission of ©SPC. All rights reserved]

**Fig. 3** Rural communities in the Solomon Islands use fossil fuels for transportation, such as motorised canoes [Source: Catherine Wilson, Inter Press Service International News Agency (IPS); Reprinted with kind permission of ©IPS. All rights reserved]

activities for women. In daily activities, small canoe boats with motors using fossil fuels also operate for passengers and cargo (Fig. 3). The remaining energy portion is then distributed to other usages.

From this case, it is evident that energy efficient and cost effective maritime transport to connect different islands within the country is necessary. The use of renewable energy for electrical propelled motor boats is also worth considering as the environment of the Pacific Islands is in favour of such renewable energy resources, including solar, wind, and wave energy. At the same time, safety issues should not be neglected. Transport operators, regardless of the size and type of vessels, must be aware of the need for safety measures, equipment, and training. Solomon Island women have been taking part in economic activities, such as selling fish and farming; however, women's participation in global economies and supply chains can be strengthened by bringing in new business models like a circular economy for a more sustainable future in the Pacific Islands.

## *6.2 A Case in Federated States of Micronesia*

Another case comes from one of the islands in the Federated States of Micronesia—Kosrae, illustrating how entrepreneurship and maritime transport are closely linked. A woman entrepreneur in Kosrae decided to start a business in waste management. Her planned business model was to collect used batteries from the island and ship them by container to an overseas recycling facility.

Even though the idea was great to support sustainability and environment, the main challenge she faced was to understand how to ship used batteries to another country by a container vessel. Her challenge can be summarised in three aspects: legal, economic, and social issues.

The first challenge was the legal aspect of starting a new business. It was not straight-forward for new business women to understand the international rules for such export. The business model for exporting used batteries involves the process of approval and authorisation from the State. The national government of the Federated States of Micronesia does not have national regulations compliant with international instruments. The absence of such national regulations technically restricts the opportunities for women's entrepreneurship in a circular economy model.

Secondly, from the economic aspect, the woman entrepreneur also faced difficulty in contacting a ship operator and understanding the constraints that shipping lines have. It was not easy for those shipping lines operating in the region to make money out of such a small-scale business. Naturally, the woman entrepreneur found it difficult to negotiate regular container exports at an affordable cost to ensure the sustainability of her business.

In addition, the third challenge appeared in the social aspect. To run the business successfully, the woman has to organise the regular collection of used batteries from households and small enterprises to ensure regular exports of a full container of used batteries. This requires a change in people's mindset to treat batteries as recycling goods. Such an attitudinal change requires a social change within the community by adopting small-scale waste management in the home.

In summary, legal, economic, and social barriers exist in terms of the realisation of Pacific Island women's entrepreneurship in a circular economy model. Interestingly, all these barriers lie in the maritime transportation system and global supply chains. It is evident that the role of maritime transport is crucial to support rural women's participation in the formal economy. Efforts by the maritime transport sector is expected to remove those barriers and create opportunities for women entrepreneurs in sustainable development.

## 6.3 *The Role of Maritime Transport and Energy from a Gender Lens*

Both cases from the Solomon Islands and the Federated States of Micronesia provide a hint of conceiving the link between maritime transportation and "Energy-Gender-Economy" issues. It is clear that maritime transportation has a vital role to play in supporting Pacific women's economic activities. Selling fish at the main market is not possible without using small canoes in the Choiseul Province of the Solomon Islands, for instance. The idea of exporting used batteries cannot be realised without accessing a maritime transport system. In other words, if maritime transportation becomes more accessible and affordable to women entrepreneurs in the Pacific Islands, it will boost opportunities for rural women to participate in formal economies. With limited examples, it is still possible to anticipate that the current status of maritime transportation in the Pacific Islands is not creating an enabling environment and various opportunities for rural women but rather challenges for them to run small- or medium-scale businesses. Heuër (2017) discusses how important it was to encourage women micro-entrepreneurs and focus on woman-to-woman sales which appeared to be a successful case of introducing new technology for energy in rural communities in Africa. By learning from good practices on energy and gender, it is assumed that maritime transport has an important role to play to integrate rural women into global supply chains and markets.

This paper argues that energy is a key driver to changing women's economic status in the Pacific Islands. The concept of energy efficiency will stimulate users to change their mindset to move towards environmentally friendly and sustainable maritime operations. It can be the use of renewable energy instead of fossil fuels and the adoption of fuel-saving techniques and methods (Ölçer et al. 2017). It is also expected to improve economic situations by diversifying energy sources instead of relying on conventional fuel types. Enabling more innovative ideas for energy efficiency will create a new opportunity for Pacific women to participate in formal economies. One such modern concept is a circular economy.

## 7 Conclusion

A literature review on energy and gender provided a good foundation for discussing the maritime energy and gender nexus which had not been developed at all at the time of the International Conference on Maritime Energy Management (MARENER 2017). This paper used the Pacific Islands as a case study to demonstrate how energy and gender are important to discuss in relation to maritime transport. It is known that the maritime sector is extremely male-dominated worldwide, as is maritime energy management. Generally, women are not considered to be decision-makers in terms of energy choice and access. In this paper, we argued how the gendered nature of maritime transport is limiting service capacity to minority users like rural women in the Pacific Islands.

It is evident that the maritime transport sector has a role to play in supporting women's participation in establishing a sustainable, energy efficient society. Energy efficiency comes from resource efficiency, which also implies a better use of human resources including women in the society. The concept of circular economy and other sustainable economic models can be used to close the gender gap and recognise the role of maritime transport to enable women to participate in the process of sustainable development. This unexplored field of research needs to be further developed to establish a better link between women's economic participation and the maritime transport and energy sectors.

## References

AusAID. (2012). *Kiribati country case study: AusAID Pacific social protection series: Poverty, vulnerability and social protection in the Pacific*. Canberra: Australian Agency for International Development (AusAID).

Becker, C. (2012). *Small island states in the Pacific: The Tyranny of distance?* Washington, DC: International Monetary Fund.

Briguglio, L. (1995). Small island developing states and their economic vulnerabilities. *World Development, 23*(9), 1615–1632.

CDP. (2017). *Missing link: Harnessing the power of purchasing for a sustainable future*. London: CDP.

Couper, A. D., Ciceri, B., & Smith, H. D. (2015). *Fishers and plunderers: Theft, slavery and violence at sea*. London: Pluto.

EC. (2015). *Factsheet on the new framework for gender equality and women's empowerment: Transforming the lives of girls and women through EU external relations (2016–2020)*. Brussels: European Commission.

ESCAP. (2010). *Sustainable development in the Pacific: Progress and challenges*. Pacific regional report for the 5-year review of the Mauritius strategy for further implementation of the Barbados programme of action for sustainable development of SIDS (MSI+5). The Economic and Social Commission for Asia and Pacific (ESCAP) Subregional Office for the Pacific, Suva.

ESMAP. (2013). *Integrating gender considerations into energy operations*. Knowledge Series 014/13. Energy Sector Management Assistance Programme. World Bank, Washington, DC.

Gokulsing, D., & Tandrayen-Ragoobur, V. (2014). Gender, education and labour market: Evidence from Mauritius. *International Journal of Sociology and Social Policy, 34*(9/10), 609–633. https://doi.org/10.1108/IJSSP-01-2013-0001

Heuër, A. (2017). Women-to-women entrepreneurial energy networks: A pathway to green energy uptake at the base of pyramid. *Sustainable Energy Technologies and Assessments, 22*, 116–123. https://doi.org/10.1016/j.seta.2017.02.020.

ILO. (2015). Labour migration in Kiribati: Briefing Note/14 April 2015. Retrieved from http://www.ilo.org/suva/areas-of-work/WCMS_407373/lang--en/index.htm. Accessed 24 November 2017.

IMO. (2013). *Busan Declaration*. IMO Regional Conference on the Development of a Global Strategy for Women Seafarers, Busan, 16–19 April 2013. http://www.imo.org/en/MediaCentre/HotTopics/women/Documents/BUSAN%20DECLARATION(Revised1)%20(3).pdf. Accessed 28 November 2017.

Karlsson, G. (2007). *Where energy is women's business: National and regional reports from Africa, Asia, Latin America and the Pacific*. Leusden: ENERGIA.

Kitada, M., Dalaklis, D., Drewniak, M., Ölçer, A. I., & Ballini, F. (2017). *Exploring the frontiers to maritime energy management research*. Paper presented at the International Association of Maritime Universities (IAMU) Conference, Varna, 12–13 October 2017 (pp. 481–490). ISBN: 978-954-8991-95-7.

Kitada, M., & Ölçer, A. I. (2015). Managing people and technology: The challenges in CSR and energy efficient shipping. *Research in Transportation Business and Management*. Special issue on Energy Efficiency. https://doi.org/10.1016/j.rtbm.2015.10.002.

Kitada, M., Williams, E., & Froholdt, L. L. (Eds.). (2015). *Maritime women: Global leadership*. Heidelberg: Springer.

Korhonen, J., Honkasalo, A., & Seppälä, J. (2018). Circular economy: The concept and its limitations. *Ecological Economics, 143*, 37–46.

Lambeth, L., Hanchard, B., Aslin, H., Fay-Sauni, L., Tuara, P., Rochers, K. D., et al. (2002). An overview of the involvement of women in fisheries activities in Oceania. In M. J. Williams et al. (Eds.), *Global symposium on women in fisheries*. Penang: ICLARM – The World Fish Center.

Ölçer, A. I., Ballini, F., Baumler, R., & Kitada, M. (2017). Maritime energy management. In I. D. Visvikis & P. Panayides (Eds.), *Shipping operations management*. Heidelberg: Springer. ISBN: 978-3-319-62364-1.

Oparaocha, S., & Dutta, S. (2011). Gender and energy for sustainable development. *Current Opinion in Environmental Sustainability, 3*, 265–271.

Walby, S. (1990). *Theorizing patriarchy*. Oxford: Basil Blackwell.

Walby, S. (1997). *Gender transformations*. London: Routledge.

World Bank. (2010). *Integrating gender considerations into energy operations*. Energy Sector Management Assistance Programme (ESMAP) Report No. 76517. World Bank, Washington, DC.

World Economic Forum. (2014). *Towards the circular economy: Accelerating the scale-up across global supply chains*. Geneva: World Economic Forum.

# Part V
# Alternative Fuels and Wind-Assisted Ship Propulsion

# Developing a Strategy for Liquefied Natural Gas Powered Transport Corridors in the Baltic Sea Region

**Josefin Madjidian, Dimitrios Dalaklis, Vytatutas Paulauskas, Lawrence Henesey, Aykut I. Ölçer, Fabio Ballini, and Momoko Kitada**

## 1 Introduction

When the discussion revolves around the topic of clean technological solutions for the shipping industry, the Baltic Sea Region (BSR) (comprising Denmark, Finland, Estonia, Germany, Latvia, Lithuania, Norway, Sweden and Russia) is clearly a leading region of the world (Dalaklis et al. 2017a). This region is also a very well-trafficked maritime corridor, since all the above mentioned countries are heavily dependent on maritime transport; it is indicative that during the last decade, ships within the Baltic Sea have increased both in number and size, currently representing the handling of up to 15% of total world cargo traffic (Stankiewicz and Vlasov 2009). Furthermore, the level of traffic continues to grow by about 5% per year. This increased level of traffic could also result in rising emissions to the air (exhaust gases) from those numerous crossing vessels. Recognising the need to protect the BSR environment, the wider region has already been assigned as a Sulphur Emission Control Area (SECA) by the International Maritime Organization (IMO).

The specific policy measure translates into the fact that since the 1st of January 2015, ships entering the Baltic Sea have been required to comply with less than 0.1 sulphur content in their fuel (IMO 2013). In turn, the effect has been a significant reduction in emissions of sulphur oxides (SOx) (−88%) and particulate matter

J. Madjidian · D. Dalaklis (✉) · A. I. Ölçer · F. Ballini · M. Kitada
World Maritime University, Malmö, Sweden
e-mail: jam@wmu.se; dd@wmu.se; aio@wmu.se; fb@wmu.se; mk@wmu.se

V. Paulauskas
Klaipeda University, Klaipėda, Lithuania
e-mail: donatasp@takas.lt

L. Henesey
Blekinge Institute of Technology, Karlskrona, Sweden
e-mail: larry.henesey@bth.se



A. I. Ölçer et al. (eds.), *Trends and Challenges in Maritime Energy Management*, WMU Studies in Maritime Affairs 6, https://doi.org/10.1007/978-3-319-74576-3_27

(PM) (−36%), compared to 2014 (Johansson and Jalkanen 2016). The Baltic Sea is also designated to become a Nitrogen Emission Control Area (NECA) as of 2021; this means that vessels built after 2021 will be required to reduce nitrogen oxides (NOx) emissions by 80% compared to the current emission levels. With the introduction of these regulations, a forced shift has been placed upon vessels/ship-owners operating in the region; changes to the current business models (and technologies involved) are necessary.

As a result, the countries surrounding the Baltic Sea are not only examining and applying different pollution reduction technologies, but also considering/introducing alternative fuels such as liquefied biogas, or methanol; research and pilot-projects are also already underway in the areas of new-builds' design and retrofits. Improvements in port infrastructure and bunkering facilities are also part of the same equation. The current analysis, aims to identify the role of Liquefied Natural Gas (LNG) in achieving the necessary compliance with the new regulations and simultaneously boost the economic development of the BSR, as part of the wider activities related to the "Go LNG project" (see Sect. 1.2). This will be accomplished by investigating the related transport flows and LNG infrastructure developments in the region under discussion. Additionally, the creation of a wider in scope value chain that incorporates all transport modalities and industries that use natural gas today should be investigated next, as a future necessary research activity.

## *1.1 LNG in the Baltic Sea Region*

When it comes to short-sea-shipping, as most operations are in SECA areas, vessels in order to comply with the newly-created regulatory framework, have the following three (3) different choices available: (a) integrating an emission abatement technology (i.e. a scrubber) in their current propulsion system, (b) avoiding changes to the current propulsion system and simply opting to use low sulphur—but more expensive—fuel such as MGO (marine gas oil) or MDO (marine diesel oil), or c) avoiding MGO and/or MDO, by opting for liquefied natural gas (LNG) (Dalaklis et al. 2017b). LNG has been viewed as the most promising alternative marine fuel technology in the short to medium term, as it can easily help to reach the environmental targets of sulphur, nitrogen and particulate matters emissions in SECAs and NECAs that have already been spelled out by IMO (Bengtsson et al. 2014; Brynolf et al. 2014; Thomson et al. 2015; Deniz and Zincir 2016). Additionally, the BSR is holding a very important experience: more of a decade of small scale LNG driven ship operation that has proven the reliability of LNG technology, including the safety aspects. The case of Norway is indicative, where the first LNG fuelled vessel (Glutra) started operating in 2000; currently, there are twenty-eight (28) vessels operating on LNG (pers. comm., Egil Rensvik, Network LNG Norway, March 2017).

Moreover, the global trade in LNG is growing rather rapidly: in 1970, global LNG trade was extremely low, around 3 billion cubic meters (bcm); by 2011, it had increased to the level of 331 bcm (The Economist 2012). As a reference, the 2015 total global consumption of "petroleum and other liquids fuel" was 93.78 million barrels per day (EIA 2016). The European Union (EU) is further considering a dramatic (50%) expansion in global supply over the next few years that will result in lower prices; this represents a major opportunity for the EU and at the same time can provide significant enhancement in relation to energy security and resilience domains (European Commission 2016). Furthermore, the number of operating LNG fuelled vessels is today also steadily increasing worldwide, with 102 of them already in operation and 108 confirmed new builds (pers. comm., Egil Rensvik, Network LNG Norway, March 2017). In order to better understand the statistics involved, 69% of the ships in operation are operating in Norway, 17% in the rest of Europe, 8% in America and 6% in Asia and the Pacific. When it comes to the confirmed bookings (as of March 2016), 58% will be operating in Europe, 25% in America, 14% in Norway and only 2% in Asia and the Pacific (DNV GL 2016).

However, the current level of progress in the deployment of LNG in the BSR (and elsewhere in the world) is often associated with the infamous "chicken-and-egg" problem, where bunker suppliers and ship-owners have both followed a "wait and see" approach (Wang and Notteboom 2015; Gerlitz and Paulauskas 2015). The lack of LNG bunkering infrastructure, and the absence of crystal clear operational standards, together with the intensive capital cost and associated risks has deterred potential stakeholders, including ship-owners and bunkering operators, to be at the forefront (Dalaklis et al. 2017a). Efforts to address these barriers, along with emphasising the above mentioned positive aspects of LNG as a marine fuel, have taken place during recent years. For example, the EU's energy plans for the future (see below, Sect. 2.1) have been amended. More specifically, in February 2016, the EU delivered an LNG Strategy, stressing the geopolitical challenges linked to ensuring secure and resilient supplies of fossil fuels and with the aim to exploit the potential of LNG (and appropriate gas storage) in order to make the EU's gas system more diverse and flexible, thus contributing to a secure, resilient and competitive gas supply (European Commission 2016). It is no coincidence that one of the countries mentioned in this strategy as being heavily dependent on a sole supplier (and therefore in need of rapid development in access to a diverse range of energy sources) is Lithuania, a partner country of the Go LNG project.

## 1.2 Go LNG Project

In a Baltic Sea ERDF part-financed INTERREG V project (2016–2019), called the Go LNG project (http://www.golng.eu/), activities have been developed with the aim to ease the implementation of the EU Clean Fuel Strategy and the Directive on the Deployment of Alternative Fuels (European Parliament 2014). The project has, among other important tasks, to provide a concept that links transport flows, LNG infrastructure development, and business models into an efficient LNG infrastructure development plan and an LNG distribution strategy for the BSR (enabling blue transport corridors). The take on this is to strengthen the interest of private parties in LNG deployment in maritime transport, Inland Waterways Transport (IWT), rail and road, and heavy industry, and to demonstrate the potential to create economic growth and jobs, while at the same time meeting policy objectives and regulations. The "Blue Transport Corridors" Strategy will be one of the outcomes of the project and is meant to provide a roadmap for promoting further LNG deployment in the BSR, both as fuel in the transport sector as well as a source of energy in other industrial sectors. This strategy will further provide stakeholders in the LNG sector an up-to-date knowledge base on relevant technical information and standards as well as policies and regulations, in the form of an online tool box and a very detailed map of bunkering facilities, along with their technical details and capacity levels. Furthermore, it will describe various well-functioning LNG solutions and business models already present in the transport sector of the region. The idea is to both create better access to technology and knowledge and build up better competences and specialisations within the BSR maritime business supply chain and LNG-related business activities, as well as showcasing the region as a competence hub in the field of LNG (Dalaklis et al. 2017a).

## 1.3 Aim

The analysis at hand will examine how the Go LNG project can develop a Blue Corridor Strategy for LNG that both supports the general policy objectives of the EU in the field of transport and energy (including the key elements of its new LNG strategy), as well as considers each partner-country's specific benefits and challenges in the introduction (and further expansion) of LNG to the market. By linking LNG infrastructure development, transport flows, business models and government policies into one cohesive and efficient LNG infrastructure development plan, the strategy under discussion will identify the existing LNG value chain of the region as well as describe the opportunities for expanding toward a wider value chain, incorporating other transport modalities and industries. In order to design a regional strategy with long-term aims, the foundation of the analysis will be provided by mapping development patterns of LNG-related activities and capabilities, regarding education, business, and infrastructure, of all the BSR countries that are involved in the Go LNG project; discussion of any other type of information that is considered valuable and necessary can also take place at a later stage. Indeed, a full exploitation of LNG opportunities requires not only the development of a well-functioning and favourable regulatory framework, but also a critical mass of investment in LNG infrastructure as well as education and cooperation. Following that, it is essential to include in the strategy business links that are currently missing, and that need to be further looked into, as well as cooperation possibilities—for a greater effect on business development and, in the longer run, the environment.

## 2 Method and Results: The Current State of Affairs and the Deployment of LNG in the Baltic Sea Region

The countries surrounding the Baltic Sea are different in many aspects concerning the development and prospects of LNG usage. Geographic location, size, number of ports, type of industries and national energy and transport strategies and/or regulations are some indicative examples. In the Go LNG project framework, the development of a questionnaire to be completed by project partners (grouped by country) was the first step in the necessary data collection effort. These questions were developed with the aim of understanding the current status of LNG deployment and the potential further LNG developments of each country, as well as to pave the way toward the identification of future potential cooperation between the countries involved. The questions were grouped into five (5) different categories: 1. Research and education, 2. Infrastructure development, 3. Policy and regulations, 4. Business development, and 5. Challenges. Following a first round, answers were gathered and new questions arose. The second round is now out for finalization. Below, some first findings are presented; certain challenges are also discussed.

### *2.1 Policy and Regulation*

In order to understand how different parts of the LNG value chain are affected by policy and law, it is necessary to gather the existing information regarding maritime transport, both on the international level as well as on national levels. The project will collate policy documents, regulatory documents, standards and norms from all levels. An important feature of the Blue Corridor Strategy is that it will also include policy and regulation regarding use of LNG for road and rail. This information will then be made available online for stakeholders in the respective sector(s). Some of the main international policies and regulations that have played a major role in the development of LNG use in the BSR are introduced next. These initiatives/legal toolboxes have already influenced and will continue to change the shipping industry's demand for different fuels, which in turn affects the fuel prices and the cost effective available technology and infrastructure.

The most significant regulation driving the introduction and continued development of LNG in the BSR is the International Maritime Organization's Convention on Prevention of Pollution by Ships (MARPOL) (IMO 2010). Annex VI represents the regulatory framework tackling exhaust gas emissions from ships. It prohibits deliberate emissions of ozone depleting substances and sets progressive reductions (tiers) in emissions of sulphur oxides ($SO_X$), nitrogen oxides ($NO_X$) and particulate matter (PM). It has also introduced designated emission control areas with more stringent standards for emissions. Today these areas are the Baltic Sea ($SO_X$ only), the North Sea ($SO_X$ only), the North American area ($SO_X$, $NO_X$ and PM), and the United

States Caribbean Sea area ($SO_X$, $NO_X$ and PM). In November 2016, IMO designated the North Sea and the Baltic Sea as NECAs, starting 1 January 2021.

When it comes to GHG from shipping, amendments to MARPOL Annex VI in 2011 added Chapter 4 which introduces two mandatory mechanisms (entered into force in January 2013), intended to ensure an energy efficiency standard for ships: (1) the Energy Efficiency Design Index (EEDI), for new ships, and (2) the Ship Energy Efficiency Management Plan (SEEMP) for all ships. Another important IMO instrument is the International Code of Safety for Ships using Gases or other Low-flashpoint Fuels, the IGF code (amendments to the International Convention for the Safety of Life at Sea (SOLAS), which entered into force 1 January 2017). The IGF code is set up to minimize the risk to the ship, its crew and the environment, having special regard to the nature of the fuels involved. It, therefore, contains mandatory provisions for the arrangement, installation, control and monitoring of machinery, equipment and systems using low-flashpoint fuels, focusing initially on LNG.

The BSR countries also need to take EU policies and regulations into account. In 2013, the EU announced a Clean Power for Transport Package, which consists of a Communication on a European alternative fuels strategy, Directive 2014/94/EU (European Parliament 2014), focusing on infrastructure and standards, and an accompanying document describing an action plan for the development of Liquefied Natural Gas (LNG) in shipping. Through the Directive 2014/94/EU, which is a common framework of measures for the deployment of alternative fuels, vital infrastructure in EU is established, setting out minimum requirements, for example, recharging points for electric vehicles and refuelling points for natural gas (both liquefied-LNG and compressed-CNG) and hydrogen, to be implemented by means of Member States' national policy frameworks, as well as common technical specifications and user information requirements.

Indeed, for many member states, the Directive 2014/94/EU has marked the starting point to react on the highest level of policy in relation to LNG developments. In November 2016, each EU country was supposed to hand in its own LNG national strategy, but as of today, all strategies have not yet been completed. Bunkering of LNG is still relatively novel and differs in several aspects from the bunkering of conventional marine fuels. As the use of LNG-fuelled vessels spreads and many more parties become involved, there is a need to standardize LNG bunkering operations. At the request of IMO, the European Commission and the Baltic and International Maritime Council (BIMCO), the world's largest international shipping association, a new standard (ISO 20519:2017) was produced. It contains requirements not covered by the IGF code, and sets requirements for LNG bunkering transfer systems and equipment used to bunker LNG fuelled vessels. However, there is still no requirement for a standard coupling for which a new working group was established (ISO 2017).

## 2.2 LNG Infrastructure, Technology and Related Services

For the various EU countries, the "TEN-T Core Network" and "Motorways of the Sea" provide the basis for the deployment of LNG infrastructure since they cover the main traffic flows and pave the way for more "network" benefits; it is necessary to provide some further information about these two important policy frameworks. As of January 2014, the EU has a new transport infrastructure policy that connects the continent between East and West, North and South. This policy aims to close the gaps between Member States' transport networks, remove bottlenecks that still hamper the smooth functioning of the internal market and overcome technical barriers such as incompatible standards for railway traffic. The specific policy promotes and strengthens seamless transport chains for passenger and freight, while keeping up with the latest technological trends. This initiative will help the European economy in its recovery and growth, with a budget of €24.05 billion up to 2020. The TEN-T programme consists of hundreds of projects (defined as studies or works), whose ultimate purpose is to ensure the cohesion, interconnection and interoperability of the trans-European transport network, as well as access to it.

Furthermore, the "Motorways of the Sea" concept aims to introduce new intermodal maritime-based logistics chains in Europe, which should improve the transport organisation within the years to come. These chains will be more sustainable, and should be commercially more efficient than road-only transport. The concept under discussion will thus improve access to markets throughout Europe, and bring relief to our over-stretched European road system. For this purpose, further use will have to be made not only of the available maritime transport resources, but also of the potential in rail and inland waterways, as part of an integrated transport chain. The concept was introduced with the 2001 Transport White Paper "European transport policy for 2010: time to decide". The European Commission proposed the development of "Motorways of the Sea" as a "real competitive alternative to land transport". This Paper also defined that the Motorways of the Sea should be part of the trans-European Network (TEN-T) and funds should be made available for its development. The EU Directive 2017/94 EU states that LNG must be available at TEN-T maritime core network ports by the end of 2025, at inland core network ports by the end of 2030, and for heavy duty vehicles, an appropriate number of fuel stations must be available along the core network by the end of 2025. Within the core network, ports that have been appointed as core ports will have to offer bunkering of LNG to their customers by the end of 2025 (Table 1). Out of the nine corridors set up by the EU, three (3) cover the Baltic Sea region and eighteen (18) ports in the BSR are core seaports that will need to offer LNG bunkering by 2025. The complete details are summarized via Fig. 1. Additionally, in Fig. 2, details of LNG availability in relation to truck transport (by road) are provided.

It is also necessary to highlight that apart from the EU master plan, which states where LNG bunkering possibilities must be available, national energy strategies as well as needs of industry may result in LNG infrastructure also being developed elsewhere. In Fig. 1, operating and planned LNG bunkering facilities and terminals

**Table 1** LNG vessels in BSR

| Vessel type | Vessel name | Vessel info | Owner |
|---|---|---|---|
| Denmark | | | |
| Ferry | Princess Isabella | | Samsö municipality |
| | MS Stavangerfjord and MS Bergensfjord | | Fjord Line |
| Tanker | Ternsund | | Terntank |
| | Fure West | | Furetank |
| Germany | | | |
| Retrofitted ferry | MS Ostfriesland | MS Ostfriesland is the first ferry in Europe using LNG as a dual-fuel propulsion system | LNG-supplier is Bomin Linde LNG GmbH |
| LNG barge | Hummel | To supply cruise ships in ports with electric power | Becker Marine Systems GmbH & Co. KG |
| Container ship | Wes Amelie | World's first container ship to be converted from heavy fuel oil to LNG. Operating on the feeder routes between the North Sea and the Baltic Sea | Wessels Rederei |
| Research vessel | Atair II | Expected to begin operations in 2020 | Federal Maritime and Hydrographic Agency |
| Lithuania | | | |
| Bunkering vessel | | Planned | |
| Norway | | | |
| Ferry | MS Stavangerfjord and MS Bergensfjord | MS Stavangerfjord—the world's first and largest cruise ferry with single LNG engine | Fjord Line |
| | | Four operating ferries serving Lofoten | Torghatten Nord |
| Supply vessel | Viking princess | Serving oilrigs. LNG powered Wärtsilä engines will during Sept 2017 be updated with batteries to a hybrid system. May with this system cut $CO_2$-emissions by 13–18%/year | Eidesvik Offshore |
| Bunkering vessel | Coralius | LNG Bunkering services for Skangas in North Sea and Baltic Sea. Ship to ship bunkering from July 2017. Co-funded by EU under TEN-T and part of the Zero Vision Tool platform | Sirius Shipping and Anthony Veder |
| LNG carrier | Coral energy | Bunkered by Coralius | Anthony Veders |
| | Pioneer Knutsen | Operates for Gasnor | Knutsen OAS Shipping |
| | | | Höeg LNG |
| | | | Golar LNG |

(continued)

**Table 1** (continued)

| Vessel type | Vessel name | Vessel info | Owner |
|---|---|---|---|
| | | | BW Gas |
| | WilForce and WilPride | Intended for international trade | Awilco LNG |
| Poland | | | |
| Bunkering vessel | | Planned for 2022 | |
| Sweden | | | |
| Cruise ship | Viking grace | Bunkered by Seagas | Viking line |
| Ferry | Visborg and Thjelvar | Among the world's most powerful and fastest LNG-fuelled RoPax-ferries, will service the route between Nynäshamn and Visby, Gotland (Sweden). OSK-ShipTech A/S (DK) naval architects | Rederi AB Gotland |
| Bunkering vessel | Seagas | | Sirius shipping |
| Cargo ship | Ternsund | Operates along North Sea-Baltic Sea | Terntank |
| LNG carrier | Stena Blue Sky, Stena Clear Sky, and Stena Crystal Sky | Transportation of LNG around the world | Stena Bulk |
| Small bulk/dry cargo vessel | LNG Sea River | The joint industry project LNG Sea River is an EU TEN-T financed project and collaboration between the Erik Thun Group, Ferus Smit and Lloyd's Register | Erik Thun AB |
| Tanker | Fure West | | Furetank |
| | Bit Viking | First product tanker in the world to use LNG as fuel, in 2011 | Tarbit Shipping |

of the BSR are displayed. With the available and planned infrastructure developments in the BSR, the Blue Corridor Strategy will, at the end of the project, connect LNG terminals, bunkering facilities and industry needs into "Blue Transport Corridors" that include all intermodal possibilities.

Apart from the planned bunkering infrastructure of the maritime TEN-T network (in place by 2025), three (3) small-scale bunkering facilities will be ready in 2018 and an additional six (6) in 2019. It should also be mentioned that one-off bunkering events have already taken place at ports that have no fixed infrastructure, e.g. in Denmark (Port of Esbjerg) and Germany (Port of Lübeck and Port of Bremerhaven). Regarding LNG terminals, three (3) will be ready in 2017 and another three (3) in 2018. In summary, the number of LNG infrastructure facilities will almost double from today's 39 (including Norway) to 77 by 2025. As can be clearly understood with the help of Fig. 2, most of the region could then be covered with LNG by truck. However, more road refueling stations will be needed to cover all possibilities.

**Fig. 1** Map of LNG terminal and bunkering infrastructure in BSR

Norway is by far the country with the most developed LNG infrastructure, at least in the maritime domain. It is also the only country of the region that has natural gas that is liquefied and exported. One large biogas (LBG) plant has been built and one more is planned; these plants will, however, mainly serve the national trucking industry. Additionally, there are two large-scale LNG import terminals in the BSR

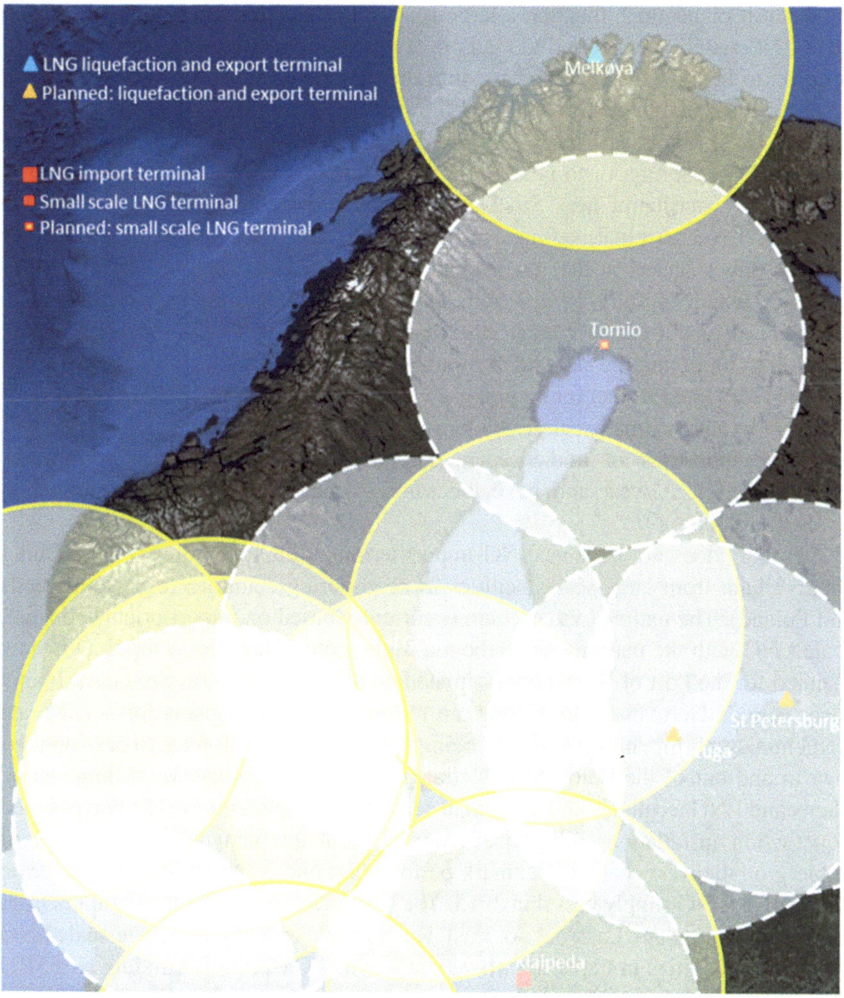


**Fig. 2** Circles with a radius of 350 km from LNG terminals (existing = yellow, planned = white striped). 350 km is the distance an LNG truck can drive before having to refuel

situated in Poland and Lithuania. From these terminals, LNG cannot only be redistributed to bunkering stations, but also re-gasified for the national gas grid. Moreover, both these facilities will have the possibility to sell and distribute LNG in the wider BSR. In Poland, the terminal is owned by the state Gaz-System, and PGNiG was awarded the contract for storage and distribution of LNG. In the near future, Gaz-System is planning to distribute LNG on a spot market basis. Small companies are importing LNG from other countries, mostly EU LNG import terminals. Poland will also have to offer small scale bunkering at the Port of Gdansk/Gdynia and the Port of Swinoujscie/Szczecin.

When it comes to Lithuania, a large-scale FSRU is now functioning, and from there LNG will be transported to the Port of Klaipeda planned bunkering facility, in addition to being re-gasified and fed into the national gas grid. No other maritime LNG facilities are planned, considering the size of the country. Furthermore, Sweden has two LNG terminals from which gas is transported by truck either to ports for small scale bunkering, or to nearby industries that use LNG as their major energy source. In Nynäshamn near Stockholm, Linde has designed and constructed a mid-scale LNG terminal, selling and distributing LNG to various municipalities without direct access to the gas grid in the eastern parts of Sweden. The LNG terminal, which is fully owned and operated by AGA, a subsidiary of Linde, commenced operations in March 2011 and is the first of its kind in the BSR. In addition to supplying the gas grid of Stockholm and industrial facilities, the terminal also provides LNG as fuel for natural gas vehicles in the region. The Port of Gävle is planning to have a small-scale LNG terminal ready by 2018 (Skangas), which will serve the industries of mid-Sweden. The aim for the terminal is to process 500,000 tons LNG/year and have the same storage capacity as the terminal in Lysekil (30,000 $m^3$).

Germany does not have an LNG import terminal due to an already functioning supply chain from large-scale facilities in neighboring countries (e.g. Netherlands and Poland). The national value chain is currently based on import-orientated small scale LNG with the main focus on the maritime sector. However, a terminal is now planned for the Port of Brunsbüttel, situated by the German North Sea coast. In any case, various German (Baltic) ports are currently hastening plans for small scale LNG bunkering for ships. Denmark, being a country that ships have to pass on their way in and out of the Baltic Sea, is strategically planning for several liquefaction plants and LNG terminals and also small scale bunkering facilities. The two planned liquefaction installations will liquefy both natural gas and biogas. The Port of Esbjerg on the west coast of Denmark performed a one-off truck-to-ship bunkering (TTS) of an LNG supply vessel in 2015. The LNG was brought by three trailers from the Port of Rotterdam. Last, but not least, Russia is planning for two large-scale LNG production facilities in the BSR, where state-owned Gazprom is building an FSRU in Kaliningrad, while the private company LNG Gorskaya is building in St Petersburg. The one in Kaliningrad is supposed to later be taken over by a larger plant in Ust-Luga (2023). It is also interesting to note that, from the previously mentioned Go LNG questionnaire, it was further found that the countries of the BSR are related to a growing number of other LNG maritime infrastructures (Table 1).

## 2.3 LNG Research, Education and Consulting

In order to facilitate fast and safe deployment of LNG in the shipping sector, not only must regulations and respective responses from the industry be on track, but also sufficient education and capacity building of people that in various and different ways must be operating LNG. Mapping the available education and current capacity building LNG portfolio in the BSR may help pinpoint areas of common interest and cooperation, as well as identify any gaps and weak points that must be addressed in the future. Most universities in the BSR offer education on energy, sustainability, logistics, business and transport. Some are of quite broad scope, while others are more specialized, for example focusing heavily towards shipping. Apart from these universities, there is a plethora of institutes that perform research in the field of maritime energy as well as consultancy companies offering both training as well as consultancy.

In the Go LNG project, a BSR LNG Competence Centre (Dalaklis et al. 2017b) is to be set up, which will also be described in the Blue Corridor Strategy. Given the relative political and economic stability of the region, cooperation between the countries has prevailed for decades and networks and consortiums have thrived. Some of these countries might be small in size, but together they create an extended pool of competence. Creating a competence centre that offers know-how on all parts and subcomponents of the LNG value chain is, therefore, cost-efficient for the whole BSR and can help those regions that are not well-developed in the field of LNG (by transferring to them the existing experience within the BSR). A first and important step toward the operationalization of the competence centre was taken in 2016, with the gathering of information that will lead to an inventory list of existing knowledge and training modules already available. The Go LNG project will catalogue and promote cooperation of institutions that are capable of providing LNG-related education and training activities, improve associated infrastructure and even create necessary models and scenarios to run optimisation exercises. For the time being, the programme of the new centre will focus on the following main categories:—LNG storage and bunkering—Cryogenic and pressure technologies—LNG as fuel for ships—LNG project management, business modelling—LNG economy purchase, contract management—LNG for road, rail and inland water transport, and industry—LNG standards and regulations, legal environment. The courses will be developed for three distinct levels/target groups: LNG value chain training, maritime basic STCW/IGF and maritime advanced STCW/IGF.

## 3 An Extended LNG Value Chain and the Future Blue Transport Corridor in the Baltic Sea Region

The Baltic Sea Region is a maritime centre with major international companies present. Regarding the deployment of LNG in the region, for the maritime industry the question is no longer if LNG could be a solution to adhering to international regulations, but rather how much of the market could be replaced by LNG and how it could best be marketed as the preferred fuel/source of energy. For the specific market to be fully exploited, and as a consequence become more financially viable, the number of sectors interacting with it needs to increase. Ideally, once the maritime

**Table 2** Gathering information on LNG related stakeholders of BSR countries

| LNG field | Country | Stakeholder | Specifics | Cooperation |
|---|---|---|---|---|
| Bunkering | | | | |
| Storage | | | | |
| Supplier | | | | |
| Distribution | | | | |
| Ships | | | | |
| End-user technologies | | | | |
| Shipbuilding | | | | |
| Shipyards | | | | |
| Consulting (risk assessment, environmental assessment etc.) | | | | |
| Classification societies | | | | |
| LNG as Power supply | | | | |
| LNG in other modalities | | | | |

industry has prepared the region with a functioning distribution chain, other industries will switch to using LNG as a principal source of energy, widening the scope of LNG end-users. One area of opportunity being explored is the spread of LNG to remote regions; due to its properties, the volumetric energy density of LNG is 2.4 times greater than that of compressed natural gas (CNG) or 1.6% that of diesel fuel, allowing larger quantities of LNG to be transported in comparison with more conventional fossil fuels.

A major part of the Blue Corridor Strategy will be devoted to fitting other modalities into the EU core port network (Baltic) as well as mapping other major industry markets, along the same network, that LNG is suitable for. As an output, the Go LNG project will perform a study on the emerging integrated LNG value chain of the region. The study will reach out to stakeholders and analyze and discuss how to implement integration of LNG business models to increase LNG distribution efficiency, demand and economy (for example by adding users and integrating technologies). Moreover, it will showcase an integrated value chain by providing business plans of existing examples. In order to perform these analyses, information on main LNG players of the region (see Table 2), value of business models, internationalization of best practices, and subsidy systems will be used. The LNG value chain study will further present the prognosis for LNG technologies and market growth in the BSR.

The SECA regulations have unquestionably generated additional costs to end-users of maritime transport and investments for ship owners and other related industries in the BSR (Bachér and Albrecht 2013), although there are certain potential socioeconomic benefits (Ballini and Bozzo 2015). However, these regulations will soon apply to other regions of the world, requiring from them on-going and functioning examples and solutions. Another important part of the value chain study will, therefore, be to transfer the challenges that have been faced by the maritime industry in the BSR into opportunities, that is, to show how additional revenue generated by complying with environmental regulations can become a source for development and innovation, subsequently resulting in new business activities, new jobs and economic growth in Europe. In order for this to be possible, the Go LNG project needs to gather information from **all** Baltic Sea countries, including efforts made to push LNG, for example the Norwegian NOx Fund (NHO 2017) and the Clean Shipping Index (CSI 2017). Both these efforts only reward companies that go beyond existing regulations (emissions).

By understanding each country's specific strategy regarding LNG development and deployment and then coupling this information with the planned TEN–T corridors of the EU, including rail and road transport, and the findings of the Go LNG extended and integrated LNG value chain (stakeholder's perspective), proposals for Blue Corridors in the BSR will be made. It is important to keep in mind that a corridor must be useful for its users, that is, it is necessary to create a transport corridor that includes all requests for well-functioning infrastructure and superstructure. These corridors will be highlighted on an online map that will build on Fig. 1, with the following additional layers:

- Available and planned LNG facilities (including core ports and bunkering barges)
- LNG refuelling stations (including TEN-T)
- LNG vessels, trucks, busses, trains
- TEN-T corridors
- National highways
- National railways
- National gas pipelines
- Cargo flows
- Industries and refineries that use LNG
- Major industries that could be potential LNG customers
- The green layer—specify LBG
- LNG in inland waterways

## 4 Discussion

The introduction of LNG is considered as a regulation-driven demand that has arisen to comply with the environmental regulations enforced by the Sulphur Emission Control Area (SECA), as introduced by IMO's MARPOL Annex VI. For the maritime industry in the BSR, the question is no longer if LNG could be a solution to adhering to international regulations, but rather how much of the market will be replaced by LNG, and, importantly, how to LNG can best be introduced as the preferred fuel/source of energy. Moreover, it needs to be noted that while the use of LNG in shipping will help to reach environmental targets of sulphur, nitrogen and particulate matters emissions, with regard to greenhouse gas (GHG) targets, the use of LNG will have to be complemented in the long run by more energy efficient engines and vessels. Therefore, one important question that remains to be addressed is how long LNG will be the best available option. Given the policies developed to keep climate change in control (e.g. Paris Agreement 2015), it is conceivable that LNG is a bridge between the world of fossil fuels and the world of renewables.

Today, there is a plethora of options of alternative fuels for future consideration: from hydrogen to methane, or even biofuel. Although some of these alternatives provide better environmental performance than LNG, unfortunately, they have limited bunker availability and the purchase-installation of the related equipment/ systems can be quite expensive. LNG is therefore a very significant stepping stone in the effort of the shipping industry to become more environmentally friendly. Promoting the use of LNG results in not only significant environmental benefits, but also substantial economic savings. The technical innovation in the production of LNG has already helped to position the specific type of fuel as one of the least expensive transportable ones, but further research activities are always welcomed.

Within the Go LNG project, the strategy that will be put forward will point toward the continued development and possibilities of LNG in the BSR. It will learn from Norwegian LNG developments and seek to acknowledge each country's specific challenges coupled with the introduction of LNG, be it economic, environmental or of a political nature. In addition, the strategy will include methods on how to ensure the sustainability of LNG infrastructure and economic growth in BSR coastal communities, for example, by demonstrating the feasibility of LBG production and distribution in a municipality. One interesting option is the mixing of natural gas with biogas before liquefaction. By doing that, the fossil derived GHG of LNG can be reduced (Sperling 2017), provided methane emissions are minimised (Brynolf et al. 2014). As the availability of biogas production increases, the dependency of the region on importing natural gas would decrease. Another way to reduce the amount of fossil GHG is to use batteries for powering ships—either complete battery driven ships or hybrid ships, using both LNG and battery. Depending on the source of electricity used to charge the batteries, these ships could be the future of short-sea-shipping.

## References

Bachér, H., & Albrecht, P. (2013). *Evaluating the costs arising from new maritime environmental regulations*. Finland: Trafi Publications. ISBN 978-952-5893-89-2.

Ballini, F., & Bozzo, R. (2015). Air pollution from ships in ports: The socio-economic benefit of cold-ironing technology. *Research in Transportation Business & Management, 17*, 92–98.

Bengtsson, S. K., Fridell, E., & Andersson, K. (2014). Fuels for short sea shipping: A comparative assessment with focus on environmental impact. *Proceedings of the Institution of Mechanical Engineers, Part M: Journal of Engineering for the Maritime Environment, 228*, 4–54.

Brynolf, S., Fridell, E., & Andersson, K. (2014). Environmental assessment of marine fuels: liquefied natural gas, liquefied biogas, methanol and bio-methanol. *Journal of Cleaner Production, 74*, 86–95.

CSI. (2017). Clean Shipping Index. https://www.cleanshippingindex.com/. Accessed March 2017.

Dalaklis, D., Madjidian, J. A., Olcer, A., Ballini, F., & Kitada, M. (2017a). Liquefied natural gas (LNG) as a marine fuel: Optimising the associated infrastructure in the Baltic Sea region. In

*Kyoto-Japan, International Association of Maritime Economists (IAME) 2017 Conference*, doi: https://doi.org/10.13140/RG.2.2.12247.98724.

Dalaklis, D., Ölcer, A., Madjidian, J. A., Ballini, F., & Kitada, M. (2017b). Bridging the LNG Knowledge Gap in the Baltic Sea region: The Go LNG initiative and the establishment of a Competence Centre. In *Proceedings of 11th International Technology, Education and Development Conference (INTED)* (pp. 1417–1425). Valencia, Spain, 6–8 March 2017, ISBN: 978-84-617-8491-2.

Deniz, C., & Zincir, B. (2016). Environmental and economical assessment of alternative marine fuels. *Journal of Cleaner Production, 113*, 438–449.

DNV GL. (2016). LNG fuelled vessels, Ship list – Vessels in operation and vessels on order. https://www.dnvgl.com/maritime/lng/ships.html. Accessed January 2017.

Energy Information Agency (EIA). Short-term energy outlook. http://www.eia.gov/forecasts/steo/report/global_oil.cfm. Accessed February 2016.

European Commission. (2016). Communication from the Commission to the European Parliament, the Council, the European Economic and Social Committee and the Committee of the Regions on an EU strategy for liquefied natural gas and gas storage. COM(2016) 49 final.

European Parliament. (2014). Directive 2014/94/EU. http://eur-lex.europa.eu/legal-content/EN/TXT/?uri=CELEX:32014L0094. Accessed January 2017

Gerlitz, L., & Paulauskas, V. (2015). *Emerging LNG value chain in the South Baltic Region.* MarTech LNG project Final report.

IMO. (2010). *MARPOL Annex VI - Prevention of Air Pollution from Ships.* London, UK: International Maritime Organization.

IMO. (2013). *Prevention of Air Pollution from Ships.* London, UK: International Maritime Organization.

ISO. (2017). ISO 20519:2017 Ships and marine technology – Specification for bunkering of liquefied natural gas fuelled vessels. https://www.iso.org/standard/68227.html. Accessed March 2017.

Johansson, L., & Jalkanen, J.-P. (2016). Emissions from Baltic Sea shipping in 2015. Baltic Marine Environment Protection Commission (Helcom).

NHO. (2017). The Business Sector's NOx Fund. https://www.nho.no/Prosjekter-og-programmer/NOx-fondet/The-NOx-fund/. Accessed March 2017.

Paris Agreement. (2015). http://unfccc.int/paris_agreement/items/9485.php. Accessed February 2017.

Sperling, K. (2017, May). How does a pioneer community energy project succeed in practice? The case of Samsø renewable energy island. Renewable and Sustainable Energy Reviews, 71, 884–897. https://doi.org/10.1016/j.rser.2016.12.116.

Stankiewicz, M., & Vlasov, N. (Eds.), (2009). Ensuring safe shipping in the Baltic. Baltic Marine Environment Protection Commission (Helcom).

The Economist. (2012). A liquid market. http://www.economist.com/node/21558456. Accessed January 12, 2017.

Thomson, H., Corbett, J. J., & Winebrake, J. J. (2015). Natural gas as a marine fuel. *Energy Policy, 87*, 153–167.

Wang, S., & Notteboom, T. (2015). The role of port authorities in the development of LNG bunkering facilities in North European ports. *WMU Journal of Maritime Affairs, 14*, 61–92.

## *Further Reading*

United Nations. (2017). Sustainable Development Goal 7: https://sustainabledevelopment.un.org/sdg7. Accessed January 2017.

# LNG Fueled Barge for Cold Ironing: Feasibility Study for the Emission Abatement in the Port of Genoa

**Michele Laviola, Massimo Figari, Marco Altosole, and Stefano Savio**

## 1 Introduction

On a global scale, by 2030 the United Nations has planned goals dealing with energy efficiency and cleaner fossil-fuel management, promoting investment in both infrastructure and technology for a more eco-friendly environment. This new scenario will involve maritime activities and international shipping as confirmed by Annex VI of IMO's pollution prevention treaty (MARPOL).

The strong interest in cutting emissions from maritime shipping is supported by a regulatory framework consisting of international regulations, European directives and national laws regarding the emissions thresholds from ships in port and possible solutions to comply with this need (Coraddu et al. 2014). The current regulations relating to such issues are:

- EU Directive 2005/33/EC, requiring a maximum sulfur content for marine fuels of 0.1% and 1.5% by mass in European ports and in exclusive economic zones respectively;
- EU Directive 2014/94/EU, requiring the installation of LNG refueling stations placed in the core TEN-T ports by 2025;
- Annex VI of the MARPOL, establishing, for pollutant emission from marine fuels, sulfur limits and NOx content up to threshold values;
- EU Directive 2006/339/EC, involving the installation of electric power stations in the port for supplying energy to ships while at berth;
- Masterplan issued by the Italian Government for promoting the use of natural gas as fuel, thus reducing the environmental impact of transport by sea and road.

The adoption of a system for supplying electrical power to ships moored in port, which allows diesel-generators to be switched off, could be an effective solution to

M. Laviola (✉) · M. Figari · M. Altosole · S. Savio
University of Genoa, Genova, Italy



A. I. Ölçer et al. (eds.), *Trends and Challenges in Maritime Energy Management*, WMU Studies in Maritime Affairs 6, https://doi.org/10.1007/978-3-319-74576-3_28

comply with the above-cited maritime rules. On the other hand, the infrastructure and upgrade technologies for supplying modern and sustainable energy services will affect the technical requirements of the ships and the related port activities. Specifically for the port of Genoa (the location considered as a study case in the present work), ships' electrical load feeding by the existing grid appears, to date, an unfeasible solution, due to the high power demand and the actual rating of the electrical distribution infrastructure in the interested area. For this reason, suitable solutions to satisfy ships' energy demands in port could be the ones proposed by the authors of this work, and characterized by the adoption of power supply floating units equipped with LNG-fueled generators.

## 2 LNG Storage for the Study Case

The LNG supply infrastructure in Italian ports is under discussion due to the EU Directives cited above. LNG refueling will be available only in a few ports that have to be identified by the authorities. For the present study the authors considered two reasonable alternatives for the port of Genoa:

- a tank directly installed in the port of Genoa: some feasibility studies about the size and layout of LNG storage tanks have already been submitted (Capocaccia 2015);
- a tank located in the port of La Spezia (50 nautical miles from Genoa) where a regasification plant has been in service for a long time (Fig. 1), using a shuttle tanker to carry the LNG from La Spezia to Genoa. Even for this case, a feasibility study has already been carried out (Moizo 2016).

**Fig. 1** Regasification plant in Panigaglia—La Spezia

## 3 Electric Energy Demand in the Old Port of Genoa

A cleaner source of electric energy to feed ships in port should be designed analyzing the energy demand in the considered area. The first step of this study dealt with the assessment of the electric energy demand of the ships at berth. Starting from collected data regarding the time spent at berth by each ship, an analysis of the calls in the port of Genoa referring to the year 2012 has been carried out.

Ships energy demand at berth has been estimated considering different algorithms according to several vessel types, namely ferries, cruise ships and containerships. For ferries, a relationship between energy demand and gross tonnage has been considered, while for cruise ships, the relationship has been based on the onboard maximum installed power. In the case of containerships, the energy demand has been evaluated taking into account the number of TEUs carried onboard. The total amount of electric energy required per year by all the ships considered has been estimated at 54 GWh, while it is equal to about 48 GWh if the energy required during the time for connection and disconnection operations (Tetra Tech, Inc. 2007) is deducted. As can be noticed from Fig. 2, the electric energy demand from ferries is the most significant, accounting for 68% of the total amount.

Figure 3 depicts the energy demand for the three ship types during the year 2012, subdivided by calendar months, while in Fig. 4 the average hourly energy required by ships is presented.

Figure 5 shows the distribution of time at berth versus the electric active power demand. As can be seen, most of the energy demand is related to electrical loads with power values below 10–15 MW, and if the histogram is analyzed numerically, it turns out that around 85% of the total energy required by the ships deals with power requests equal to or below 10 MW, the average being about 6 MW.

## 4 Power Supply Technical Solutions

If the goal is to satisfy the energy demand of all the ships calling the old port of Genoa, the first option identified is represented by a fixed power supply barge moored in port, connected to a dedicated network for energy distribution at each

Fig. 2 Electric energy demand in port for several ship types—comparison

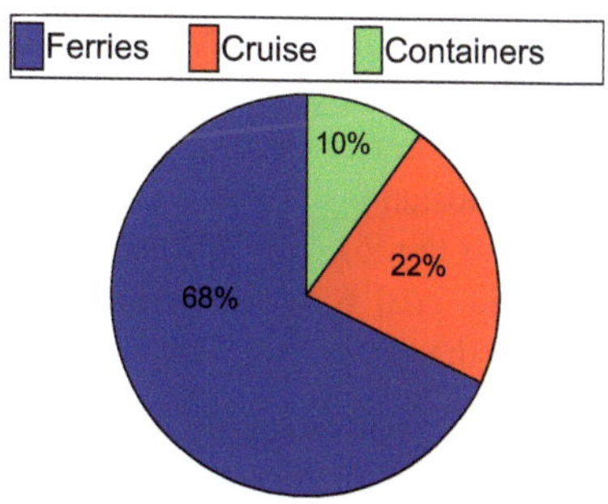

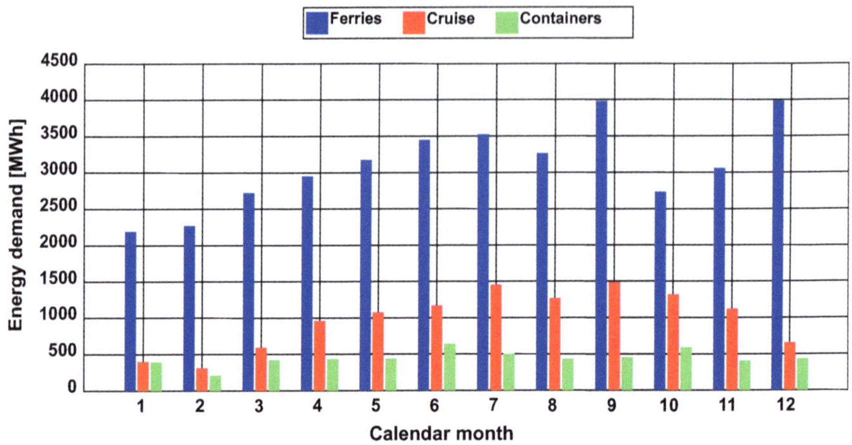


**Fig. 3** Monthly electrical energy demand

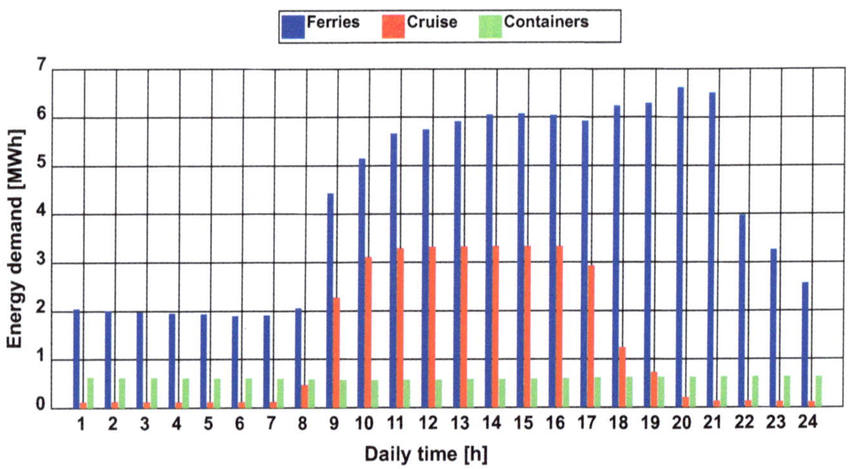


**Fig. 4** Average hourly electrical energy demand

berth. In such a way, it would be possible to create a shore-to-ship electrical supply system for distributing electric energy from the main energy source (power supply barge) to the final users. Just as an example, a possible layout of the electrical grid is represented in Fig. 6, where the location of the power supply barge (PSB) is marked with a white rectangle. Electrical power produced by the barge is conveyed to four main substations (green labels in Fig. 6) and then distributed to 14 terminals (red labels in Fig. 6).

If the goal is to satisfy the energy demand for only a limited number of ships calling the old port of Genoa, two technical solutions are identified:

- a towed power supply barge;
- a self-propelled power supply vessel.

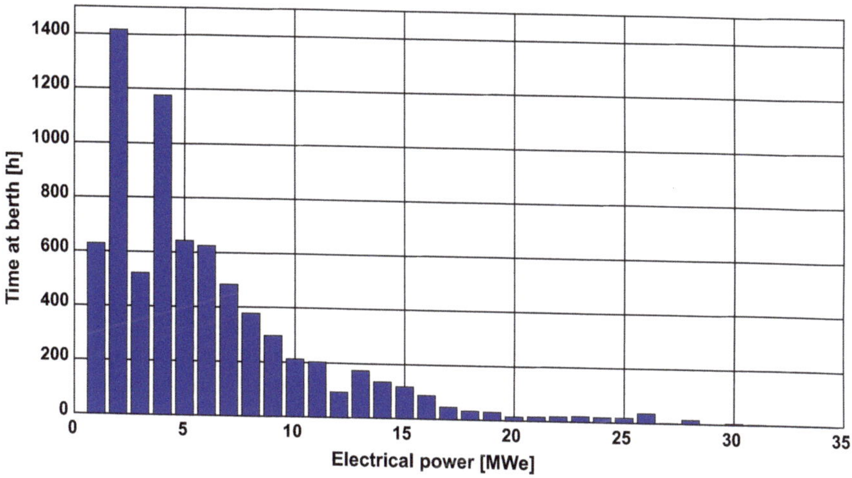


**Fig. 5** Distribution of time at berth vs electric power demand

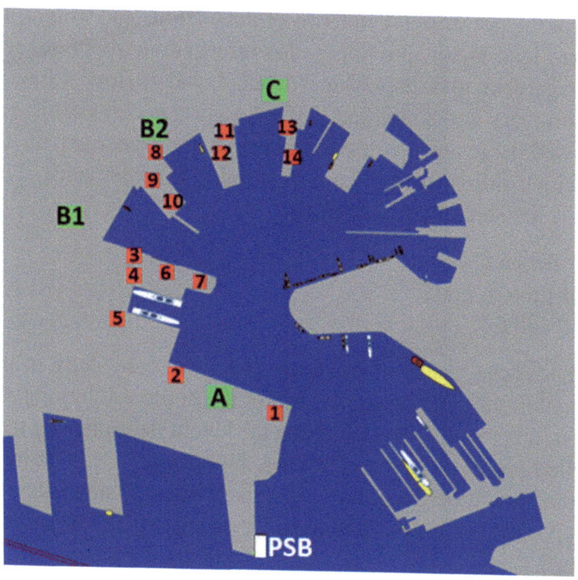


**Fig. 6** Port electrical infrastructure. Proposed layout for the old Port of Genoa

Regarding these two options, in the first stage of the study, it could be argued that the towed barge requires less initial investment compared to the self-propelled vessel, but it requires the assistance of one or more harbor tugs.

In particular, the self-propelled vessel could be a valid solution in case the LNG storage tank is located in the port of La Spezia.

Obviously, these two solutions obviate the need to install an electrical grid ashore because the electric energy produced on-board the power supply barge/vessel can directly supply one or two berthed ships.

Starting from the input data and the considerations previously described and taking into account the three options described in this chapter, a design procedure aimed at exploring feasible solutions has been carried out by the authors. The procedure and the results are detailed in the following section.

# 5 Power Plant Layout Identification

In order to carry out the study, a dedicated MATLAB® code has been developed in-house by the authors. As far as LNG fueled generators are concerned (Altosole et al. 2017), a database has been implemented and populated with the main features of the generation units produced to date by three different manufacturers (named C, R,W) in the considered power range. Figure 7 shows a simplified scheme of input data and output results of the code. For each of the three options considered (fixed barge, towed barge and self-propelled vessel), given a configuration in terms of number and power rating of the generation units, the code provides the main characteristics of the power plant, in terms of dimensions and weight, fuel consumption and average load factor, providing also information about generators contemporaneity management to satisfy ships demand and to lower fuel consumption. Moreover, preliminary and simplified cost estimation is computed.

As far as towed barge and self-propelled vessel are concerned, the code automatically selects which ship to feed in case of simultaneous demands. To this aim, the best candidate is the ship characterized by:

- high power demand;
- long layover time in port;
- frequent port calls.

Table 1 provides the results of the analysis for the case study of the old port of Genoa. In particular, the main characteristics of the optimal solution for each floating unit are reported. It is worth mentioning that the layout definition for the towed barge and self-propelled vessel has been identified, introducing a constraint (12 MW) on the maximum electric power generation. The fixed barge, instead, has the capacity to supply all the possible users simultaneously, according to the results previously reported in Fig. 5.

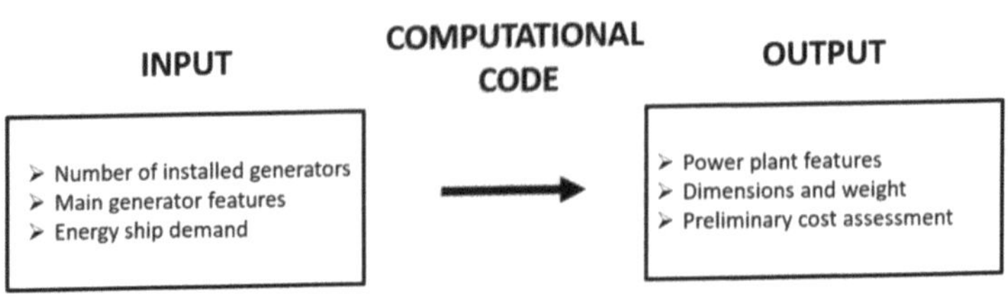


**Fig. 7** Code input data/output results

**Table 1** Main characteristics of the possible solutions

| | Towed barge | Self-propelled vessel | Fixed barge |
|---|---|---|---|
| Generators # and power [MW] | 6 × 2.1 | 6 × 2.1 | 9 × 4.2 |
| Energy [GWh/year] | 9.3 | 9.3 | 47.9 |
| Efficiency [%] | 46.6 | 46.6 | 45.9 |
| Load factor [–] | 0.80 | 0.80 | 0.66 |
| LNG capacity [m$^3$] | 200 | 200 | 500 |
| LNG consumption [tons/year] | 1528 | 1616 | 8307 |
| Displacement [t] | 2435 | 2477 | 5000 |
| Length [m] | 48 | 58 | 64 |
| Beam [m] | 12.5 | 12.5 | 18.2 |
| Draught [m] | 4.3 | 4.3 | 4.4 |

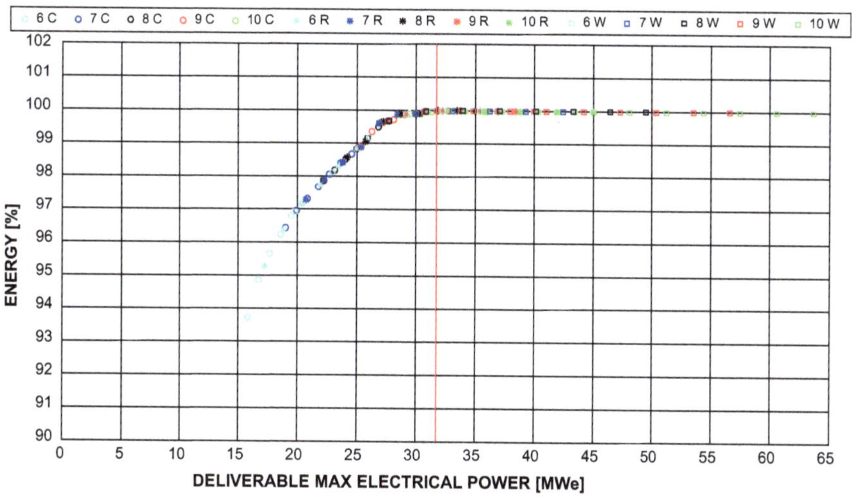


**Fig. 8** Percent energy coverage vs installed electrical power—fixed barge

In the following, a deeper analysis of the detailed results is presented just for the fixed barge option, as it is the technical solution that can satisfy the energy demand of all the ships calling the considered port area. Figure 8 shows the achievable percent energy coverage as a function of the rated electrical power available for the different layouts investigated. Each dot represents one of the generation configurations, which range from 6 to 10 units of different manufacturers and different rated powers. Considering a maximum power request of about 31.5 MW (as reported in Fig. 5), all the configurations (dots) to the right of the vertical red line in the current figure allow 100% coverage of energy demand at the port.

Regarding the barge size and weight, as can be noticed from Figs. 9 and 10 respectively, taking into account a power request of 31.5 MW, the length ranges from about 50 to 80 m, while the displacement ranges from 4000 to about 7000 tons, depending on the layout selection.

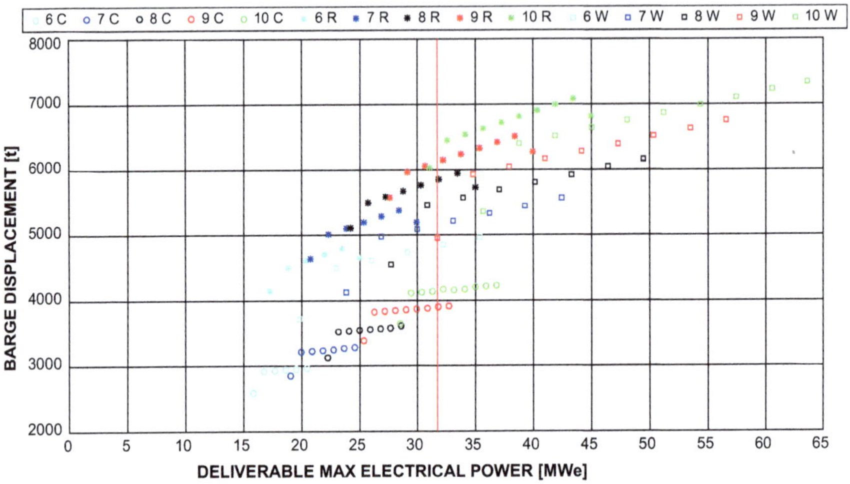


**Fig. 9** Length of barge vs installed electrical power—fixed barge

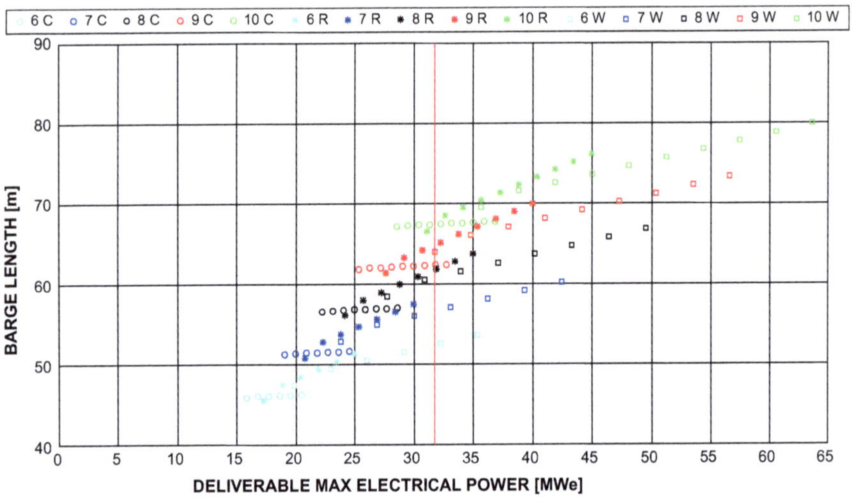


**Fig. 10** Deadweight of barge vs installed electrical power—fixed barge

The estimation of the barge displacement $W$ has been carried out considering the sum of four items, as given in Eq. (1):

$$W = W_H + W_O + W_{GG} + W_T. \tag{1}$$

where:

- the weight of hull $W_H$ and the weight of outfitting $W_O$ has been derived from available data of similar units;

- the weight of gas-fueled generators $W_{GG}$ and LNG on-board tank $W_T$ has been derived from the product catalogue.

The evaluation of the main dimensions of the barge has been carried out considering the footprint of the generators and an additional clearance as found in Potapov (2012).

As far as the preliminary cost assessment is concerned, a first estimation of the investment cost has been performed, neglecting the cost figure due to the current discount rate. In particular, the investment cost has been computed considering the sum of four items, as reported in Eq. (2):

$$CapEx = C_H + C_O + C_{GG} + C_{T\&E}. \tag{2}$$

where:

- $C_H$ is the cost of hull, estimated through the correlation found in Vernengo and Rizzuto (2014);
- $C_O$ is the cost of the outfitting, estimated through the technical information collected in Peckham (2013);
- $C_{GG}$ is the cost of the gas-fueled electric generator, estimated using the information found in Masaki et al. (2014);
- $C_{T\ \&\ E}$ is the cost of LNG tank and gas processing equipment, estimated consulting technical information in HEC and CCDTT (2013).

The specific cost of energy $EC$ has been estimated through Eq. (3):

$$EC = \frac{\frac{CapEx}{LT} + OpEx}{E}. \tag{3}$$

where:

- $LT$ represents the lifetime of the power barge considered for economic investigations: 30 years is an appropriate value (Engblom 2014);
- $E$ is the energy consumed per year by ships in port;
- $OpEx$ represents the yearly operational costs, estimated by relationship in Eq. (4):

$$OpEx = O_{fuel} + O_{maint} + O_{lo} + O_{crew}. \tag{4}$$

The cost of fuel ($O_{fuel}$) has been obtained considering a unit value of 307 €/ton, while the cost related to the maintenance of generators ($O_{maint}$), lubricating oil ( $O_{lo}$) and crew ($O_{crew}$) have been found in (V.T.P. Engineering n.d.).

The specific cost of energy is reported in Fig. 11; taking into account a power demand of 31.5 MW and continuous operations, it shows that a cost of about 0.11 € per kWh could be achieved selecting a fixed barge equipped with nine generators from manufacturer W and a rated power of the single unit of 4.2 MW.

It is not easy to predict how many ships could be equipped with an on-board system able to receive energy from the shore connection, so the variation of the

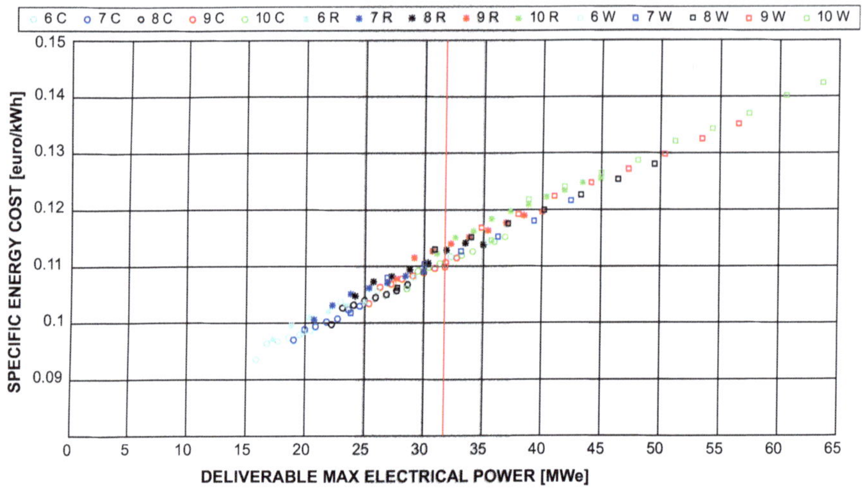


**Fig. 11** Specific energy cost vs installed electrical power—fixed barge

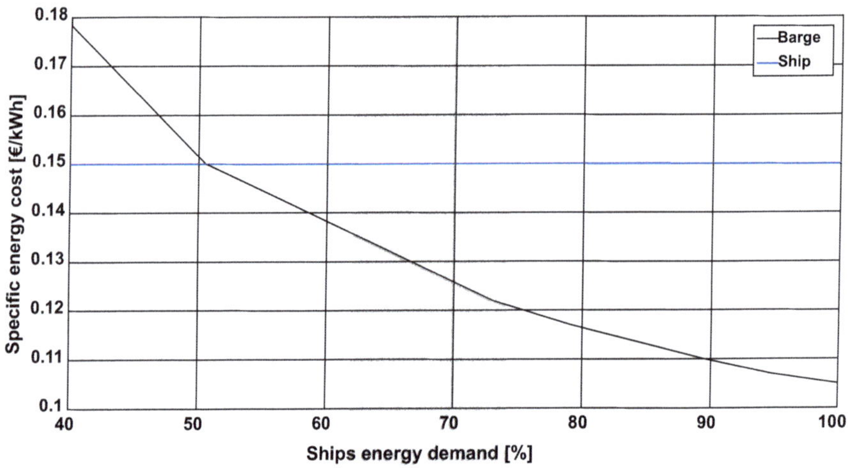


**Fig. 12** Fixed barge specific energy cost vs percent ships energy demand

specific cost of energy provided by the fixed barge, as a function of the electric energy demand from ships, has been analyzed, and the results are shown in Fig. 12. It proves that if 50% (or more) of the total energy required by the ships is received from the barge, the proposed solution could be an effective option. An average specific cost of 0.15 €/kWh has been considered for electric energy produced on-board by ships (Levander and Sipilä 2008).

To further cut the specific cost of energy produced by the barge, it would be necessary to increase the amount of energy delivered by such a system. Table 2

**Table 2** Specific cost of energy [€/kWh] vs ships and ashore activities energy request

| | Energy yearly delivered to ships [GWh] | | | | | | |
|---|---|---|---|---|---|---|---|
| $\alpha_{LF}$ | 17.1 | 24.2 | 35.0 | 37.9 | 42.9 | 45.4 | 47.9 |
| 0 | 0.190 | 0.150 | 0.122 | 0.117 | 0.110 | 0.107 | 0.105 |
| | (0) | (0) | (0) | (0) | (0) | (0) | (0) |
| 0.2 | 0.100 | 0.096 | 0.091 | 0.090 | 0.089 | 0.088 | 0.087 |
| | (35.5) | (34.1) | (31.9) | (31.3) | (30.4) | (29.9) | (29.4) |
| 0.4 | 0.084 | 0.083 | 0.081 | 0.081 | 0.080 | 0.080 | 0.080 |
| | (71) | (68.1) | (63.8) | (62.7) | (60.7) | (59.7) | (58.8) |
| 0.6 | 0.077 | 0.077 | 0.076 | 0.076 | 0.076 | 0.076 | 0.076 |
| | (106.5) | (102.2) | (95.7) | (94) | (91.1) | (89.7) | (88.2) |
| 0.8 | 0.0739 | 0.0737 | 0.0735 | 0.0734 | 0.0733 | 0.0733 | 0.0732 |
| | (142) | (136.2) | (127.7) | (125.3) | (121.4) | (119.5) | (117.6) |
| 1 | 0.0713 | 0.0713 | 0.0713 | 0.0713 | 0.0713 | 0.0713 | 0.0713 |
| | (177.4) | (170.3) | (159.6) | (156.6) | (151.8) | (149.4) | (146.9) |

provides the specific cost of energy produced by the identified configuration of the fixed barge to satisfy different requests from both the ships and other ashore industrial activities in port. To this aim the land factor $\alpha_{LF}$ of Table 2 allows the identification of the latter amount of energy delivered by the barge generators ($\alpha_{LF}$ times the amount of energy still remaining, once the ships demand has been satisfied, taking into account the available gross total based on the installed electrical power). In such a context, the table also reports, for the different scenarios, the yearly value of the energy delivered to the ashore industrial activities [GWh] within brackets.

The results of the analyzed case studies show that, if full power is continuously delivered all year, the minimum cost is achieved, until a value of about 0.07 € per kWh. It is fundamental to note that this value cost of energy is very near to the cost of energy provided by the electrical grid in Italy.

## 6 Reduction of Air Pollutant Emission

The calculation of air pollutant emissions by ships at berth in the old port of Genoa has been performed through the emission factor as reported in the literature (Yorke engineering 2007). Table 3 reports the amount of the main pollutants for the different scenarios investigated in this study (all the ships supplied by the fixed barge, or the maximum ships that can be supplied by the towed/self-propelled barges), as well as for the actual situation where berthed ships use MGO as fuel for diesel generators.

It is to be noticed that a lower reduction is achieved by means of the towed barge and self-propelled vessels because of the lower electrical energy supplied by gas-fueled generators.

**Table 3** Air pollutant emission for different scenarios [tons/year]

| | Using MGO as fuel | Cold ironing (fixed barge) | Cold ironing (barge towed or self-propelled vessel) |
|---|---|---|---|
| NOx | 829 | 257 | 718 |
| PM 10 | 25 | 3.0 | 21 |
| PM 2.5 | 13.7 | 1.6 | 11.4 |
| SOx | 122 | 15 | 101 |

## 7 Conclusion

The current chapter presents technical solutions to feed ships' electric systems in port by means of LNG supplied power generation systems. Three different solutions have been identified and analyzed for a case study related to the old port of Genoa. From the study, it turns out that a generation plant installed on a fixed barge could be a valid option for supplying energy to all the ships at berth. Taking into account that in the considered port some issues do not allow ships' electrical equipment to be directly fed by the national grid, specifically the local high power demand and the actual rating of the electrical distribution infrastructure, the fixed barge option could be a good solution to satisfy the overall ships' energy request, lowering the pollutant emissions in port. An alternative could be represented by the adoption of the towed barge and/or self-propelled vessel, which require a lower capital investment and are characterized by smaller dimensions and weight. On one hand, due to their size, such latter solutions can offer flexibility of movement in port and allow ships to be fed by approaching each of them. On the other hand, they can supply energy just to a reduced number of ships and do not achieve the same performances as the fixed barge solution, in terms of air pollutant reduction in port. In any case, whatever solution is adopted, ashore connection of ships for energy purposes through solutions fueled by LNG seems an effective alternative to make the port and the surrounding environment more ecofriendly. The introduction of this fuel in the maritime sector could also represent a pull factor for all ship-owners interested in switching to greener energy for moving cargo and passengers in this new era where it seems that environmental aspects, as stated by MARPOL Annex VI, will be the benchmark for the maritime sector.

## References

Altosole, M., Benvenuto, G., Campora, U., Laviola, M., & Zaccone, R. (2017). Simulation and performance comparison between diesel and natural gas engines for marine applications. *Proceedings of the Institution of Mechanical Engineers Part M: Journal of Engineering for the Maritime Environment, 231*(2), 690–704.

Capocaccia, F. (2015). *Dal Green Shipping Summit il punto su LNG*. Genova: TTM.

Coraddu, A., Figari, M., & Savio, S. (2014). Numerical investigation on ship energy efficiency by Monte Carlo simulation. *Proceedings of the Institution of Mechanical Engineers Part M: Journal of Engineering for the Maritime Environment, 228*(3), 220–234.
Engblom, K. (2014). Features and parameters of various power plant technologies. *Wärtsilä Technical Journal.*
HEC and CCDTT. (2013). LNG as Ship Fuel, Effects on ship design, operations and supporting infrastructure.
Levander, O., & Sipilä, T. (2008). LNG auxiliary power in port for container vessels. *Wärtsilä Technical Journal, 2*, 42–47.
Masaki, A., Hiroyuki, K., Tetsugo, F., Shota, O., & Kazuyoshi, H. (2014). Economic analysis of trans-ocean LNG fueled. *Journal of Marine Science and Technology, 19*, 470–478. https://doi.org/10.1007/s00773-014-0262-5.
Moizo, A. (2016). Bunkeraggio LNG a Panigaglia, assegnato il bando, Ship2shore.
Peckham, J. (2013). Becker Touts 'World's First Hybrid LNG Barge' for Cold-Ironing Cruise Ships. Diesel fuel News.
Potapov, V. (2012). *DF integrated propulsion system and LNG pack – Technical developments, benefits and operational experience.* Saint Petersburg: Wartsila. http://www.korabel.ru/filemanager/OTHER/0/0/7.pdf.
Tetra Tech Inc. (2007). Draft use of shore-side power for ocean-going vessels white paper.
V.T.P. Engineering COLD IRONING. Una nuova soluzione eco-compatibile per l'alimentazione delle navi in porto.
Vernengo, G., & Rizzuto, E. (2014). Ship synthesis model for the preliminary design of a fleet of compressed natural gas carriers. *Ocean Engineering, 89*, 189–199.
Yorke engineering. (2007). Port of San Diego: Cold ironing study.

# Decision Framework for Shipowners to Comply with Air Emission Reduction Measures: A Case Study of Methanol as a Fuel

**Aditya Srivastava, Aykut I. Ölçer, and Fabio Ballini**

## 1 Introduction

The maritime transport carries a large portion of global trade which is a complex and dynamic sector. The maritime logistic system provides benefits to the customer but causes an adverse impact on the environment. Air pollution caused by maritime transport affects human health, climate, flora fauna, and oceans (Andersson et al. 2016). These effects are not reflected in transport prices. Hence, a comprehensive based approach is needed to achieve environmental, health and climate benefits while supporting the continuous positive growth of the maritime sector and profitability to ship-owners (OECD Council Working Party on Shipbuilding 2010).

When it comes to compliance with regulations, decision makers are facing a finite number of technological solutions, which are overlapping (Han 2010). Ship-owners are suffering through a substantial economic burden, owing to fluctuation in fuel prices. Ship-owners should make compliant strategies keeping in mind externalities caused by emissions to enable a sustainable environment. The idea of developing a decision framework for shipowners to reach the sustainability goals of the United Nations (UN) is the main driving force behind the paper (Sustainable Development goals 2015). Rozmarynowska (2012) shows that the new regulations for $SO_x$ and $NO_x$ emissions will cause a modal backshift and will, consequently, increase the costs of freight. Andersson et al. (2016) argue the need for alternative measures with new mandatory $SO_x$, $NO_x$ tier III and Monitoring Reporting and Verification (MRV) regulations coming in the future. The ship-owner should make use of available

A. Srivastava (✉)
Supply Chain, Maersk Line, Mumbai, India
e-mail: s16077@alumni.wmu.se

A. I. Ölçer · F. Ballini
World Maritime University, Malmö, Sweden
e-mail: aio@wmu.se; fb@wmu.se



A. I. Ölçer et al. (eds.), *Trends and Challenges in Maritime Energy Management*,
WMU Studies in Maritime Affairs 6, https://doi.org/10.1007/978-3-319-74576-3_29

technologies, environmental assessment and emission reduction techniques to meet their objectives and ensure their profitability (Corbett and Fischbeck 2000). In this regard, a comparison between the technical measures, operational measures, and an alternative fuel is made. The analysis will assist shipowners to select one or numerous measures in a cost-effective manner while making their strategy. The decision framework is exemplified by a case study of adopting Methanol as an alternative fuel in a ROPAX vessel operated in the Northern European region.

Ölçer and Ballini (2015) discussed a decision-making framework for evaluating the trade-off solutions of cleaner seaborne transportation for Copenhagen port by using cold ironing technology. In this study, the economic perspective of the technology was taken into consideration. The paper used the fuzzy logic technique for decision making for shipowners. Banawan et al. (2010) proved that the conversion process of a ship's main engine from conventional fuel to an alternative fuel has both environmental and economic benefits. The emission reduction in $NO_x$, $SO_X$, $CO_2$, and PM was demonstrated. Seddiek and Elgohary (2014) analysed various methods which would reduce emissions. The results showed a valuable percentage reduction by using Selective catalytic reduction (SCR) and scrubber systems. The study concluded that use of Liquified Natural Gas (LNG) as a fuel would give better results environmentally and economically. Elgohary et al. (2014) compared alternative marine fuels for their environmental benefits; the main emphasis was on LNG as a future marine fuel. Jiang et al. (2010) used the voyage based model to calculate the external cost for a ship. In the study, they did a cost-benefit analysis of savings in air pollution and climate change costs. In conclusion, environmental benefit is sufficient to increase the benefit-cost ratio. A similar methodology is used in the case study.

## 2 Methodology

Emission factors of Carbon dioxide ($CO_2$), Methane ($CH_4$), Nitrous oxide ($N_2O$), Nitrogen oxides ($NO_X$), Sulphur dioxide ($SO_2$), Particulate matter ($PM_{10}$), Non Metallic volatile organic compounds (NMVOC), and Carbon monoxide (CO) of Marine gas oil (MGO) and Methanol are used for calculating air emissions (Brynolf 2014). Total emissions for each gas are calculated by summing up emissions during standby, manoeuvring and cruise speed. The equation used for calculating emissions from MGO is given below (Banawan et al. 2010):

$$E = P * LF * efd * T \tag{1}$$

The vessel uses blended oil after switching fuel, of which 15% is MGO used as pilot fuel and the remaining 85% includes Methanol (Stefenson 2014). The emission calculation is modified slightly based on the ratio of mixing of the fuel. In this case, the following equation is used for calculation of air emissions:

$$E = P * LF * (0.15efd + 0.85efm) * T \tag{2}$$

where "efd" is the emission factor (g/kWh) in MGO mode and "efm" is the emission factor (g/kWh) in Methanol mode, "T" is the engine running time in hours, and "P" is engine power at the maximum continuous rating (kW), "LF" (%) is load factor of the engine.

The environmental benefit can be calculated by substituting the emission factors for $CO_2$, $CH_4$, $N_2O$, $NO_X$, $SO_2$, $PM_{10}$, NMVOC, and CO for both MGO and Methanol. Financial analysis is done for the vessel used in the case study. Net Present Value (NPV) is calculated for the next 15 years. The NPV is the sum of the present values of all cash flows from the project (including initial investment), with the cash flows being discounted at 10%. NPV is calculated by the following equation (Juhász 2011):

$$NPV(i, N) = \sum\nolimits_{t=0}^{N} \frac{Rt}{(1+i)^t} \tag{3}$$

Internal Rate of Return (IRR) is the discount rate that sets the NPV equal to zero. The following equation is used for calculating the IRR (Juhász 2011):

$$NPV(i, N) = 0 = \sum\nolimits_{t=0}^{N} \frac{Rt}{(1+i)^t} \tag{4}$$

where i = discount rate, N = total number of periods, Rt = Net cash flow, and t = time of cash flow. The payback period of a project is found by counting the number of years needed before cumulated forecast cash flows equal the initial investment. There are fluctuations in fuel prices and capital investment during different market conditions. Therefore, it is necessary to conduct a sensitivity analysis for the useful life of the vessel. Monte Carlo simulation is used for the sensitivity analysis. Furthermore, four different scenarios were analysed where internal rate of return was calculated. Additionally, the paper uses a voyage based model for calculation of external costs which depend on different routes and vary from ship to ship. Detailed information was taken for the estimation of external costs such as the type of ship, engine type, sailing, berthing and manoeuvring distance. The following equation is used for the calculation:

$$C_{ij} = E_{ij} * MC_i \tag{5}$$

where i represents four types of ship emissions, including $NO_x$, $SO_x$, $CO_2$ and PM. $C_{ij}$ represents the external costs (Euro) of emission $i$ from the ship, $E_{ij}$ represents the total amount (kg) of emission from the ship and $MC_i$ represents the marginal external costs (Euro per kg) of emission.

The alternative measures available to ship-owners are technical measures, operational measures or use of alternative fuel. Identification of criteria for choosing an optimal measure for ships was collated by interviewing 15 experts from different ship owning companies and experts in the maritime industry. Operational expenses

(OPEX), Capital investment (CAPEX), Payback period and Carbon tax were the most important criteria in the decision framework for ship-owners.

Ship owners' decision-making is first formulated by an m x n decision matrix D where there are m alternatives $A_i$ (i = (1, 2, … ,m)) and n selection criteria $C_j$ (j = (1, 2, . . . ,n)). The evaluation value $x_{ij}$ for alternative i with respect to criterion j is obtained from various data sources. Expert judgements were used to assess the criteria for a particular abatement measure.

$$D = \begin{matrix} . & C_1 & C_j & C_n \\ A_1 & x_{11} & x_{1j} & x_{1n} \\ A_i & x_{i1} & x_{ij} & x_{in} \\ A_m & x_{m1} & x_{mj} & x_{mn} \end{matrix} \tag{6}$$

The weight for each criterion is estimated by experts using utility scores ranging from 1 (equal importance) to 9 (extreme importance). The weights of each criterion reflect the preferences of ship owners in the selection of criteria used for investing in air emission reduction measures. All the measures available are normalised so that the elements are unit free. Vector transformation is used to normalise the value:

$$R_{ij} = \frac{x_{ij}}{\sqrt{\sum_{i=1}^{m} x_{ij}^2}} \tag{7}$$

where $R_{ij}$ is the normalised value of $x_{ij}$. This helps in easier comparisons across different criteria. A weighted normalised TOPSIS matrix $V_{ij}$ is evaluated with the weightage of criteria obtained from AHP where $W_j$ is the average utility score assigned by the experts to the weight of $C_j$.

$$V_{ij} = \frac{W_j}{\sqrt{\sum_{i=1}^{m} W_j}} * R_{ij} \tag{8}$$

The positive ideal solution vector ($S^p$) and negative ideal solution ($S^n$) vector is obtained by involving the best and worst normalised scores for each criterion respectively and is defined as follows:

$$\begin{aligned} S^p &= \left[V_1^+, \ldots\ldots, V_j^+, \ldots, V_n^+\right] \\ S^n &= \left[V_1^-, \ldots\ldots, V_j^-, \ldots, V_n^-\right] \end{aligned} \tag{9}$$

Distances from alternatives to the positive and negative ideal solutions can be measured by using separation measures for each alternative available and are evaluated as follows:

$$S_i^+ = \sqrt{\sum_{j=1}^{n} (V_{ij} - V_J^+)^2} \quad i = 1, 2, \ldots..m \tag{10}$$

$$S_i^- = \sqrt{\sum_{j=1}^{n} (V_{ij} - V_J^-)^2} \quad i = 1, 2, \ldots..m \tag{11}$$

To rank the emission control techniques available to ship-owners in the current market situation, a relative closeness index (*CI*) based on the separation measures is developed. The air emission reduction measure can be ranked as per the increasing value of the closeness index.

$$CI_i = \frac{S_i^-}{S_i^+ + S_i^-} \tag{12}$$

## 3 Case Study of Methanol Conversion of M.V. Stena Germanica

Methanol as a future marine fuel is the best alternative fuel for ship-owners and is available within existing infrastructure (Haraldson 2015). A case study is done to determine the economic and environmental benefits borne by Stena Lines after the application of methanol as the main fuel for a ROPAX vessel called M.V. Stena Germanica. The vessel has four Wärtsilä engines with a total of 32,000 horsepower for propulsion. The vessel was running with MGO before switching fuel to Methanol. The converted engines are of dual fuel type, which means that Methanol is the primary fuel, but they can still run on MGO as a backup. MV Stena Germanica has no emissions during the port stay as it uses cold ironing technology during the port stay.

The most common method of producing methanol is from synthesis of natural gas. The Carnol process (thermal decomposition of methane for methanol synthesis by using renewable source of energy) and direct oxidation of methane are also some of the methods to form methanol. Fossil fuel availability is decreasing with high energy demand, hence the production of methanol from biomass is a good option for the future. Biomass and coal can be gasified for the production of the synthesis gas. Methanol can also be produced from forest industry residuals, by carbon capture technology and from renewable feedstocks (Matzen et al. 2015). Green Methanol development (Bio- and $CO_2$ captured Methanol) is leading towards zero emissions (Lasselle 2015).

The criteria for a marine fuel should include logistics, safety, security and strategic aspects. The IGC[1] and IBC[2] code contain guidelines about transportation and carriage of methanol on the vessel (Freudendahl 2015). Methanol is highly toxic for humans and exhibits a highly corrosive nature to metals (Tinnerberg 2015). Therefore, a vessel requires more sophisticated construction in the fuel system; for example, inerting of the tank is necessary for safe transportation. Carriage of methanol is quite common in the land industry. There are many guidelines for the transportation of methanol (ISO 8217: 2012).

When comparing Liquified Natural Gas (LNG) and Methanol from the well head to the propeller, it was found that Methanol conversion cost is almost comparable to scrubber technology (Stefenson 2014). The conversion cost of M.V. Stena Germanica was around 350 Euro/kW, whereas the cost for BIT Viking to convert to LNG use was 1000 Euro/kW (Stefenson 2014). Therefore, there is a clear distinction between LNG and Methanol considering the economic perspective. The fluctuation of the methanol prices (Methanex 2015a, b) and MGO fuel prices (Clarkson 2015) in the last 14 years has been considerable. The prices of methanol were about 50% less from 2008 to 2014, and during the last 3 years, the prices of methanol were found to be close to MGO price; thus, it makes methanol a better substitute. On the other hand, the price of renewable methanol is quite high in the current market situation as the production cost is quite high (Matzen et al. 2015).

The data collection for the case study is done by conducting interviews with officials of Stena Lines and by studying various Stena Lines publications. The voyage route of the vessel is from Göteburg to Kiel and the total time required is 14 days. 45,000 cars or lorries are lifted from the road every year, and around 1300 persons can travel. Before conversion, MGO consumption by Stena Germanica was 11,000 tonnes/year. As the energy content of Methanol is 2.167 times less than that of MGO, methanol consumption was calculated to be 23,800 tonnes. As per the officials, operational expenses before and after conversion did not change.

## 3.1 Environmental Benefit

Appendix 2 shows the machinery particulars and emission calculations for $CO_2$, $CH_4$, $N_2O$, $NO_X$, $SO_2$, $PM_{10}$, NMVOC, and CO. Emission factors used for specific gases are converted to g/kWh (Srivastava 2016). As per the calculations of emissions of individual gases, it is apparent that the shift from MGO to methanol has resulted in a reduction of $CO_2$, $CH_4$, $N_2O$, $NO_X$, $SO_2$, $PM_{10}$, NMVOC, CO by 5%, 85%, 85%,

[1]International Code of the Construction and Equipment of Ships Carrying Liquefied Gases in Bulk (IGC Code).

[2]The IBC Code provides an international standard for the safe carriage in bulk by sea of dangerous chemicals and noxious liquid substances.

**Table 1** Calculations of air emissions

| | |
|---|---|
| Air emissions when using MGO | 43576.91 $tCO_2e$ |
| Air emissions when using Methanol | 40623.21 $tCO_2e$ |
| Total reduction in air emissions gases in 1 year | 2953.70 $tCO_2e$ |

99%, 85%, 52%, 85% and 85%, respectively. Table 1 gives the reduction in air emissions caused by fuel switching. The total reductions in $CO_2$ eq(equivalent) are evaluated to be around 2953.79 tonnes per year.

## 3.2 *Economic Benefit*

Factors included in the financial analysis are cargo capacity, passengers, duration of time at sea and other parameters. The interviewee stated that the conversion costs of 24 megawatts ROPAX ferry MV Stena Germanica were around 22 million Euro, which includes investment in an internal storage tank on shore (7 million Euro), conversion of a bunker barge (2 million Euro) and complete conversion from MGO engine to dual fuel engine (13 million Euro) (Stefenson 2015). The earnings were taken from the consolidated income statement (Financial report Stena AB 2016). All the earnings were divided by the total capacity of passengers and lane metres of all the vessels operated by Stena Lines to determine the revenue, expenses and profit earned per person and per lane metre. Table 2 gives the estimated cash flow by multiplying the capacity of the vessel with the values obtained per passenger and per lane metre to give the revenue, profit, and OPEX for M.V. Stena Germanica. A cash flow analysis was made for the next 15 years to determine the feasibility of the use of methanol (Srivastava 2016).

Other assumptions required for the calculation are discussed briefly. The discount rate or hurdle rate is assumed to be 10%.[3] The carbon equivalent pricing is the current carbon tax value in Sweden. The tax rate is assumed to be 22%[4] as per the national Swedish tax regime. Useful asset life is assumed to be 15 years[5] after the conversion. The depreciation method used for the vessel is straight line depreciation. Macroeconomic assumptions made in the calculation are exchange rate and inflation, which are taken as 0.93 Euro/USD[6] and 1.02%,[7] respectively. NPV, IRR and Payback period were calculated and compared for both cases as shown in Table 3.

A graphic representation of after-tax IRR for the next 15 years for MV Stena Germanica after switching fuel is shown in Fig. 1. One of the main findings of the

[3]Discount rate is chosen as 10%.

[4]Tax rate is taken from tax yearbook of Sweden 2014 http://www.skatteverket.se/.

[5]Useful asset life is taken from STENA AB 2015 financial report.

[6]Exchange rate is taken from Sveriges Riksbank.

[7]The Riksbank's target is to maintain inflation at a rate of 2% when measured by CPI.

**Table 2** Revenue, expenses, and profit for MV Stena Germanica

| Capacity | Passenger 1300 | Lane metre 4000 | |
|---|---|---|---|
| Income statement | Mn Euros | Mn Euros | Total in Mn Euros |
| Revenue | 11.428 | 35.162 | 46.589 |
| Expenses | 8.483 | 26.101 | 34.583 |
| Profit | 2.945 | 9.061 | 12.006 |

**Table 3** NPV, IRR and Payback period for MV Stena Germanica for MGO and Methanol

| Fuel | After-tax IRR | NPV (Euros) | Payback year |
|---|---|---|---|
| MGO | 38.9% | 5,01,32,044.5 | 4 |
| MeOH | 40.3% | 5,25,21,783.1 | 4 |

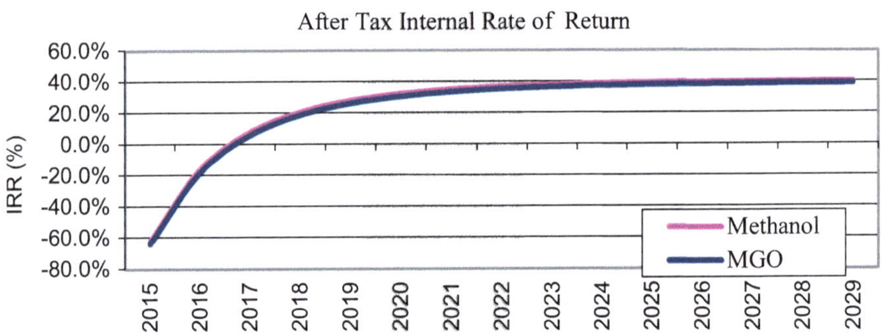


**Fig. 1** IRR of a project with MGO and Methanol

project was that the payback period was determined to be around 4 years. Secondly, the NPV is more than zero and, lastly, the IRR is above the discounted rate for the next 15 years.

## 3.3 *Sensitivity Analysis and Possible Scenarios*

Monte Carlo simulation is used for the sensitivity analysis, which runs 250 iterations of the model randomly varying the value of the fuel prices, CAPEX, OPEX and carbon tax. The percentage change of the fuel prices, CAPEX and Carbon tax were kept between 40 and 50% of the current prices. After running the simulation, Fig. 2 is generated which shows the value of the possible IRR outcomes.

Furthermore, a scenario analysis is done only for the change in fuel prices. Four scenarios are taken to explain the project feasibility. IRR is calculated for the following scenario of fuel prices of MGO and methanol; the result is shown in Table 4. When conducting a Scenario analysis for fluctuation in fuel oil prices, the

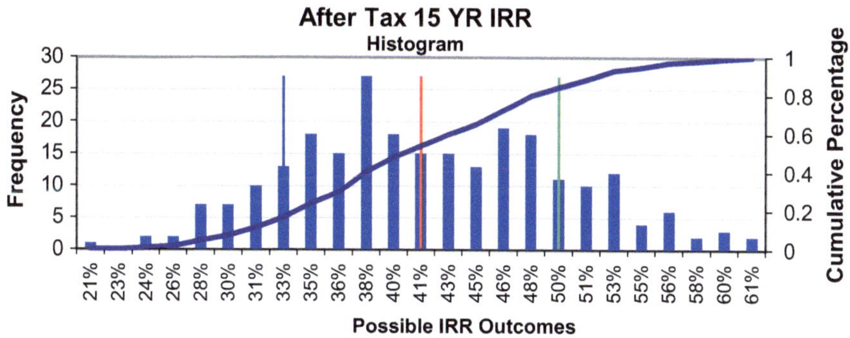


**Fig. 2** Sensitivity analysis for change in fuel price, CAPEX and Carbon tax

**Table 4** Scenario analysis

| Scenario analysis | MGO €/mmbtu | MeOH €/mmbtu | IRR |
|---|---|---|---|
| Scenario 1 | 5.14 | 12.75 | 51.27% |
| Scenario 2 | 14.89 | 13.20 | 22.57% |
| Scenario 3 | 7.24 | 8.58 | 45.03% |
| Scenario 4 | 10.20 | 14.10 | 36.34% |

CAPEX, OPEX and carbon tax were assumed constant at present value. IRR was found positive in all scenarios.

## *3.4 Calculation of External Cost for M.V. Stena Germanica*

For the calculation, the total time of sailing is divided into berthing, manoeuvring, and free sailing time from Kiel to Göthenburg. Berthing time is taken as 2 h at each port, and manoeuvring time is taken as 12 h for each trip at each port. Rest of the time is taken as the free sailing time. The total running time for the marine engines on board MV Stena Germanica is estimated to be around 8400 h. The total power includes the power of the main and auxiliary engines used for sailing. Air pollution cost factors are taken from the HEATCO project for calculation of external costs (Maibach et al. 2008). Appendix 4 gives the values of the all the external costs caused by the emissions. Emissions calculated at the free sailing stage are far greater than the levels at the manoeuvring and the berthing stages. This imbalance will become even more with the increase of sailing time and distance between ports. The air pollution costs are calculated by adding the costs of $NO_2$, $SO_x$, and PM whereas the costs of $CO_2$ are taken as climate change costs. The overall externalities are evaluated to be around 9,100,278 Euro when using MGO, which includes 1,276,697 Euro for climate change costs and 7,823,581 Euro for air pollution costs. Similarly, for methanol, overall externality came in around 2,683,946 Euros which includes

1,206,741 and 1,477,205 as climate change and air pollution costs, respectively. Changes in the air pollution costs and climate costs because of fuel switching amounts are evaluated as 6,346,376 and 69,956 Euro, respectively. The differences in the external costs are considerable and place emphasis on the need for ship emissions' control for ship owners. The primary focus should be on reduction of PM as marginal external costs around the coastal area for PM are high.

## 4 Decision-Making Criteria for Ship-Owners Through AHP and Ranking of Measures Through TOPSIS (Technique for Order Preference Using Similarity to Ideal Solution)

To evaluate the decision-making criteria for developing a framework for air emission reduction measures, different criteria were chosen. The Analytical Hierarchy Process (AHP) technique is used to calculate the weights of criteria and eventually to process for subjective evaluation. Industrial visits and interviews were carried out to collate the data from maritime professionals of three shipping companies operating in the Nordic region, Gulf countries and South East region and other experts, including naval architects, technical managers and professors. Interviewees whose role affected the policy making or decision making in the organisation with respect to selecting energy efficiency measures were selected. The interviews were carried out through open ended questions and focus was placed on ranking the importance of decision making criteria while choosing air emission reduction measures. The relative ranking of the criteria can be measured by pair wise comparison. The weight and ranking of each criterion are given in Table 5 and Appendix 3. The consistency ratio is less than 0.10, which proves that the result is consistent and acceptable. After finding of all the alternatives and criteria related to air emission reduction measures, a TOPSIS matrix is developed. The relative closeness index is evaluated for the alternatives as shown in Table 6.

Appendix 1 shows the measures available for air emission reduction measures and the closeness to ideal solution through TOPSIS. The best alternative emerges as technical measures while the worst is alternative fuel. The closeness index shows technical measures are closer to the ideal positive solution as compared to the other alternatives because the decision criteria are only based on cost and benefit analysis.

**Table 5** The weight and ranking of decision making criteria through AHP

| Criteria | Weight | Rank | | |
|---|---|---|---|---|
| CAPEX (C1) in euros | 0.58 | 1 | AHP Result | Value |
| OPEX (C2) in euros | 0.12 | 3 | Consistency Index | 0.07 |
| Payback period (C3) years | 0.25 | 2 | Random Index | 0.90 |
| Carbon tax (C4) in euros | 0.05 | 4 | Consistency Ratio | 0.07236 |

**Table 6** Ranking of measures available to ship-owners through technique for order preference using similarity to ideal solution (TOPSIS)

| Alternatives | C1 | C2 | C3 | C4 | CI | Ranking |
|---|---|---|---|---|---|---|
| Alternative fuel (Methanol) | 22,000,000 | 25,000 | 4 | 374,656 | 0.36 | 3 |
| Technical (Waste Heat Recovery)[a] | 5,554,839 | 20,000 | 10 | 210,000 | 0.88 | 1 |
| Operational (Scrubber)[b] | 5,483,870 | 182,500 | 3 | 60,000 | 0.64 | 2 |

[a]EE appraisal tool for IMO

[b]Investigation of appropriate control measures (abatement technologies) to reduce black carbon emissions from international shipping (2012)

Ship-owners hesitate to select alternative fuel and other measures because of uncertainty of fuel prices and high CAPEX. The fuel cost is expected to increase because of new sulphur and nitrogen regulations. There is a resistance by ship owners toward change as profitability is not guaranteed and, in many countries, bunker facilities are not available for alternative fuels. Since the ship-owners' decision criteria exclude health cost analysis, future scenarios and impact assessment, the benefits are difficult to analyse. The results of the case study are an effort to estimate the benefits that ship-owners can avail of by switching to alternative fuel.

### 4.1 *Proposed Decision Framework*

The decision framework analyses the criteria a ship-owner must consider while choosing abatement measures for emission reduction. In the current situation, ship-owners focus on CAPEX, OPEX, fuel prices, and shorter payback period. The decision framework designed by the author, shown in Fig. 3, gives various approaches for a ship-owner to avoid air emissions and maintain their cash flow. In the framework, there is an emphasis on the implementation of carbon tax for a sustainable environment. It can be seen clearly how carbon tax will push ship-owners to comply with abatement measures for long-term benefits and social benefits.

## 5 Conclusion

The paper proposes a decision framework for ship-owners to make a compliant strategy to reduce air emissions. The framework is validated by using a case study of a ROPAX vessel. The environmental benefit of 2953.70 tonnes of $CO_2$ eq per year was evaluated after switching to methanol.

The economic benefits of the case study are reduced OPEX, positive NPV, and IRR evaluated positive, and the payback period is 4 years. Reduction in climate change cost was calculated at 69,956 Euro, whereas the reduction in health cost was evaluated at 6,346,376 Euros. It is apparent that switching to methanol is a cost effective measure for the ship-owners in the long term. Methanol can comply with

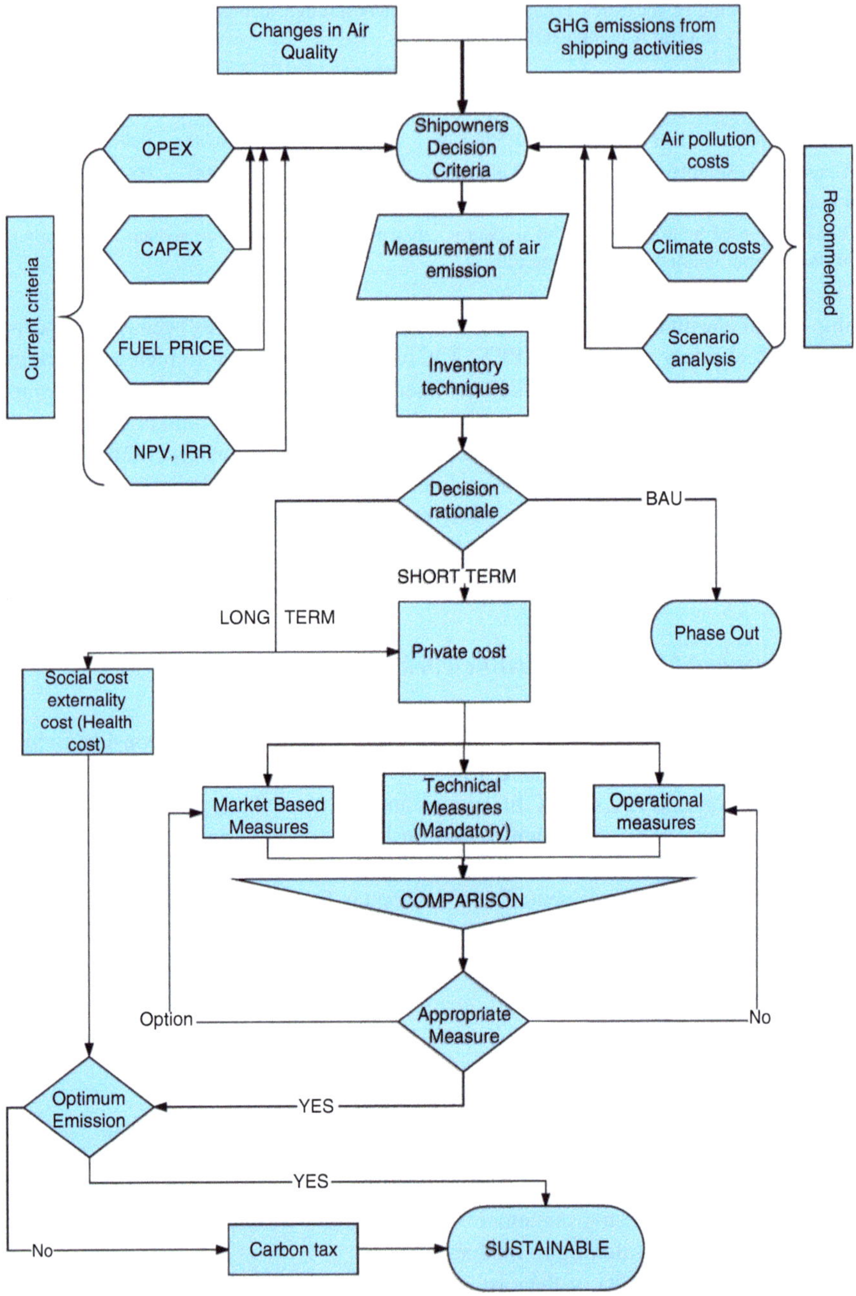


**Fig. 3** Decision framework for ship-owners towards zero emission and sustainability

stricter sulphur and $NO_x$ Tier III regulations and its high capital cost is justified in the long term if the technology is mature and available. As the new MRV regulation came into force in January 2018, because of switching to a new and clean fuel, Stena AB Lines will benefit. Proposed criteria to include in the decision framework for ship-owners while making energy efficiency investments are *health cost, climate change cost* and *future scenarios* to move towards a *sustainable environment.*

**Acknowledgements** We would like to express our very sincere gratitude to Stena lines and Stena officials especially to Ms. Catherine Lee who is Business Development Director of Stena Lines Group and Mr. Per Stefenson who is Marine Standards Advisor for Stena Rederi AB Technical Division for their valuable input and providing the first author with data for successful completion of the paper. Every result described in this paper was accomplished with the help and support of fellow colleagues and collaborators. We would like to thank all the people who contributed in some way to the work described in this paper.

## Appendix 1: Measures available for air emission reduction through TOPSIS

| Measures | $S_i^+$ | $S_i^-$ |
|---|---|---|
| Technical (Waste Heat Recovery) | 0.442 | 0.060 |
| Operational (Scrubber) | 0.407 | 0.226 |
| Alternative fuel (Methanol) | 0.224 | 0.407 |

## Appendix 2: Machinery particulars and emissions for M.V. Stena Germanica

| Machinery particulars | | | | | | | | Emission in tonnes | | | | | |
|---|---|---|---|---|---|---|---|---|---|---|---|---|---|
| | MGO | | | 85% MeOH + 15% MGO | | | | MGO | | | 85% MeOH + 15% MGO | | |
| | ME | AE 1 | AE 2 | ME | AE 1 | AE 2 | Gases | ME | AE 1 | AE 2 | ME | AE 1 | AE 2 |
| Model | 8ZAL40S | 6L26 | 9L26 | 8ZAL40S | 6L26 | 9L26 | CO2 | 35603 | 2781 | 4172 | 33945 | 2652 | 3978 |
| No. of installed | 4 | 1 | 1 | 4 | 1 | 1 | CH4 | 0 | 0 | 0 | 0 | 0 | 0 |
| MCR (kW) | 5760 | 1800 | 2700 | 5760 | 1800 | 2700 | N2O | 2 | 0 | 0 | 0 | 0 | 0 |
| MCR total | 23040 | 1800 | 2700 | 23040 | 1800 | 2700 | NOX | 732 | 57 | 86 | 20 | 2 | 2 |
| Load factor at sea | 0.70 | 0.70 | 0.70 | 0.70 | 0.70 | 0.70 | SO2 | 23 | 2 | 3 | 3 | 0 | 0 |
| LF at manoeuvring | 0.70 | 0.70 | 0.70 | 0.70 | 0.70 | 0.70 | NH3 | 0 | 0 | 0 | 0 | 0 | 0 |
| 14 days for one trip | | | | | | | PM10 | 5 | 0 | 1 | 3 | 0 | 0 |
| 25 (assuming 15 days of repair and maintenance) trips in 1 year | | | | | | | NMVOC | 28 | 2 | 3 | 4 | 0 | 1 |
| 8400 running hours in 1 year | | | | | | | CO | 63 | 5 | 7 | 10 | 1 | 1 |
| | | | | | | | Total | 36457 | 2848 | 4272 | 33985 | 2655 | 3983 |

## Appendix 3: Ranking of criteria for ship-owners through AHP

| Criteria | CAPEX | OPEX | Payback period | MBM | Criteria | CAPEX | OPEX | Payback period | MBM | Total | Ranking |
|---|---|---|---|---|---|---|---|---|---|---|---|
| CAPEX | 1.00 | 5.00 | 4.00 | 7.00 | CAPEX | 0.63 | 0.54 | 0.73 | 0.41 | 2.30 | 1.00 |
| OPEX | 0.20 | 1.00 | 0.33 | 3.00 | OPEX | 0.13 | 0.11 | 0.06 | 0.18 | 0.47 | 3.00 |
| Payback period | 0.25 | 3.00 | 1.00 | 6.00 | Payback period | 0.16 | 0.32 | 0.18 | 0.35 | 1.01 | 2.00 |
| Carbon tax | 0.14 | 0.33 | 0.17 | 1.00 | Carbon tax | 0.09 | 0.04 | 0.03 | 0.06 | 0.21 | 4.00 |
| Total | 1.59 | 9.33 | 5.50 | 17.00 | Total | 1.00 | 1.00 | 1.00 | 1.00 | 4.00 | |

## Appendix 4: Calculation of external costs for M.V. Stena Germanica

| Voyage | | Berthing | Manoeuvring | Free sailing | Berthing | Manoeuvring | Free sailing | |
|---|---|---|---|---|---|---|---|---|
| Time in hours | | 300 | 50 | 3850 | 300 | 50 | 3850 | |
| Port | | Kiel | Kiel | Kiel | Gothenburg | Gothenburg | Gothenburg | |
| Air pollution cost factor | $CO_2$ | 0.03 | 0.03 | 0.03 | 0.03 | 0.03 | 0.03 | HEATCO[a] project |
| | $NO_X$ | 12.7 | 12.7 | 12.7 | 4.1 | 4.1 | 4.1 | HEATCO project |
| | $SO_2$ | 10.9 | 10.9 | 10.9 | 4.2 | 4.2 | 4.2 | HEATCO project |
| | PM | 227.6 | 227.6 | 33.6 | 231.3 | 231.3 | 17.0 | HEATCO project |
| MGO (external cost Euros) | $CO_2$ | 45596.3 | 7599.4 | 585152.8 | 45596.3 | 7599.4 | 585152.8 | |
| | $NO_X$ | 396625.6 | 66104.3 | 5090028.2 | 128044.5 | 21340.7 | 1643237.4 | |
| | $SO_2$ | 10666.2 | 1777.7 | 136883.0 | 4109.9 | 685.0 | 52743.9 | |
| | PM | 52125.6 | 8687.6 | 98754.6 | 52972.9 | 8828.8 | 49965.1 | |
| | Total external cost | 505013.7 | 84168.9 | 5910818.6 | 230723.7 | 38453.9 | 2331099.3 | |
| MeOH (external cost Euros) | $CO_2$ | 43097.9 | 7183.0 | 553089.7 | 43097.9 | 7183.0 | 553089.7 | |
| | $NO_X$ | 74036.8 | 12339.5 | 950138.6 | 23901.6 | 3983.6 | 306737.7 | |
| | $SO_2$ | 0 | 0 | 0 | 0 | 0 | 0 | |
| | PM | 20376.4 | 3396.1 | 38604.1 | 20707.6 | 3451.3 | 19531.8 | |
| | Total external cost | 137511.0 | 22918.5 | 1541832.3 | 87707.1 | 14617.9 | 879359.1 | |

[a]Handbook on estimation of external costs in the transport sector, Maibach et al. (2008)

# References


Andersson, K., Brynolf, S., Lindgren, J. F., & Wilewska-Bien, M. (Eds.). (2016). *Energy efficiency and fuel changes to reduce environmental impacts in shipping and the environment: Improving environmental performance in marine transportation* (pp. 295–339. Web. 9 Jan. 2017). Berlin: Springer. https://doi.org/10.1007/978-3-662-49045-7_10.

Banawan, A. A., Gohary, M. M., & Sadek, I. S. (2010). Environmental and economical benefits of changing from marine diesel oil to natural-gas fuel for short-voyage high-power passenger ships. *Proceedings of the Institution of Mechanical Engineers, Part M: Journal of Engineering for the Maritime Environment, 224*(2), 103–113. https://doi.org/10.1243/14750902jeme181.

Brynolf, S. (2014). *Environmental assessment of present and future marine fuels*. Thesis. Chalmers University of Technology, N.p.: n.p., n.d. Doktorsavhandlingar Vid Chalmers Tekniska Högskola. Ny Serie. 2014. Web. 18 June 2016. <https://trid.trb.org/view.aspx?id=1367598>.

Clarkson Intelligence network. (2015). Retrieved July 07, 2016. Clarkson Research Services Limited, Commodity Quay, St Katharine Docks, London. https://sin.clarksons.net/.

Corbett, J. J., & Fischbeck, P. S. (2000). Emissions from waterborne commerce vessels in the United States continental and inland waterways. *Environmental Science & Technology, 34*(15), 3254–3260.

Elgohary, M. M., Seddiek, I. S., & Salem, A. M. (2014). Overview of alternative fuels with emphasis on the potential of liquefied natural gas as future marine fuel. *Proceedings of the Institution of Mechanical Engineers, Part M: Journal of Engineering for the Maritime Environment, 229*(4), 365–375. https://doi.org/10.1177/1475090214522778.

Financial Report Stena AB. (2016). Stena Raderi AB, Gothenburg. http://www.stena.com/en/Documents/ENG_STAB_%C3%85R_2016_Del2_.pdf.

Freudendahl, U. (2015). Methanol as ship fuel. Handling and safety matters. In *MOT-2030 Workshop*, Lund. http://www.lth.se/fileadmin/mot2030/filer/7._Freudendahl_-_Methanol_as_ship_fuel.pdf.

Han, C. (2010). Strategies to reduce air pollution in shipping industry. *The Asian Journal of Shipping and Logistics, 26*(1), 7–29. https://doi.org/10.1016/s2092-5212(10)80009-4.

Haraldson, L. (2015). Methanol as marine fuel. Retrieved July 7, 2016, from http://www.sintef.no/globalassets/marintek/cimac2015/f5-Methanol-as-marine-fuel---lennart-haraldsson---Wärtsila.pdf.

IMO Secretary-General Welcomes Adoption of Sustainable Development Goals. N.p., 27 Sept. 2015. Web. 18 June 2017. http://www.imo.org/en/mediacentre/pressbriefings/pages/41-sdgs.aspx.

ISO 8217: 2012 "Petroleum products-Fuels (class F)-Specifications of Marine Fuels".

Jiang, L., Kronbak, J., & Chridtensen, L. P. (2010). *External costs of maritime shipping: A voyage-based methodology*. Maritime Research, Department of Environmental and Business Economics. Retrieved June 21, 2016.

Juhász, L. (2011). Net present value versus internal rate of return. *Economics & Sociology E&S, 4* (1), 46–53. https://doi.org/10.14254/2071-789x.2011/4-1/5.

Lack, D. A., Thuesen, J., & Elliot, R. (2012). *Investigation of appropriate control measures (abatement technologies) to reduce black carbon emissions from international shipping*. Study report and prepared by Litehauz, ERRIAZ. http://www.imo.org/en/OurWork/Environment/PollutionPrevention/AirPollution/Documents/Air%20pollution/Black%20Carbon.pdf.

Lasselle, S. (2015). Use of methanol as fuel. Methanol as marine fuel: Environmental benefits, technology readiness, and economic feasibility (Rep. No.2015-1197, Rev. 2).DNVGLdistribution. http://www.imo.org/fr/OurWork/Environment/PollutionPrevention/AirPollution/Documents/Report%20Methanol%2021.01.2016.pdflimit-1/exceedance-of-air-quality-limit-5 (last accessed 06.03.14).

Maibach, M., Schreyer, C., Sutter, D., Van Essen, H. P., Boon, B. H., Smokers, R., et al. (2008). *Handbook on estimation of external costs in the transport sector*. Delft: CE Delft. http://www.euskalyvasca.com/pdf/estudios/2008/20080314_costes_externos_.pdf.

Matzen, M., Alhajji, M, & Demirel, Y. (2015). *Technoeconomics and sustainability of renewable methanol and ammonia productions using wind power-based hydrogen*. OMICS International. OMICS International, 06 July. Web. 18 Nov. 2016. https://doi.org/10.4172/2090-4568.1000128.
Methanex. (2015a). *Methanex monthly average regional posted contract price history*. https://www.methanex.com/sites/default/files/.
Methanex. (2015b). *Methanex posts regional contract methanol prices for North America, Europe and Asia* [Online]. Available: https://www.methanex.com/our-business/pricing. Accessed 27 July 2016.
Oecd Council Working Party On Shipbuilding (Wp6). (2010). *Environmental and climate change issues in the shipbuilding industry*. Retrieved May 7, 2016, from https://www.oecd.org/sti/ind/46370308.pdf.
Ölçer, A., & Ballini, F. (2015). The development of a decision making framework for evaluating the trade-off solutions of cleaner seaborne transportation. *Transportation Research Part D: Transport and Environment, 37*, 150–170.
Rozmarynowska, M. (2012). LNG jako alternatywne paliwo dla statków-aspekty techniczne, ekologiczne, ekonomiczne i regulacyjne. *Logistyka, 5*.
Seddiek, I. S., & Elgohary, M. M. (2014). Eco-friendly selection of ship emissions reduction strategies with emphasis on SOx and NOx emissions. *International Journal of Naval Architecture and Ocean Engineering, 6*(3), 737–748. https://doi.org/10.2478/ijnaoe-2013-0209.
Srivastava, A. (2016). *Compliant strategy for shipowners towards sustainable maritime transport: A decision framework for air emission reduction measures*. http://commons.wmu.se/cgi/viewcontent.cgi?article=1518&context=all_dissertations.
Stefenson, P. (2014). *The use of biofuel in the marine sector or Methanol, the marine fuel of the future*. European Biofuels Technology Platform Brussels 15 October 2014. Brussels. http://www.marinemethanol.com/publications/category/4-methanol-as-a-marine-fuel.
Stefenson, P. (2015). *Email correspondence*. Stena Raderi AB, Gothenburg.
Tinnerberg, H. (2015). Methanol as a fuel exposure and possible health risks. In *MOT 2030 Workshop*. Lund. transport/shipping/index_en.htm.

## Further Reading

Franc, P., & Sutto, L. (2013). Impact analysis on shipping lines and European ports of a cap-and-trade system on CO2 emissions in maritime transport. *Maritime Policy & Management, 41*(1), 61–78. https://doi.org/10.1080/03088839.2013.782440.
Nielsen, C., & Chess, C. (2012). Vessel emission study: Comparison of various abatement technologies to meet emission levels for ECA's. In *9th Annual Green Ship Technology Conference* (p. 13). http://www.finaval.com/site/images/documenti/news/2014/articoloNOV14paper.pdf.
Stott, P. (2013). A retrospective review of the average period of ship ownership with implications for the potential payback period for retrofitted equipment. *Proceedings of the Institution of Mechanical Engineers, Part M: Journal of Engineering for the Maritime Environment, 228*(3), 249–261. https://doi.org/10.1177/1475090213486096.

# Commercial Wind Propulsion Solutions: Putting the 'Sail' Back into Sailing

Gavin Allwright

## 1 Introduction

Can the shipping industry create a twenty-first century business model that is both profitable, resilient and delivers on all aspects of the triple bottom line, concerning economic, social, and environmental values? With the help of commercial wind propulsion solutions, that question becomes much easier to answer. It will help future proof the fleet at a time of great transition, disruption and risk.

Our modern world economy was founded on 'wind ships'. With the growing pressure on the shipping industry to implement more ambitious targets for all emissions, wind offers an abundant, free energy source that can be harnessed once again using refined, automated technologies for vessels of all classes and all sizes. Harnessing this renewable resource helps to de-risk the future by securing a significant percentage of operational expenses, fixing it at zero for the lifetime of the vessel.

Wind energy delivers economic and energy security against the volatility of the fuel markets, and more certainty in the face of emissions regulations—the European Commision's (EU's) 2020 sulphur directive (EU Directive 2012/33/EU), the Paris Agreement and the International Maritime Organization's (IMO's) Marine Environment Protection Committee (MEPC) 70th work on the industry's 'fair share' of Green House Gases (GHGs). Whether regulation leads to a carbon tax, carbon cap, rationing or more stringent enforcement of mandated efficiency increments, introducing 'free' renewable energy into the asset's operating budget creates a more predictable and future-proofed business model.

The shipping industry emits the same amount of $CO_2$ as the 6th ranked country (between Germany and Japan), and emissions are predicted to rise. Global population is predicted to grow by at least 2 billion by 2050, creating additional demand on

G. Allwright (✉)
International Windship Association (IWSA), London, UK
e-mail: secretary@wind-ship.org



A. I. Ölçer et al. (eds.), *Trends and Challenges in Maritime Energy Management*, WMU Studies in Maritime Affairs 6, https://doi.org/10.1007/978-3-319-74576-3_30

resources and the global supply chain. IMO projects a doubling of the world fleet by 2050, and world trade growth will lead to a 50–250% growth in emissions (IMO 2015). With land-based emissions set to decline dramatically in the coming decade, a 'business as usual' approach in shipping will see its share of global emissions grow to 17% of global $CO_2$ emissions in 2050, if left unregulated (ENVI 2015).

The burning of Heavy Fuel Oil (HFO) and its health effects are widely acknowledged as unacceptable—SOx, NOx, and PM have a huge impact locally and throughout the coastal regions of countries, where much of the population is generally clustered. The potential health cost savings for national governments run into tens of billions and far outweigh the remediation costs. There is a real need to join up those dots at a policy level (Danish EcoCouncil 2011).

## 2 Wind Propulsion Technology Variations

With over 30 technology providers and projects working closely with classification societies, various shipping companies, designers and engineering companies to produce commercially viable systems, this sector is growing and becoming a viable option for the industry. These technologies can be broken down into eight main categories from Flettner rotors to traditional soft sail, dynarig systems, hard sail variations to kites, suctions wings, turbines and hull form variants.

Even in the face of lower oil prices, there has been continued research and development (R&D) and many of the technologies are either in late stage development or market ready. Retrofits of wind propulsion technology can deliver 10–30% fuel savings, and new builds up to 50%. Along with these significant fuel savings comes long-term economic certainty through securing a non-commodity fuel source for propulsion. This is particularly important bearing in mind that ships designed and built today will need to operate in a net zero emissions world by 2050, if we are to keep within the now ratified Paris Agreement to limit global temperature increases to 1.5 °C.

Shipping is in transition, much like the land-based energy sector over the last decade; however, the scale and speed of change required in the maritime industry mean that we face an ever-steepening change curve. The big question is what is the shipping industry 'transitioning' to? Liquefied Natural Gas (LNG) is viewed as a significant move in the industry, but is seen as a 'transition fuel' at best and the market has yet to fully embrace the fuel, with weak uptake – c.100 vessels rather than the projection of approximately 1000 vessels by 2020 from DNV GL (which has been subsequently almost halved), with dual fuel engines and LNG-ready vessels being the preferred industry options. However, LNG is still a fossil fuel, has significant climate impact and ships built today using LNG will be redundant before the end of their lives. This market uncertainty can only be improved by taking a large percentage of that difficult fuel choice out of the equation.

It is time to unleash the phenomenal talent of our naval architects, ship designers and engineers to design—and more importantly build—the bold, ambitious zero emissions vessels of the future. Small mandated incremental changes are insufficient to tackle the climate challenge, or the inevitable attentions of the rest-of-the-world demanding shipping do its 'fair-share'.

The recently released EU commissioned report, Study on the analysis of market potentials and market barriers for wind propulsion technologies for ships (CE Delft 2016), states that; 'In 2030, the market potential could amount to around 3,700–10,700 installed systems on bulkers and tankers, associated with approximately 3.5–7.5 million tonnes $CO_2$ savings and 6,500–8,000 direct and 8,500–10,000 indirect jobs.'

Significant market and non-market barriers still exist, and much work is already underway to address these challenges. Lloyd's Register also produced a barriers and solutions report focusing on wind propulsion and a return on investment (ROI) analysis (LR 2015) and there have been a number of academic studies by UCL Energy Institute backing these claims (Rehmatulla et al. 2017).

## 3 The International Windship Association

The International Windship Association (IWSA) was established in late 2014 and has brought together over 50 members and registered supporters, including 22 technology providers; key universities, including TU Delft, Fraunhofer ISI, Emden Leer University, and University of Tokyo, and NGOs. The founding members identified the need to strengthen the message that wind propulsion is a credible, viable and increasingly necessary propulsion solution and that the industry needed a central, one-stop supplier of information regarding this developing trend.

The association works to facilitate and promote wind propulsion for commercial shipping worldwide and to bring together all parties in the development of a windship sector to shape industry and government attitudes and policies. We have five key work streams, focusing on issues from communication, policy, technical and certification, finance and market transformation and the strength of the association is in our collaborative approach. We have engaged with a number of EU research and development programs, the United Nations (UN) and the Organisation for Economic Co-operation and Development (OECD) events and reports, numerous industry conferences and panel discussions worldwide and will continue to do so. Our recent 'Wind Propulsion Innovation Awards' and both the 'Blue Week Natural Propulsion Seminars' and last year's Shipbuilding, Machinery and Marine Technology (SMM) 'Wind Propulsion Innovation Forum' have attracted support from many key industry players and been welcomed warmly by ship owners, classification societies, the media, technology providers and research institutes.

# 4 Development Stage and Market Opportunities

Wind propulsion technologies are moving from design and late stage development into commercial deployment. Many of the technologies are ready for sea trials and appropriate vessels are being sought. The UN-affiliated International Renewable Energy Agency (IRENA)—Renewable Energy Option in Shipping: Technology Brief (Mofor et al. 2015) presented a snapshot of the stage each wind propulsion technology had achieved. However, in the last year we have seen accelerated progress. The Sustainable Shipping Initiative which includes Maersk and Cargill, LR, ABS, Unilever and ABN-AMRO, among others, has predicted a substantial number of wind propulsion demonstrator vessels by 2030. Furthermore, Lloyds Register analysed the potential market for wind propulsion, identifying 6000 vessels or 10% of the fleet as having high retrofit potential. New entrants such as Google, which registered a commercial wind propulsion patent recently, validate the market opportunity. The coming 12–18 months will see further strides in development and an increase in commercial uptake and test vessels.

# 5 Technical Options Overview

## *5.1 Rotors*

Commercial Flettner Rotors generate propulsion from harnessing the Magnus Effect, created when wind passes over a revolving cylinder. They were first developed in the 1920s on a number of ships, including the 3000 dwt MV Barbara. In 2010, Enercon (Ger) began trials of the 12,800 dwt prototype E-Ship 1 with four Flettner rotors and it has been in operation worldwide for the last 5 years. In 2015, it was joined by the 9700 dwt MV Estraden, a Bore Shipping RoRo vessel servicing a North Sea route with 2 Norsepower (Fin) rotors fitted. These vessels will soon be joined the MV Fehn Pollux when it commences trials of the Eco Flettner system. There are a number of other companies working on variants of the standard rotor design, including US-based Magnuss, with a retractable telescopic rotor system.

## *5.2 Fixed Sail*

Fixed-sails are essentially rigid 'wings' on a rotating mast. Current proposals include use on large ships, notably the 84,000 dwt bulker Wind Challenger. With successful land-based tests of their telescoping reefed masts completed, the Japanese consortium, including University of Tokyo, NYK Lines, and MOL are now moving toward sea trials. The EffShip (Swe) and the Wallenius Wilhelmsen (Nor) S/V Orcelle project also propose retractable rigid system designs. 'Wingsails' are a variation tested in the 1980s and recently updated with companies such as Oceanfoil (UK) and Windship International (UK). Ocius, Detlev and Ecomarine all offer rigid solar/sail hybrid rigs; the former technology is currently in use on ferries in Australia and China, and 10 retractable Detlev sails are proposed as auxiliary systems for the Japanese Peace Boat's 55,000 dwt Ecoship cruise liner project and the Ecomarine system, in its late stage development in Japan. Other rigid sail designs derive from systems used in the leisure yacht and racing industry, including Wind+Wing Technologies (US), Bound4Blue (Esp), Propelwind (Fra) and Blue Technology (Den).

## 5.3 *Automated Soft Sail*

These are conventional soft-sails operated with a push-button system from a free-standing mast. This technology is proven, requires no additional crew to operate, meets all health and safety needs, has no obstacles to accessing holds, and is suitable for both retrofit and new builds of small to medium sized tankers, bulkers and general cargo vessels. There are a number of cruise vessels using soft sails along with a number of smaller commercial vessels. New fully automated rig designs based on the 'Dynarig' system, proven on the 90 m superyacht, the Maltese Falcon, include the Smart Green Shipping Alliance (UK) (includes LR, Rolls Royce) 8000 dwt bulker, and the 7000 dwt Ecoliner (Neths), which has been attracting interest from Volkswagen recently. A number of other designs have been developed over the years, most notably Seagate Sails (Ita), Neoline (Fra) and the German Autares system, PINTA rigs and the INDOSAIL rigs used successfully for years on the Rainbow Warrior II and the coastal freighter Murata Jaya900.

## 5.4 *Suction Wing*

In the 1980s, Capt. Jacques Cousteau, and his team developed the 'TurboSail', a non-rotating fan-driven design which also harnesses the Magnuss effect. The promising system was extensively tested on his research vessel Alcyone and is being further refined by Crain Technologies (Fra).

## 5.5 *Kite Sails*

Google released a patent in late 2016 for an energy kite sail for use on commercial vessels. The energy generated is used to power an electro-dialysis system to extract $CO_2$ from seawater to produce hydrogen gas. A refinery system uses the $CO_2$ and $H_2$ through the Fischer-Tropsch process to produce the synthetic diesel fuel C11H24. It then converts this synthetic fuel into ethanol by using syngas fermentation as alternative fuel. The market leader in kite technology is Skysails GmbH. In 2008, MS Beluga Skysails was the world's first commercial container cargo ship partially powered by a 160-$m^2$ kite and in the years that followed eight more commercial vessels were equipped with the technology. Kite sails are attached to the bow of the vessel, which then operate at altitude to maximise wind speeds.

### 5.6 *Turbines*

Design and testing of wind turbines for use on commercial vessels is underway by Inerjy (US) and Propit (Swe). There have also been a number of tests of ancillary power production with turbines scaled up from leisure yacht use, and further technical development can be expected.

### 5.7 *Hull Form*

Norwegian Lade AS's Vindskip design concept is for a hybrid merchant vessel powered by an LNG-fuelled engine and an aerodynamic hull that functions as a giant sail. Tank testing and CFD simulation have shown significant potential fuel savings.

## 6 Barriers to the Uptake of Wind Propulsion

There have been a number of studies into the barriers facing the uptake of wind propulsion technologies including recent ones from UCL (Rehmatulla et al. 2017) and included in Lloyds Register report (LR 2015): Wind-powered Shipping: Review of commercial, regulatory and technical factors affecting uptake of wind-assisted propulsion. These focus on both market and non-market challenges to the sector and some of the barriers are familiar to all technology innovation sectors but many are particularly challenging in the shipping industry. Indeed, IWSA was established to challenge many of these barriers and collaborate across the industry to find ways to reduce and ultimately do away with many of these.

LR's report highlights seven key barriers outside of the basic ROI calculations that are currently challenging. Industry structure is a key one, especially the issues around the split incentive and recently highlighted market failures to value more efficient ships with higher charter rates, though more efficient vessels are likely to secure charters in the first place (Prakash et al. 2016). New mechanisms are being developed to deal with this, with the potential for technology to be sold or leased as a service. SSI2040 and Carbon War Room are working on funding schemes that share the cost burden. Numerous efficiency indexes such as the Clean Shipping Index seek to lock in the value of efficiency ratings.

The lack of external drivers pushing the shipping industry to make change is also a structural failure. The logistics chain still has very little influence on shipping uptake of efficiency innovation. Ports give incentives but these are still small and the IMO still needs to nudge its member States to the need for serious regulatory changes. This has been a major factor missing over the last few years and has not yet taken up the slack of record fuel price lows.

Perception and the need to substantiate claims made around levels of savings are also very important and will only be fully dealt with by a substantial increase in demonstrator vessels. IWSA is undertaking much work in this area with engagement with researchers, policy makers, the industry and the media to update the market on developments, increase understanding of the pipeline of projects and to collate research and industry reports.

The cost of bringing demonstration vessels and retrofit technologies to market is a major barrier facing projects. IWSA seeks to increase collaboration between our members to share best practices, support funding applications and assist in finding shipping companies that are willing to be first movers and engage with new wind technologies. In an industry that has limited capacity to embrace change, there is increased need for both our members and the IWSA network as a whole to also tackle the operational and technical barriers that the industry identifies. These include issues around route specific savings rather than savings on all routes, compliance issues around navigation, safety, and cargo handling. Many of these issues are being dealt with by classification societies and some of these issues are being championed and solved by other innovation sectors, such as autonomous shipping's use of sensors and cameras to overcome navigation and line of sight issues. Automated wind propulsion systems, sophisticated route modelling and weather routing software and operations management are also challenging many of these (IWSA 2017).

# 7 Brief Case Studies

## 7.1 *Flettner Rotors*

As mentioned, the Flettner rotor or rotor sails were invented over 80 years ago and there are now a number of technology providers that are competing to bring their products to market. The early mover was Enercon, with its E-ship 1 prototype vessel that has been at sea proving the viability of rotors for the past 6 years. The vessel is currently being operated commercially by Auerbach Shipping. This new built vessel was specifically designed to maximise the efficiency gains from the 4 rotors and has delivered up to 20% savings on selected routes over that period (Enercon 2013).

The EcoFlettner project, a collaboration of 15 companies based in the E-ship 1 homeport of Leer has been working on a new generation of motor-rotor-sailing ships with an eye on the short sea trade. The base vessel is a 4000 dwt multipurpose vessel. A prototype is being completed at present and will be tested aboard the Fehn shipping MV Fehn Pollux in the coming months.

The Finnish company, Norsepower, also presented up to date details of their product, financing options and savings during the RINA conference last year. They have already confirmed details of their first installation of two rotors aboard the MV Estraden operated by Bore Shipping, with an average net saving of 6.1% reported, with an ROI of 5 years with MGO at US$400/ton. Bore has reported that they are very happy with the retrofit, which was completed without disruption to their operations, and with the simplicity of using the fully automated system. Computers adjust the rotor rotation speed and optimise the thrust and the collection of data from these operations is enabling Norsepower to produce accurate estimates of savings on other vessels and routes. Norsepower has recently secured additional funding from the EU and Tekes, the Finnish government's funding agency for innovation, to develop a rotor 30 m in height and 5 m in diameter, with a maximum main-engine equivalent power output of more than four megawatts (MW).

Other designs are also embracing this rotor development trend, with Dutch C-Job designers moving ahead with plans for the Eco-Freighter, a design with four rotors, and the Deltamarin's Delta Challenger concept vessel with six rotors. Both concepts look to utilise Norsepower rotors, whereas the Wind Hybrid Coaster will use 2 Eco Flettners.

Magnuss, a US-based company developing rotor technology has spent the last few years focused on the development of its Magnuss VOSS™ or Vertically-variable Ocean Sail System, a mechanical rotor sail that can be retracted and stowed out of the way. They have been working with engineers at MIT and from the shipping industry on this promising innovation. The VOSS system is now class approved by Lloyd's Register and case studies have been completed including a Rio Tinto case study. After extensive simulation testing, models have been specifically designed for tanker and dry bulk carriers (ideally suited for panamax sizes and above) and the company is currently working on bringing the technology to sea trial with a demonstration vessel.

## *7.2 UT Wind Challenger*

The Wind Challenger Project was launched by the prestigious leading University of Tokyo and Japanese major shipping companies in 2009, to find a way to utilise maximum ocean wind power for the main propulsion of a large cargo vessel. The Joint Industry Project includes NYK Line, Mitsui O.S.K. Lines, Ltd., Oshima Shipbuilding Co., Ltd., Tadano Ltd., ClassNK, and Tokyo Keiki Inc.

The vessel is equipped with very large rigid sails (Height: 50 m, Breadth: 20 m, Thickness: 4 m, Area: 1000 $m^2$) on the upper deck as a main propulsor which is made of advanced light material such as CFRP composite or aluminium alloy. This rigid sail, whose horizontal section is crescent shaped, has a retractable steel spar and totally enclosed wing of composite material, and also has a vertically telescopic reefing and self-rotating mechanism to meet the wind velocity and direction. The four rigid sails (total sail area 4000 $m^2$) are expected to generate enough forward thrust for a 84,000 dwt bulk carrier to maintain a cruising speed of 14 knots, which has been confirmed by wind tunnel testing and full scale CFD simulation. A larger 180,000 dwt version is also being designed. The power management system between sails and engines has also been studied.

A routing case study has been completed on the route between Yokohama (Japan) and Seattle (USA) and approximately 30% of propulsion energy was achieved from wind power, on average. The cost performance study also produced good results and once the sail unit costs are brought down through economies of scale, the ROI will be increasingly attractive (Ouchi et al. 2013).

## *7.3 Smart Green Shipping Alliance*

The Smart Green Shipping Alliance (SGSA) has an ambitious trajectory of development for commercial shipping. Minimising carbon starts on naval architects drawing boards. Designing for 100% reuse reduces carbon by 29%, according to the research led by Gilbert et al. (2016). Building the vessel for extreme longevity, looking into condition-based monitoring, new materials and operation approaches to extend use of the asset over time, further reduces embedded carbon. Powering the vessel with 100% renewable energy, allowing fuel cost to be predictable over the lifetime of the ship, means that we can envisage today an ultra-low carbon, future-proof resilient asset through combining existing and proven technologies in new ways.

Once the collective creativity of the shipping industry realises the potential in maritime renewable energy, we can anticipate exponential growth mirroring that of onshore renewable energy. New financial products have triggered an onshore renewable revolution. The SGSA is creating innovative business models for the maritime sector that reflect the fundamentally different economics enabled by renewable energy—the asset is decoupled from volatile commodity fossil fuels and harnesses, through technology, freely and infinitely available power.

The final piece in the jigsaw is the replication for the maritime sector of the analysis tool that underpinned the development of onshore wind. 'Virtual Met Mast', created by the UK's Met Office weather service, enables all contracting parties—asset developer, financier, end-user—to be able to confidently predict the value of wind to any project, to effectively monetise the value of renewable energy and give all contractual parties confidence.

Of course maritime analysis is inherently more complex. The windy hill where we might develop a fleet of wind turbines is static, unlike hybrid-wind powered ships which move through waves and currents as well as variable meteorological conditions. But by mobilising the power of big data, the SGSA is, with the Met Office, developing a tool called 'Tradewind' to perform a similar function to Virtual Met

Mast. It can analyse new vessel designs against 30 years hind cast data in wind, waves and currents, enabling a reliable valuation of the contribution of wind on any ship, on any route. Monetising the value of the wind and making the unpredictable predictable enable rapid design optimisation.

## 8 Conclusion

Pressure is increasing on the shipping industry to make significant reductions in emissions, especially $CO_2$. Ambitious targets can be reached if the industry embraces wind propulsion technologies, many of which are in late development or market ready, and class certified. They are the only cluster of technologies that provide significant savings, with reasonable ROI's today and can be deployed as retrofits or new builds using existing technological know-how and materials.

We have seen that there is a growing wind technology basket, and that diversity in itself is one of the keys to robust development. The creation of the IWSA was also an important step forward to help consolidate and strengthen the viability and credibility of the sector across the industry and in policy circles.

The uptake of wind technologies will be driven by both market and non-market forces. The market will drive the sector primarily through the ROI, with fuel prices and the increased credibility and affordability of the equipment itself, which in turn will develop quickly as the number of demonstrator vessels and retrofit installations rise. Regulatory and policy decisions, restrictions and levies on emissions (SOx, NOx, PM and $CO_2$) and incentives to embrace low carbon solutions will all play their part; however, the critical transition will be in the mind-set of the industry. Embracing a hybrid, multi-fuel, multi-solution future designed for a resilient and profitable heterogeneous fleet is an important part of that journey. This reframes the challenge question, not 'Why should we look at wind propulsion solutions?' rather, 'How far can wind technologies take us along the decarbonisation voyage?'

## References

CE Delft. (2016). Study on the analysis of market potentials and market barriers for wind propulsion technologies for ships. http://www.cedelft.eu/?go=home.downloadPub&id=1891&file=CE_Delft_7G92_Wind_Propulsion_Technologies_Final_report.pdf. Accessed 07 January 2017.

Danish EcoCouncil. (2011). Cleaner Shipping – focus on air pollution, technology and regulation. http://ecocouncil.dk/documents/publikationer/1323-130730-shipping-uk/file. Accessed 14 December 2016.

Enercon. (2013). E-Ship: A Wind-Hybrid Commercial Cargo Ship 4th Conference on Ship Efficiency, Hamburg, 23–24. September 2013. http://www.ship-efficiency.org/onTEAM/pdf/06-STG_Ship_Efficiency_2013_100913_Paper.pdf. Accessed 05 January 2017.

ENVI Committee. (2015). Emission Reduction Targets for International Aviation & Shipping – Study for ENVI Committee, European Parliament. http://www.europarl.europa.eu/RegData/etudes/STUD/2015/569964/IPOL_STU(2015)569964_EN.pdf. Accessed 17 December 2016.

Gilbert, P., Walsh, C., & Hodgson, P. (2016). The role of material efficiency to reduce CO2 emissions during ship manufacture: a life cycle approach. *Marine Policy*. doi: https://doi.org/10.1016/j.marpol.2016.04.003. Publication link: 8b76150b-9fa7-48c5-8da7-fe59d9b4055e.
International Maritime Organisation (IMO). (2015). *Third IMO Greenhouse Gas Study 2014*. London: IMO Publications.
International Windship Association (IWSA). (2017). Research centre – listings of many of the publicly available research papers. http://wind-ship.org/en/research. Accessed 29 October 2017.
Lloyd's Register (LR). (2015). Wind-powered Shipping: Review of commercial, regulatory & technical factors affecting uptake of wind-assisted propulsion. http://www.lr.org/en/news-and-insight/publications/windpoweredshipping.aspx. Accessed 07 January 2017.
Mofor, L., Nuttall, P., & Newell, A. (2015). *Renewable energy options for shipping: Technology brief*. Bonn: IRENA.
Ouchi, K., Uzawa, K., Kanai, A., & Katori, M. (2013). Wind challenger the next generation hybrid sailing vessel. In *Third international symposium on marine propulsors smp'13, Launceston, Tasmania, Australia*, May 2013. http://www.marinepropulsors.com/proceedings/2013/11B.3.pdf. Accessed 05 January 2017.
Prakash, V., Smith, T., Rehmatulla, N., Mitchell, J., & Adland, R. (2016). Revealed preferences for energy efficiency in the shipping markets. http://shippingefficiency.org/sites/shippingefficiency.org/files/Revealed%20preferences%20for%20energy%20efficiency%20in%20shipping%20markets.pdf. Accessed 04 January 2017.
Rehmatulla, N., Parker, S., Smith, T., & Stulgis, V. (2017). Wind technologies: Opportunities and barriers to a low carbon shipping industry. *Marine Policy, 75*, 217–226.

## Further Reading

S@IL Program– Engineering Report - EU Intereg IV. (2015). http://www.nsrsail.eu/wp-content/uploads/2015/11/SAIL_FinalReport_Engineering.pdf. Accessed 12 December 2016.
S@IL Program– Roadmap for Sail Transport - EU Intereg IV. (2015). http://www.nsrsail.eu/wp-content/uploads/2015/12/Roadmap-SAIL-Transport-WEB-Bonduelle-WP4.pdf. Accessed 12 December 2016.
Sustainable Shipping Initiative (SSI) - Roadmap Vision 2040. (2016). http://www.ssi2040.org/wp-content/uploads/2016/03/SSI-Roadmap-Simple-web.pdf. Accessed 07 January 2017.

# Part VI
# Marine Renewable Energy

# A Multipurpose Marine Cadastre to Manage Conflict Use with Marine Renewable Energy

Séverine Michalak

Notwithstanding their potential benefit as a non-carbon-emitting energy source, the number and the size of marine renewable energy (MRE) farms increases conflict uses, creating a kind of private occupation of the sea space. On one hand, 3344 offshore wind turbines garnish the European marine spaces, without including those under construction.[1] The United States forecasts 22 GW in offshore wind by 2030 and 86 GW by 2050.[2] On the other hand, these fixed units gathered in industrial production farms reach huge sizes, extended by safety zones. Dong Energy offshore wind projects *Hornsea* 1, 2, 3 and 4, located off the coast of Yorkshire in the United Kingdom, will respectively cover 407, 462, 696 and 845 $km^2$ in the *Hornsea* zone of 4735 $km^2$ dedicated to offshore wind energy.[3]

Successful development of MRE will result from a good balance between a certain degree of private use required by MRE parks and the legal regime of the sea space. Despite its shape-shifting characteristics, horizontally firstly, from territorial sea to high sea regime, and vertically then, from seabed to sea surface, the main legal principle of common use governs the sea space.

Aware of economic and legal issues of private occupation required by coastal and marine activities, some States, such as Canada,[4] built the concept of a multipurpose

[1]The total of the current projects is 21.7 GW (http://www.offshorewind.biz/2016/08/02/infographic-european-offshore-wind-trends-and-statistics-in-2016/). Accessed on 2 August 2016.

[2]Wind Vision: a new ERA for wind power in the United States, Department of Energy (DOE) Report, USA, 2015. https://www.energy.gov/sites/prod/files/WindVision_Report_final.pdf.

[3]http://www.4coffshore.com/windfarms/hornsea-project-two-united-kingdom-uk1u.html.

[4]Canada has the longest coastline (7.1 millions of $km^2$). More than 60% of the territory under the jurisdiction of AATC (Association des Arpenteurs des Terres du Canada) is on water. Canada has created 5 LOMA (Large Ocean Management Area) to help marine planning (Ng'ang'a et al. 2001).

S. Michalak (✉)
University of Nice Sophia-Antipolis, Nice, France
e-mail: severinemichalak@gmail.com



A. I. Ölçer et al. (eds.), *Trends and Challenges in Maritime Energy Management*, WMU Studies in Maritime Affairs 6, https://doi.org/10.1007/978-3-319-74576-3_31

marine cadastre (MMC) in the 1990s, followed by the United States,[5] Australia,[6] and New Zealand.[7] More recently, resulting from MRE development, the Netherlands,[8] Belgium,[9] United Kingdom,[10] Germany,[11] Sweden,[12] Poland,[13] Israel[14] and Greece[15] have also started to examine the concept of MMC. It is interesting to notice that studying the MMC idea is encouraged in Resolution n°4 of the 16th United Nations Regional Cartographic Conference for Asia and Pacific (UNRCC-AP) about Cadastral and SDI. In France, some authors have written about this subject,[16] recently brought up to date by surveyors.[17] The Council of European Geodetic Surveyors (CGLE) coordinates the set-up of MMCs in several European Union Member States.[18]

The MMC seems to be an efficient tool to determine better way to allocate exclusive rights to ocean energy developers, in accordance with other users rights. The United-States are among the pioneers with their marinecadastre.gov website, which has been set clearly to promote offshore renewable energy, as a complement to marine spatial planning.

What does the MMC consist of? We will try to underline its distinctive features compared to the land cadastre and to provide some definitions (Sect. 1). In a land

[5]American marine cadastre can be seen on www.marinecadastre.gov. The USA are among the precursors thanks to their OPIS (*Ocean Planning Information System*) (Fulmer 2007; Fowler and Treml 2001, pp. 493–507).

[6]Australia has set up the *Australian Marine Spatial Information System* (AMSIS), a system of interactive cartography helping the decision making. It contains more than 80 information layers about marine boundaries, bathymetry, physical and environmental data, legal interests, fishing and navigation. (https://www.ga.gov.au/amsis/index.jsp) (Binns et al. 2005).

[7]Knight (2010), pp. 25–32.

[8]Barry et al. (2003), pp. 313–324.

[9]De Latte (2016).

[10]Dumashie (2013), p. 84.

[11]www.geoseaportal.de.

[12]Andersson (2016).

[13]Dawidowicz and Zrobek (2014), pp. 127–144.

[14]Srebro (2015).

[15]Arvanitis et al. (2016).

[16]Boissery and Falque (2009), pp. 7–10; Boissery et al. (2012), p. 4.

[17]The "Portail du Littoral et de la Mer", that will be a development of the "Portail Géofoncier", is being experimented, in collaboration with the Agence de l'Eau Rhône Méditerranée Corse, the ICREI and the Association France International Expertise Foncière, in 2 test zones: a first zone of 40 km between Carry-le-Rouet and La Ciotat, on the French maritime public domain and marine uses, and a second zone of 4 km at the level of Sète ("Marseille, zone test du premier 'cadastre marin'", www.laprovence.com, 11 May 2016). See http://www.sogefi-sig.com/portfolio-item/portail-du-littoral-et-de-la-mer/.

[18]"Portail du littoral et de la mer": Initiative taken in Athens during the PCC (*Permanent Committee on cadastre*) Conference on May 2014, agreed in Rome on December 2014 and launched by CGLE (Comité de liaison des géomètres experts), EULIS (*European Land Information Service*), ELRA (*European Land Registry Association*), and Euro Geographics (Schmidt 2016).

cadastre, the right to exclude others constitutes the main characteristic of property ownership.[19] In the sea space, the degree of exclusivity required by MRE farms varies from country to country (Sect. 2). Given the lack of exclusive use in the sea space legal regime, it would be interesting to analyze the way the presence of MRE farms matches with the public rights and the innocent passage right in the territorial sea and other States' rights in the exclusive economic zone (EEZ) (Sect. 3).

# 1 Distinctive Characteristics and Definitions of the Multipurpose Marine Cadastre (MMC)

After highlighting the characteristics of the MMC (Sect. 1.1), we will try to give some definitions (Sect. 1.2).

## *1.1 Distinctive Characteristics of MMC*

The FIG in its *Statement on the cadastre* gives this definition of a land cadastre: "a parcel based and up-to-date land information system containing a record of interests in land (e.g. rights, restrictions and responsibilities). It usually includes a geometric description of land parcels linked to other records describing the nature of the interests, the ownership or control of those interests and often the value of the parcel and its improvements".[20]

Cadastres at sea already exist. In France, for instance, the development of aquaculture has required the creation of specific marine cadastres[21] and the VALPENA programme aims to map the presence of fishing vessels in the areas where MRE farms are likely to be set-up.[22] However, the new idea consists of gathering all marine activities in a single marine cadastre, called a Multipurpose Marine Cadastre (MMC). The efficiency of the MMC depends on its ability to respect all specific features of the sea space, such as the impossible mapping of physical parcels, and its three or four dimensional aspect, leading to overlap of sometimes incompatible activities. Indeed, "few marine activities take place on the "surface" of the water. Nearly everything marine takes up the water column. Most marine rights, such as aquaculture, mining, fishing, and mooring rights and even

[19]Bergel et al. (2010).

[20]http://www.fig.net/resources/publications/figpub/pub11/figpub11.asp.

[21]Article 1st of Decree of 29 February 2012 about administrative management of marine farming permits (NOR: AGRM1202806A) (Mongruel et al. 2006, p. 137 et s).

[22]GIS VALPENA is a collaborative platform federating University of Nantes and seven Regional Committees for Maritime Fishing and Marine Fish Farming (CRPMEM) gathering more than 3200 ships.

navigation have an inherently 3D nature, which makes a 2D definition of these rights legally inadequate. Where and how do these rights overlap?(...) Any two marine rights are likely to intersect not at the surface of the water, but at some point, far below, in the water column or even within the seabed. Controlling, and regulating marine activity and, consequently, a facilitating ocean governance requires a more accurate portrayal of rights in the water column, by using a 3D representation of these rights in a "*marine cadastre*".[23]

The production of marine renewable electricity itself needs this three or four-dimensional space. The MRE installations are built or fixed to the seabed. The harnessing of MRE resources takes place in the water column, at the water surface or in the airspace above for wind energy. Land space also presents a three-dimensional aspect, given subsoil mining and geothermal exploitation and air space use. However, the communication between these three dimensions is less easy than in marine space. Besides, consents granted are temporary, given that sea space is, as a priority, dedicated to public use.

The MMC is also distinguished by a shared sovereignty regime, with the innocent passage of foreign vessels in the territorial sea and other states' rights in the EEZ.

## *1.2 Definition(s) of the Multipurpose Marine Cadastre*

The MMC is defined as a tool aimed to "create a sustainable and equitable management system for spatially governed marine rights, restrictions and responsibilities" or as "the inventory, the positioning, the recording and updating of regulation, activities, uses, rights and easements located in the marine and coastal space, and the persons responsible for".[24] An MMC is also defined as a "marine information system, encompassing both the nature and spatial extent of interests and property rights, with respect to ownership and various rights and responsibilities in the marine jurisdiction".[25] These property rights are related to resource use and not to private appropriation as *ownership.*[26]

MMC represents therefore a permanent tool for listing all rights at sea. First, the rights held by the coastal state according to the international law of the sea, whether it is the exercise of sovereignty in the territorial sea,[27] or its sovereign rights in the EEZ.[28] Internal public property law then varies from State to State. In France, the

[23]Ng'ang'a et al. (2004), p. 447.

[24]Paper presented by Mazuyer F at 1st Mediterranean Union of Surveyors (UMG) Congress in Marrakech, 17–19 April 2015.

[25]Nichols et al. (2000), pp. 415–424; another definition is: "*A marine cadastre is a system to enable the boundaries of maritime rights and interests to be recorded, spatially managed and physically defined in relationship to the boundaries of other neighboring or underlying rights and interests*".

[26]Kirat (2012), p. 59.

[27]UNCLOS Article 2.

[28]UNCLOS Article 56. Coastal State has sovereign rights regarding the production of energy from the water, currents and wind.

legal regime of the seabed, which is part of the maritime public domain,[29] is different from the water column regime, called *res communis*,[30] and from the overhead airspace,[31] thereby supporting the 4D aspect of the sea space. The MMC also lists public rights, such as freedom of navigation, and other States' rights, such as innocent passage in the territorial sea,[32] and the rights exercised by the third States in the EEZ.[33] It lists, finally, restrictions such as marine protected areas or MRE farms safety zones, and private use consents. MMC could become, like the land cadastre, given huge economic issues related to marine activities and "Blue Growth",[34] an efficient tool to manage financial resources from these activities. Indeed, it would facilitate "a fair taxation system, allowing taxation of both sea-related property and sea-generated income. In the EEZ, taxation of property and income would obviously be possible following prior registration of the offshore units under a national flag, ideally pursuant to international conventions concluded at the initiative of the IMO and/or the Comité Maritime International (CMI)".[35]

MMC is a complementary tool for maritime spatial planning (MSP), subject of the directive 2014/89.[36] According to Article 3, MSP consists of analyzing and organizing "human activities in marine areas to achieve ecological, economic and social objectives". MSP aims to, according to Article 8, establish plans "which identify the spatial and temporal distribution of relevant existing and future activities and uses, in their marine waters, in order to contribute to the objectives set out in Article 5".

MSP will be set up more efficiently and at a lower cost if data collected thanks to MMC are used.[37] In the case of existing zoning, MMC facilitates recording and checking of all consents and rights granted in the marine area.[38] MMC is the main tool of integrated ocean management strategy as the unifying element of all different

[29]The EEZ and continental shelf are not considered as public properties. Article L2111-4 of CG3P has only included in the French maritime public domain the seabed and subsoil, but neither the continental shelf, nor the water column above.

[30]Article 714 Civil Code.

[31]The French doctrine is divided regarding the question of the existence of an air public domain (Godfrin and Degoffe 2012, p. 68); Bettio, n° 21: "French State property of the air space is only agreed regarding the space overhanging public domain dependencies (Council of State, 23 Oct. 1963, SNCF: RDP 1963, p. 1191, conclusions Braibant)".

[32]UNCLOS Articles 17–32.

[33]UNCLOS Article 58.

[34]http://ec.europa.eu/maritimeaffairs/policy/blue_growth/index_fr.htm.

[35]Balla and Wouters (2017).

[36]Directive 2014/89/EU of the European Parliament and of the Council of 23 July 2014 establishing a framework for maritime spatial planning.

[37]Cf. The directive 2007/2/EC of the European Parliament and of the Council of 14 March 2007 establishing an Infrastructure for Spatial Information in the European Community (INSPIRE). Cf. (https://data.ifremer.fr/Tout-savoir-sur-les-donnees/Cadre-juridique-de-diffusion/Directive-Inspire).

[38]Arvanitis et al. (2016).

competitive uses.[39] In the United States, MMC has proved its efficiency in the development of the *Territorial Sea Plan of Oregon*,[40] whose part 5A describes proceedings related to MRE projects. MMC is then a tool, among others, in the global MSP process.[41]

In many countries, MRE development has increased interest in the marine cadastre concept, given growing private space required by MRE farms. Yet private use is not the norm in the sea space, contrary to the land space. We will now analyze how this private use is translated in different countries.

## 2 The Different Manifestations of Private Use by MRE Developers: Some Case Studies

Private use presents itself in different ways, such as navigation restriction or ban on MRE farms because of maritime safety. Private use also exists for strategic reasons, as part of sea space management, and regarding other fixed activities, such as aquaculture or oil and gas exploitation. For instance, Germany, Belgium and Portugal dedicate some marine areas to MRE activity, almost in an exclusive way. We will now study the cases of some States in more detail.

### *2.1 United Kingdom*

United Kingdom leads the offshore wind energy market.[42] Wide areas dedicated to offshore wind were selected a long time ago as part of *Rounds*,[43] thanks to the *Marine Resource System* (MaRS) GIS.[44] In *Round* 3[45] some *Zones Development Agreements* (ZDA) were selected, in which exclusive rights were granted to developers, allowing them to choose, themselves, suitable areas to implement farms, after excluding oil and gas, and aquaculture concessions, *Round* 1 and 2 areas, and rail

[39]https://www.acls-aatc.ca/en/node/40: cf. *Canada's Ocean Action Plan* and *Canada's Oceans Act.*

[40]Harte et al. (2010), p. 26.

[41]We can mention also the *Marine Scaping* for marine activities structuring (Toonen and Van Tatenhove 2013, pp. 43–52).

[42]*The European offshore wind industry. Key trends and statistics 2016.* Published in January 2017, www.windeurope.org.

[43]https://www.thecrownestate.co.uk/media/5644/ei-round-3-offshore-wind-site-selection-at-national-and-project-levels.pdf.

[44]The Crown Estate uses this tool as part of seabed leases granting.

[45]The Round 3 includes nine development zones located around the United Kingdom. Seven of them are located in the English and Welsh waters.

navigation.[46] Article 36 B of the *Electricity Act* 1989 guarantees public right of navigation[47] and ensures priority to main navigation rails.

Public rights might be restricted by the *Energy Act* of 2004 (article 99) and the new article 36 A in the *Electricity Act* of 1989, in case of negative interactions with MRE farms, protected by safety zones (*Energy Act* of 2004 articles 95 & 96). Safety zones are usually fixed at 500 m distance around MRE farms under construction, at 500 m also around fixed platforms for staff accommodation and extensive maintenance works and 50 m around operating wind turbines.[48] UNCLOS has set a maximum limit for safety zones around installations in the EEZ of 500 m (art. 60.5).

Article 96 bans vessel access to safety zones, except if expressly allowed, to prevent harmful effects to farms building or maintenance from vessels, and bans also access to every prohibited activity according to Article 95. Fishing vessel access, for instance, is allowed if a 50 m area around turbines is respected, but they may be restricted or forbidden for safety reasons.[49] Public rights may be restricted if justified by *the public interest,* and the British government has recognized that "*the loss of access to fishing grounds by fishermen might in individual cases give rise to an entitlement to compensation*".[50] The UK supports compensation for the restriction or loss of sea space user's rights.[51]

The two hats worn by the Crown may lead to conflicts of interest. Indeed, the Crown, as public rights' trustee, has to protect them. But the Crown, through the Crown Estate, also owns the seabed and grants leases to MRE developers. That is why the Scottish Parliament voted the *Land Reform (Scotland) Act* in 2003[52] creating *access rights* directly held by the public that they can invoke in court. Scotland, given its huge targets in MRE,[53] wishes its marine and land spaces management to be decentralized because of overly commercial management led by the Crown Estate, lacking protection of public interests.

---

[46] UK had declared a Renewable Energy Zone (REZ) before declaring a EEZ. The Exclusive Economic Zone Order *2013* acknowledging the EEZ in UK has come into effect on 31 March 2014, abrogating in article 1(2) the REZ Order Act.

[47] The Crown's ownership of the seabed, and every title granted on its basis, shall respect public rights, such as fishing or sailing (Todd 2012, pp. 667–672: public rights shall be respected also in EEZ by MRE developers) (A-G for British Columbia v A-G for the Dominion of Canada, 1914, AC 153,173).

[48] Safety zone statement; Pursuant to regulation 6(1), b(ii) of the Infrastructure Planning (Applications: Prescribed forms and Procedure) Regulations 2009, Dogger Bank Creyke Beck August 2013, Forewind. The Electricity (Offshore Generating Stations) (Safety Zones) (Application Procedures and Control of Access) Regulations 2007 (SI No 2007/1948).

[49] Christie et al. (2014), pp. 254–261.

[50] Plant (2004), pp. 868–881.

[51] Todd (2012), p. 671.

[52] Land Reform (Scotland) Act 2003. The Bill for this Act of the Scottish Parliament was passed by the Parliament on 23rd January 2003 and received Royal Assent on 25th February 2003.

[53] In the context of independence attempts, the Scottish minister of Energy, Fergus Ewing, has approved four offshore wind projects of 2.3 GW, reproaching to UK to give priority to nuclear sector www.greenunivers.com. Accessed 16 October 2014.

## 2.2 Germany

Germany has planned offshore wind energy (OWE) development through three areas. *Priority zones* (*Vorrangebiete*) grant *exclusivity* to OWE farms. Any other kind of exploitation is banned if incompatible with OWE. Three priority zones have been defined in the North Sea and two in the Baltic Sea.[54] *Reserved zones* (*Vorbehaltsgebiete*) confer *priority* and not *exclusivity* to OWE, in relation to other activities that are allowed as part of preliminary planning. Finally, within *appropriate zones* (*Eignungsgebiete*) OWE is allowed *a priori*, without any arbitration with other users.

Most of the OWE parks are located in the EEZ, far from the strong potential conflict use areas, thanks to the low bathymetry in the German EEZ. Sailing and fishing are forbidden inside OWE farms under construction. For farms in operation, rules vary from park to park. For instance, any access to EnBW Baltic 1 is allowed for vessels over24 m and for trawler ships. Moreover, access is allowed provided that visibility is at least 500 m.[55]

## 2.3 Belgium

By a royal decree of 20 March 2014[56] establishing a marine spatial plan for Belgium, a special area dedicated for MRE activities subject to consents has been allocated. In this zone MRE activity is expressly prioritized in relation to other activities. The royal decree of 11 April 2012[57] clarifies the safety zones around MRE farms. The maximum distance allowed is 500 m around the farm but also around each turbine, and around a group of farms. No access is allowed, except by warships, research vessels, submarine pipelines and other exceptional cases.

## 2.4 Portugal

Portugal has, since 2007,[58] created a special legal régime for wave energy, by allowing private use of the water column of the territorial sea. The decree-law n°5/2008 of 8 January 2008, creating a pilot zone for wave energy, allows in article

---

[54]Prall (2009).

[55](BSH M5) 01/16.

[56]Section 3 article 8 § 1 and 2.

[57]Royal decree of 11 April 2012 establishing a safety zone around artificial islands, installations and devices for power generation from water, currents and the wind in maritime regions under Belgium jurisdiction (n°2012011172).

[58]Law n°57/207 of 31 August 2007.

4 other activities in the pilot zone provided they respect the high priority granted to wave energy. The pilot zone is managed as part of the public service delegation and wave energy consents are granted after a tender process.

## 2.5 France

Exclusive rights are granted to MRE developers laureates in a tender process. The laureate wins the exclusive right to harness MRE resources (wind, waves, current...) in the dedicated zone. But they may boast this exclusive right only in relation to their direct MRE competitors, and not to other users of the sea space, such as fishers. The legal regime applying to the private use of the maritime public domain by MRE developers is called "concession d'utilisation du domaine public maritime en dehors des ports".[59] The disadvantage of this regime is the obligation to justify the public interest in the MRE. Yet, this public interest is very often disputed. MRE activity needs the creation of a special legal regime, like aquaculture.

Regarding other nomadic activities, such as navigation, the maritime prefect, in virtue of their role in the respect of public order in the marine space,[60] can prohibit the set-up of MRE farms in the traffic separation scheme (TSS) and submit it to the respect of maritime easements (submarine cables, pipelines...). The Local Nautical Commission or the Great Nautical Commission may be consulted before setting any signal traffic.[61] Rules regarding the establishment of safety zones vary from farm to farm. Freedom of sailing may be restricted for administrative or environmental reasons, provided these restrictions do not exceed what is required by the pursued objectives.[62] Once MRE farm set-up is allowed, freedom of sailing may be restricted just because of close proximity of an MRE farm, or restricted inside its safety zone.

The Great Nautical Commission has recommended, as part of the offshore wind farm project located in Courseulles-sur-Mer,[63] the establishment of a safety zone of 500 m around each building area. During the operation phase, they propose to ban vessels exceeding 25 m in length from sailing inside the farm, to prohibit vessels from accessing a safety zone of 50 m around each wind turbine, to ban the parking or anchorage inside the whole farm, except in case of damage to the ship, to ban also diving and to limit passenger ships navigation to 500 m from the MRE farm. Fishing will also be limited.[64]

---

[59]Cf. *supra.*

[60]Decree n°2004-112 of 6 February 2004 related to the organization of State Action at Sea (JORF n°32 of 7 February 2004, p. 2616), article 1.

[61]Decree n°86-606 of 14 March 1986 (JO 19 March 1986) modified by the decree n°88-199 of 29 February 1988 (JO 2 March 1988).

[62]Rezenthel (2007), p. 684; Rezenthel (2012), p. 736.

[63]Great Nautical Commission (2015).

[64]Regulation about fishing isn't yet established.

Insofar as private use of space, which is the norm in the land cadastre, constitutes an exception in the sea space, it would be interesting to analyze the way private use matches with public rights or other state rights in the territorial sea and the EEZ. Indeed, establishing an MMC requires, after listing all the rights existing in the sea, to know their hierarchy. For instance, to what extent may the public rights be restricted by MRE farms? How might MRE farms be advantaged while respecting innocent passage? MMC study demands also a dynamic approach and requires the redesign of legal backgrounds of marine spatial planning and conflict use management.

## 3 The Private Use of MRE to the Test of the Rights of States Under International Law

We will focus now on the articulation between the presence of MRE farms and other States regarding, firstly, the innocent passage right in the territorial sea (Sect. 3.1) and, secondly, the other States' rights in the EEZ such as the freedom of navigation and of laying of sub-marine cables (Sect. 3.2).[65]

### *3.1 Innocent Passage Right in the Territorial Sea*

The coastal State cannot hamper the innocent passage of foreign ships through the territorial sea except in accordance with this Convention.[66] When does this passage cease to be innocent? According to article 19.2.k of UNCLOS, "Passage of a foreign ship shall be considered to be prejudicial to the peace, good order or security of the coastal State if in the territorial sea it engages in any of the following activities: (...) any act aimed at interfering with any systems of communication or any other facilities or installations of the coastal State". This question has been raised about offshore oil rigs. A coastal state is free to allow the construction of installations in its territorial sea provided these installations do not interfere with innocent passage of foreign ships.[67] We could apply this situation to MRE works. While Nordic countries set up their offshore wind farms in the EEZ, France will establish them in its territorial sea, as the Saint-Nazaire, Fécamp or Courseulles-sur-Mer projects, except in the Mediterranean Sea where the floating offshore wind projects will take place in the EEZ, given significant conflicts of use in the territorial sea.

According to article 22, "the coastal State may, where necessary having regard to the safety of navigation, require foreign ships exercising the right of innocent

[65]High sea legal regime won't be studied here, given the hypothetical MRE projects.

[66]UNCLOS Article 24.

[67]Esmaeli (2001), pp. 72–73.

passage through its territorial sea to use such sea lanes and traffic separation schemes as it may designate or prescribe for the regulation of the passage of ships". The General Secretary of IMO (International Maritime Organization) has encouraged the Member States to make their vessel traffic management systems more compatible with new uses, such as marine renewable energy projects, while ensuring the security of navigation.[68] The evaluation of security reasons falls within the coastal State's discretionary power and cannot be controlled except in exceptional cases such as abusive or excessive evaluation.[69] That is why the coastal state may not prohibit totally the navigation of ships inside MRE farms. At most, it could, on the basis of article 25.3, suspend temporarily the exercise of foreign ships' innocent passage inside the MRE safety zone.

MRE economic and strategic stakes carry weight in the exercise of a Coastal State's discretionary power. Could we classify MRE parks and their safety zones as *Particularly Sensitive Sea Areas* (*PSSA*)[70]? A *PSSA* is "an area that needs special protection through action by IMO because of its significance for recognized ecological or socio-economic or scientific reasons and which may be vulnerable to damage by international maritime activities". These zones are usually located in dense maritime traffic areas and may meet one of three criteria. Except ecological and scientific criteria, MRE areas might fulfil the third, socio-economic, criterion, which may concern touristic interest, or strong dependence level of the local population on the marine environment, or marine resources exploitation, such as MRE resources.

## 3.2 *The Rights of Other States in the EEZ*

We will analyze the freedom of navigation in relation to MRE safety zone establishment (Sect. 3.2.1) and to the removal of MRE installations (Sect. 3.2.2) and the distinguish between the freedom of laying of sub-marine cables and MRE farms (Sect. 3.2.3).

### 3.2.1 The Safety Zones

UNCLOS grants priority to freedom of navigation. Article 60.7 prohibits installations and safety zones in the EEZ in case of risk to the use of recognized sea lanes

[68] Address of the secretary-General (2012) at the Opening of the Fifty-Eighth Session of the Sub-Committee on Safety of navigation. www.imo.org/MediaCentre/SecretaryGeneral/Secretary-GeneralsSpeeches/Meetings/Pages/Sub-Committee-on-Safety-of-Navigation-(NAV)-58th-session.aspx. Accessed 2 July 2012.

[69] Scalieris (2011), p. 56.

[70] www.imo.org; Lefebvre-Chalain (2012), p. 416 onwards.

essential to international navigation.[71] In the absence of such risk, the coastal State may establish safety zones around installations to the limit of 500 m (art. 60.5). It may take *appropriate measures* in these safety zones to ensure the safety of both navigation and of the installations (art. 60.4).

UNCLOS is quite unclear about these *appropriate measures* and allows the coastal State to decide what kind of vessel is allowed, or not, to access the safety zone. The Coastal State's power shall be exercised in a well-proportioned way, and the safety zones "shall be designed to ensure that they are reasonably related to the nature and function of (...) installations" (art. 60.5). Any conflict arising between the coastal State and any other State shall be resolved on the basis of equity, according to article 59 of UNCLOS.

### 3.2.2 The Removal of MRE Installations

Article 60.3 of UNCLOS orders the removal of abandoned installations in respect of generally accepted international standards of fishing, of the protection of the marine environment and the rights and duties of other States. A partial removal is allowed, provided appropriate publicity. The French decree related to the establishment of installations in the French EEZ (article 17) allows a partial removal if necessary to ensure the protection of the marine environment and provided security of navigation is not threatened.[72] The principle of total removal has been softened by the London Convention on the Prevention of Marine Pollution by Dumping of Wastes and Other Matter,[73] confirmed by the negotiations of the Third United Nations Conference on the law of the Sea. The British negotiators, yet fervent defenders of the freedom of navigation, adopted the point of view of oil companies, given the huge amount of total removal of offshore platforms. France, as a nuclear power State, was less in favor of this softening because of consequences to fishing or navigation safety. France has finally joined the United Kingdom. We can guess that partial removal of MRE installations will also be preferred for economic or safety reasons, more than environmental ones. Indeed, in the UK, the cost of removal of offshore wind turbines as part of *Rounds* 1 and 2 has been estimated at 335 million pounds.[74]

---

[71]Esmaeli (2001), p. 240.

[72]Decree n°2013-611 of 10 July 2013 related to regulation applied to artificial islands, installations, devices in the exclusive economic zone, on the continental shelf in the economic zone and in the ecological protection zone, and to the drawing of submarine cables and pipelines.

[73]Adopted on 13 November 1972 and come into force on 30 August 1975.

[74]This is equivalent to about 460 millions euros, related to mostly offshore wind at work, in building or authorized as part of *Round 1* and 2. Estimation based on the hypothesis of the set-up of 7.2 GW by 2020 as part of *Round 2* (Climate Change Capital, Offshore Renewable Energy Installation Decommissioning Study Final Report, 17 November 2010).

### 3.2.3 The Freedom of Laying of Sub-marine Cables

Among the freedoms enjoyed by other States, we can find the freedom of laying of sub-marine cables in the EEZ (art. 58 UNCLOS) or on the continental shelf (article 79). Article 79.3 states that the drawing of pipelines on the continental shelf shall be agreed by the coastal State. Interference can occur between MRE farms and the presence of telecommunication cables laid in the same area.[75] Given that MRE technologies are more recent, MRE farms will probably be established where cables already exist. Offshore wind turbines set-up close to a submarine cable increase the probability of exposing it to damage caused by scraping of the seabed resulting from sedimentary erosion around the offshore wind structure.[76] MRE farms may also make access more difficult for ships in charge of the repairing telecommunication cables. According to article 79.5 UNCLOS, "States shall have due regard to cables or pipelines already in position" when they lay sub-marine cables or pipelines.

In UK, the *Crown Estate, Subsea Cables UK*, the *European Subsea Cables Association* and *Renewable UK* have put forward some recommendations about the respect of distances between MRE farms and sub-marine cables. If the distance is more than one nautical mile, no proximity agreement is required. If less, the parties involved shall find an agreement based on a minimal distance of 750 m. A distance superior to 500 m is considered dangerous.[77]

The *International Cable Protection Committee* (ICPC) has recommended, when the bathymetry is at least 40 m, a minimal distance of 1100 m, that it can be extended or reduced according to geographical or bathymetric circumstances, or other parameters required by ships that fix cables.[78]

Interference might also occur between the repair services of MRE electric cables and telecommunication cables. According to the *North American Submarine Cables Association, "For an undersea telecommunications cable owner planning to install a new undersea telecommunications cable that will cross an energy export cable, installation costs (for negotiating multiple cable crossing agreements) and risks both to the cable and to its commercial agreements in the event of delay have dramatically increased*". They also recommend finalizing "*reasonable crossing agreements*" to reduce such risks.

The UNCLOS gaps related to the degree of priority that shall be granted to MRE will probably lead States to make more and more rules on a case-by-case basis, following recommendations of international and regional organizations.

---

[75]Nielsen and Davenport (2014), p. 371 onwards.

[76]Comments of the North American Submarine Cable Association (NASCA) (2012).

[77]Subsea Cables UK (2012).

[78]ICPC (2010).

# 4 Conclusion

A multipurpose marine cadastre (MMC), as a complementary tool for Marine Spatial Planning (MSP) and a base layer of the Marine Spatial Data Infrastructure (MSDI),[79] would be efficient to clarify the rights, restrictions and responsibilities existing at sea, and to facilitate transfer of titles, and establishment of mortgages. Besides, it would give the opportunity to redesign a property rights regime more suitable to the sea space, taking into account the growing private use of the sea and its necessary combination with common use principle. It may be appropriate to encourage a right to *include* more than a right to exclude, to highlight specific features of MMC in comparison with land cadastre, and to support synergies between activities, such as aquaculture inside MRE farms, preventing MRE developers from being accused of a potential "*crime of enclosure*".[80]

The set-up of an MMC meets, nevertheless, some constraints, such as scattering of data related to rights, restrictions and responsibilities (granting of permits and concessions, delimitation of marine areas, environmental data), held by a multiple range of actors. Aside from the international law of the sea, the establishment of an MMC depends mostly on each internal legal system, and constitutes, therefore, a challenge for global ocean governance.

## References

Andersson, M. (2016). *Marine cadastre. The Swedish case.* Paper presented at the Common vision conference 2016, Migration to a smart world, Amsterdam, 6 June 2016.

Arvanitis, A., Giannakopoulou, S., & Parri, I. (2016). *Marine cadastre to support marine spatial planning*. Paper presented at the Common vision conference 2016, Migration to a smart world, Amsterdam, 5–7 June 2016.

Balla, E., & Wouters, R. (2017). Marine cadastre in Europe: State of play. Paper prepared for presentation at the "2017 World Bank Conference on land and poverty". The World Bank, Washington DC, 20–24 March 2017.

Barry, M., Elema, I., & Van der Molen, P. (2003). Ocean governance and the marine cadastre: The Netherlands North Sea. *Geomatica, 57*(3), 313–324.

Bergel, J. -L., Bruschi, M., & Cimamonti, S. (2010). Traité de droit civil, Les biens, LGDJ, 2nd éd. 2010 n°95 et s.

---

[79] The Marine Spatial Data Infrastructure includes seabed topography, geology, marine infrastructure, resources utilization, administrative and legal boundaries, areas of conservation, marine habitats and oceanography (International Hydrographic Organization 2011). https://www.iho.int/iho_pubs/CB/C-17_e1.1.0_2011_EN.pdf

[80] Cf. Supreme Court of India Civil appeal n°1132/2011 of 28 January 2011: the small lake in the District of Patiala that offered varied functions has been qualified by the Supreme Court of "collective good", judging that the real estate company who wanted to monopolize it was guilty of the "*Crime of enclosure*" (Parance and de Saint Victor 2014, p. 13).

Binns, A., Rajabifard, A., Collier, P. A., & Williamson, I. (2005). Developing the concept of a marine cadastre: An Australian case study. Department of Geomatics, University of Melbourne, Australia.

Boissery, P., & Falque, M. (2009). Prospective de gestion du domaine public maritime et cadastre marin. *Les Annales de la Voirie, 138*, 7–10.

Boissery, P., Valentini-Poirier, C. A., & Falque, M. (2012). Territorialiser la mer. Un cadastre marin pour le domaine public maritime, ICREI, Etudes foncières n°160, novembre-décembre 2012, p. 4.

Christie, N., Smith, K., Barnes, R., & Elliott, M. (2014). Co-location of activities and designations: A means of solving or creating problems in marine spatial planning? *Marine Policy, 43*, 254–261.

Comments of the North American Submarine Cable Association (NASCA). (2012). Before the Bureau of Ocean Management, US. Department of the Interior In the Matter of Atlantic OCS Proposed Geological and Geophysical Activities, Mid-Atlantic and South Atlantic Planning Areas Draft Programmatic Environmental Impact Statement, OCS EIS/EA BOEM 2012-005, p. 20. Retrieved May 30, 2012, from http://www.wiltshiregrannis.com/siteFiles/News/F7FE3C69FE76D28CD58D601E45D5C56B.pdf (Statement of NASCA).

Dawidowicz, A., & Zrobek, R. (2014). Multipurpose water-marine cadastre in Poland – Development directions. *Acta Adriatica, 55*(2), 127–144.

De Latte, G. (2016). *Marine cadastre. Legal framework. UNCLOS and EU Legislation*. Paper presented at the Common vision conference 2016, Migration to a smart world, Amsterdam, 5–7 June 2016.

Dumashie, D. (2013). The marine cadastre: A rights based approach to marine conservation in England. In M. Falque & H. Lamotte (eds.), *Biodiversité, droits de propriété, économie et environnement* (p. 84). Bruylant.

Esmaeli, H. (2001). *The legal regime of offshore oil rigs in international law*. Aldershot, England: Ashgate/Dartmouth.

Fowler, C., & Treml, E. (2001). Building a marine cadastral information system for the United States: A case study. *Computers, Environment and Urban Systems, 25*, 493–507.

Fulmer, J. (2007). *The Multipurpose Marine Cadastre Web Map*. Paper presented at 2007 ESRI Survey & Engineering GIS Summit, San Diego, CA, 16–19 June 2007.

Godfrin, P., & Degoffe, M. (2012). Droit administratif des biens (10e éd, p. 68). Sirey. Paris: Dalloz.

Great Nautical Commission. (2015). Report after meeting on 8 April 2015 about offshore wind farm project off the coast of Courseulles-sur-Mer. http://www.shom.fr/fileadmin/data/DMI/PLANS/Programmes_GHO/PV-CONSLUSIONS_GCN_parc-eolien-Courseulles_8.04.2015_Vf_01.pdf

Harte, M., Campbell, H. V., & Webster, J. (2010). *Identification of OCS renewable energy space-use conflicts and analysis of potential mitigation measures* (p. 26). Oregon State University, 31 March 2010.

ICPC. (2010). Recommendation n°13 Proximity of Wind Farm Developments and Submarine Cables. Retrieved September 27, 2010, from https://www.iscpc.org/publications/recommendations/

Kirat, T. (2012). *Economie du droit* (p. 59). Paris: Editions La Découverte.

Knight, P. (2010). Beyond the horizon: Making way for offshore resource management in New Zealand. *Journal of the New Zealand Institute of Surveyors, 300*, 25–32.

Lefebvre-Chalain, H. (2012). La stratégie normative de l'organisation maritime internationale (OMI), PUAM, n°585 onwards, p. 416 onwards.

Mongruel, R., Perez Agundez, J. A., Girard, S., Bailly, D., & Nassiri, A. (2006). Analyse des stratégies d'occupation du domaine maritime côtier: le cas des concessions conchylicoles. Rapport final du projet ASTRODOME, Publications Amure, série Rapport, n° R-11-2008, Programme National Environnement Côtier, Novembre 2006, p. 137 et s.

Ng'ang'a, S., Nichols, S., Sutherland, M., & Cockburn, S. (2001). Toward a multidimensional marine cadastre in support of good ocean governance. New spatial information management tools and their role in natural resource management. Paper presented at the International Conference on spatial information for sustainable development, Nairobi, Kenya, 2–5 October 2001.

Ng'ang'a, S., Sutherland, M., Cockburn, S., & Nichols, S. (2004). Toward a 3D marine cadastre in support of good ocean governance: A review of the technical framework requirements. *Computers, Environment and Urban Systems, 28*, 447.

Nichols, S., Monahan, D., & Sutherland, M. (2000). Good governance of Canada's offshore and coastal zone: Towards an understanding of the marine boundary issues. *Geomatica, 54*(4), 415–424.

Nielsen, W. F., & Davenport, T. (2014). Submarine cables and offshore energy. In D. R. Burret, R. C. Beckman, & T. M. Davenport (Eds.), *Submarine cables. The handbook of law and policy* (p. 371 onwards). Leiden, the Netherlands: Martinus Nijhoff Publishers.

Parance, B., & de Saint Victor, J. (2014). « Commons, biens communs, communs »: une révolution juridique nécessaire. In B. Parance (ed.), *Repenser les biens communs* (p. 13). Paris: CNRS Editions.

Plant, G. (2004). Offshore renewable energy development and the Energy Bill 2004. *JPL*, 868–881.

Prall, U. (2009). *Legal frame for the use of offshore wind energy in Germany*. Paper presented at the 3rd Seminar on EIA for offshore wind farms vis –à vis Natura 2000 and other land uses: "Which kind of legal frame we need?" Riga, Latvia, 15–16 April 2009.

Rezenthel, R. (2007). La navigation de plaisance: une conciliation de la liberté et de l'intérêt général, DMF: 684.

Rezenthel, R. (2012). La navigation de plaisance: une liberté fondamentale conciliable avec la protection de l'environnement, DMF: 736.

Scalieris, E. (2011). L'exercice du pouvoir discrétionnaire de l'Etat côtier en droit de la mer Pédone Paris, p. 56.

Schmidt, N. (2016). Marine cadastre: A support to the blue economy? Paper presented at CGLE General Assembly, Tirana, 18 March 2016.

Srebro, H. (2015). Implementation of marine cadastre in Israel, International Federation of Surveyors, Article of the Month, September 2015.

Subsea Cables UK. (2012). Guideline n°6 on the Proximity of Offshore Renewable Energy Installations and Submarine Cable Infrastructure in Waters, Issue 4. The Crown Estate. Retrieved August 2012, from https://www.thecrownestate.co.uk/media/5658/ei-km-in-pc-cables-082012-proximity-of-offshore-renewable-energy-installations-submarine-cable-infra structure-in-uk-waters-guideline.pdf

Todd, P. (2012). Marine renewable energy and public rights. *Marine Policy, 36*, 667–672.

Toonen, H. M., & Van Tatenhove, J. (2013). Marine scaping: The structuring of marine practices. *Ocean and Coastal Management, 75*, 43–52.

# Ocean Energy: Seeking the Balance Between States' Exclusive Rights of Exploitation and Marine Biodiversity Conservation

Wellington Tebar

## 1 Introduction

Environmental preservation is considered a fundamental condition for the maintenance of life, at all its levels, on this planet. It is important to say, however, that the emergence of an ecological conscience can be attached to the growing awareness that the natural resources are finite. In other words, nature's newly discovered value cannot be attributed, at least not in a legal perspective, to a sudden recognition of its intrinsic value (Mendes 2000); otherwise, it could imply the inevitable recognition of rights to other living beings.

It can explain, in part, the nomenclature adopted to represent this new way of facing economic growth: "sustainable development". There are several (different) definitions for the expression "sustainable development".[1] However, perhaps the

[1]Baroni (1992), in a study designed to investigate a few definitions that were given to the term "sustainable development", concluded that some authors define it as they believe it should be, or as they would like it to be. On the other hand, there are those who "confuse sustainable development with ecological sustainability—which refers only to the capacity that natural resources have to reproduce themselves". Differently, some authors recognize that the current model of economic growth leads to an unsustainable exploitation of natural resources and, for that reason, it must be limited. Finally, there are those who replace the traditional notion of economic development by the modern notion of sustainable development, so that the new adjective attached to it ("sustainable") implies that development itself was not able to increase the well-being of citizens and to reduce the poverty. Taking into account all these point of views, Frey (2001) has summarised the several concepts of sustainable development presented by the specialized literature. He has based his analysis on three dimensions: the model of political-administrative relation, the role of the State in economy regulation and the democratic potential regarding decision making in environmental affairs. As a result, the author has proposed the following systematization, divided into three groups

W. Tebar (✉)
Study Center for the Law of the Sea "Vicente Marotta Rangel" of the University of São Paulo, School of Law (CEDMAR-USP), São Paulo, Brazil
e-mail: cedmar.usp@gmail.com



A. I. Ölçer et al. (eds.), *Trends and Challenges in Maritime Energy Management*, WMU Studies in Maritime Affairs 6, https://doi.org/10.1007/978-3-319-74576-3_32

most commonly known can be found in the report of the World Commission on Environment and Development "Our Common Future" (1987) (Brundtland Report), elaborated in preparation of the relevant discussions that took place in Rio de Janeiro, in 1992, at the United Nations Conference. According to this document, "sustainable development is development that meets the needs of the present without compromising the ability of future generations to meet their own needs".

It is important to remark that the definition set by the report is not immune to critics, because it does not define precisely the needs of the present or the future generations.[2] Nevertheless, it has succeeded in drawing specific attention to new methods of promoting economic development, which encompass the commitment to preventing the depletion of the world's natural resources and to avoiding environmental damage (Barbosa 2008). In this respect, since the emergence of the sustainable development concept, environmentally friendly concepts were adopted in reference to this new way of facing economic growth, such as "green economy"[3]and "blue economy",[4] among others.

The term "blue economy" is particularly suitable for the scientific objective of this paper, as it refers to the use of the seas and the oceans to achieve the goals of sustainable development. Here, the investment and the development of renewable energies that take place at maritime zones can be mentioned. More precisely, within all types of energy that can fulfill this characteristic, special focus has been given to ocean energy.

However, the possibility of ocean energy exploitation faces several challenges: the development of the needed technologies; the legal framework (concerning the analysis of whether a State owns the right to exclusively exploit resources within its national jurisdiction according to both domestic and international law); the regulatory framework (structure for the project, sitting and permitting procedures, intellectual property issues); and the environmental aspects (environmental attributes, environmental impact assessment, marine biodiversity conservation).

---

of ideas: (1) the free market notion, which sees on the market the regulatory strength of sustainable development; (2) the ecological-technocrat notion of economic planning, which considers the State regulatory action on economy the key point to ensure the prevalence of citizens well-being in the development process; and (3) the political notion of democratic participation, which considers the society, through social mobilization, the key element to the achievement of sustainable development. Ultimately, it can be concluded that there is no consensus on how to achieve sustainability through development, due to the highly different ideological components of the many definitions given to the aforementioned term by its authors. As a result, this lack of conceptual clarity leads to inefficient policy making, especially in international contexts. At this regard, Cf. Lélé (1991), pp. 607–221.

[2]Although, some indicative elements can be found in the mentioned document, such as: the increase of employment, the guarantee of minimum consumption standards, more food production (but based on ecological policies), the change of energy consumption patterns (to more greener standards), housing, water supply, sanitation, and health care, among others.

[3]For an overview of general aspects, Cf. A guidebook to the Green Economy (2012).

[4]For an overview of general aspects, Cf. Blue Economy (2015).

Addressing all of these questions would be extremely problematic because the scientific object of this paper could become excessively wide. Therefore, in order to maintain scientific coherence, this paper is divided into three parts. Firstly, it addresses the precise definition of "ocean energy" (in contrast with the concept of "marine renewable energy").

Secondly, it focuses exclusively on the international legal framework, especially with regard to the provisions of the United Nations Convention on The Law of The Sea (hereinafter UNCLOS). In fact, UNCLOS is the primary legal document to be taken into consideration, since it sets a static legal framework of maritime zones and (to them) correlated jurisdiction(al rights) of exploitation. In the second part, the legal regimes of maritime zones (drawing special attention to the zones subject to State's national jurisdiction), are addressed in order to analyze whether the coastal State has exclusive rights of exploitation over certain maritime zones, or whether it must consider the (possible) rights of third Parties (namely those related to navigation and fishery).

Thirdly, this paper analyzes whether UNCLOS sets any provision that implies the existence of a duty to preserve biodiversity while developing industrial activity in the ocean. In this sense, biodiversity is addressed as an externality regarding the industrial exploitation of ocean energy. This paper also addresses the legal tool for taking biodiversity into account in the decision making process (sitting and permitting procedure), which is the environmental impact assessment (EIA).

## 2 Ocean Energy

Ocean energy can be considered as a renewable energy. Taking into account that the objective of this paper could become excessively wide, it is important to define what types of renewable energies will be addressed and which international legal framework will be considered. Effectively, can every type of energy produced in (or by) the ocean be categorized as ocean energy? Which international statute will be considered to resolve potential disputes between States triggered by the exploitation of ocean energy?

Firstly, it is important to remark on the difference between "marine renewable energy" and "ocean energy". Although both forms of energy can be considered as renewable, the first one is a broader term used to designate all forms of energy derived from industrial activity that takes place in one of the maritime zones. In this sense, offshore wind can be considered as a marine renewable energy (Iddri 2015). On the other hand, the second term is used to designate all forms of energy which provide the generation of electricity, at any level of ocean power, such as the energy generated by the movement of waves and tides, or even its thermal energy and salinity gradient (Batey 2010).

Secondly, in the event of a dispute between States, considering that the installation of the infrastructure necessary for the exploitation of ocean energy will be set in the ocean domain, any dispute between States, resulting from that, should be examined in light of UNCLOS provisions.[5] Therein will be addressed the extension of the States' jurisdiction(al powers) in the development of industrial activities in the ocean, according to the legal regimes of maritime zones described in UNCLOS.

Finally, one can argue that, in a legal international perspective, there is not much of a difference between addressing offshore wind energy and ocean energy, as far as both could be analyzed according to UNCLOS provisions. Indeed, this could be true, depending on the nature of the dispute between the States concerned. However, the distinction is important for several reasons. For example, beyond the conceptual relevance, the environmental issues to be addressed and the legal issues to be considered could be different, claiming, therefore, for specific solutions.

## 3 United Nations Convention on the Law of the Sea

The United Nations Convention on the Law of the Sea (UNCLOS), concluded in 1982, in Montego Bay, regulates the legal discipline to be applied to the (exploitation of) natural resources located in the ocean domain. In fact, there are at least five main legal regimes that govern all aspects concerning marine resources: Territorial Sea, Exclusive Economic Zone, Continental Shelf, Area and High Seas. Hereinafter, this paper will analyze the extension of the States' jurisdiction(al powers) in the development of industrial activities in the ocean, according to the legal regimes of maritime zones described in UNCLOS. In other words, when developing its offshore operations, can the State interfere or affect the rights of third parties (other States)?

### *3.1 Territorial Sea*

The legal regime of the territorial sea is distinguished by the remarkable presence of sovereignty (Art.2, §1). This sovereignty extends itself to the air space over the territorial sea as well as to its bed and subsoil and must be exercised according to the provisions set by UNCLOS and other rules of international law (Art.2, §2 and §3). It can be concluded, then, that States have full entitlement over the territorial sea. It means that States retain exclusive rights of exploitation of any kind of industrial

[5]It is well known that, in an international perspective, there are few regional legal frameworks that regulates, event thought they are not yet fully developed, the specific features of ocean energy. It can be mentioned, here, the initiatives of the European community law, such as the action plan to support the development of blue energy within Europe 2020 Strategy and beyond.

activity, so that the coastal States can refuse any kind of unlawful interference in their jurisdictional waters.

Other States merely enjoy the right of innocent passage through the territorial sea (Art.17), as long as it is not prejudicial to the peace, good order or security of the coastal State (Art.19). In fact, coastal States can lawfully regulate innocent passage (Craig 2012) with respect to, among other subjects: protection of navigational aids and facilities and other facilities or installations; the conservation of living resources of the sea; the preservation of the environment of the coastal State and the prevention, reduction and control of pollution thereof.

## 3.2 Exclusive Economic Zone

The legal regime of the Exclusive Economic Zone (hereinafter EEZ) is, indeed, interesting. UNCLOS assigns *sovereign rights* to the States for the specific purposes of exploring and exploiting, conserving and managing the natural resources, whether living or non-living, of the waters superjacent to the seabed and of the seabed and its subsoil, and with regard to other activities for the economic exploitation and exploration of the zone (Art.56, §1, "a"). On the other hand, UNCLOS also assigns *jurisdiction* to the States, with regard to: (1) the establishment and use of artificial islands, installations and structures; (2) the marine scientific research; (3) the protection and preservation of the marine environment (Art.56, §1, "b").[6]

For the purpose of this paper, it is important to consider that the States have *sovereign rights* regarding the *production of energy from the water, currents and winds* (Art.56, §1, "a"), as well as exclusive rights to construct and to authorize and to regulate the construction, operation and use of installations and structures for the purposes provided for in article 56 (Art.60, §1, "b"). It can be concluded, then, that States have full entitlement over the EEZ in this particular matter. It means that States retain exclusive rights of exploitation of any kind of ocean energy.

On the other hand, it can be mentioned that in exercising its rights and performing its duties under UNCLOS in the exclusive economic zone, the coastal State shall have due regard to the rights and duties of other States and shall act in a manner compatible with the provisions of UNCLOS (Art.56, §2).

---

[6]It is important to remark that UNCLOS assigns *sovereign rights* and *jurisdiction(al rights)* to the States, instead of *sovereignty*, in referring to the prerogative of powers exercised by coastal States over their respective EEZ. This can be explained by the fact that, actually, the EEZ constitutes a third genre (EEZ is said to be *sui generis*), since it does not fit into the standard legal discipline of the territorial sea, nor the high seas (Gavouneli 2007), but retains legal implications of the both mentioned maritime zones. Then, it can be implied that although the EEZ is subject to the prerogative of power exercised by coastal States (typical feature of the territorial sea), such prerogative is limited by the rules of international law, such as the freedom of navigation (typical feature of the high seas). In other words, meanwhile the *full sovereignty* is exercised over the territorial sea and the *absence of sovereignty* is the characteristic that drives the high seas, only *sovereign and jurisdictional rights* are exercised over the EEZ (Gavouneli 2007).

## 3.3 Continental Shelf

Regarding the legal regime of the continental shelf, UNCLOS assigns *sovereign rights* to the States for the purpose of exploring it and exploiting its natural resources (Art.77, §1), consisting of the mineral and other non-living resources of the seabed and subsoil together with living organisms belonging to sedentary species, that is to say, organisms which, at the harvestable stage, either are immobile on or under the seabed or are unable to move except in constant physical contact with the seabed or the subsoil (Art.77, §4), so that no one may undertake these activities without the express consent of the coastal State (Art.77, §2).

For the purpose of this paper, it is important to consider that the infrastructure necessary for the *production of energy from the water and currents* in the EEZ will probably be installed in the continental shelf (because, in general, the water column above the continental shelf corresponds to the domains of EEZ), as provided for in Art.60 of UNCLOS. It means that (the legal framework of) States sovereign powers exercised over the EEZ will be partially and symmetrically extended to the continental shelf (specifically in the matter of ocean energy exploitation).

It can be concluded, then, that States have full entitlement over the continental shelf in this particular matter, so that States retain exclusive rights of exploitation of any kind of industrial activity related to that. It does not mean, however, that the States own the water column or the continental shelf, but solely that States possess the necessary rights to proceed to exploitation of the natural resources. In other words, States retain exclusive rights in the sense that third Parties cannot exploit such resources without the consent of the coastal State (Jayakumar 2013).

Symmetrically, the exercise of the rights of the coastal State over the continental shelf (by installing the aforementioned infrastructure, for example) must not infringe or result in any unjustifiable interference with navigation and other rights and freedoms that other States can exercise in the water column above (EEZ), so that the coastal State shall act in a manner compatible with the provisions of UNCLOS (Art.78, §2).

## 3.4 Area

The legal regime of the Area (that comprehends the seabed, the ocean floor and the subsoil, beyond the limits of national jurisdiction) is guided by the principle of common heritage of the world (Art.136), so that all of its resources (that comprehends all solid, liquid or gaseous mineral resources *in situ* in the Area at or beneath the seabed, including polymetallic nodules—Art.133, "a") are vested in mankind as a whole, on whose behalf the Authority shall act (Art.137, §2). This means that no State shall claim or exercise sovereignty or *sovereign rights* over any part of the Area or its resources (Art.137, §1), nor shall any State or natural or juridical person claim, acquire or exercise rights with respect to the minerals recovered from the Area except in accordance with the rules set by the Convention (Art.137, §3).

It can be concluded, then, that if any exploitation of ocean energy takes place in the Area, it can be done by the authority and the results of such exploitation shall be governed by the principles set by UNCLOS, such as common heritage of mankind (Art.136), transfer of technology (Art.144), share of benefits (Art.140), protection of the marine environment (Art.145), among others.

### 3.5 *High Seas*

At last, concerning the legal regime of the high seas, UNCLOS still retains the fundamental right of freedom for flagged vessels, so that the high seas are open to all States. However, this is not an unconditioned freedom, since it must be exercised under the conditions laid down by UNCLOS and by other rules of international law (Art.87, §1) (Zanella 2013). In fact, no State may validly purport to subject any part of the high seas to its sovereignty (Art.89) and shall be reserved for peaceful purposes (Art.88).

It can be concluded, then, that if any exploitation of ocean energy takes place in the high seas, it can be done by any State, as long as there is no unjustifiable interference with navigation and other rights and freedoms that other States can exercise on the high seas (Art.87, §1).

## 4 Ocean Governance and Marine Biodiversity Conservation

The protection of the animal and vegetal species, as well as their *habitats* and the correct functioning of terrestrial and marine ecosystems, is not a recent concern. What is new is the term "biodiversity" that, in the past, was not employed. Nevertheless, now, there is a great variety of international instruments, of regional and universal scope, that promotes, punctually, the protection of biodiversity, directly[7] or indirectly.[8]

[7] It can be mentioned: The Convention on Wetlands (1971); The Convention concerning the Protection of World Cultural and Natural Heritage (1972); The Bern Convention on the Conservation of European Wildlife and Natural Habitat (1973); The Convention on the Conservation of Migratory Species of Wild Animals (1979); The Convention on the Conservation of Antarctic Marine Living Resources (1980); The Convention on Biological Diversity (1992).

[8] It can be mentioned: The Convention for the Protection of the Natural Resources and Environment of the South Pacific Region (1976); The United Nation Convention on the Law of the Sea (1982); World Charter for Nature (1982); The United Nations Framework Convention on Climate Change (1992). It has been said that the there is an indirect protection, because in these international

Considering that, for the purpose of this paper, eventual industrial activities will be set in the ocean domain, the legal analysis will remain based on the Constitution of the Seas. In this respect, it has to be said that the UNCLOS sets general and geographically oriented duties regarding environmental protection (Craig 2012).

Geographically oriented duties mean that UNCLOS sets specific duties of environmental protection according to the legal regimes of maritime zones. It can be mentioned, for example, the duties provided for in article 67 (EEZ), regarding the responsibility of coastal States for the management of the *catadromous* species in whose waters that species spend the greater part of their life cycle.

On the other hand, general duties mean that the provisions regarding the protection of environment shall be applied to all maritime zones. It seems to be consensual that articles 192 and 194, §5, of UNCLOS, which restate customary international law, constitute part of the legal basis that imposes to States general duties to achieve the protection and preservation of the marine environment, that are extended even to the high seas,[9] even though UNCLOS also provides for rights of exploitation (Art.193).

By setting these duties for the States to comply with environmental protection, UNCLOS stimulates the improvement of ocean governance regimes, since third States shall respect the governance boundaries established by coastal States. However, at the same time, it can be said that "these provisions underscore a fundamental tension that runs through UNCLOS, and, indeed, through marine governance generally, between environmental protection and the right to use marine resources" (Craig 2012).

So, the challenge to be faced by marine governance approaches and tolls refers to the achievement of the balance between the States' (exclusive) rights of exploitation and their duty to protect the environment, whereas while developing industrial activities in the maritime zones, States must take into account measures to protect biodiversity. There are a few approaches that can be undertaken to achieve this goal, such as place-based marine governance.

According to Craig (2012), "the basic premise of place-based marine governance is that regulators can delineate a particular area of the ocean (large or small) and create a governance regime for that area that simultaneously addresses all the values to be protected and all activities of concern. Specifically, this governance regime can identify key habitats, functions, and ecosystems of the area that it seeks to protect, restore, and maintain; prioritizes among uses of the area, such as privileging fishing over recreation or vice-versa; outlaw (or limit) certain uses (such as fishing) and

---

instruments, the biodiversity has been introduced as a privileged requirement to be considered when carrying out the Environmental Impact Assessment. Regarding European community law, the introduction of biological diversity as a requirement to be considered when carrying out the Environmental Impact Assessment is also present. It can be mentioned: The EIA Directive 85/337/CEE, that applies to a wide range of defined public and private projects, and the Directive 2001/42/CEE, that applies to a wide range of public plans and programmes. In this respect, Cf. Billet (2008), pp. 62–63.

[9]In fact, these general duties constitute the main argument for justifying the creation of Marine Protected Areas in the High Seas. In this respect, Cf. Scovazzi (2003).

activities (such as polluting or habitat destroying activities); and limit or regulate access to prevent overexploitation or overuse or to allow a damaged area to recover".

In fact, the legal framework set by UNCLOS allows the adoption of a place-based marine management approach, since it provides for the coastal States the necessary rights to govern the maritime zones under their respective jurisdictions (Craig 2012), and, at the same time, imposes general and specific duties to preserve the marine environment. In other words, States are allowed to exploit the natural resources in the ocean domain, but they shall comply with the duties regarding biodiversity preservation. There are many methods that can be used to implement a place-based marine management, such as Integrated Coastal Management[10] (ICM) and Ecosystem-Based Management[11] (EBM). Within these approaches, there are specific tolls, such as the implementation of Marine Protected Areas[12] (MPAs) and Ocean Zoning[13] (OZ).

The common ground to all of these approaches and tolls is that they are all part of a long and complex decision making process (sitting and permitting procedure). For this reason, for the purpose of this paper, just two important stages of this long process will be taken into account, namely the environmental impact assessment and public participation. In effect, the environmental impact assessment is a necessary tool to take environmental values into consideration, since it evaluates the environmental impacts of industrial activities. On the other hand, public participation is critically important, whereas the socioecological dimension of the governance regime must be considered in the decision making process (Craig 2012).

## 5 Environmental Impact Assessment and Decision Making

The Environmental Impact Assessment is a well-known tool in International Environmental Law. In this regard, UNCLOS was the first Convention of universal scope to prescribe the need for States to assess the potential effects that could derive from the exploitation of activities in the marine environment when the States have

---

[10]For an overview, Cf. Secretariat of the Convention on Biological Diversity (2015), 108 pages, especially pp. 32–46 (for the concept).

[11]For an overview, Cf. United Nations Environment Programme (UNEP)/Global Programme of Action for the Protection of the Marine Environment from Land-based Activities (GPA). Ecosystem-based management: markers for assessing progress (2006), especially pp. 4–7 (for the concept). Highlighting the importance of this concept, it can be pointed out that the Convention for the Protection of the Marine Environment of the North-East Atlantic ("Ospar Convention") has explicitly mentioned, in its Annex V, the "integrated ecosystem approach" as a necessary tool "to protect the maritime area against the adverse effects of human activities so as to safeguard human health and to conserve marine ecosystems".

[12]For an overview, Cf. Commonwealth Department of Environment and Heritage of Australia. The benefits of Marine Protected Areas (2003). Especially p. 4 (for the concept).

[13]For an overview, Cf. Agardy (2010), especially pp. 09–10 (for the concept—including explaining the relationship and the difference between ocean zoning and land zoning).

"reasonable grounds for believing that planned activities under their jurisdiction or control may cause substantial pollution of or significant and harmful changes to the marine environment" (Art.206).

Indeed, while UNCLOS has multiplied the possibilities for the use of the seas, it has also prescribed specific duties to the States, such as the management and protection of marine biological resources (Cot 2010). In this sense, the Environmental Impact Assessment guarantees that no activity will be implemented when there is a potential threat of serious harm to biodiversity (Long 2013). It is important to remark that the term "threat" encompasses not only the pollution, but any potential cause of serious damage to the environment (Massop 2013), such as a harmful interference in the *habitat* of a certain species, for example, by the installation of the infrastructure to be used in ocean energy exploitation.

It is true that ocean energy is a type of renewable energy, so it is relatively nonpolluting and causes fewer adverse environmental impacts in comparison with other forms of energy, such as the combustion of fossil fuels (Starrs 1993). However, in order to comply with UNCLOS provisions, before implementing an industrial activity in the ocean (to exploit its energy), the States must assess the adverse impacts that the enterprise may cause to the marine environment.

Here, it is important to remark on the "strategic environmental impact assessment", which involves States cooperation (especially maritime bordering States) with the purpose of mapping the most profitable areas (from an energetic point of view) and, at the same time, the less vulnerable ones (from an ecological and financial point of view). This would allow, for example: (1) the adoption of the best exploitation strategies; (2) the prevention of conflicts among States (transboundary harms); and (3) the anticipation of the effects of climate change regarding ocean currents (movement and temperature) (Pelc and Fujita 2002).

In addition, it is important to remark that even in the areas beyond national jurisdiction, the States should comply with the principles of international customary law, acting with due diligence with regard to the approval of industrial projects, in order to avoid serious harm to the environment (Massop 2013). In effect, the precautionary principle has become an important principle of international customary law, especially with regard to the protection of biodiversity in the field of ocean governance (Marffy-Mantuano 2010).

The Environmental Impact Assessment, if correctly executed, can reduce, diminish or even eliminate the environmental stressors derived from the industrial activity. However, it must be remarked that the scientific knowledge about the potential environmental impacts of the activity to be implemented is always limited because it depends on the monitoring and data gathering of the specific sites where the industrial activity would be exploited (Long 2013).

In a context characterized by uncertainty, the precautionary principle could be managed in a way to obstruct any kind of environmental interference. In this respect, the absence of accurate data should not automatically lead to the banishment of the industrial activity, since the risks (to the environment) are not completely predictable. Indeed, in an environmental perspective, there is no such thing as zero risk (Silva 2005). In other words, it can not be neglected that uncertainty is a factor that generates instability, but it is also true that it can not be fully eliminated, so it must be managed in order to avoid its utilization as a paralyzing tool (Amado Gomes 2007).

Therefore, to avoid this problem, a multi-phased sitting and permitting procedure should be implemented with, after a technically oriented environmental impact assessment,[14] the imposition of adequate constrictive measures. These measures could be revised through time according to a standardizing model of information gathering, regarding the type of industrial activity developed and the seriousness of the inherent risk. This proposition would allow the development of the industrial activity, while its environmental impacts would be carefully monitored, which would enable, in the future, after the information gathering, a review of the measures previously imposed (Gollier et al. 2000). In this respect, a banishing measure could be recommended if the constant data evaluation reveals serious impacts to the marine environment.

## 6 Public Participation and Decision Making

The need to establish a multi-phased sitting and permitting procedure regarding the industrial activities to be implemented in the ocean domain was previously mentioned. One crucial part of this procedure refers to public participation, as it comprehends the sociological dimension of the ocean governance regime. In fact, considering that many people shall be affected, directly or indirectly, by the development of industrial activity in ocean areas, the opportunity to influence policy making must be given to them. In this sense, public participation (in a collective or individual approach) is the key to legitimizing the public decision making process (Kannen et al. 2013), bringing to it democratic values (Prieur 1999). It can be concluded, then, that public participation is a condition of validity of the decision making process.[15]

For that reason, it is extremely important to consider the effectiveness of public participation in the decision making process, which will only be achieved by the implementation of specific strategies and tools. Above all, the relationship between the citizens and the sea must be strengthened, mainly through environmental

[14]For example, in a context of uncertainty, there are very interesting propositions in the literature, such as the concept of a probabilistic risk assessment, that could be applied at several risk scenarios. For an overview, Cf. Malsch (2013), pp. 215–224.

[15]It is important to remark that considering public participation as a condition of validity of the decision making process does not mean that decision maker is fully bounded by the public will. It means that the public intervention must be taken into account (Schrage 1999), but it is not binding. It can be explained by the fact that decision makers figure the risk out in a different way compared with the public in general. In fact, as decision makers are assisted by experts, who bear specialized knowledge, they take into consideration not solely the potential risks, damages and losses, but also the potential benefits which could result from the exploitation of certain industrial activity (Sunstein 2013).

education and guarantee of access rights to environmental information (Prieur 1999), in order to promote people's environmental awareness. The "ocean citizens" (Fletcher and Potts 2007) are more prepared to take environmentally responsible decisions and, at the same time, influence future public policies regarding ocean energy exploitation. These very environmentally oriented citizens can also play a decisive role in the organizational capacity of civil society through, for example, the establishment of non-governmental organizations (NGOs), which could lead to an increase in its influence on policy making (Lane 2003).

Besides that, for public participation to be effective, it is important to define precisely the "public concerned" (Munn and Ricketts 2015). Naturally, the participation of the whole community (local, regional and international) is welcomed. However, on particular occasions, wide and uncoordinated public participation could be costly and ineffective. Therefore, policy makers must assess the community's particular needs, in order to choose the appropriate form of public participation and decision making process (Irvin and Stansbury 2004).

For instance, if the exploitation takes place in the territorial sea, a wide range of participation (as it would be onshore) is expected, since there shall be a great number of people directly affected by the industrial enterprise. On the other hand, if the exploitation takes place in another maritime zone (effectively offshore), then the public participation will be mainly addressed to the fishing community, as it could face several problems directly caused by the industrial activity, such as: (1) adverse effects on fish community structure through changes in species composition; (2) exclusion of fishing within deployment areas; (3) adverse effects on migratory species; among others (Boehlert and Gill 2010), requiring, then, a particular approach.

Lastly, it is important to highlight that, in an internal perspective, the extent of public participation in policy making depends on the legal system of each State, which reflects the environmental culture and perception of its citizens. The States should guarantee to the civil society full access to environmental information and value the dialogue among all stakeholders, such as economic actors and citizens, in order to try to achieve an environmentally responsible decision by *consensus*. In this sense, the decision making process would become more democratic and transparent (Prieur 1999).

Naturally, the internal legal systems of each State can be influenced by international legal frameworks. In this regard, in an international perspective, the UNECE Convention on Access to Information, Public Participation in Decision-making and Access to Justice in Environmental Matters (known as—and hereinafter—Aarhus Convention), adopted on 25 June 1998, at Aarhus, Denmark, plays an important role in guarantying the participation and effective influence of the citizens regarding the formulation of public policies in the environmental domain.

The access rights to environmental information are provided for in article 4 (Access to Environmental Information) and 5 (Collection and Dissemination of Environmental Information). On the other hand, the right to public participation in the decision making process is provided for in the articles 6 (Public Participation in Decisions on Specific Activities), 7 (Public Participation Concerning Plans, Programmes and Policies relating to the Environment) and 8 (Public Participation during the Preparation of Executive Regulations and/or Generally Applicable Legally Binding Normative Instruments). Finally, in order to ensure environmental transparency, the right of access to Justice is provided for in article 9 (Access to Justice).

In this international perspective, set by the Aarhus Convention, the provisions regarding access to environmental information, public participation and access to Justice in the environmental domain can give rise to legal and political implications within the framework of national legislation. In effect, the Aarhus Convention creates personal rights for the individuals regarding participation in public regulation (Bétaille 2010). In addition, article 8 of the Aarhus Convention sets general duties to States in order to enable effective public participation and to promote "administrative democracy" (Monediaire 2010). Finally, the Aarhus Convention creates personal rights for all citizens regarding access to all environmental information, ensuring environmentally responsible participation of civil society in policy making, and personal rights of access to Justice, in order to overcome any obstacle to access to environmental information.

## 7 Conclusion

In the past few years, the ocean as a potential resource of energy has caught the attention of the industrial sector. Being a renewable energy, its exploitation allows the environment to naturally recover, at the same time promoting diminished dependence on pollutant forms of energy, such as the combustion of fossil fuels.

The development of renewable energies, especially ocean energy, is a recent concern. In this respect, although there is a great incentive for the development of these resources, by, for example, the elimination of domestic regulatory barriers in several States (Starrs 1993), their legal regime, mostly in an international perspective, remains inhomogeneous. Perhaps, with the expansion of the use of renewable energies (and the inevitable disputes between States resulting from that), the regulation will become part of the international political agenda.

Regarding ocean energy, although there is an (almost) complete lack of international regulation, its exploitation is possible. However, a few conditions must be satisfied. Firstly, States can exploit the ocean energy resource within their national jurisdiction, conserving exclusive rights resulting from industrial activity, but they must not unjustifiably interfere with the rights of third Parties (namely those related to navigation and fishery).

Secondly, the exploiting aspiration must be submitted to a complex decision making process (sitting and permitting procedure). This paper has highlighted just two stages of this long process, namely the environmental impact assessment and public participation. In effect, it is absolutely necessary to assess the potential effects that could derive from the exploitation of activities in the marine environment through a technically oriented environmental impact assessment. In addition, public participation is critically important as a condition of validity of the entire decision making process.

# References

## *Doctrine*


Agardy, T. (2010). *Ocean zoning: Making marine management more effective*. London: Earthscan.

Amado Gomes, C. (2007). *Risco e Modificação do Acto Autorizativo Concretizador de Deveres de Proteção do Ambiente*. Dissertation, University of Lisbon.

Barbosa, G. S. (2008). O Desafio do Desenvolvimento Sustentável. *Revista Visões, 1*(4), 1–11.

Baroni, M. (1992). Ambiguidades e Deficiências do Conceito de Desenvolvimento Sustentável. *Revista de Administração de Empresas, 32*(2), 14–24.

Batey, D. L. (2010). Intellectual property – Protections and pitfalls. In *The law of ocean and tidal energy: A guide to business and legal issues* (pp. 1–5). Stoel Rives Ocean Energy Team.

Bétaille, J. (2010). La Contribution du Droit aux Effets de la Participation du Public: de la Prise en Considération des Résultats de la Participation. *Revue juridique de l'environnement, 35*(2), 197–217.

Billet, P. (2008). La place de la diversité biologique dans les évaluations environnementales, l'information du public e la participation. In *Revue juridique de l'environnement, Limoges, n. spécial. Biodiversité et Évolution du Droit de la Protection de la Nature: Réflexion Prospective* (pp. 57–67).

Boehlert, G. W., & Gill, A. B. (2010). Environmental and ecological effects of ocean renewable energy development: A current synthesis. *Oceanography, 23*(2), 68–81.

Cot, J. P. (2010). L'Océan Partagé. In *Annuaire du Droit de la Mer* (Vol. Xv, pp. 13–21). Paris: Institut du Droit Economique de la Mer. Pedone.

Craig, R. K. (2012). *Comparative ocean governance: Place-based protections in an era of climate change. New horizons in environmental and energy law*. Cheltenham, UK: Edward Elgar.

Fletcher, S., & Potts, J. (2007). Ocean citizenship: An emergent geographical concept. *Journal of Coastal Management, 35*(4), 511–524.

Frey, K. (2001). A Dimensão Político-Democrática nas Teorias de Desenvolvimento Sustentável e suas Implicações para a Gestão Local. *Ambiente & Sociedade, 9*, 115–143.

Gavouneli, M. (2007). *Functional jurisdiction in the law of the sea. Publications on ocean development* (Vol. 62). Boston: Martinus Nijhoff.

Gollier, C., Jullien, B., & Treich, N. (2000). Scientific progress and irreversibility: An economic interpretation of the "Precautionary Principle". *Journal of Public Economics, 75*, 229–253.

Irvin, R. A., & Stansbury, J. (2004). Citizen participation in decision making: Is it worth the effort? *Public Administration Review, 64*(1), 55–65.

Jayakumar, S. (2013). The continental shelf regime under the United Nation Convention on the law of the sea: Reflections after thirty years. In M. H. Nordquist, J. N. Moore, A. Chircop, & R. Long (Eds.), *The regulation of continental shelf development: Rethinking international standards* (pp. 3–12). Leiden: Center for Oceans Law and Policy. Martinus Njhoff Publishers.

Kannen, A., Kremer, H., Gee, K., & Lange, M. (2013). Renewable energy and marine spatial planning: Scientific and legal implications. In M. H. Nordquist, J. N. Moore, A. Chircop, & R. Long (Eds.), *The regulation of continental shelf development: Rethinking international standards* (pp. 153–178). Leiden: Center for Oceans Law and Policy. Martinus Njhoff Publishers.

Lane, M. B. (2003). Participation, decentralization, and civil society: Indigenous rights and democracy in environmental planning. *Journal of Planning Education and Research, 22*(4), 360–373.

Lélé, S. (1991). Sustainable development: A critical review. *World Development, 19*(6), 607–221.

Long, R. (2013). Offshore wind energy development and ecosystem-based marine management in the UE: Are the regulatory answers really blowing in the wind? In M. H. Nordquist, J. N. Moore, A. Chircop, & R. Long (Eds.), *The regulation of continental shelf development: Rethinking international standards* (pp. 15–52). Leiden: Center for Oceans Law and Policy. Martinus Njhoff Publishers.

Malsch, G. (2013). Environmental regulation and probabilistic risk assessment. In M. H. Nordquist, J. N. Moore, A. Chircop, & R. Long (Eds.), *The regulation of continental shelf development: Rethinking international standards* (pp. 215–224). Leiden: Center for Oceans Law and Policy. Martinus Njhoff Publishers.

Marffy-Mantuano, A. (2010). Governance Internationale de la biodiversité marine dans une perspective de développement durable. In *Annuaire du Droit de la Mer* (Vol. Xv, pp. 177–205). Paris: Institut du Droit Economique de la Mer. Pedone.

Massop, J. (2013). The legal framework for the regulation of safety and environmental issues on the outer continental shelf. In M. H. Nordquist, J. N. Moore, A. Chircop, & R. Long (Eds.), *The regulation of continental shelf development: Rethinking international standards* (pp. 179–193). Leiden: Center for Oceans Law and Policy. Martinus Njhoff Publishers.

Mendes, P. S. (2000). *Vale a pena o Direito Penal do Ambiente?* Lisboa: AAFDL.

Monediaire, G. (2010). La Participation du Public à L'élaboration des Actes Réglementaires de L'état et de ses Établissements Publics en Matière d'Environnement: La Nécessité d'une Loi Grenelle 3? *Revue juridique de l'environnement, 35*(1), 223–239.

Munn, J., & Ricketts, S. (2015). Public participation in environmental decision-making, plan-making and executive regulations-Articles 6, 7 and 8 of the Aarhus Convention. In *The Aarhus Convention: A guide for UK lawyers* (pp. 121–139). Oxford: Hart.

Pelc, R., & Fujita, R. M. (2002). Renewable energy from the ocean. *Marine Policy, 26*, 471–479.

Prieur, M. (1999). La Convention d'Aarhus: instrument universel de la démocracie environnementale. *Revue juridique de l'environnement, 24*(1), 09–29.

Schrage, W. (1999). La Convention sur l'Accès à l'Information, la Participation du Public au Processus décisionnel et L'accès à la Justice em Matière D'environnement. *Revue juridique de l'environnement, 24*, 5–7.

Scovazzi, T. (2003). *Marine protected areas on the high seas: Some legal and policy considerations*. Duban, South Africa. World Parks Congress, Governance Session.

Silva, C. P. (2005). *Verde cor de Direito: Lições de Direito do Ambiente*. Coimbra: Almedina.

Starrs, T. (1993). Solar, wind and geothermal energy. In C. Campbell-Mohn, B. Breen, J. William Futrell, J. M. McElfish Jr., & P. Grant (Eds.), *Environmental law: From resources to recovery* (pp. 482–526). St. Paul, Minn.: Environmental Law Institute. West Publishing Co..

Sunstein, C. R. (2013). *Beyond the precautionary principle. Public law and legal theory* (Working paper n°. 38) (pp. 1–48). Chicago.

Zanella, T. V. (2013). Liberdades e Restrições à Navegação Marítima em Alto Mar no Direito Internacional. *Revista do Instituto de Direito Brasileiro, 2*(10), 11795–11824.

## Online Documents

*A guidebook to the Green Economy* (2012). Report prepared by Cameron Allen and Stuart Clouth for the UN Division for Sustainable Development. Available at: <https://sustainabledevelopment.un.org/content/documents/GE%20Guidebook.pdf>. Access in: 02/01/2017.

*Blue Economy* (2015). UNEP Regional Seas Report and Studies No. 195. Available at: <http://www.unep.org/ecosystemmanagement/water/regionalseas40/Portals/50221/BlueEconomy_SSSIC_screen_rev.pdf>. Access in: 02/01/2017.

*Ecosystem-based management: markers for assessing progress* (2006). United Nations Environment Programme (UNEP)/Global Programme of Action for the Protection of the Marine Environment from Land-based Activities (GPA). The Hague. Available at: <http://www.unep.org/pdf/GPA/Ecosystem_based_Management_Markers_for_Assessing_Progress.pdf >. Access in: 02/01/2017.

*Integrated Coastal Management for the Achievement of the Aichi Biodiversity Targets: Practical Guidance for Implementation Based on Experience and Lessons Learned from Coastal and Ocean Governance in the Seas of East Asia* (2015). Secretariat of the Convention on Biological Diversity. Montreal, Technical Series No. 76, 108 pages. Available at: <https://www.cbd.int/doc/publications/cbd-ts-76-en.pdf>. Access in: 02/01/2017.

*Ocean energy: key legal issues and challenges* (2015). In Issue Brief – oceans and coastal zones, n°04/15. IDDRI, Sciencespo, pp. 1–4. Available at: <http://www.iddri.org/Publications/Collections/Syntheses/IB0415.pdf>. Access in: 02/01/2017.

*Our Common Future: Report of the World Commission on Environment and Development* (1987). Chapter 2: Towards Sustainable Development. United Nations. Available at: <http://www.un-documents.net/wced-ocf.htm>. Access in: 02/01/2017.

*The benefits of Marine Protected Areas* (2003). Commonwealth Department of Environment and Heritage of Australia. Available at: <https://www.environment.gov.au/resource/benefits-marine-protected-areas-discussion-paper>. Access in: 02/01/2017.

# Learning from Humpback Whales for Improving the Energy Capturing Performance of Tidal Turbine Blades

**Weichao Shi, Mehmet Atlar, and Rosemary Norman**

## 1 Introduction

Learning from Mother Nature is a never ending process for modern technology development. According to the research development (Fish and Battle 1996; Fish et al. 2011; Miklosovic et al. 2004), the protrusion/tubercles on the leading edges of the flippers of humpback whales have shown the ability to improve the performance of the flippers in terms of improving the lift-to-drag ratio as well as delaying the stall. This finding has shown the potential to enhance the performance of tidal turbine blades.

This giant species of marine mammal generally ranges from 12 to 16 m. They first attracted the attention of marine biologists for their unusual feeding behaviour. A study conducted from the 1970s to the 1980s into their feeding behaviour (Hain et al. 1982) showed their most inventive technique which is known as bubble net feeding. First a group of humpback whale swims in a shrinking circle and then they blow bubbles below a school of prey. With the shrinking ring of bubbles, the school of fish can be confined in an ever-smaller cylinder. Finally, they swallow their prey in the bubble net.

W. Shi (✉)
School of Marine Science and Technology, Newcastle University, Newcastle upon Tyne, UK

Department of Naval Architecture, Ocean & Marine Engineering, University of Strathclyde, Glasgow, UK
e-mail: weichao.shi@strath.ac.uk

M. Atlar
Department of Naval Architecture, Ocean & Marine Engineering, University of Strathclyde, Glasgow, UK

R. Norman
School of Marine Science and Technology, Newcastle University, Newcastle upon Tyne, UK

 
A. I. Ölçer et al. (eds.), *Trends and Challenges in Maritime Energy Management*,
WMU Studies in Maritime Affairs 6, https://doi.org/10.1007/978-3-319-74576-3_33

With such unique preying habits, humpback whales surprised researchers with agile moves like sharp turns and tight circular swimming. Their unusually long flippers show spectacular hydrodynamic design with NACA 63 series like cross section. But, in contrast, their leading edge does not seem to be either geometrically or hydrodynamically smooth; instead, it has large protuberances which lead to the research reviewed here.

Since then, research studies in both numerical simulations and experimental investigations have been conducted to investigate the effect of application on 2D and 3D foils. Based on these studies, foil based devices equipped with leading-edge tubercles have shown maintained lift coefficients after the stall (Kim et al. 2016; Hansen et al. 2016; Bolzon et al. 2016; Aftab et al. 2016; Wei et al. 2015). Based on the above findings, the tubercles might be able to provide tidal turbine blades with the following hydrodynamic advantages:

1. Better lift-to-drag ratio: This particular characteristic might be able to enhance the maximum power coefficient to help the turbine generate more power and more use of the available energy;
2. Delaying the stall angle: This factor might be able to help the turbine to operate more stably without the sudden loss of the lift which may cause fluctuation in power generation, hence lowering the power quality. This may in turn cause vibration and fatigue problems that reduce the lifetime of a tidal turbine.
3. Maintaining the lift under stall: This is another factor which might be very effective in the starting process of fixed pitch tidal turbines, as this process relies purely on the initial torque of the turbine to overcome the frictional torque and to start spinning. Since the initial torque is very small due to the high angle of attack, maintaining the lift under high angle of attack is crucial.

The above reviewed developments and technical observations have motivated this study to develop this biomimetic concept further with a view to improving the performance of tidal turbines. Despite the benefits of such potential, it has been found that there is an "extremely limited amount of research available for the application of the leading-edge tubercles on a tidal turbine". This will require further research, in particular experimental research, to help in understanding the physics of the leading-edge tubercles and whether they can improve turbine performance.

Within this framework, this project first focused on a tubercle design and optimization, which is described in Sect. 2. By conducting fundamental experimental and numerical investigations on the representative tidal turbine blade, the leading-edge tubercles demonstrated significant benefits, which are further elaborated in Sect. 3. Following the confidence built on the results of the representative turbine blade, a series of model testing campaigns with the three tidal turbine models to investigate the effect of leading-edge tubercles on the turbines has been conducted and results have been presented and discussed in Sect. 4. In Sect. 5, the findings in the representative turbine blade and the turbine models are liaised to summarize the major impact of applying leading-edge tubercles onto tidal turbines.

# 2 Tubercled Leading Edge Design and Optimisation

## 2.1 *Description of Target Tidal Turbine*

The target tidal turbine blade was designed by using the S814 profile cross-section from the NREL series (Barltrop et al. 2006; Wang et al. 2007), as shown in Fig. 1. The key parameters are given in Table 1.

## 2.2 *Tubercle Profile Design*

The research on the two dimensional foil sections was more focusing on the optimisation of the sinusoidal shape tubercle profiles defined by various parameters because of the close pattern with the specimen (Van Nierop et al. 2008; Swanson and Isaac 2011; Stanway 2008; Hansen et al. 2009, 2011, 2012, 2016). However, the current literature does not display any useful application of the leading-edge tubercles on thick and highly cambered tidal turbine foil sections like the S814 section used in this study.

Therefore, combined with the experience from the previous studies, the sinusoidal form of tubercles was selected as the basis shape to conduct the numerical optimisation process. Via systematically changing two variants, the Height (H) and the Wavelength (W), of these protrusions, various sizes of leading-edge

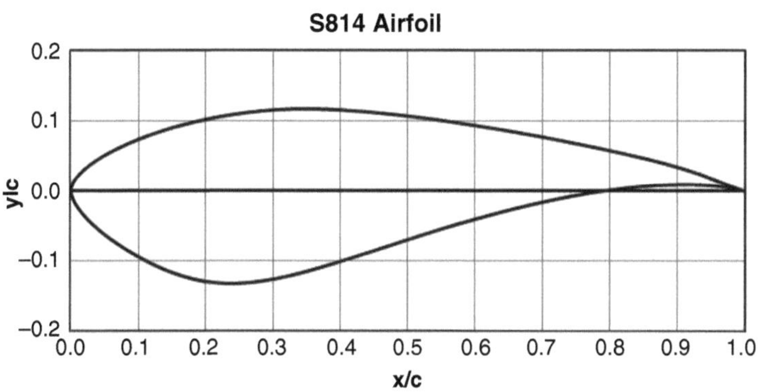


**Fig. 1** Cross-section profile of S814 (Shi et al. 2016)

**Table 1** Main particulars of the tidal stream turbine model (D = 400 mm) (Shi et al. 2016)

| r/R | 0.2 | 0.3 | 0.4 | 0.5 | 0.6 | 0.7 | 0.8 | 0.9 | 1 |
|---|---|---|---|---|---|---|---|---|---|
| Chord length (mm) | 64.35 | 60.06 | 55.76 | 51.47 | 47.18 | 42.88 | 38.59 | 34.29 | 30 |
| Pitch angle (deg) | 27 | 15 | 7.5 | 4 | 2 | 0.5 | −0.4 | −1.3 | −2 |

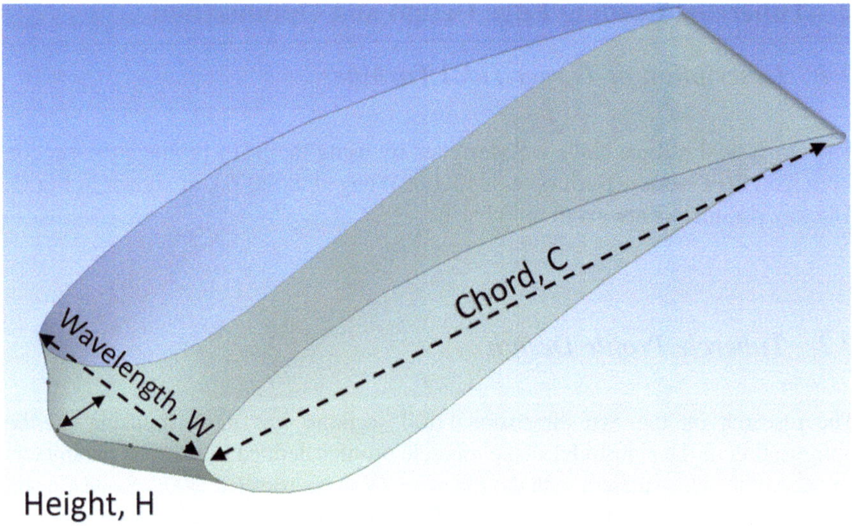


**Fig. 2** Definition of 2D foil with a sinusoidal tubercle (Shi et al. 2016)

tubercles can be generated. The definitions of these parameters are shown in Fig. 2. The leading-edge profile is defined as in Eq. 1.

$$h = \frac{H}{2} cos \left[ \frac{2\pi}{W} s - \pi \right] + \frac{H}{2} \tag{1}$$

where $h$ is the local protrusion height of the leading-edge tubercles compared to the reference one which is the smooth leading-edge profile, mm; $s$ is the spanwise location of foil section, mm.

## 2.3 *Optimization Methodology*

Numerical optimization of leading-edge tubercles for a 2D foil section S814 (with the representative profile section of S814) is the pathway to find the optimum design. With the changing variants, the models were built, meshed and evaluated all in the integrated environment of ANSYS-Workbench, which includes ANSYS-Designmodeller as a geometry generator, ANSYS-Meshing as a mesh generator and ANSYS-CFX as a CFD code for performance evaluation.

## 2.4 *Optimum Design of Tubercles for S814*

The study was documented in detail as in reference (Shi et al. 2016). With the finding of the optimisation study, a sinusoidal leading-edge tubercle with the size of *0.1C*

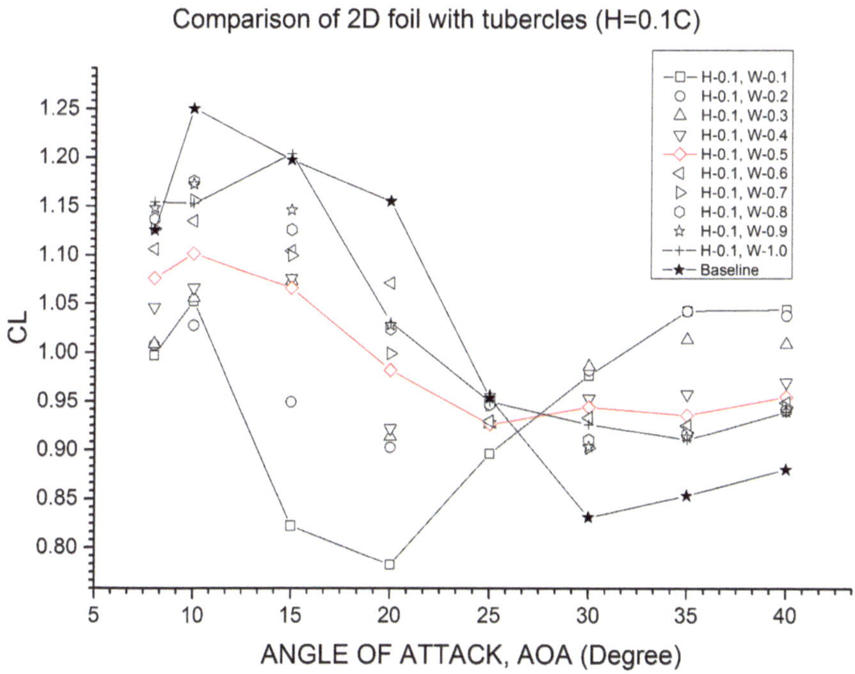


**Fig. 3** Optimisation result of 2D foil regarding to the lift coefficients with the varying wavelength (W) at a constant height (H = 0.1 C) (Shi et al. 2016)

high and *0.5C* wide presented a compromised design with the maintained lift over the varied angles of attack (marked red in Fig. 3) at the price of slightly lowering the maximum $C_L$ (defined in Eq. 2).

$$C_L = \frac{Lift}{\frac{1}{2}\rho V^2 A} \tag{2}$$

where the *Lift* is the force acting on the foil perpendicular to the incoming flow; $\rho$ is the density of the water, kg/m$^3$; $V$ is the inflow velocity, m/s; $A$ is the reference area of the foil, m$^2$, which is $C$ (Chord) * $W$ (Wavelength).

## 3 Applied on a Representative Turbine Blade

Having conducted the numerical optimization of leading-edge tubercles for a 2D foil, the next task of this research was to apply the optimum tubercle shapes onto a 3D straightened turbine blade to study the effect on the tidal turbine blades (Shi et al. 2016).

This straightened three dimensional blade was based on the blade of the target tidal turbine which had the same distribution of the chord length but with a constant pitch. Via changing the angles of attack (AOAs), performance of the representative 3D blade is similar to the tidal turbine operating at varying tip speed ratios (TSRs).

## 3.1 Experimental Investigation of the Blade with the Optimised Tubercles

The representative turbine blades were tested and measured in the Emerson Cavitation Tunnel (ECT), which is a medium size propeller testing water channel, for performance evaluation and flow measurement. Various leading-edge profiles have been tested by taking advantage of the 3D printed interchangeable leading-edge modules which allow different combinations of tubercled and smooth leading-edge modules to systematically change the tubercle coverage as shown in Fig. 4.

The force measurement of the lift and the drag has been conducted with the comparisons of the lift-to-drag ratios. The Foil "0001", 1/4 leading-edge tubercle coverage has shown the highest lift-to-drag ratio in a range of AOAs, which has indicated to be the most efficient configuration, which can be clearly seen in Fig. 5. Compared to the reference, Foil "0000", Foil "0001" shows a performance improvement from 0° to 26° AOA and also displays more than 10% improvement in the maximum lift-to-drag ratio at 5° AOA. Therefore, Foil "0001" may offer potential improvement in the performance of a tidal turbine over a wider range of tip speed ratios.

On the other hand, Foil "1111" displayed the highest performance improvement at 16° AOA in terms of lift-to-drag ratio. Therefore, it might result in the most prominent effect of leading-edge tubercles over the reference turbine.

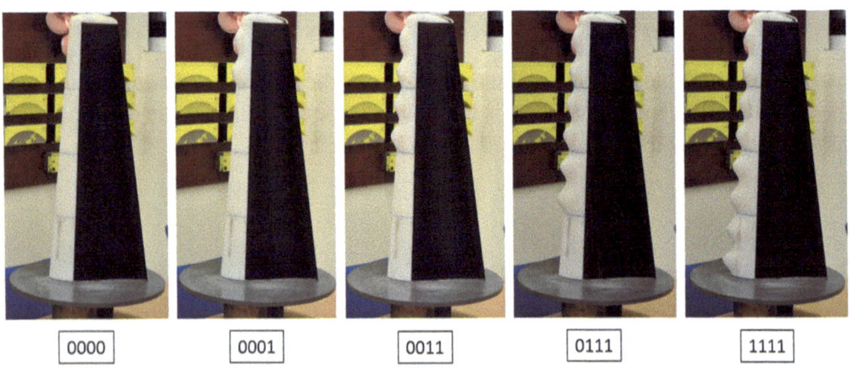


**Fig. 4** Tested 3D hydrofoil models with interchangeable leading-edge parts

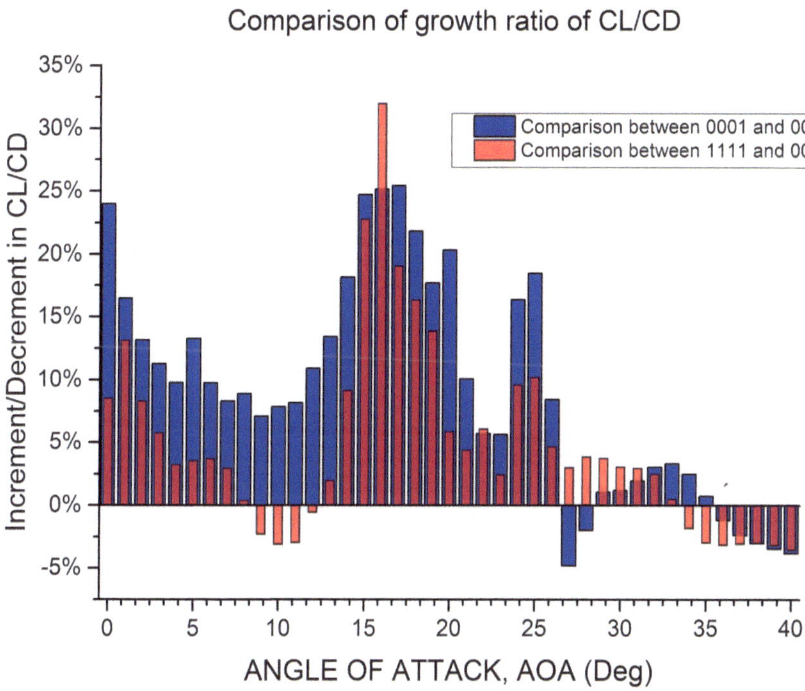


**Fig. 5** Performance enhancement of Foil "1111" and Foil "0001" in terms of lift-to-drag ratio (Shi et al. 2016)

**Fig. 6** Typical tip vortex cavitation generated by Foil "0000" (Diameter is around 10 mm) (Shi et al. 2016)

In the flow visualization study, a significant tip vortex cavitation has been generated by Foil "0000" as shown in Fig. 6; while no tip vortex cavitation has been observed in the test for Foil "1111". Combined with the understanding of the relationship between cavitation and vortex: the larger tip vortex cavitation is, the stronger tip vortex is; Foil "1111" presented a better performance over the reference Foil "0000" in terms of lowering the strength of tip vortex and restricting the development of tip vortex.

### 3.2 *Numerical Simulation of the Blade with the Optimised Tubercles*

Following the experimental investigation, CFD simulations were conducted as a complement to provide further details of the flow field and, hence, to understand the effect of tubercles. The simulations were conducted by Star-CCM+ to solve the incompressible Navier–Stokes equations with Detached Eddy Simulation (DES) models in order to take into account both vortex resolution and computational economy.

The CFD simulation domain was described and dimensioned according to the experimental environment, i.e. the measuring section of the ECT, as shown in Fig. 7. Taking advantage of the automatic polyhedral mesher, 21 million mesh was generated.

The simulation has shown a good agreement with the experimental data. The comparison of the CFD predicted flow separation patterns at AOA = 16° for Foil "0000" and Foil "1111" are shown in Fig. 8 by the iso-surface representations for the axial velocity (Vx) equal to 0 m/s. It can be clearly seen that the iso-surface of the reference hydrofoil was all connected and formed into a huge cloud, while the one of the tubercled hydrofoil was separated and containerised by the effect of the leading-edge tubercles. In addition, the separated volume was much smaller than that observed for Foil "0000".

Further demonstration of the vortex activity on the foil surfaces in Fig. 9 shows that the vorticity [i] distribution on the suction side indicates a pair of contra-rotating vortices generated from the tubercles along the span of the hydrofoil. By scrutinizing the details of the vorticity [i] distribution at the tip of the hydrofoil, the secret of why the leading-edge tubercles always perform better was revealed. The contra-rotating vortices generated by these tubercles were interacting with the tip vortex by stopping it from spreading on the surface of the hydrofoil. As shown in Fig. 10, a vortex generated by a tubercle next to the tip, which shows a positive vorticity [i] along a

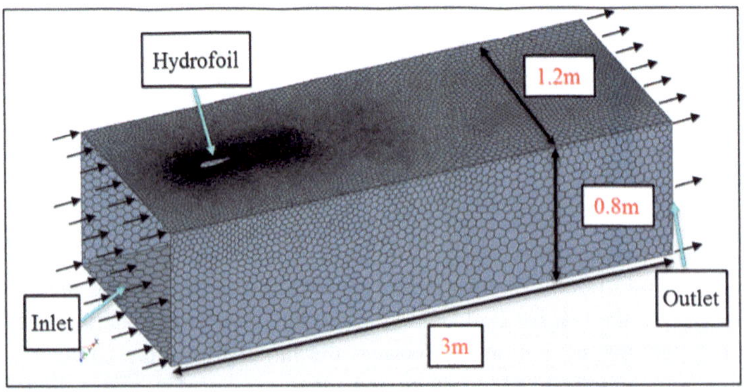


**Fig. 7** Computation domain and boundary conditions

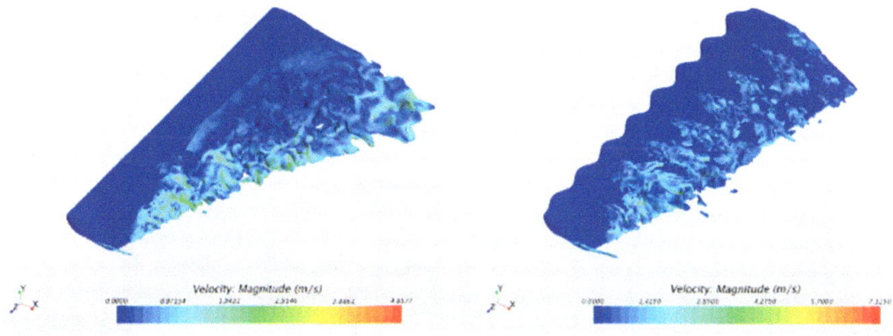


**Fig. 8** Comparative flow separation patterns of two hydrofoils (Iso-surface: Vx = 0 m/s; AOA = 16°)

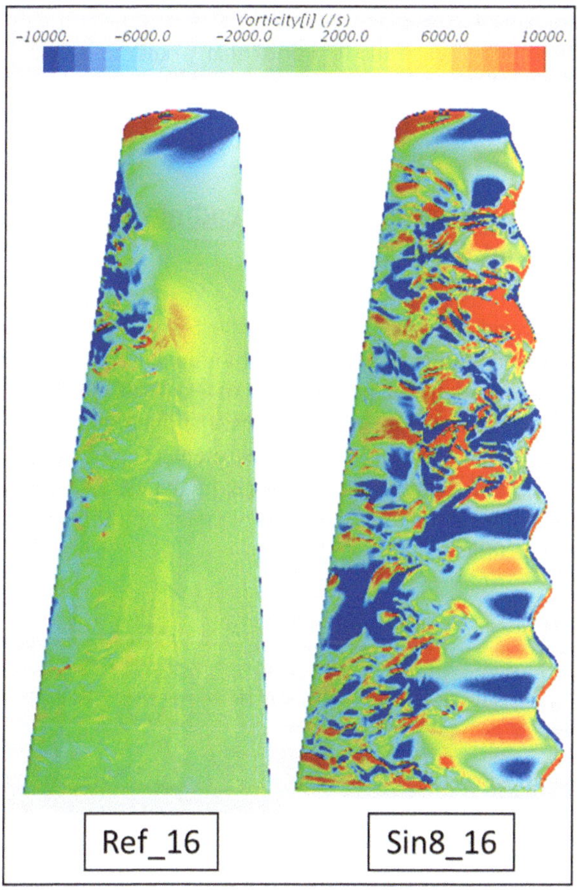


**Fig. 9** Comparative vorticity [i] distributions along two hydrofoils on the suction side

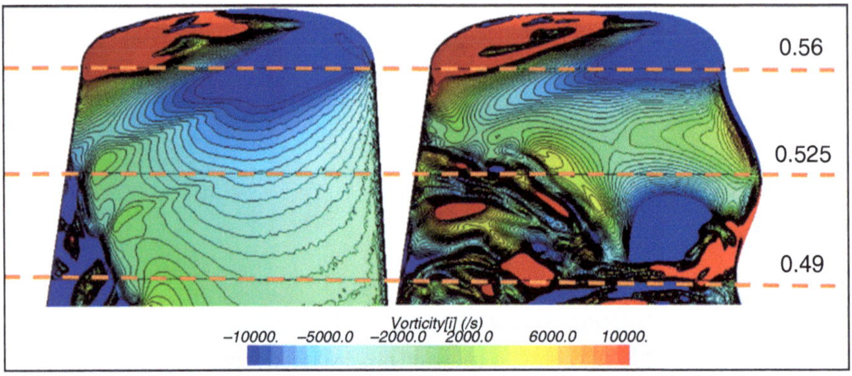


**Fig. 10** Comparative details of vorticity [i] distribution at the suction side and tip of two hydrofoils

span position of Z = 0.525–0.56, cancelled the strong negative tip vortex. As a result, the detrimental effect of the tip vortex that would be responsible for performance loss was greatly weakened.

## 4 Experimental Investigations of Leading-Edge Tubercles As Applied on Tidal Turbine

With the confidence gained from the studies of the foil section and the representative turbine blade, it was investigated whether this biomimetic concept could be effective for improving the hydrodynamic performance of a state-of-the-art tidal turbine with the actual models. The test campaign includes: (1) open water performance tests; (2) cavitation observation tests; (3) noise measurement tests and (4) detailed flow measurement tests, which were all conducted in the ECT, using state-of-the-art model testing equipment. During the test, various non-dimensional parameters were used, listed as follow:

$$\text{Power Coefficient}: Cp = \frac{Q\omega}{\frac{1}{2}\rho A_T V^3} \tag{3}$$

$$\text{Thrust Coefficient}: C_t/10 = \frac{T}{\frac{1}{2}\rho A_T V^2}/10 \tag{4}$$

$$\text{Tip Speed Ratio}: TSR = \frac{\omega r}{V} \tag{5}$$

$$\text{Reynolds Number}: Re_{0.7r} = \frac{C_{0.7r}\sqrt{\left(V^2 + (0.7\omega r)^2\right)}}{\nu} \tag{6}$$

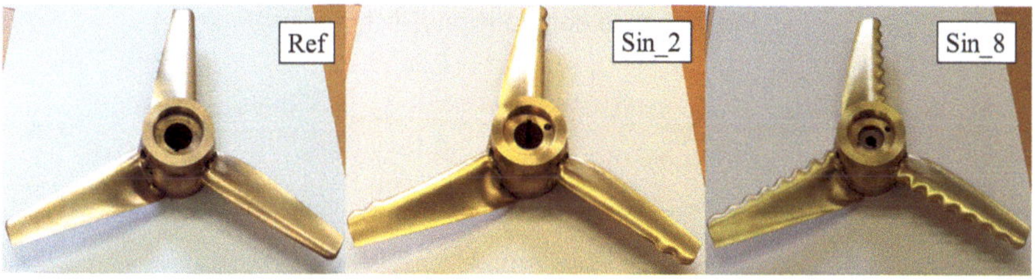


**Fig. 11** Tested turbine models

$$\text{Cavitation Number}: Cav_{0.7r} = \frac{P_{0.7r} - P_v}{\frac{1}{2}\rho\sqrt{\left(V^2 + (0.7\omega r)^2\right.}} \tag{7}$$

where $Q$ refers to the torque of the turbine, Nm; $T$ refers to the thrust, N; $\omega$ refers to the rotational speed, rad/s; $A_T$ refers to the swept area of the turbine and equal $\pi D^2/4$, $m^2$; $\rho$ refers to the tunnel water density, $kg/m^3$; $V$ refers to the incoming velocity, m/s, $D$ refers to the turbine diameter, m; $C_{0.7r}$ refers to the chord length of the turbine at 0.7 radius, m; $\nu$ refers to the kinematic viscosity of the water, $m^2/s$; $P_{0.7r}$ refers to the static pressure at the upper 0.7 radius of the turbine, Pa; $P_v$ refers to the vapour pressure of the water, Pa.

Based on the results from Sects. 2 and 3, three turbine models equipped with pitch adjustable blade design with three different leading-edge profiles were manufactured in brass, as shown in Fig. 11. The reference turbine, Model "Ref", refers to the one with a smooth leading-edge profile. The Model "Sin_2", with only two leading-edge tubercles at the tip, as shown in the centre of Fig. 11, is supposed to be the most efficient design according to the representative blade tests. Finally, the one with eight leading-edge tubercles, named "Sin_8", presented the maximum lift coefficient in the representative blade tests.

## 4.1 *Open Water Hydrodynamic Performance Tests*

The typical performance difference caused by the tubercles can be observed in Fig. 12. The plots of comparison of the power coefficients (Cp) and the thrust coefficients (Ct/10) have been presented in the top two plots. The comparison analysis between the turbines with tubercles and the reference turbine has been conducted and presented in the bottom two plots showing the percentage difference relative to the reference turbine.

The leading-edge tubercle has the capability to improve the turbine's performance in terms of generating higher torque. In the low TSR region (0.5–2.5), when the turbine blades are suffering from stall and under flow separation, the turbine with leading-edge tubercles can produce more torque compared to the reference one. Model "Sin_8" presented the best performance, with 40% higher Cp achieved. At the same time, higher thrust has also been observed during these conditions caused by tubercles, with 2–4% increment for "Sin_2" and 6–8% for "Sin_8".

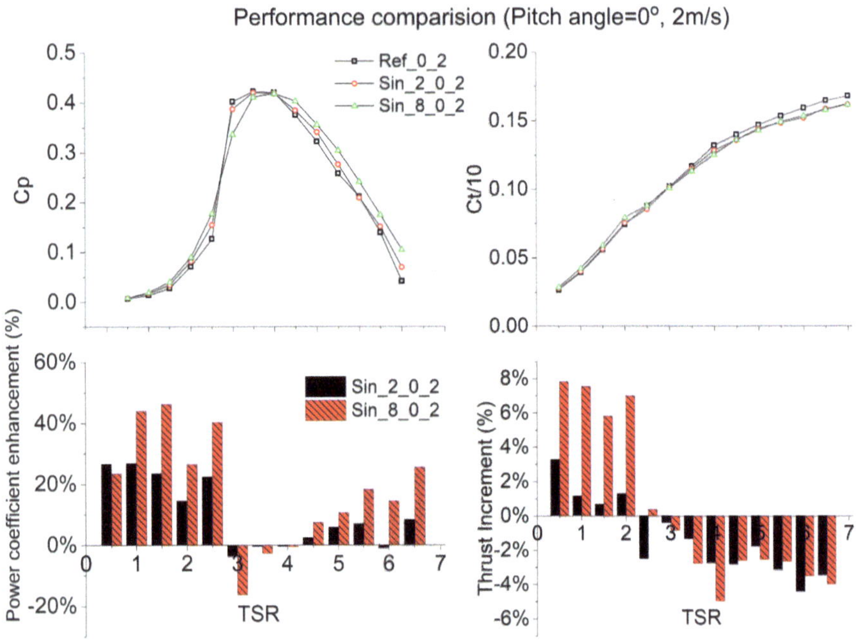


**Fig. 12** Performance comparison (Pitch = 0°, 2 m/s) (Shi et al. 2016)

With the increasing TSR, the turbine models gradually reached their maximum efficiency, with Cp = 0.43 at TSR = 3.5 for Model "Ref" and Model "Sin_2" and a similar level of Cp at a slightly delayed TSR (TSR = 4). Apart from the delayed TSR reaching the highest power coefficient, the tubercled turbine does not have clear benefits/drawbacks regarding the maximum power coefficient.

With a continual increase in TSR further beyond the maximum power coefficient region, turbines reached a condition wherein the blades were experiencing smaller angles of attack. Model "Sin_2" and Model "Sin_8" presented 15–20% more torque but around 4% less thrust. The experimental setup and the detailed result have been documented in the reference (Shi et al. 2016), while the result of the other condition is very similar to the result shown in Fig. 12.

## 4.2 *Cavitation Observation Tests*

Following the performance test, cavitation observation tests were continued. As two typical types of cavitation, blade tip vortex cavitation as well as cloud cavitation with a misty appearance, either at the back or face side, were observed during the test.

The cavitation diagram is presented in Fig. 13, which shows the cavitation occurrence of the turbine in different blade pitch angles (indicated by the last digit of the legend label), a range of cavitation numbers and a range of TSRs. The blade

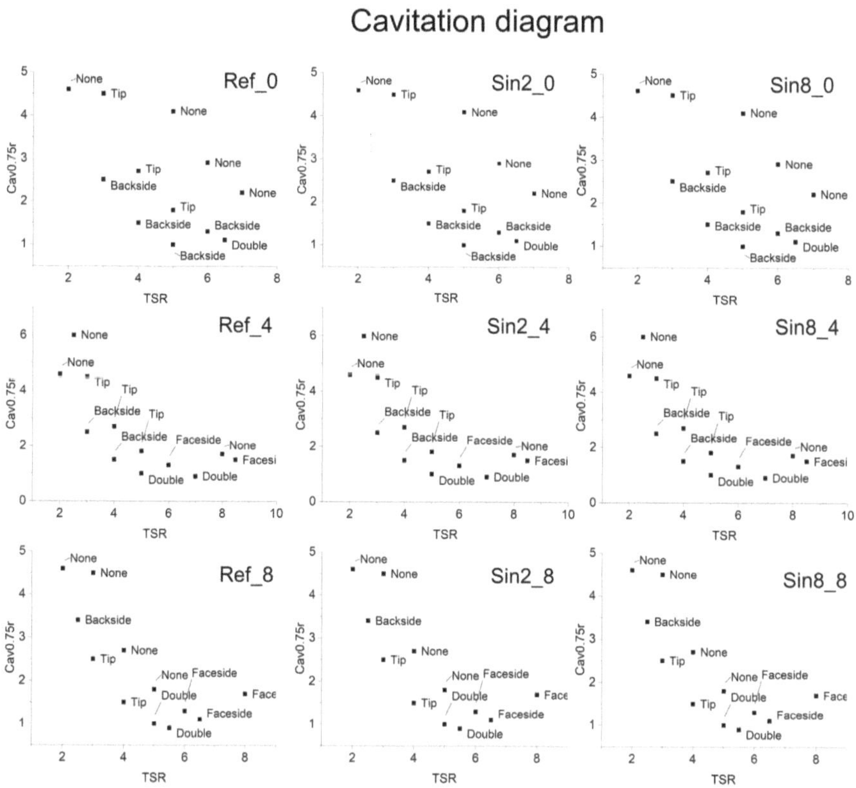


**Fig. 13** Cavitation diagram regarding to TSR and cavitation number (Shi et al. 2016)

cavitation is developed in a typical sequence as follows: first the tip vortex cavitation would be developed, which was the main type of tidal turbine cavitation; after the occurrence of tip vortex cavitation and with the decreasing cavitation number, unsteady cloud cavitation would be generated on either side of the blades depending on the TSRs. This kind of unsteady cloud cavitation will cause a significant detrimental effect on the turbine efficiency, while the tip vortex cavitation will not.

Regarding the cavitation inception, a detailed cavitation inception test was not conducted. But from the cavitation diagram and the observation during the test, the inception of cavitation can be triggered earlier by the tubercled turbine blades compared to the reference turbine blades. But the diameter of tip vortex cavitation did not show any significant difference.

On the other hand, the detailed comparison showed that the cloud cavitation can be limited by the tubercles within the trough area between two adjacent tubercles, as shown in Fig. 14. This has been proved to have the benefit of reducing the cavitation extent as well as intermit the cavitation development in order to prevent the formation of continuous large extent cavitation.

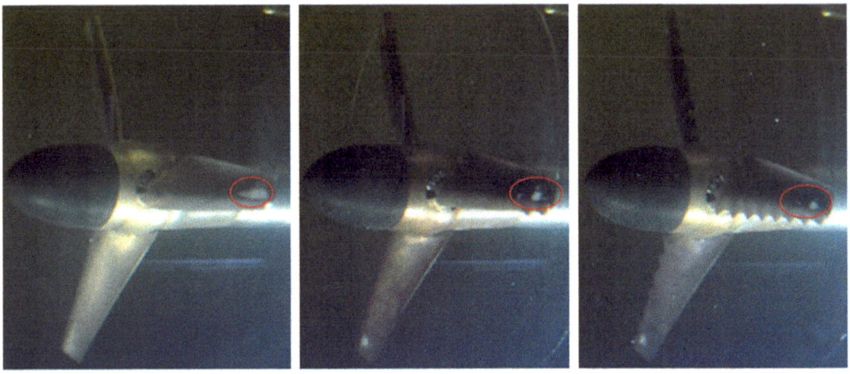

**Fig. 14** Detailed comparison of cavitation pattern (Pitch angle = +8°; V = 3 m/s; TSR = 6; $Cav_{0.75r} = 1.3$) (Shi et al. 2016)

## 4.3 Noise Measurement Tests

Meanwhile, with the cavitation observation test, a noise measurement test was also conducted to monitor the underwater radiated noise level. The noise measurement result in Fig. 15 was analysed with the cavitation inception diagram in Fig. 13 to correlate the cavitation development with the noise level.

It can be seen that prior to the generation of cavitation, three turbine models displayed a similar level of total noise including background noise, which can be noticed from the top three plots in Fig. 15 (TSR $\leq$ 2). However, once the cavitation occurred, the noise levels of the turbines with tubercles (3 $\leq$ TSR $\leq$ 5) were generally higher than those of the reference turbine, which is because of the early generation of the tip cavitation by the tubercled turbines.

Whereas at TSR = 6 all the turbines were suffering from cloud cavitation, as shown in Fig. 13, Model "Ref" generated a higher level of noise compared to the tubercled ones at the higher frequency end from 3 kHz. This is because the occurrence of the cloud cavitation on the blade surface was constrained by tubercles. As shown in Fig. 14, a much smaller extent of blade cavitation can be noted with the increasing numbers of tubercles.

## 4.4 Detailed Flow Measurement Tests

Having investigated the effect of the leading-edge tubercles on various performance characteristics of the tidal turbine blades, significant differences were observed requiring further explanation. The most fundamental and perhaps effective explanations could be made by conducting a detailed flow investigation around the turbine blades and slipstream of the turbines, preferably by using a state-of-the-art Stereo Particle Image Velocimetry (SPIV) system.

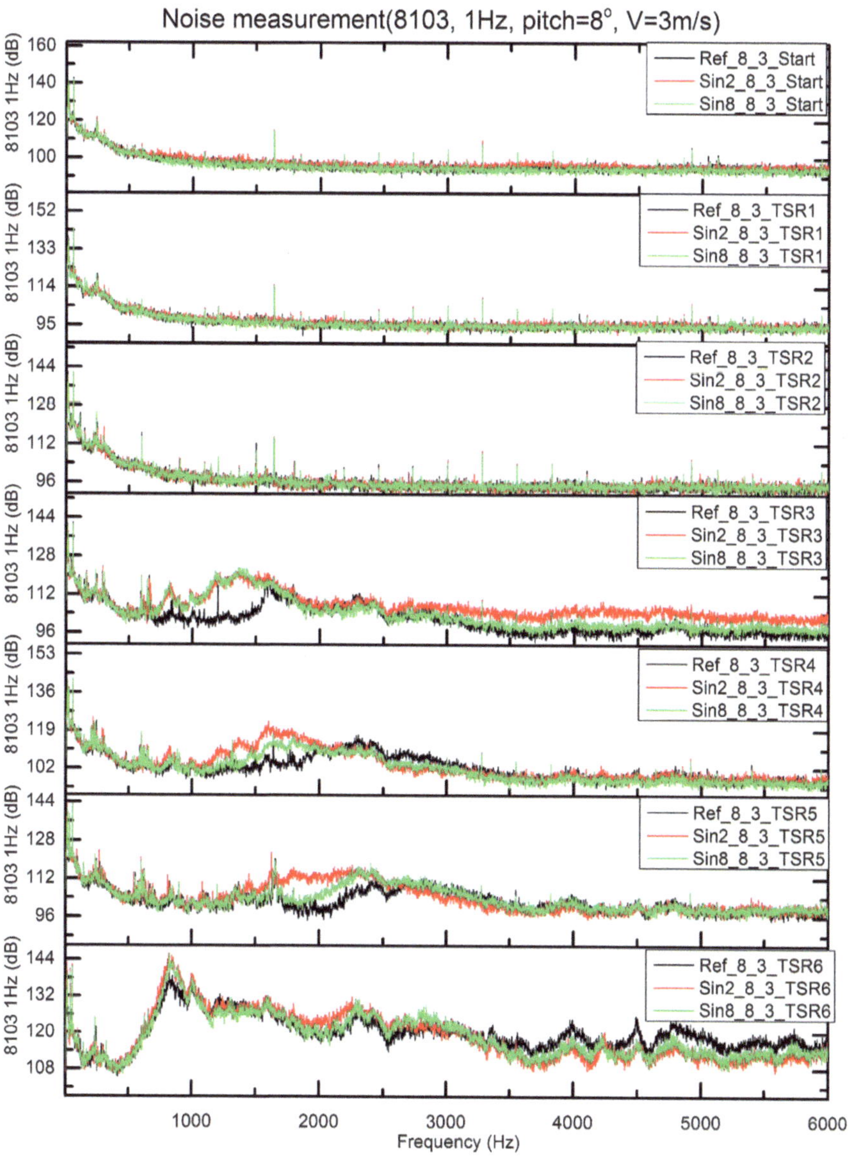


**Fig. 15** Comparative total noise levels measured in 1 Hz band (Pitch angle = 8°, V = 3 m/s)

Based on the measurement results of the blade cross-section under the "Stall condition", while all the tested turbine blades suffered from severe flow separations, the turbine blades with the leading-edge tubercles were able to maintain the flow to be more attached to their surfaces at certain positions, as seen in Fig. 16. This

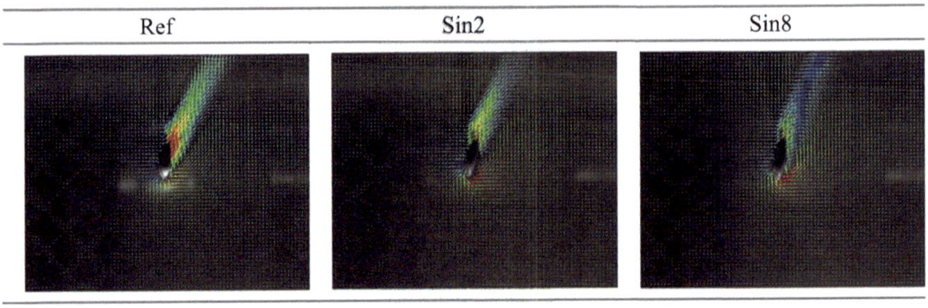


**Fig. 16** PIV measurement of the cross-section at 0.9r of the three models at TSR = 2 and V = 4 m/s

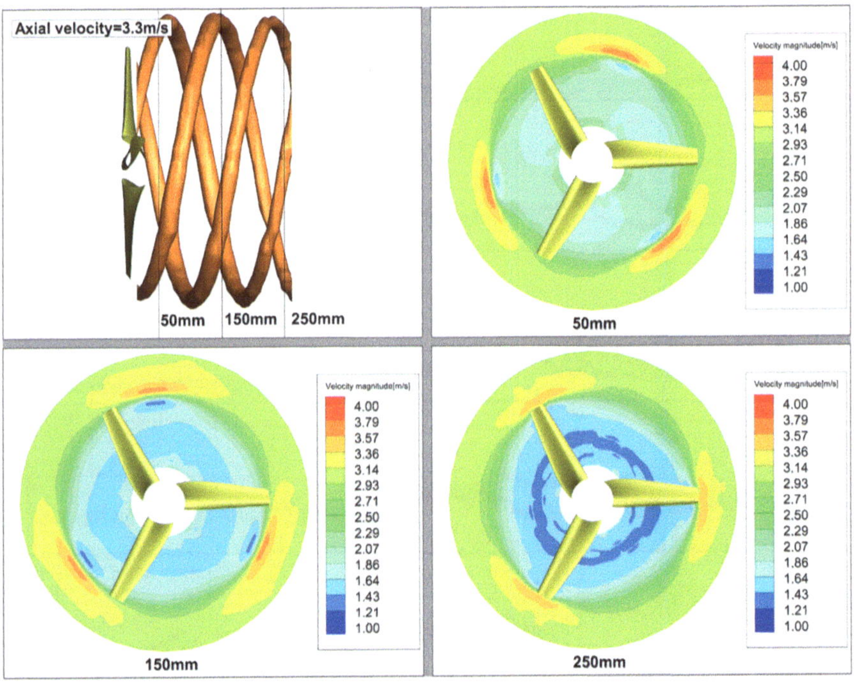


**Fig. 17** Velocity distribution for the reference turbine at varied sections, 50, 150 and 250 mm downstream of model turbine at TSR4 with the iso-surface of axial velocity = 3.3 m/s

provided the turbine with additional torque for starting, as shown in the observed Cp benefits in lower TSRs.

On the other hand, the flow structure measured downstream of the turbine revealed that the turbines with tubercles can induce a higher induction factor, which resulted in lower velocities in the wake field as shown in the comparison of Figs. 17, 18 and 19. This effect is positively related to the number of tubercles, which result in higher power coefficients (2.2% for Sin_2 and 4.3% for Sin_8) and also

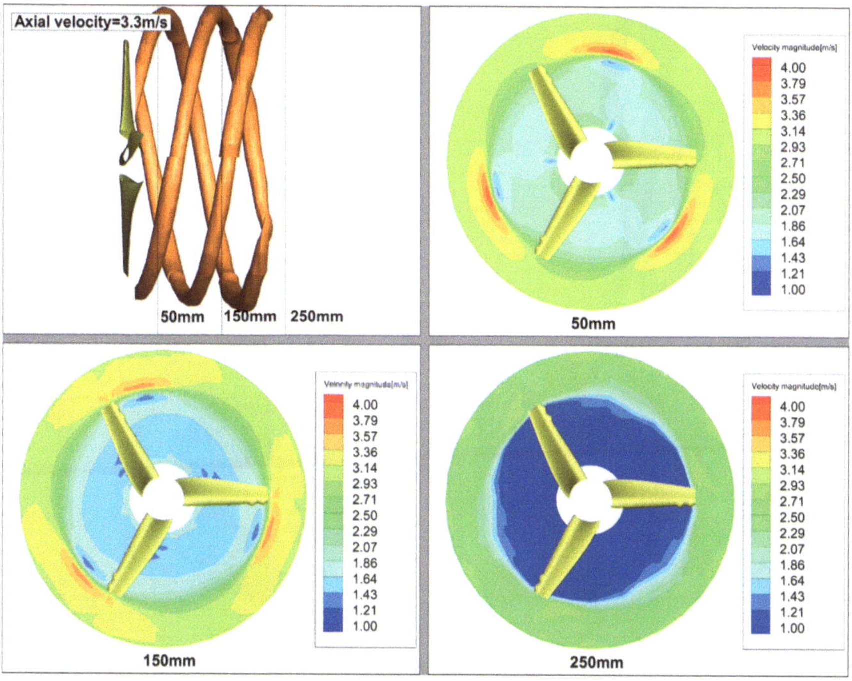


**Fig. 18** Velocity distribution for the Sin_2 turbine at varied sections, 50, 150 and 250 mm downstream of model turbine at TSR4 with the iso-surface of axial velocity = 3.3 m/s

higher thrust coefficients (4.6% for Sin_2 and 7.7% for Sin_8) at this specific condition with the optimum TSR (TSR = 4 and V = 3 m/s).

PIV measurements in Figs. 17, 18 and 19 also revealed that the tip vortex was weakened and its axial trajectory was shortened by the tubercles in the "Optimum working condition" (TSR = 4 and V = 3 m/s). This was further confirmation of the reduced 3D effects of the actual turbine blades by the tubercles, reflecting the findings in Sect. 3. This finding is also positively related to the number of tubercles.

# 5 Conclusions

Humpback whales have been living in the ocean for millions of years, but time does not wash their rough tubercles away. Natural selection has kept their non-streamlined feature for a reason. By going through such a comprehensive set of experimental investigations of tidal turbine models with and without leading-edge tubercles, the comparative hydrodynamic efficiencies in the current were investigated including the cavitation and underwater noise performances. These investigations were supported with comprehensive flow analyses using state-of-the-art experimental tools for further exploration and understanding of the effects of

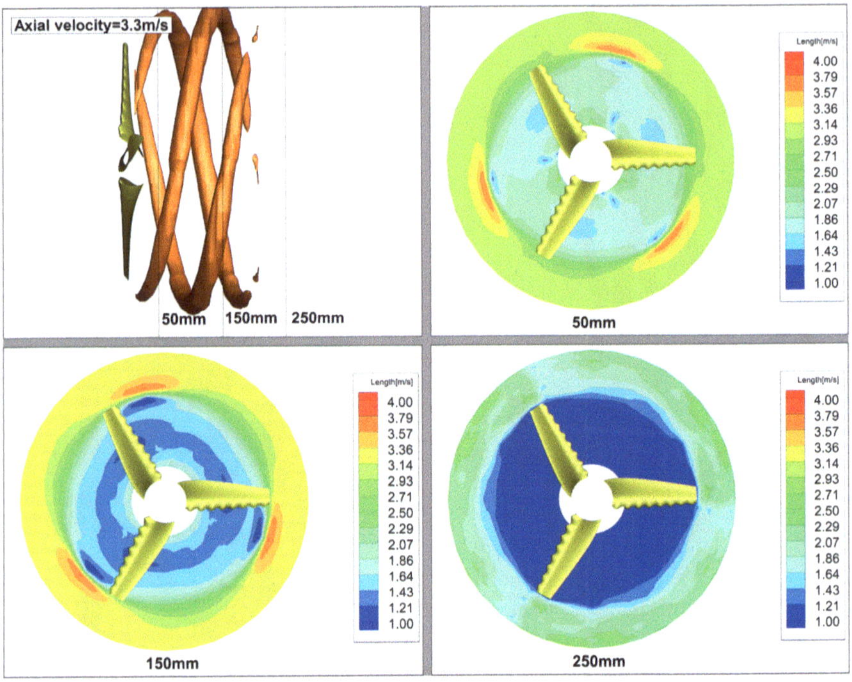


**Fig. 19** Velocity distribution for the Sin_8 turbine at varied sections, 50, 150 and 250 mm downstream of model turbine at TSR4 with the iso-surface of axial velocity = 3.3 m/s

tubercles on tidal turbines as well as presenting bench mark data for future research and developments.

Studies conducted in this paper proved the beneficial effects on various hydrodynamic performances of these turbines with leading edge tubercles, mainly with improved performance in low TSRs without influencing the maximum power coefficient, limited cavitation development and, hence, lowered noise level. Based on this model scale demonstration, this bionic concept is ready for the next full-scale applications.

**Acknowledgements** This research is funded by the School of Marine Science and Technology, Newcastle University and China Scholarship Council. Hence the financial support obtained from both establishments is gratefully acknowledged. The Authors would like to thank all the team members in the Emerson Cavitation Tunnel for the help in testing and sharing their knowledge. And the last but not the least, special acknowledgement to the team in Strathclyde University. Thanks for the support for the last stage of this study.

## References


Aftab, S. M. A., Razak, N. A., Mohd Rafie, A. S., & Ahmad, K. A. (2016). Mimicking the humpback whale: An aerodynamic perspective. *Progress in Aerospace Sciences, 84*, 48–69.

Barltrop, N., Varyani, K. S., Grant, A., Clelland, D., & Pham, X. (2006). Wave-current interactions in marine current turbines. *Proceedings of the Institution of Mechanical Engineers, Part M: Journal of Engineering for the Maritime Environment, 220*, 195–203.

Bolzon, M. D., Kelso, R. M., & Arjomandi, M. (2016). Tubercles and their applications. *Journal of Aerospace Engineering, 29*, 04015013.

Fish, F. E., & Battle, J. M. (1996). Hydrodynamic design of the humpback whale flipper. *Journal of Morphology, 225*, 51–60.

Fish, F. E., Weber, P. W., Murray, M. M., & Howle, L. E. (2011). The tubercles on humpback Whales' flippers: Application of bio-inspired technology. *Integrative and Comparative Biology, 51*, 203–213.

Hain, J. H. W., Carter, G. R., Kraus, S. D., Mayo, C. A., & Winn, H. E. (1982). Feeding-behavior of the humpback whale, Megaptera-Novaeangliae, in the western North-Atlantic. *Fishery Bulletin, 80*, 259–268.

Hansen, K., Kelso, R., & Doolan, C. (2012). Reduction of flow induced airfoil tonal noise using leading edge sinusoidal modifications. *Acoustics Australia, 40*, 172–177.

Hansen, K. L., Kelso, R. M., & Dally, B. B. (2009). The effect of leading edge tubercle geometry on the performance of different airfoils. In *Exhft-7*. Krakow, Poland.

Hansen, K. L., Kelso, R. M., & Dally, B. B. (2011). Performance variations of leading-edge tubercles for distinct airfoil profiles. *AIAA Journal, 49*, 185–194.

Hansen, K. L., Rostamzadeh, N., Kelso, R. M., & Dally, B. B. (2016). Evolution of the streamwise vortices generated between leading edge tubercles. *Journal of Fluid Mechanics, 788*, 730–766.

Kim, J. W., Haeri, S., & Joseph, P. F. (2016). On the reduction of aerofoil–turbulence interaction noise associated with wavy leading edges. *Journal of Fluid Mechanics, 792*, 526–552.

Miklosovic, D. S., Murray, M. M., Howle, L. E., & Fish, F. E. (2004). Leading-edge tubercles delay stall on humpback whale (Megaptera Novaeangliae) flippers. *Physics of Fluids, 16*, L39–L42.

Shi, W., Atlar, M., Norman, R., Aktas, B., & Turkmen, S. (2016). Numerical optimization and experimental validation for a tidal turbine blade with leading-edge tubercles. *Renewable Energy, 96*, 42–55.

Shi, W., Atlar, M., Rosli, R., Aktas, B., & Norman, R. (2016). Cavitation observations and noise measurements of horizontal axis tidal turbines with biomimetic blade leading-edge designs. *Ocean Engineering, 121*, 143–155.

Shi, W., Rosli, R., Atlar, M., Norman, R., Wang, D., & Yang, W. (2016). Hydrodynamic performance evaluation of a tidal turbine with leading-edge tubercles. *Ocean Engineering, 117*, 246–253.

Stanway, M. J. (2008). *Hydrodynamic effects of leading-edge tubercles on control surfaces and in flapping foil propulsion.* Msc, Massachusetts Institute of Technology.

Swanson, T., & Isaac, K. M. (2011). *Biologically inspired wing leading edge for enhanced wind turbine and aircraft performance*. AIAA.

Van Nierop, E. A., Alben, S., & Brenner, M. P. (2008). How bumps on whale flippers delay stall: An aerodynamic model. *Physical Review Letters, 100*, 054502.

Wang, D., Atlar, M., & Sampson, R. (2007). An experimental investigation on cavitation, noise, and slipstream characteristics of ocean stream turbines. *Proceedings of the Institution of Mechanical Engineers Part A-Journal of Power and Energy, 221*, 219–231.

Wei, Z. Y., New, T. H., & Cui, Y. D. (2015). An experimental study on flow separation control of hydrofoils with leading-edge tubercles at low Reynolds number. *Ocean Engineering, 108*, 336–349.

# CFD Simulation of a Passively Controlled Point Absorber Wave Energy Converter

**Minghao Wu, Weizhi Wang, Johannes Palm, and Claes Eskilsson**

## 1 Introduction

Point absorbers are a special class of wave energy converters (WECs), defined by having a geometry that is small in relation to the wavelength. Point absorbers are oscillating bodies and typically subjected to large amplitude motions in the resonance region. Seminal analytical studies of the power capture of point absorbers include Budal and Falnes (1975) and Evans (1976). In order to increase the power capture different control strategies have been proposed, such as reactive control, latching, and declutching; see e.g. Falnes (2002), Hals et al. (2011) and Korde and Ringwood (2016) for discussions and comparisons of hydrodynamic control of WECs. Recently, a new passive control strategy based on negative spring arrangement was presented: The WaveSpring technology (Todalshaug 2014). The WaveSpring technology is today developed and used in the CorPower Ocean buoy, soon to be deployed at EMEC's scale test site in Orkney, Scotland (EMEC 2017).

It is well known that the motion response of point absorbers in the resonance region is subjected to nonlinear effects and that linear radiation-diffraction methods over-predict the response amplitude for steeper waves. Recent CFD studies of point absorbers in the resonance region have also illustrated this. Yu and Li (2013) compared a CFD model with a linear model and showed that the linear model

M. Wu · W. Wang · J. Palm
Department of Shipping and Marine Technology, Chalmers University of Technology, Göteborg, Sweden

C. Eskilsson (✉)
Department of Civil Engineering, Aalborg University, Aalborg Øst, Denmark

Department of Shipping and Marine Technology, Chalmers University of Technology, Göteborg, Sweden
e-mail: cge@civil.aau.dk

 
A. I. Ölçer et al. (eds.), *Trends and Challenges in Maritime Energy Management*,
WMU Studies in Maritime Affairs 6, https://doi.org/10.1007/978-3-319-74576-3_34

over-predicted the power production of a self-reacting point absorber in heave. Palm et al. (2016) presented a coupled mooring analysis using CFD for a point absorber undergoing six degree-of-freedom (DoF) motion, and the model accurately recreated the strong wave height dependence of the response amplitude. These nonlinear effects are even more highlighted for point absorbers using phase control. Indeed, in a recent study Giorgi and Ringwood (2016) implemented latching control for a heaving sphere in a CFD-based numerical wave tank and compared to a standard linear method. The results show that the linear model overestimates the amplitude of motion, and, as a result, the extracted power. Importantly, the optimal latching duration given by the linear model will not maximize the performance of the actual device.

So while the advantage of CFD is obvious—nonlinear and viscous effects, overtopping, large amplitude waves and motions, etc. are all included—the major disadvantage of using CFD for modelling WECs is that it is very computationally expensive, especially when irregular sea states are simulated. Eskilsson et al. (2015) computed the overtopping discharge of the Wave Dragon wave energy converter and reported a computational effort in the order of 150,000 CPU hours for a 3-hour sea state simulation using a computational mesh of 14 million cells. However, in a recent technical review regarding the applicability of the CFD-based numerical tank test of offshore floating structure design (Kim et al. 2016), it is concluded that the CFD-based numerical tank test shows the benefits of lower cost, shorter project period and higher flexibility compared with the physical tank test.

In light of the findings of Giorgi and Ringwood (2016), we expect that the passively controlled CorPower buoy also will require CFD simulations in order to obtain reliable values for motion response and power production. In this paper, the main objective is to establish a three-dimensional numerical wave tank based on the open-source CFD solver OpenFOAM and test its feasibility of modelling the CorPower WEC motion in waves. The simulation will consider both the pretension mooring force and the nonlinear PTO (power-take-off) force and will support 6 DoF motion. The nonlinear effects due to the viscous flow and the nonlinear motion excited by WaveSpring phase control system are to be investigated. The results of motions and forces will be compared with the laboratory test data under the same settings.

## 2 The CorPower Point Absorber

CorPower Ocean AB is a Swedish technology company focusing on the research and development of the high-efficiency wave energy converter. The CorPower WEC prototype adopts the point-absorber type and the phase control technology to minimize the size of the buoy and maximize the efficiency. The shape and the inner structure of the buoy is shown in Fig. 1. The inside PTO gear extracts the kinematics energy and convert to electricity.

**Fig. 1** CorPower Ocean WEC prototype. Courtesy CorPower Ocean AB

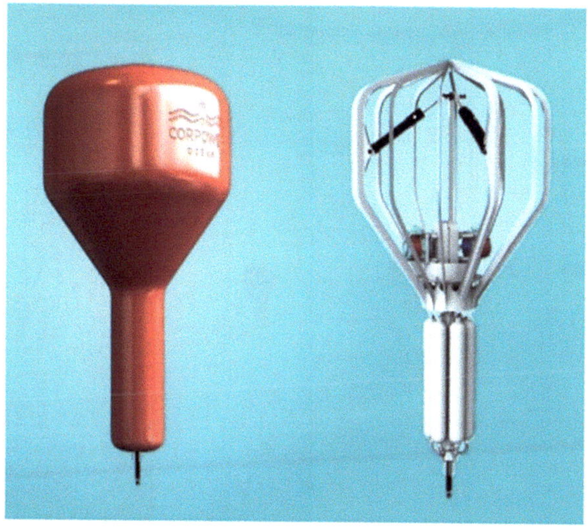

The application of the WaveSpring phase control system invented at Norwegian University of Science and Technology (Todalshaug 2014) gives a more efficient wave energy conversion. WaveSpring is a passive control system and, in contrast to conventional latching control, requires fewer sensors and active loops since the incident wave information is not utilized by the control system. The working principles of the WaveSpring system are described in Todalshaug et al. (2016). In short, pressurized pneumatic cylinders (see Fig. 1) provide a spring force which is contrary to the heave hydrostatic force causing the motion to be amplified when the buoy is off the equilibrium position. The net vertical force along the central rod, $F_{WS}$ is expressed as

$$F_{WS}(z) = N_c A_c P(z) \frac{z}{\sqrt{l_0^2 - z^2}}, \tag{1}$$

where $N_c$ is the number of cylinders, $A_c$ is the area of a cylinder, $l_0$ is the initial length of the cylinder at its horizontal position and $P(z)$ is the cylinder pressure. Here $z$ is the relative WEC excursion along the center rod.

This negative spring arrangement is able to amplify the power capture of the buoy and increase the power absorption by widening the response bandwidth. In the experimental study described in the next section it was found that the average power increased by a factor 3 by passive control provided by the WaveSpring system (Todalshaug et al. 2016).

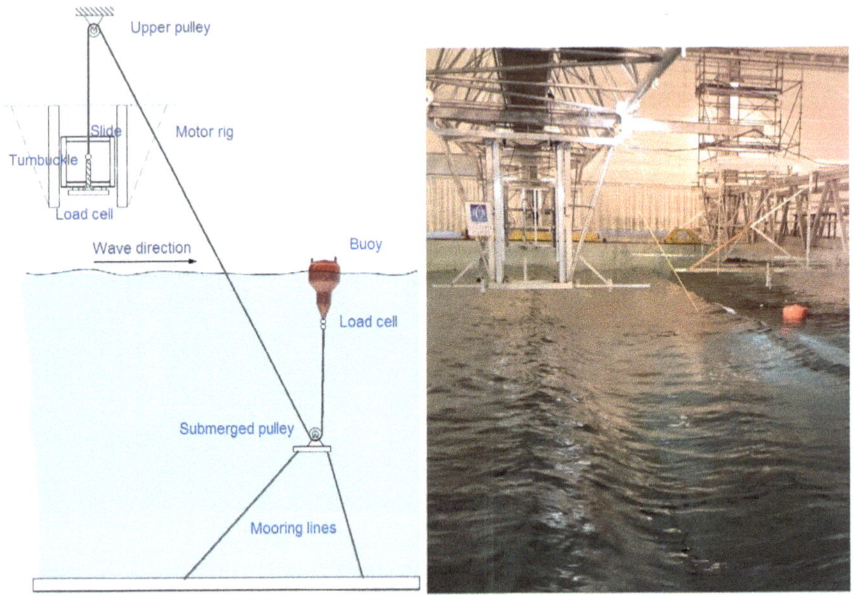


**Fig. 2** Tank test configuration (*left*) and photo of experiment (*right*). Reprinted from International Journal of Marine Energy, Vol 15, Todalshaug et al., Tank testing of an inherently phase-controlled wave energy converter, Pages 68–84., Copyright (2016), with permission from Elsevier

## 3 Physical Experiments

The tank tests of the CorPower WEC model were carried out in the Hydrodynamic and Ocean Engineering Tank at Ecole Centrale de Nantes in France during November 2014. A detailed description of this series tank test is presented in Todalshaug et al. (2016) The system consists of a buoy model in scale 1:16, a rope, two pulleys and the motor rig, as shown in Fig. 2. The rope connected to the buoy is used as a mooring line to restrain the buoy, while the bottom mooring lines are used to fix the submerged pulley. The mooring line is pre-tensioned to keep the buoy in its equilibrium position. The motor rig is connected to a load control system and can approximate the on-board machine force as long as the angle between the buoy axis and the mooring line is small. The experimental parameters are listed in Table 1.

Various sensors are used for capturing the transient motion, wave height and forces. The wave surface elevation is measured by five resistive wave probes. The six-degree-of-freedom motion of the buoy is measured by Qualisys optical position sensors. Two force sensors are installed at both the buoy end and the motor rig end of the mooring line.

In the experiments a wide array of different regular and irregular waves is tested, but here only one wave is investigated: a regular wave with amplitude 0.156 m and a wave period of 2.25 s. This is a fairly linear incident wave, having a steepness of 0.02.

**Table 1** Experimental parameters

| Item | Value | Item | Value |
|---|---|---|---|
| Buoy diameter | 0.525 m | Water depth | 5 m |
| Buoy height | 1.125 m | Rope length | 15.26 m |
| Buoy mass | 19.75 kg | Depth at mooring point | 3.09 m |
| Buoy displacement[a] | 0.0708 $m^3$ | Rope and sheave mass | 1.85 kg |
| Buoy waterline[a] | 0.98 m | Mooring line stiffness | 202 N/m |
| Moment of inertia | 2.38 $kgm^2$ | Pretension force | 500 N |

[a]At equilibrium position

# 4 Numerical Modelling

The numerical model is based on incompressible Reynolds Averaged Navier-Stokes (RANS) equations with the single fluid (mixture) approximation. The two-phase problem is handled using the volume of fluid (VOF) approach.

The VOF-RANS equations are solved using the interDyMFoam solver in the open-source finite volume framework OpenFOAM (OpenFOAM 2017; Weller et al. 1998). The interDyMFoam solver supports dynamic mesh motion to allow for the motion of the WEC and waves are generated/absorbed by the waves2Foam package (Jacobsen et al. 2012). This combination is a fairly popular choice of CFD-based numerical wave tank in the wave energy community, see Davidson et al. (2015) and the references therein.

In order to model the CorPower buoy the WaveSpring force (1) was implanted as a restraint inside body motion library of OpenFOAM. Also, due to the set-up of the motor rig, a simplified mooring made up of a linear pre-tensioned spring was used in the simulations.

## 4.1 *Mesh and Numerical Settings*

The fluid domain is described in a Cartesian coordinate system with the origin point at the bottom of the WEC buoy. The positive *x*-direction is to the wave-propagation direction, the *y*-direction points to aside and the *z*-direction is vertically upward. The length of the domain is 48 m. The bottom as well as the side walls are modelled as slip conditions. To avoid the reflection from the bank, the distance from the floating buoy to the side wall is set to 4 m, which gives an 8.6 m full width. Two relaxation zones are set at two ends of the domain for wave generation (inlet) and cancellation (outlet). Each relaxation zone is set to cover two wavelengths. Figure 3 outlines the computational domain used.

The grids are generated with the snappyHexMesh utility. In order to capture the near-body flow and the wave surface, several levels of grid refinement are used to increase the grid resolution near the buoy surface and the air-water interface. To keep a good grid quality, the background mesh is created as uniform hexahedral cells,

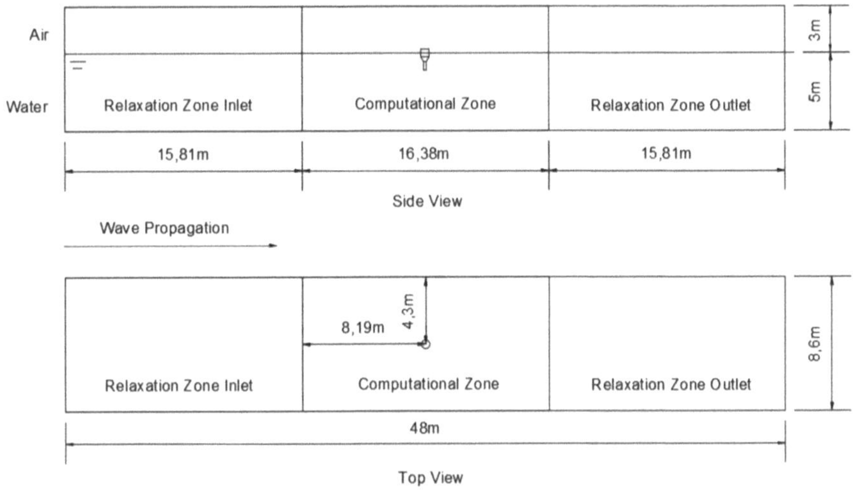


**Fig. 3** Numerical wave tank dimensions: side view (*top*) and top view (*bottom*)

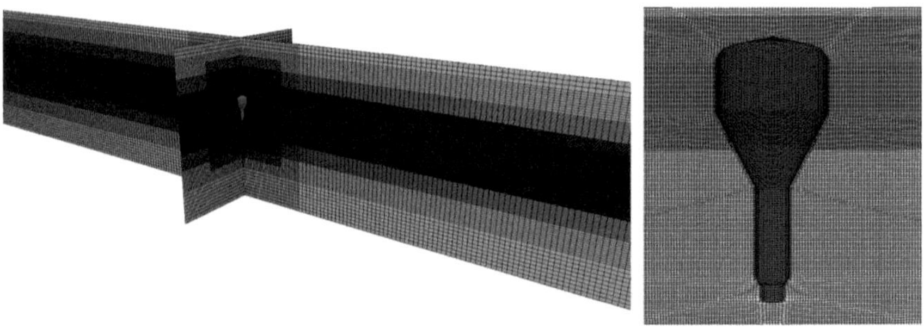

**Fig. 4** Computational mesh: global view (*left*) and close up at the buoy (*right*)

with a cell aspect ratio approximately equal to unity after refinement. Figure 4 shows the computational mesh. The total cell number is 10.9 million cells. There are roughly 40 cells per wave length along the $x$-direction and 30 cells per wave height in the $z$-direction. By adding five layers to the near-wall region, the $y+$ value is roughly 50.

The $k$-$\omega$ SST model (Menter 1994) is used together with wall functions. For oscillatory flow, it is advisable to use $y+$ values below one, but in order to lessen the computational effort, wall functions were employed in this study.

The second-order van Leer scheme is used for the divergence terms in the Navier-Stokes equations while the first-order upwind scheme is used in the turbulence equations. The diffusion terms are treated with second-order central difference scheme. The solution is advanced in time using the implicit Crank-Nicholson scheme. Ten outer iterations are performed per time-step to ensure that the motion solver is converted properly.

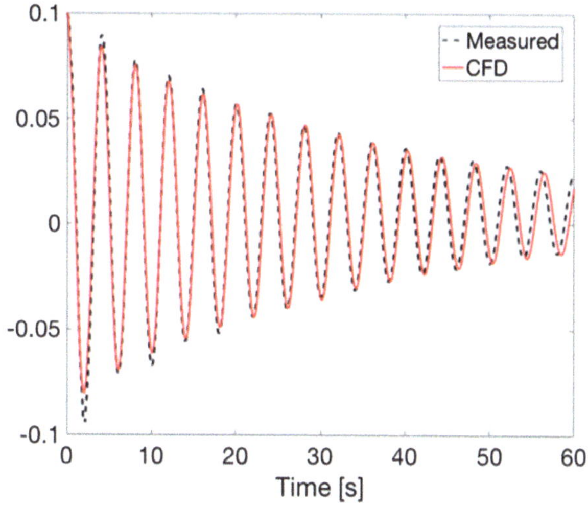


**Fig. 5** Surge decay test. Experimental data is presented in dashed black line while computational results are printed in solid red lines

**Table 2** Surge decay test

| | Experimental | Numerical | Relative error |
|---|---|---|---|
| Average period | 3.98 s | 4.03 s | 1.3% |
| Damping ratio | 9.55E-02 | 8.75E-02 | 8.3% |

Average surge decay period and non-dimensional linear damping factor

# 5 Test Cases

## *5.1 Surge Decay Test*

The surge decay of the buoy in calm water is evaluated numerically and compared to the tank test experiment. The mooring line is simplified as a spring with a linearized stiffness coefficient, which is calculated by dividing the pretension by the mooring line length. The anchor of the mooring line is fixed at the submerged pulley in the test. The resulting time series are plotted in Fig. 5 and the obtained surge natural period and damping ratios in Table 2. The computed and measured natural surge period are in general good agreement. The initial release position is a bit uncertain from the experimental tests, which explains the amplitude difference in the first oscillation and also the larger difference in damping ratio.

## *5.2 Regular Waves with Linear Damper Only*

In this case, the motion with only the linear damper PTO system is simulated. The linear damping coefficient is set to 628 kg/s, which is the optimum value obtained for unconstrained heave. The set-up of this case corresponds to a Keulegan–Carpenter

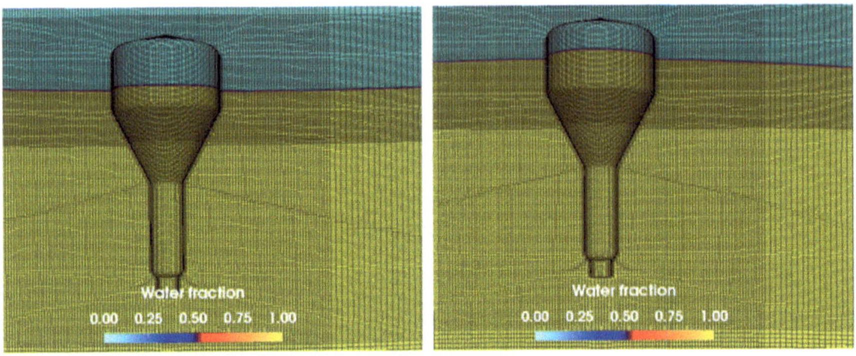


**Fig. 6** Buoy and mesh motion response in regular waves with linear damper: At $t = 48.5$ s (*left*) and $t = 49.5$ s (*right*)

number (KC) of 1.8, giving that the flow is inertia dominated (Sumer and Fredsøe 1997). The incident waves are given by 5th order Stokes theory.

The simulation is run for 50 s, by which time a steady periodic motion of the WEC has been established. Figure 6 shows the perspective global view of the wave surface and the side view of local mesh motion around the buoy at simulation time $t = 48$ s. Under this relatively large PTO damping coefficient, the buoy heave amplitude is small so that the wave surface is well captured in the mesh refined region near the waterline. No overtopping is observed according to the post-processing.

The motion and force logs of both the CFD simulation and experiments are plotted in Fig. 7, including surge, pitch, heave and the PTO-mooring system force. The response is seen to be linear; there are no higher harmonics visible in the time series. Generally, there is good agreement between numerical and experimental response amplitude. The problem is the heave (the motion peak is over-predicted) and the pitch (negative pitch is under-predicted) responses. These two modes are coupled and it was found that these motions are sensitive to the value of the pre-tension, which is a bit uncertain from the experimental tests. In addition, the moment of inertia is estimated based on approximate gyration radius data, which might explain some of the error in the pitch motion.

### 5.3 *Regular Waves with WaveSpring System*

As mentioned above, in order to increase the efficiency of the power capture the WaveSpring system is operated together with the linear damper PTO system. This introduces highly nonlinear factors in the numerical modelling. The wave case of Sect. 5.2 is repeated with the WaveSpring system included. The WaveSpring system in this test is made up of three pistons, each with cross section of 5.82E-04 $m^2$, initial length 0.25 m and initial volume 1.59E-04 $m^3$.

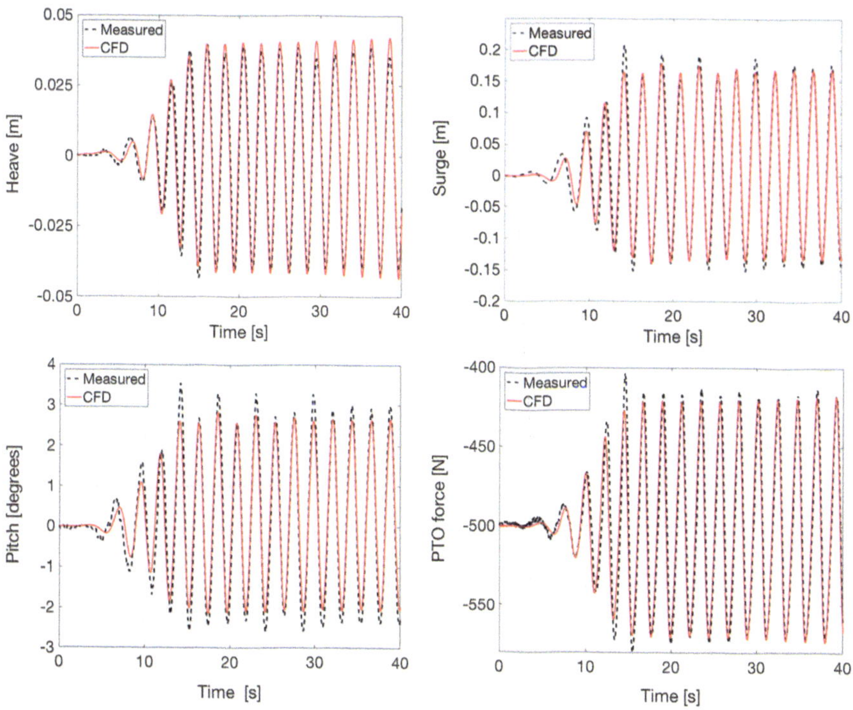


**Fig. 7** WEC motion response in regular waves with linear damper: Heave (*top left*), surge (*top right*), pitch (*bottom left*) and PTO force (*bottom right*). Experimental data is presented in dashed black line while computational results are printed in solid red lines

Figure 8 illustrates the buoy motion and mesh deformation at four instances, sampled during two wave periods. The motion is highly exaggerated compared to the linear damper only case. The buoy is close to becoming fully submerged. Please note that there are instances where it is very close that the water surface becomes positioned outside the refinement zone, due to the large motion of the buoy.

The motion and forces time series are shown in Fig. 9. The experimental test is aborted after 50 s in order to avoid waves reflected from the basin walls interfering with the WEC. It is clear that the experimental test has not reached a steady periodic behavior. Clearly 50 s is too short a time for the initial transients in pitch and surge to stabilize. The magnitudes of the motion response are far larger than the linear damper case. It is also clear that higher order terms are present in all heave, pitch and surge motions. The passive phase control clearly enhances the nonlinearity and amplifies the motion response.

The numerical simulations show a reasonable fit to the experimental results. Please note that the incident waves in the numerical model do not stop at 50 s, so after 50 s the numerical and experimental data should differ. Clearly, the majority of the nonlinear effects are captured in the numerical model and the WaveSpring modules appears to work as expected. Generally, the responses are slightly under

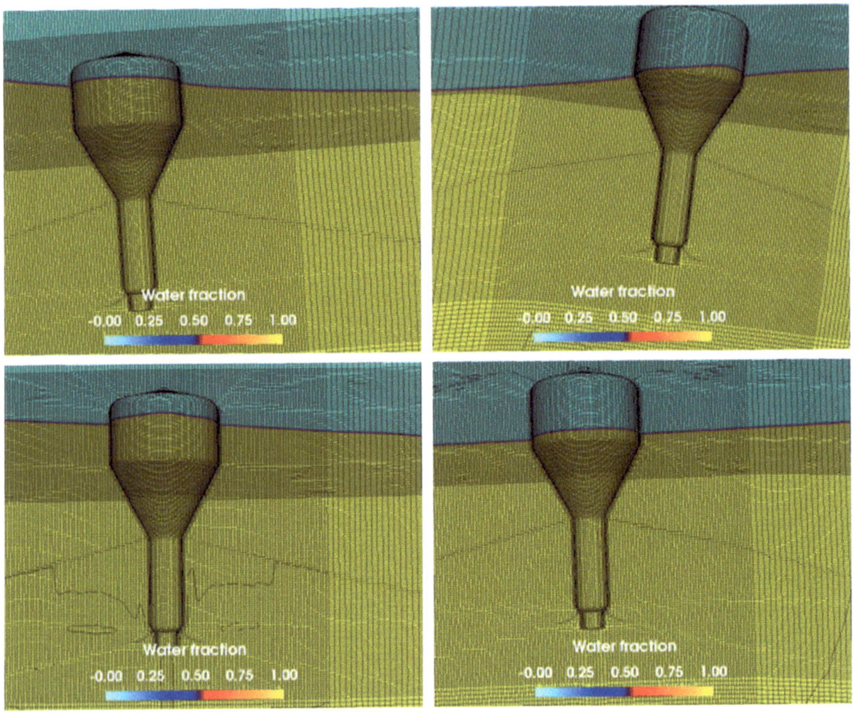


**Fig. 8** WEC motion response in regular waves with WaveSpring system. At $t = 30.75$ s (*top left*), $t = 32.25$ s (*top right*), $t = 33.5$ s (*bottom left*) and $t = 34.5$ s (*bottom right*)

predicted. Probably the mesh should be even more refined for this nonlinear case; another possible reason for the under prediction could be the use of wall functions.

## 6 Concluding Remarks

In this paper, a CFD-based numerical wave tank is used to compute the WEC motion and forces of a passively controlled point-absorber designed by CorPower Ocean AB. The negative spring arrangement of the WaveSpring system has been implemented into the six degrees-of-freedom solver and three computational cases are carried out:

- WEC surge decay test with a 0.1 m initial horizontal offset.
- WEC motion in waves when only PTO system operates.
- WEC motion in waves when both PTO system and WaveSpring system operates.

The simulated results showed generally good agreement with the laboratory data and the relative errors between CFD and experimental results are acceptable, albeit

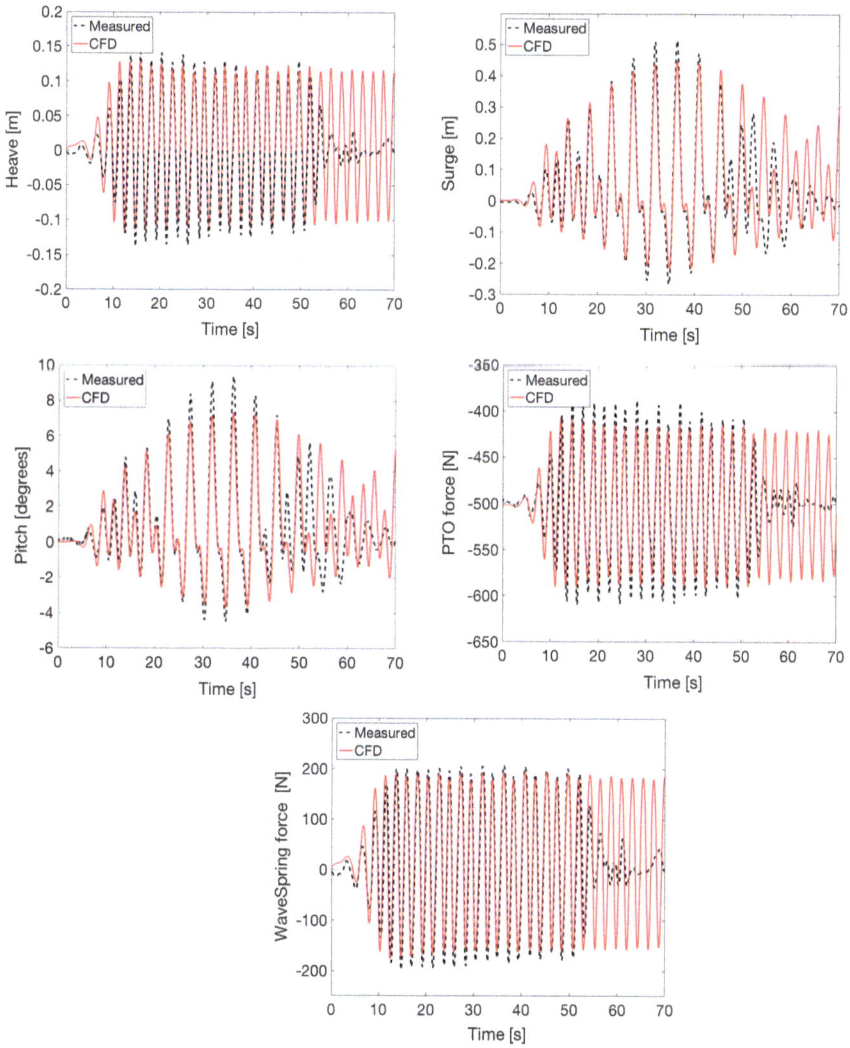


**Fig. 9** WEC motion response in regular waves with WaveSpring system: Heave (*top left*), surge (*top right*), pitch (*middle left*), PTO force (*middle right*) and WaveSpring force (*bottom*). Experimental data is presented in dashed black line while computational results are printed in solid red lines

there are differences, of course. There are typical large problems associated with validation of CFD results. As CFD lacks semi-empirical tuning parameters it is more sensitive to input data than traditional radiation-diffraction methods. For example, Palm et al. (2016) illustrated that by shifting the center of gravity just 3 mm for a 0.5 m diameter cylinder (well inside the uncertainty range of the experimental data) the decay period was significantly changed. In the case of the CorPower buoy tests there are uncertainties associated with the pre-tension, with the moment of inertia, as well as the fact that the WaveSpring case was not run until a steady periodic state had been established. This makes detailed validation very hard, but this problem is frequently occurring since WEC tank experiments are typically not performed with enough accuracy to be used as validation cases—it is simply not the main purpose of the experimental test campaigns.

For the strongly nonlinear motion excited by the WaveSpring technology, the simulation appears to capture the key system characteristics; the exaggerated motion response as well as the high-order motion responses are clearly visible. This provides confidence in the continued use of the CFD-based wave tank for further tuning of the WaveSpring/buoy design, as well as providing validation data for radiation-diffraction models. Further work is needed to examine the effects of using different turbulence models, in order to gain confidence in future full-scale WEC modelling using CFD.

**Acknowledgements** The experimental data was gratefully provided by CorPower Ocean AB. We especially thank Dr. Jørgen Hals Todalshaug at CorPower for the invaluable assistance with regard to the WaveSpring implementation. Support for this work was given by the Swedish Energy Agency under grant no. P40428-1. The simulations were made on resources from Chalmers Centre for Computational Science and Engineering, provided by the Swedish National Infrastructure for Computing.

## References

Budal, K., & Falnes, J. (1975). A resonant point absorber of ocean-wave power. *Nature, 256*, 478–479.

CorPower. (2017). CorPower Ocean AB. http://www.corpowerocean.com. Accessed 02 February 2017.

Davidson, J., Cathelain, M., Guillemet, L., Huec, T., & Ringwood, J. (2015). Implementation of an OpenFOAM numerical wave tank for wave energy experiments. In *Proceedings of the 11th European Wave and Tidal Energy Conference*, Nantes, 6–11 September 2015.

EMEC. (2017). The European Marine Energy Centre Ltd. http://www.emec.org.uk/about-us/wave-clients/corpower-ocean. Accessed 02 February 2017.

Eskilsson, C., Palm, J., Kofoed, J. P., & Friis-Madsen, E. (2015). CFD study of the overtopping discharge of the Wave Dragon wave energy converter. In C. Guedes Soares (Ed.), *Renewable Energies Offshore. 1st International Conference on Renewable Energies Offshore, Lisbon, November 2014* (p. 287). Croyden: CRC Press.

Evans, D. (1976). A theory for wave-power absorption by oscillating bodies. *Journal of Fluid Mechanics, 77*(1), 1–25.

Falnes, J. (2002). Optimum control of oscillation of wave-energy converters. *International Journal of Offshore and Polar Engineering, 12*(2), 147–155.

Giorgi, G., & Ringwood, J. V. (2016). Implementation of latching control in a numerical wave tank with regular waves. *Journal of Ocean Engineering and Marine Energy, 2*(2), 211–226.

Hals, J., Falnes, J., & Moan, T. (2011). A comparison of selected strategies for adaptive control of wave energy converters. *Journal of Offshore Mechanics and Arctic Engineering, 133*, 031101.

Jacobsen, N.-G., Fuhrman, D.-R., & Fredsøe, J. (2012). A wave generation toolbox for the open-source CFD library: OpenFoam®. *International Journal for Numerical Methods in Fluids, 70*, 1073–1088.
Kim, J.-W., Jang, H., Baquet, A., O'Sullivan, J., Lee, S., Kim, B., et al. (2016). Technical and economical readiness review of CFD-based numerical wave tank for offshore floater design. In *Offshore Technology Conference*, Houston, 2–5 May 2016.
Korde, U. A., & Ringwood, J. V. (2016). *Hydrodynamic control of wave energy devices*. Cambridge: Cambridge University Press.
Menter, F. R. (1994). Two-equation eddy-viscosity turbulence models for engineering applications. *AIAA Journal, 32*, 1598–1605.
OpenFOAM. (2017). OpenCFD Ltd. http://www.openfoam.com. Accessed 02 February 2017.
Palm, J., Eskilsson, C., Moura Paredes, G., & Bergdahl, L. (2016). Coupled mooring analysis for floating wave energy converters using CFD: Formulation and validation. *International Journal of Marine Energy, 16*, 83–99.
Sumer, B. M., & Fredsøe, J. (1997). *Hydrodynamics around cylindrical structures*. Singapore: World Scientific Publisher.
Todalshaug, J. H. (2014). Wave Energy Convertor. Patent WO2015107158 A1, January 20, 2014.
Todalshaug, J. H., Ásgeirsson, G. S., Hjálmarsson, E., Maillet, J., Möller, P., Pires, P., et al. (2016). Tank testing of an inherently phase controlled wave energy converter. *International Journal of Marine Energy, 15*, 68–84.
Weller, H., Tabor, G., Jasak, H., & Fureby, C. (1998). A tensorial approach to computational continuum mechanics using object-oriented techniques. *Computers in Physics, 12*(6), 620–631.
Yu, Y.-H., & Li, Y. (2013). Reynolds-Averaged Navier-Stokes simulation of the heave performance of a two-body floating-point absorber wave energy system. *Computers and Fluids, 73*, 104–114.

# A Framework to Improve the Coexistence of Maritime Activities & Offshore Wind Farms

**Raza Ali Mehdi, Jens-Uwe Schröder-Hinrichs, Aykut I. Ölçer, and Michael Baldauf**

## 1 Introduction

### *1.1 Background*

The offshore wind sector (OWS) has experienced significant growth in the last few decades. Since the inception of the very first offshore wind farm (OWF)—Vindeby—in 1991, there have been significant technical, technological and strategic improvements with regards to offshore wind energy generation. This is best evidenced by the rise in the numbers and sizes of both offshore winds turbines (OWTs) and OWFs. And although OWFs are still not on par with their on-shore equivalents in terms of pure cost, scientists and engineers are confident that the price gap between the two can be closed substantially over the next few years with increasingly reliable OWTs (Siemens 2014). Subsequently, with unrelenting improvements in materials, structures, aerodynamics, hydrodynamics—as well as in installation, operations, maintenance and decommissioning strategies (Maegaard et al. 2013a, b; Mehdi et al. 2016)—the sustained, strong growth of the OWS is predicted to continue for the foreseeable future (GWEC 2016; EWEA 2016).

R. A. Mehdi (✉)
Maritime Risk & System Safety (MaRiSa) Research Group, World Maritime University, Malmö, Sweden
e-mail: rm@wmu.se

J.-U. Schröder-Hinrichs · M. Baldauf
MaRiSa Research Group, Maritime Safety and Environmental Administration (MSEA), WMU, Malmö, Sweden
e-mail: jus@wmu.se; mbf@wmu.se

A. I. Ölçer
World Maritime University, Malmö, Sweden
e-mail: aio@wmu.se

 
A. I. Ölçer et al. (eds.), *Trends and Challenges in Maritime Energy Management*, WMU Studies in Maritime Affairs 6, https://doi.org/10.1007/978-3-319-74576-3_35

In the grand scheme of things, the developments in the OWS are good news for our society: with an increasing number of people becoming aware of the harsh realities of climate change, as well as strong climate policy instruments such as the UN 2030 SDG Goals and various EU directives, the continued growth of the OWS is more welcome than ever before—from both a bottom-up and top-down perspective.

That being said, the growth of the OWS is not without its challenges. Apart from the sheer technological and financial challenges posed by larger, more complex turbines in harsher environmental conditions, the OWS is also under sharp scrutiny from environmental experts, who are particularly concerned about the potential impacts that OWFs can have on avian and marine species. Furthermore, there are also legal and policy-related hurdles that the OWS has yet to overcome. But perhaps the strongest challenge to the growth of the OWS comes from the maritime industry. Indeed, the shipping industry—which proudly defends its status as a traditional user of the sea—often views the growth of the OWS quite warily. Although maritime accidents in the vicinity of OWFs have been largely avoided thus far, there are concerns that the sustained growth of the OWS may have a significant negative impact on navigational safety in the long run. There are also concerns that the placement and location of certain OWFs may reduce navigational *efficiency*, and increase ship emissions due to measures such as re-routeing.

## 1.2 Problem Description

It is evident that the impacts of the OWS are complex and multi-dimensional—a fact that is well recognized by the coastal states that approve and consent OWFs. In fact, most governmental authorities have clear guidelines and specifications that require a thorough assessment of the environmental, societal, and maritime impacts before granting a license for any OWF in their waters.

However, these assessments—whilst being quite thorough—are often conducted in a very *segregated* manner: the different impacts are usually assessed and evaluated by different consultative bodies and authorities. This means that it is difficult to get an overall, 'integrated' picture of the various different impacts associated with OWFs. Consider, for example, a maritime navigational risk assessment which has recommended re-routeing as a risk-control option to mitigate the navigational risks in the vicinity of an OWF. This measure may inadvertently lead to financial costs, environmental impacts (e.g. dredging a new channel), and even reduced energy efficiency (e.g. through increased ship emissions offsetting the 'clean' energy produced by the OWF). With the current segregated approaches, however, *only* the primary navigational risk aspect of the re-routeing would be evaluated—whilst the secondary and tertiary impacts on the environment and energy efficiency would likely be ignored. Similar problems may also be encountered when proposing measures to mitigate the environmental impact, or improve energy efficiency from an OWF.

As a solution to the current segregated approach, the present work aims to describe an *iterative* optimization framework, wherein the different impacts associated with OWFs are evaluated in an integrated manner. By describing a step-by-step approach, the authors envision that the proposed framework can allow stakeholders to overcome the shortcomings of the existing 'boxed' approach to OWF approval.

## 2 Literature Review

The evolution and growth of the OWS over the last two-and-a-half decades has resulted in an abundance of literature pertaining to the design, installation, operation, maintenance, decommissioning, and impact assessment of OWTs and OWFs.

A recent publication (Ostachowicz et al. 2016) for instance, describes various tools, methods and models that can be used to design and optimize various OWT components, improve maintenance strategies and enhance OWF layouts. Similarly, the work of Mehdi and Schröder-Hinrichs (2016) and Deeb et al. (2017) provides a comprehensive review of models and methods that are used for maritime navigational risk assessment (NRA) around OWFs—whilst also proposing a novel harmonized NRA framework. The interaction between OWFs and maritime navigation has also been explored in a *legal* context recently by Chircop and L'Esperance (2016).

The societal impacts of OWFs are also quite well explored (Lacroix and Pioch 2011). Authors such as Devine-Wright (2005), Ladenburg (2011) and Chen et al. (2015), for example, have investigated the perceptions of societies and coastal communities with regard to OWFs.

There is also substantial research on the environmental impacts of OWFs. The impacts of OWFs on avian species have been addressed by authors such as Desholm and Kahlert (2005), Dierschke et al. (2006), Hüppop et al. (2006), Degraer and Brabant (2009), Beiersdorf and Radecke (2014) and New et al. (2015). Likewise, the impacts of OWFs on marine life have been investigated in the works of Köller et al. (2006), Degraer and Brabant (2009), Andersson (2011), Bailey et al. (2014) and Verfuss et al. (2016).

There have also been studies focussing on the lifecycle environmental (Weinzettel et al. 2009; Wang and Sun 2012) and economic (Blanco 2009; Snyder and Kaiser 2009) impacts of OWFs which consider different aspects of OWTs from inception to decommissioning.

In an attempt to integrate the different impacts of OWFs, authors such as Samoteskul et al. (2014) have also explored how vessel re-routeing can reduce the costs associated with OWFs.

In addition to OWF-specific literature, there have also been significant advances in research dealing with general maritime (e.g. Boukani 2016) and marine environmental aspects. This, however, is a topic beyond the scope of the present work—and is thus not explored in greater detail here. It is nevertheless important to consider this fact, because most developments in research are still quite 'segregated'. Nonetheless, the existing literature—both OWF specific and general—does serve as a crucial starting point for the present work, wherein we identify various tools, methods and models that complement each other in a manner that can be used to create an integrated framework.

# 3 Proposed Framework

The literature review indicates that despite initial efforts, and sophisticated existing models and methods, there is still a research gap when it comes to an *integrated* approach to assessing the multi-dimensional impacts of OWFs. As a solution, the present work proposes a theoretical framework that integrates the various different impact calculations in a coherent manner. The proposed framework is shown in Fig. 1.

The first step in the framework is the design of individual wind turbines. This step, in itself, is iterative as the design of different OWT components has to be optimized both individually and in relation to other components. A typical design process for an OWT may start with the blade design, including materials and aero-elastic optimization. Depending on the size and the potential load that the blades have to bear, one may then design a suitable gearbox unit and other nacelle components. This may be followed by the design of the OWT tower which needs to bear the sustained aerodynamic and mechanical loads from the blades and nacelle assembly over an extended period of time. An adequate support structure is also needed to support the entire blade-nacelle-tower assembly. Together, the tower and support structure must also be capable of bearing hydrodynamic loads in often-harsh sea-conditions. The installation, maintenance, and decommissioning strategies for the various components must also be optimized. The aforementioned publication edited by Ostachowicz et al. (2016) describes various tools that can be used to optimize the design and installation/maintenance/decommissioning strategies for the various OWT components; these tools would be ideally suited to fulfil the first step of the framework.

Following the design of an OWT, one can focus on optimizing the design of an OWF in terms of layout and location. The layout considerations are primarily driven by aerodynamic efficiency requirements, which in turn are influenced by the wake effects from the OWTs (Mehta 2016; Giebel and Hasager 2016).

For OWF design, it is also important to consider the wind resource—i.e.—the direction, speed, consistency, duration, etc. of wind flow in an area. This consideration can have an influence on both the layout, as well as the location, of the OWF. The design and layout of OWTs, and the location of an OWF also has an influence on the installation (Sarker and Ibn Faiz 2017; Boquist 2015), maintenance (Martínez García et al. 2016; Dalgic et al. 2015) and decommissioning (Gjødvad and Ibsen 2016) costs which are important factors to consider.

Thus, the design, layout and location of the OWTs together determine the financial cost and the energy debt of an OWF. Often, the fabrication, installation, maintenance and decommissioning operations are only considered in terms of

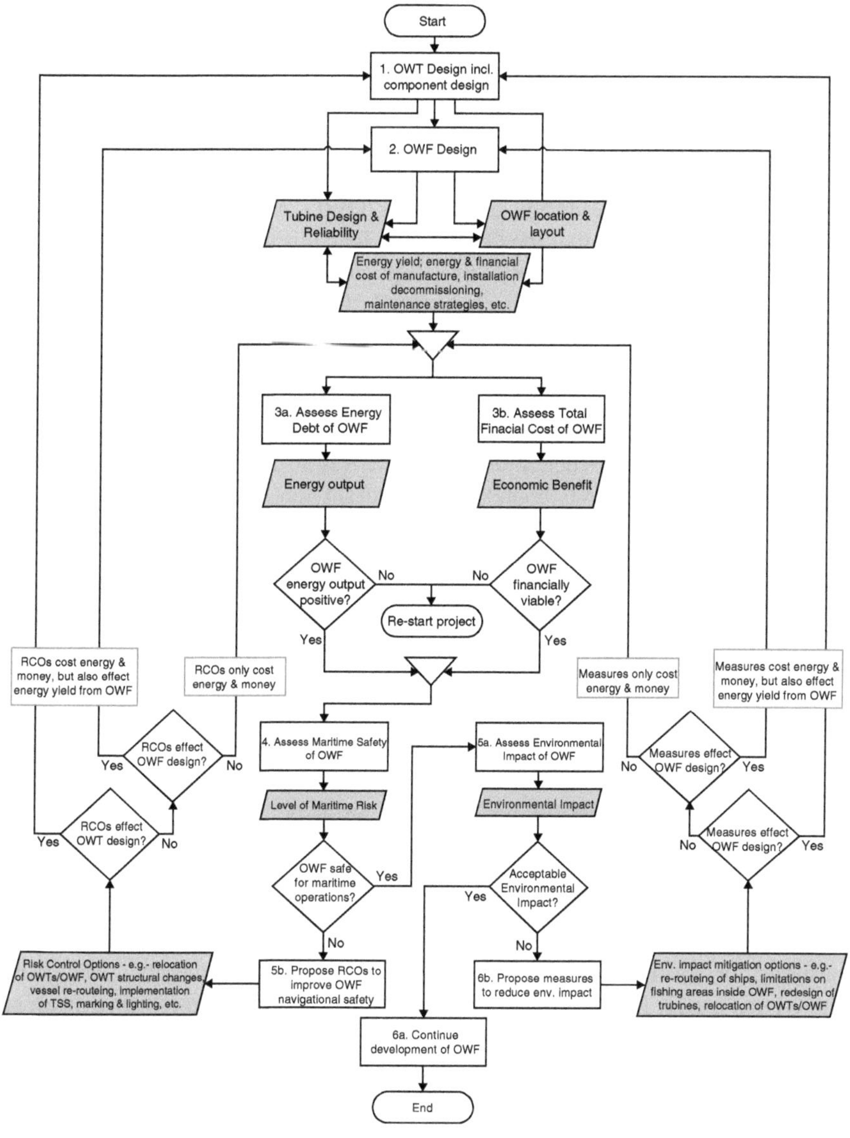


**Fig. 1** The proposed integrated framework

*financial* costs—whilst the energy consumption figures are ignored. For the present framework, it is also important to consider that the energy cost of these operations, and various tools—as mentioned in Sect. 2—can be used for both the financial and energy lifecycle analyses in step 3a and step 3b. If the OWF is found to be both financially profitable, and energy positive, the project can proceed.

Step 4 of the framework requires the users to assess the maritime safety impacts that an OWF may pose. There are several tools (Mehdi and Schröder-Hinrichs 2016) which can be used at this stage to quantify the risk of various navigational incidents arising due to the presence of an OWF. If the risk is found to be above a predetermined level, stakeholders may recommend that the OWF owners implement certain risk control measures in step 5a. These risk control measures may or may not have an influence on the design of OWT or OWF. A common requirement in Germany, for instance, is that the OWT tower be collision friendly—i.e.—in case of a drifting side-on collision between a ship and a turbine, the ship's hull must sustain damage *below* a certain threshold (BSH 2015). This of course, can have an influence on the design of the OWT: having a more 'elastic' or flexible tower may limit the loads that the tower can bear, and thus have an effect on the size of the blades and the energy that can be generated. There have also been cases where the layout of an OWF has been changed due to maritime safety concerns. In Belgium, for example, the authorities determined that there was insufficient distance between a shipping lane and an OWF boundary; this meant that the OWF owner had to eliminate a row of OWTs to ensure the minimum safe-passing-distance between turbines and vessels was achieved (Mehdi et al. 2017). Again, such a decision can have a significant influence on the energy yield and profitability of an OWF. Thus, if it is the case that a risk control option affects either the design of an OWT, or that of an OWF, the users of the present framework are required to loop back to step 1 or 2 respectively, to ensure that the OWF is still viable in terms of finance and energy.

It is also possible to have risk control measures that *do not* affect the design of an OWT or OWF. Examples of such measures include painting, lighting, traffic separation schemes, shore-based monitoring of vessels, and re-routeing of shipping lanes. Each of these measures, however, has an associated financial and energy cost which must be considered in the present framework. Perhaps the most 'expensive' of such measures is vessel re-routeing: such a measure may require vessels to take a significantly long detour around an OWF, increasing the operational cost as well as fuel emissions. Whilst the increased financial cost may be passed on the consumer, or absorbed by the ship owners, the increased emissions cannot be ignored that easily. Indeed, if over the lifecycle of an OWF, a re-routeing measure produces more emissions than can be offset by the electricity generation from the OWF, it may be prudent to scrap the entire project altogether. Examples of re-routeing around OWFs can be found in Belgium and the Netherlands; both countries have submitted several joint FSA studies to the IMO for these routeing measures to be implemented (IMO 2015a, b). The present framework, therefore, requires users to loop back to steps 3a and 3b after step 4, to ensure that a risk control measure does not nullify the financial and energy benefits of an OWF to an extent where it is imprudent to continue with the project. This loop requires the use of further tools, such as ship emission calculators (Boukani 2016), which focus on assessing the impact of speed reduction of ships on cost and emission figures as long as relevant input is fed into the developed calculator; the inputs to this tool include basic data about the ship and the journey (e.g. age and size of ship, year of build, distance of journey, ship speeds and fuel price), costs (e.g. crew provisions, lubricants, P&I insurance, wages), and engine consumption (based on speed, load and ballast condition of vessel). The outputs from the tool, which are presented as graphs and tables, include optimum voyage speed and time, and total gas emissions in terms of CO, $CO_2$, $NO_2$, $NO_x$, $SO_2$.

If maritime safety is at an adequate level, and the OWF is still profitable and energy positive, the users can proceed to step 5b of the framework. This step requires users to assess the environmental impact of OWFs. This includes an assessment of both the impacts related to the OWF itself, as well as an environmental impact assessment of the risk control measures implemented to ensure maritime safety in the vicinity of the wind farm. In other words, the environmental impact step in the present framework requires a consideration of direct OWT-related factors, as well as any indirect *change* in maritime activities or the marine environment *due* to the OWTs. Tools that can be used for this step can be derived from such works as listed in Sect. 2, and primarily involve an impact assessment in the context of marine and avian species. If the impacts are found to be unacceptable, stakeholders can propose mitigation measures; as with the risk control options for maritime safety, these mitigation measures may have an influence on OWT or OWF design, which would mean that the financial costs and energy yield would have to be recalculated through step 1 and/or 2. Alternatively, the measures may also have an influence on maritime safety, in which case step 4 would need to be revisited. If the environmental impact mitigation measures do not effect OWT design, OWF design or maritime safety, users are still required to loop back to step 3a and 3b to calculate the total cost and energy debt of the project, whilst taking into account the introduction of the measures.

Once all the different impacts have been assessed, four different parameters can be plotted on a heat-map as shown in Fig. 2.

Each of the four parameters represents the four primary aspects that are covered in this framework. The background heat-map should be scaled to represent the acceptability limits for each of the different parameters. Maritime risk should ideally be a combination of the peak navigational risk probability and consequence value. The economical cost should be a monetary value representing the total costs associated with the project including the fabrication of OWT components, installation, maintenance and decommissioning costs, the costs of maritime safety risk control options, and the cost of environmental impact mitigation measures. The energy debt should be an energy value (in Watts); it should factor in the total energy yield from the OWF and subtract the total energy cost associated with the turbines, the OWF, the navigational risk control options, and the environmental protection measures. There is no present consensus on how to represent the various environmental impacts on a uniform scale for the present work.

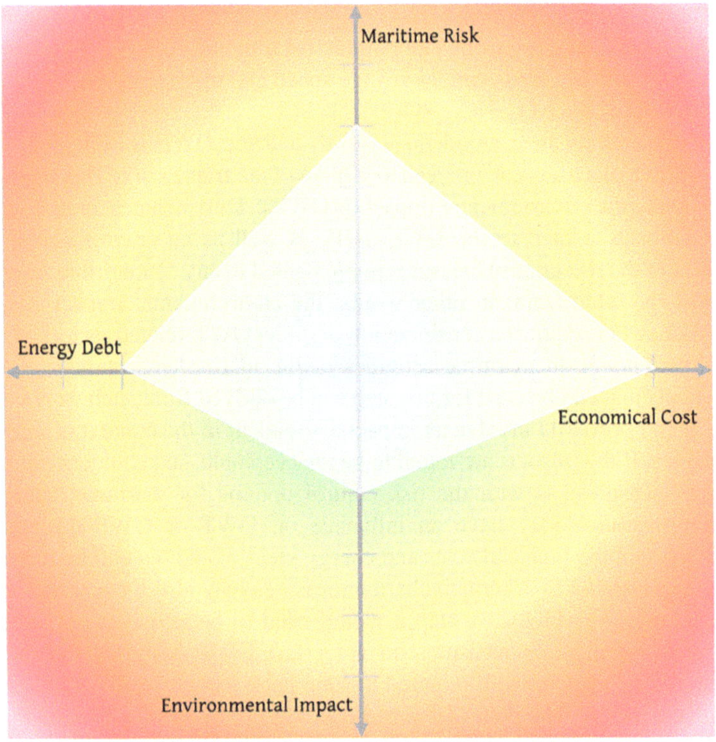


**Fig. 2** Envisioned output of the proposed framework

## 4 Discussion

One of the main discussion points of the present work is who the *users* of this model might be. Each step requires the use of specialised tools, models and methods which may not be available to, or even useable by, all different stakeholders. Indeed, it may be unwise to suggest that one authority or individual perform all the different calculations, as each stage requires specialist knowledge of a subject matter. It is thus envisioned that each of the impacts is calculated separately by specialists, but that the relevant data flow and information exchange is managed through the proposed framework by either the responsible *Coastal State*, or by an *OWF developer* if the state chooses to delegate this responsibility.

The information exchange would require, for instance, the OWF planners to share spatial information about the OWF, and OWT designers to share structural information about the OWTs, with maritime stakeholders; in turn, the maritime authorities may revert with modified spatial files or structural files in case the design of the OWF or OWT needs to be changed. The proposed framework, in a way, provides a uniform approach about what data needs to be shared, when, and between whom. Data that is commonly needed for more than one assessment—e.g.—meteorological data used for OWF design, as well as navigational risk calculations—can be provided by the primary user of the framework to all different parties. This would ensure that different models, methods and tools utilize uniform data sets.

With marine space becoming increasingly congested, the need for this framework is quite urgent. Spatial planners from various coastal states need to be able to determine what the best use of space is—and for this, they need as much information as they can get, in as coherent a manner as possible. Stakeholders rely on tools such as marine spatial planning (MSP) in order to balance the needs of different marine users. The MSP approach is advocated by regional and international bodies such as the EU (EU 2014, 2016) and the IMO. MSP can be defined as 'a public process of analysing and allocating the spatial and temporal distribution of human activities in marine areas to achieve ecological, economic, and social objectives that are usually specified through a political process. It is a continuing, iterative process that learns and adapts over time' (Ehlers and Douvere 2009). From the definition, it is clear that an important part of MSP is understanding the impacts of different projects and activities in a holistic manner. The framework presented in this work allows stakeholders to do exactly that: by encouraging a graphical plot of the various impacts of OWFs (or indeed any offshore renewable energy installations (OREIs)), this framework ensures that any decisions with regard to use of marine space, are on the back of cross-sectoral, risk-based approach.

The framework also proposes the incorporation of cost-benefit analysis calculations in each iteration. For these calculations, the guidelines of IMO on the Formal Safety Assessment (FSA) can be instrumental. The FSA is defined as 'a rational and systematic process for assessing the risks associated with shipping activity and for evaluating the costs and benefits of IMO's options for reducing these risks' (IMO 2017). FSA has been successfully used by IMO to introduce regulations and proactive risk-control measures over the last few decades. Coastal states who wish to introduce risk-control measures (e.g. altering shipping lanes to make way for other marine uses) are also encouraged to submit FSA studies—and many have done so already to ensure safety of navigation around OWFs, for instance. Enhanced version of the FSA cost-benefit formulae can thus be used to assess the cross-sectoral impacts—allowing the framework to further complement the MSP process as well.

# 5 Future Work and Conclusions

## *5.1 Future Work*

One area which has not been explicitly covered in the present framework is the so-called societal impact of OWFs. With OWFs in Europe moving further away from shore, at a greater distance from coastal communities, the societal impacts are not considered an urgent issue for the present work. Nonetheless, this is a factor which is still quite relevant in countries where there is still near-shore capacity—and the authors intend to include it in future iterations of the framework.

In the current paper, the authors have also refrained from providing elaborate descriptions of the various tools, models and methods that can be used at each assessment stage. A recommendation of such tools would require a thorough study of the different options that are available, before an informed decision can be made. This again, is a task which the authors hope to accomplish in future research. That being said, the publications mentioned in the literature review section do contain excellent tools which can be used in initial (and perhaps even advanced) versions of this framework.

## 5.2 *Conclusions*

The current work provides a description of a theoretical framework, which can be used by decision makers to optimize the use of increasingly limited sea-space due to OWFs. In developing this framework, the authors present a brief discussion about the various aspects of OWFs, and highlighted the complex interconnectivity between these different aspects—an element that is often not adequately considered in current approval procedures.

It is envisaged that through such a framework, a better mutual understanding of the multi-dimensional impacts of OWFs can be instilled within the many different stakeholders that are involved in the design, approval, consenting and general lifecycle of OWFs. And with this better understanding—as well as a more coherent, holistic information exchange promoted by the presented framework—it is anticipated that decision makers can make better, more well-informed decisions about optimum use of sea space.

**Acknowledgements** This research was supported by a Marie Curie Initial Training Network Grant within the 7th European Community Framework Programme. The authors of this work gratefully acknowledge support for this research under the project No. 309395 MARE-WINT provided by the EU.

## References

Andersson, M. H. (2011) *Offshore wind farms – Ecological effects of noise and habitat alteration on fish*. Dissertation, Stockholm University

Bailey, H., Brookes, K. L., & Thompson, P. M. (2014). Assessing environmental impacts of offshore wind farms: Lessons learned and recommendations for the future. *Aquatic Biosystems*. https://doi.org/10.1186/2046-9063-10-8.

Beiersdorf, A., & Radecke, A. (Eds.). (2014). *Ecological research at the offshore Windfarm alpha ventus: Challenges, results and perspectives*. Wiesbaden: Springer.

Blanco, M. I. (2009). The economics of wind energy. *Renewable and Sustainable Energy Reviews, 13*, 1372–1382.

Boquist, P. (2015). *Offshore wind power investment model using a reference class forecasting approach to estimate the required cost contingency budget*. Dissertation, Uppsala University.

Boukani, L. N. (2016). *Ship energy management and development of software for vessels' optimum economic speed.* Dissertation, World Maritime University.

BSH. (2015). Minimum requirements concerning the constructive design of offshore structures within the Exclusive Economic Zone (EEZ). In *BSH Books*. Available via BSH. http://www.bsh.de/de/Produkte/Buecher/Standard/index.jsp. Accessed 18 August 2017.

Chen, J.-L., Liu, H.-H., Chuang, C.-T., & Lu, H.-J. (2015). The factors affecting stakeholders' acceptance of offshore wind farms along the western coast of Taiwan: Evidence from stakeholders' perceptions. *Ocean Coast Manage, 109*, 40–50.

Chircop, A., & L'Esperance, P. (2016). Functional interactions and maritime regulation: The mutual accommodation of offshore wind farms and international navigation and shipping. *Ocean Yearbook, 30*, 439–487.

Dalgic, Y., Lazakis, I., Dinwoodie, I., McMillan, D., & Revie, M. (2015). Advanced logistics planning for offshore wind farm operation and maintenance activities. *Ocean Engineering, 101*, 211–226.

Deeb, H., Mehdi, R. A., & Hahn, A. (2017). A review of damage assessment models in the maritime domain. *Ships Offshore Structures, 12*(1), 31–54.

Degraer, S., & Brabant, R. (Eds.). (2009). *Offshore wind farms in the Belgian part of the North Sea: State of the art after two years of environmental monitoring*. Brussels: Royal Belgian Institute for Natural Sciences.

Desholm, M., & Kahlert, J. (2005). Avian collision risk at an offshore wind farm. *Biology Letters, 1*(3), 296–298.

Devine-Wright, P. (2005). Beyond NIMBYism: Towards an integrated framework for understanding public perceptions of wind energy. *Wind Energy, 8*, 125–139.

Dierschke, V., Garthe, S., & Mendel, B. (2006). Possible conflicts between offshore wind farms and seabirds in the German sectors of North Sea and Baltic Sea. In J. Köller, J. Köppel, & W. Peters (Eds.), *Offshore wind energy: Research on environmental impacts* (pp. 121–143). Heidelberg: Springer.

Ehlers, C., & Douvere, F. (2009). *Marine Spatial Planning: a step-by-step approach toward ecosystem-based management. Intergovernmental Oceanographic Commission and Man and the Biosphere Programme* (IOC Manual and Guides No. 53, ICAM Dossier No. 6). UNESCO Publications. Available via UNESCO. http://unesdoc.unesco.org/images/0018/001865/186559e.pdf. Accessed 17 May 2017.

EU. (2014). DIRECTIVE 2014/89/EU OF THE EUROPEAN PARLIAMENT AND OF THE COUNCIL of 23 July 2014 establishing a framework for maritime spatial planning. *Official Journal of the European Union*. Available via EU. http://eur-lex.europa.eu/legal-content/EN/TXT/HTML/?uri=CELEX:32014L0089&from=EN. Accessed 17 May 2017.

EU. (2016). Political declaration on energy cooperation between the North Seas Countries. *European Union Documents*. Available via EU. http://europa.eu/rapid/attachment/IP-16-2029/en/Political%20Declaration%20on%20Energy%20Cooperation%20between%20the%20North%20Seas%20Countries%20FINAL.pdf. Accessed 17 May 2017.

EWEA. (2016). The European offshore wind industry - key trends and statistics 2015. https://www.ewea.org/fileadmin/files/library/publications/statistics/EWEA-European-Offshore-Statistics-2015.pdf. Accessed 1 March 2017.

Giebel, G., & Hasager, C. B. (2016). An overview of offshore wind farm design. In W. Ostachowicz, M. McGugan, J.-U. Schröder-Hinrichs, & M. Luczak (Eds.), *MARE-WINT: New materials and reliability in offshore wind turbine technology* (pp. 337–346). Cham: Springer.

Gjødvad, J. F., & Ibsen, M. D. (2016). ODIN-WIND: An overview of the decommissioning process for offshore wind turbines. In W. Ostachowicz, M. McGugan, J.-U. Schröder-Hinrichs, & M. Luczak (Eds.), *MARE-WINT: New materials and reliability in off-shore wind turbine technology* (pp. 403–419). Cham: Springer.

GWEC. (2016). Global Wind Report 2015 – Annual market update. http://www.gwec.net/publications/global-wind-report-2/global-wind-report-2015-annual-market-update/. Accessed 1 March 2017.

Hüppop, O., Dierschke, J., Exo, K.-M., Fredrich, E., & Hill, R. (2006). Bird migration and off-shore wind turbines. In J. Köller, J. Köppel, & W. Peters (Eds.), *Offshore wind energy: Research on environmental impacts* (pp. 91–116). Heidelberg: Springer.

IMO. (2015a). Routeing measures and mandatory ship reporting systems. Report on the Formal Safety Assessment (FSA) related to proposals for new and amended routeing measures off the Netherlands- Belgian Coast between West Hinder, North Hinder and Maas West traffic separation schemes Submitted by Belgium and the Netherlands. NSCR 3/INF.3 In: IMO Docs. Available via IMO. https://docs.imo.org/. Accessed 18 August 2017.

IMO. (2015b). Routeing measures and mandatory ship reporting systems. New traffic separation scheme and amendments to existing traffic separation schemes "In the Approaches to Hook of Holland and at North Hinder" Submitted by Belgium and the Netherlands. NSCR 3/3/2 In: IMO Docs. Available via IMO. https://docs.imo.org/. Accessed 18 August 2017.

IMO. (2017). Formal Safety Assessment. http://www.imo.org/en/OurWork/Safety/SafetyTopics/Pages/FormalSafetyAssessment.aspx. Accessed 10 June 2017.

Köller, J., Köppel, J., & Peters, W. (Eds.). (2006). *Offshore wind energy: Research on environmental impacts*. Heidelberg: Springer.

Lacroix, D., & Pioch, S. (2011). The multi-use in wind farm projects: More conflicts or a win-win opportunity? *Aquatic Living Resources, 24*, 129–135.

Ladenburg, J. (2011). Attitude and acceptance of offshore wind farms – The influence of travel time and wind farm attributes. *Renewable and Sustainable Energy Reviews, 15*, 4223–4235.

Maegaard, P., Krenz, A., & Palz, W. (Eds.). (2013a). *Wind power for the world: The rise of modern wind energy: Pan Stanford series on renewable energy* (Vol. 2). Boca Raton, FL: CRC Press.

Maegaard, P., Krenz, A., & Palz, W. (Eds.). (2013b). *Wind power for the world: International reviews and developments: Pan Stanford series on renewable energy* (Vol. 3). Boca Raton, FL: CRC Press.

Martínez García, I. E., Sánchez Sánchez, A., & Barbati, S. (2016). Reliability and preventive maintenance. In W. Ostachowicz, M. McGugan, J.-U. Schröder-Hinrichs, & M. Luczak (Eds.), *MARE-WINT: New materials and reliability in offshore wind turbine technology* (pp. 235–272). Cham: Springer.

Mehdi, R. A., Ostachowicz, W., & Luczak, M. (2016). Introduction. In W. Ostachowicz, M. McGugan, J.-U. Schröder-Hinrichs, & M. Luczak (Eds.), *MARE-WINT: New materials and reliability in offshore wind turbine technology* (pp. 1–9). Cham: Springer.

Mehdi, R. A., & Schröder-Hinrichs, J.-U. (2016). A theoretical risk management frame-work for vessels operating near offshore wind farms. In W. Ostachowicz, M. McGugan, J.-U. Schröder-Hinrichs, & M. Luczak (Eds.), *MARE-WINT: New materials and reliability in offshore wind turbine technology* (pp. 359–400). Cham: Springer.

Mehdi, R. A., Schröder-Hinrichs, J.-U., van Overloop, J., Nilsson, H., & Pålsson, J. (2017). A critique of navigational risk assessment processes for offshore wind farms. Submitted to Energy Research & Social Science.

Mehta, D. (2016). Large Eddy simulation of wind farm aerodynamics with energy-conserving schemes. In W. Ostachowicz, M. McGugan, J.-U. Schröder-Hinrichs, & M. Luczak (Eds.), *MARE-WINT: New materials and reliability in offshore wind turbine technology* (pp. 347–358). Cham: Springer.

New, L., Bjerre, E., Millsap, B., Otto, M. C., & Runge, M. C. (2015). A collision risk model to predict Avian fatalities at wind facilities: An example using golden eagles, Aquila chrysaetos. *PLoS ONE*. https://doi.org/10.1371/journal.pone.0130978

Ostachowicz, W., McGugan, M., Schröder-Hinrichs, J.-U., & Luczak, M. (Eds.). (2016). *MARE-WINT: New materials and reliability in offshore wind turbine technology*. Cham: Springer.

Samoteskul, K., Firestone, J., Corbett, J., & Callahan, J. (2014). Changing vessel routes could significantly reduce the cost of future offshore wind projects. *Journal of Environmental Management, 141*, 146–154.
Sarker, B. R., & Ibn Faiz, T. (2017). Minimizing transportation and installation costs for turbines in offshore wind farms. *Renewable Energy, 101*, 667–679.
Siemens, A. G. (2014). A macro-economic viewpoint: what is the real cost of offshore wind? Available via Siemens AG. http://www.energy.siemens.com/hq/pool/hq/power-generation/renewables/wind-power/SCOE/Infoblatt-what-is-the-real-cost-of-offshore.pdf. Accessed 1 March 2017.
Snyder, B., & Kaiser, M. (2009). Ecological and economic cost-benefit analysis of offshore wind energy. *Renewable Energy, 34*, 1567–1578.
Verfuss, U. K., Sparling, C. E., Arnot, C., Judd, A., & Coyle, M. (2016). Review of offshore wind farm impact monitoring and mitigation with regard to marine mammals. In A. N. Popper & A. Haw-kins (Eds.), *The effects of noise on aquatic life II. Advances in experimental medicine and biology* (Vol. 875, pp. 1175–1182). New York, NY: Springer.
Wang, Y., & Sun, T. (2012). Life cycle assessment of CO2 emissions from wind power plants: Methodology and case studies. *Renewable Energy, 43*, 30–36.
Weinzettel, J., Reenaas, M., Solli, C., & Hertwich, E. G. (2009). Life cycle assessment of a floating off-shore wind turbine. *Renewable Energy, 34*, 742–747.

## Further Reading

Gibson, E., & Howsam, P. (2010). The legal framework for offshore wind farms: A critical analysis of the consents process. *Energy Policy, 38*, 4692–4702.
Wright, G., Mehdi, R. A., & Baldauf, M. (2016). 3-dimensional forward looking sonar: Off-shore wind farm applications. In: *Proceedings of the 24th European Navigation Con-ference (ENC 2016)*, European Group of Institutes of Navigation, Helsinki 30 May – 2 June 2016.

# Editorial Conclusion

**Aykut I. Ölçer, Momoko Kitada, Dimitrios Dalaklis, and Fabio Ballini**

From the very first meeting of the committee responsible for the organisation of the International Conference on Maritime Energy Management (MARENER 2017) until the publication of this cumulative research effort, a significant period of time has elapsed. During these two and a half years, a growing interest in the field of Maritime Energy Management (MEM), both inside and outside the maritime industry, has been observed. There is a very simple explanation why: the sustainable future of the maritime industry is strongly related to energy efficient design and operation of ships, as well as numerous other emerging maritime energy-related areas of study. In this regard, the World Maritime University (WMU) has responded well in advance to the specific needs of the maritime industry with five different tangible outcomes: (a) the "operationalisation" of the Maritime Energy Management research group (MarEner); (b) the launch of a postgraduate diploma course in MEM; (c) the introduction of a Master of Science (MSc) programme in MEM; (d) the conduct of the MARENER 2017 conference; (e) the publication of this book—"Trends and Challenges in Maritime Energy Management".

This book includes 34 chapters divided into six different but interrelated categories: (1) regulations, (2) ship design, (3) ship and port operations, (4) economic and social aspects, (5) alternative fuels and wind-assisted ship propulsion, and (6) marine renewable energy. These chapters reflect the vast majority of the contemporary trends and challenges which the maritime industry is observing in energy management. In this cumulative research effort the breadth of knowledge and research necessary to understand a holistic picture of MEM is highlighted. However, the task is not over yet; there is now an open invitation for a wide range of scholars, policy makers, and industry experts to share their knowledge and experience to promote a low carbon and energy efficient future for the maritime sector. It is,

A. I. Ölçer (✉) · M. Kitada · D. Dalaklis · F. Ballini
World Maritime University, Malmö, Sweden
e-mail: aio@wmu.se



A. I. Ölçer et al. (eds.), *Trends and Challenges in Maritime Energy Management*,
WMU Studies in Maritime Affairs 6, https://doi.org/10.1007/978-3-319-74576-3_36

therefore, our aspiration as editors that this book will help establish a basis for further development of MEM research, not only in developed countries but also in developing ones.

To conclude the current effort, this chapter will address what comes next in the MEM domain and how to participate in the development of future MEM research. The following sections present three important developments: the Maritime Energy Management Research Strategy; the MARENER 2017 Conference evaluation study; and the role of MEM in Research and Innovation for Sustainable Maritime Transportation.

## 1 Maritime Energy Management Research Strategy

During the MARENER 2017 Conference, WMU's research strategy in the MEM field was launched (see the Annex). The essence of this strategy is captured in its vision: "*To become the world's leading University in the research field of maritime energy management to play a vital role in transforming the maritime world to achieve a sustainable, low carbon and energy efficient future by delivering research of global excellence*". It is also important to point out that the mission of WMU in relation to MEM research is to act as the world's leading institute in the respective research field by delivering research of global excellence through addressing the needs of maritime and ocean communities.

As stated earlier, interdisciplinary research is the pivot point of the MEM research strategy. In order to achieve the strategic goals, the following actions are identified:

- Advancing the knowledge in the maritime energy management field by conducting world-class foundamental and applied research in the thematic areas shown in the Annex over a life-cycle perspective, from shipping to oceans, and from ports to shipyards;
- Contributing to the capacity building and the goals of the International Maritime Organization (IMO) and its Member States and relevant United Nations (UN) bodies, and in particular to the UN Sustainable Development Goals (SDGs) 7, 12 & 13;
- Fostering strategic relationships with other universities, governmental organisations, companies and other maritime stakeholders across the world to advance research;
- Setting a global research agenda in the maritime energy management field through addressing maritime community's needs; and
- Undertaking research at an international standard through scholarly publication.

To operationalise the strategy, key thematic areas have been identified as "Energy Efficiency"; "Regulatory Framework"; "Renewable Energy"; "Human Factors"; "Economics" and "Innovation". These key thematic areas will be viewed from a life-cycle perspective: "Design – Production – Operation – Recycling". Further, the thematic areas and the life-cycle perspective have three focused applications, which are, "Ships", "Onshore facilities (ports and shipyards)", and "Oceans". The combinatorial interaction between key thematic areas, life-cycle perspective, and focus applications is crucial for the effective implementation of the MEM research strategy.

Strategic directions in MEM are reasonably represented in this book, covering all the key elements of the MEM strategy. Nevertheless, some research areas are weaker than the others. For example, economic and social aspects as well as port and shipyard operations are research areas that must be strengthened in the future.

## 2 MARENER 2017 Evaluation Study

After the MARENER 2017 conference, an online evaluation survey was conducted from 30 January to 14 February 2017. In total, 55 non-WMU participants responded; the overall response rate was 46%. The conference participants were both academics and maritime experts related to the maritime energy management field. A summary of the MARENER 2017 evaluation study was reported in an information paper to the 71st session of the IMO's Marine Environment Protection Committee (MEPC)[1] on the Agenda item 5: Air Pollution and Energy Efficiency.

According to the survey, a high percentage of respondents agreed with the following:

- MARENER 2017 informed the challenges and opportunities in the maritime industry to achieve the United Nation's (UN) Sustainable Development Goal (SDG) 7.
- MARENER 2017 provided a forum for all stakeholders to discuss ways of achieving an energy efficient and low carbon future for the maritime industry.
- There is value in WMU offering future conferences focused on maritime energy management.

The purpose of the study was not only to evaluate the MARENER 2017 conference itself but also to identify gaps in current MEM research. The analysis of the survey data revealed that the perceived challenges in MEM are not always common; different countries tend to face different problems. This implies that socio-economic issues can be underlying the challenges associated with MEM. In the survey, 12.7% ($n = 7$) of respondents expressed the importance of the human element, for example, education and training, social and economic impact on sustainable energy, gender, and awareness of energy efficiency in the industry. Another aspect that emerged from the survey was the business and commercial elements (10.9%, $n = 6$), that can also affect MEM, such as split incentives, applications of energy systems, integrated logistics, and practical management issues.

[1]MEPC 71/INF.13 (Information document on the International Conference on Maritime Energy Management held in Malmö, Sweden, from 24 to 25 January 2017).

Furthermore, even in the research areas where a good number of contributions have been made, for example, ship design and operations, or port facilities, it is necessary to further develop the specific research topics. The survey revealed that 20.0% (n = 11) of respondents emphasised the need for building specialised expertise within the focused area of MEM. This is somehow expected, considering that various on-going research efforts on MEM are still in the middle of development or immature. Hence, further development of new ideas and their applications into practice is expected. Practical applications of theories were mentioned by the respondents (16.4%, n = 9). Such opinions include practical approaches from maritime companies; commercial experiences and their relation to academic research; actual examples of implementing regulations; applicability of maritime energy technologies in developing countries; ship owners' perspectives in decision-making in terms of the application of new technologies; viewpoints from energy efficiency equipment manufacturers; and more technical, efficient and economically feasible methods.

Last but not least, research addressing the UN SDGs was also noted as important (7.3%, n = 4). Future MEM research must contribute to bridging the gap between developed and developing countries, which, the Editors believe, was very valuable feedback. SDG Goal 7 "Ensure access to affordable, reliable, sustainable and modern energy for all" has a close link to other SDGs, such as Goal 12 (Responsible consumption and production) and Goal 13 (Climate Change). Hence, future MEM research should consider an intersectional approach, addressing several SDGs.

## 3 The Role of MEM in Research and Innovation for Sustainable Maritime Transportation

Four of the ten goals in the IMO's Sustainable Maritime Transportation System[2] are attributed and linked to energy management: Energy efficiency and ship-port interface; Energy supply for ships; Technical co-operation; and New technology and innovation. Following the Paris Agreement at the 21st Session of the Conference of the Parties to the United Nations Framework Convention on Climate Change (COP 21), energy efficiency has been receiving more attention by the maritime industry. With growing concern about climate change and its environmental impacts, it is inevitable that the maritime industry has to play an active role in reducing $CO_2$ and other emissions from ships.

Nonetheless, research in the MEM domain is relatively new compared to other mature disciplines, like safety research. Much of the available data on energy are general and it is difficult to find data specific to the maritime sector; in some cases, it does not exist at a national level, especially in developing countries. This clearly

[2] IMO (2013).

highlights the need for MEM experts who can contribute to appropriately "localising" the global concerns in MEM in depth.

With rapid advancement of modern technologies, there seems to be plenty of opportunities but, at the same time, a great deal of uncertainty in the maritime industry. The MARENER 2017 conference can be considered as WMU's first call for a large international forum to voice our concerns and coping strategies through education and research.

It is also worth noting that contemporary maritime energy problems are so widely spread through various disciplines, such as naval architecture, marine engineering, ocean engineering, law and policy, economics, social sciences, and environmental sciences. This is why interdisciplinary research is necessary and crucial for MEM; hence, the MARENER 2017 facilitated a dialogue among various stakeholders from different disciplines.

Increasing expectations for alternative fuels and (marine) renewable energy research are evident. This indicates that "innovation" is certainly another important keyword for sustainable maritime transportation and for zero-emission ships of the future. Whilst conventional ideas and methods are still valid, new approaches and methodologies should be sought in order to create both technological and social innovation in the maritime world towards a low carbon and energy efficient future.

"*A vision without a plan is a wish. A plan without a timetable is a wish list.*[3]" In this regard, the Editors sincerely believe that the MEM research field will gain a new momentum and re-defined vision. The publication of this book and the research strategy are expected to serve both as an underlying foundation and a trigger for further developments.

## References

IMO. (2013). *World maritime day: A concept of a sustainable maritime transportation system.* London: International Maritime Organization.

Oung, K. (2013). *Energy management in business: The manager's guide to maximising and sustaining energy reduction.* London: Routledge.

[3]See Chapter 3, page 38 of Oung, K. (2013).

# Annex

 
A. I. Ölçer et al. (eds.), *Trends and Challenges in Maritime Energy Management*,
WMU Studies in Maritime Affairs 6, https://doi.org/10.1007/978-3-319-74576-3

# Maritime Energy Management Research Strategy

## OUR VISION

To become the world's leading University in the research field of maritime energy management and to play a vital role in transforming the maritime world to achieve a sustainable, low carbon and energy efficient future by delivering research of global excellence.

## OUR STRATEGY

This vision will be reflected by MarEner (Maritime Energy Management and Marine Technology) Research Group's interdisciplinary research, whereby WMU will be appreciated for:

*Advancing* the knowledge in the maritime energy management field by conducting world-class fundamental and applied research in the thematic areas given below over a life-cycle perspective, from shipping to oceans, and from ports to shipyards;

*Contributing* to the capacity building and the goals of the IMO and its Member States and relevant UN bodies, in particular to UN SDGs 7, 12 & 13;

*Fostering* strategic relationships with other universities, governmental organizations, companies and other maritime stakeholders across the world to advance research;

*Setting* a global research agenda in the maritime energy management field through addressing maritime community's needs; and

*Undertaking* research of an international standard through scholarly publication.

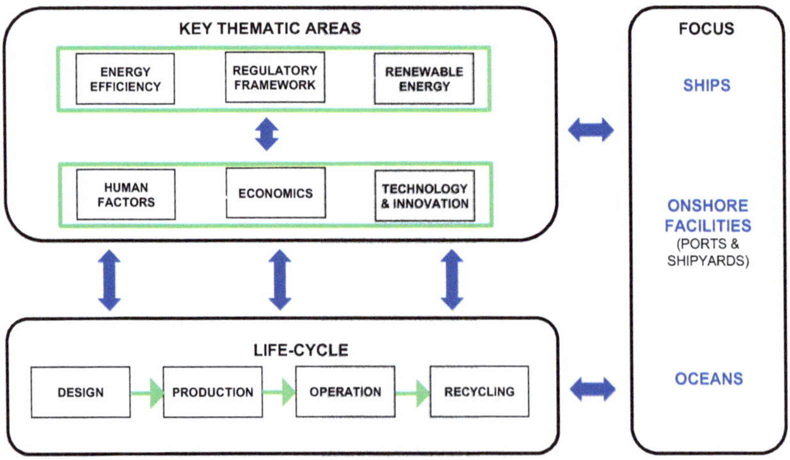


WORLD MARITIME UNIVERSITY
Fiskehamnsgatan 1, 211 18
PO Box 500, SE-201 24
Malmö, Sweden
www.wmu.se/research

WMU WORLD MARITIME UNIVERSITY

MIX
Papier aus verantwortungsvollen Quellen
Paper from responsible sources
FSC® C105338

If you have any concerns about our products,
you can contact us on
**ProductSafety@springernature.com**

In case Publisher is established outside the EU,
the EU authorized representative is:
**Springer Nature Customer Service Center GmbH**
**Europaplatz 3, 69115 Heidelberg, Germany**

Printed by Libri Plureos GmbH
in Hamburg, Germany